Methods in Microbiology
Volume 33

Recent titles in the series

Volume 24 *Techniques for the Study of Mycorrhiza*
JR Norris, DJ Reed and AK Varma

Volume 25 *Immunology of Infection*
SHE Kaufmann and D Kabelitz

Volume 26 *Yeast Gene Analysis*
AJP Brown and MF Tuite

Volume 27 *Bacterial Pathogenesis*
P Williams, J Ketley and GPC Salmond

Volume 28 *Automation*
AG Craig and JD Hoheisel

Volume 29 *Genetic Methods for Diverse Prokaryotes*
MCM Smith and RE Sockett

Volume 30 *Marine Microbiology*
JH Paul

Volume 31 *Molecular Cellular Microbiology*
P Sansonetti and A Zychlinsky

Volume 32 *Immunology of Infection*, 2nd edition
SHE Kaufmann and D Kabelitz

Methods in Microbiology

Volume 33
Functional Microbial Genomics

Edited by

Brendan Wren

*Department of Infectious and Tropical Diseases,
London School of Hygiene and Tropical Medicine,
London*

and

Nick Dorrell

*Department of Infectious and Tropical Diseases,
London School of Hygiene and Tropical Medicine,
London*

ACADEMIC PRESS
An imprint of Elsevier Science

Amsterdam   Boston   London   New York   Oxford   Paris
San Diego   San Francisco   Singapore   Sydney   Tokyo

This book is printed on acid-free paper.

© 2002 Elsevier Science Ltd. All rights reserved

No part of this publication may be reproduced or transmitted in any form or by any means, electronic or mechanical, including photocopy, recording, or any information storage and retrieval system, without permission in writing from the Publisher.

The appearance of the code at the bottom of the first page of a chapter in this book indicates the Publisher's consent that copies of the chapter may be made for personal or internal use of specific clients. This consent is given on the condition, however, that the copier pay the stated per copy fee through the Copyright Clearance Center, Inc. (222 Rosewood Drive, Danvers, Massachusetts 01923), for copying beyond that permitted by Sections 107 or 108 of the U.S. Copyright Law. This consent does not extend to other kinds of copying, such as copying for general distribution, for advertising or promotional purposes, for creating new collective works, or for resale. Copy fees for pre-2002 chapters are as shown on the title pages. If no fee code appears on the title page, the copy fee is the same as for current chapters.
0580-9517/02 $35.00

Explicit permission from Academic Press is not required to reproduce a maximum of two figures or tables from an Academic Press chapter in another scientific or research publication provided that the material has not been credited to another source and that full credit to the Academic Press chapter is given.

Academic Press
An Elsevier Science Imprint
84 Theobald's Road, London, WC1X 8RR, UK
http://www.academicpress.com

Academic Press
An Elsevier Science Imprint
525 B Street, Suite 1900, San Diego, California 92101-4495, USA
http://www.academicpress.com

A catalogue record for this book is available from the British Library

Cover photograph: Adapted from Chapter 3, figure 2, courtesy of D. Ussery.

ISBN 0-12-521533-9 (Hardback)
ISBN 0-12-787765-5 (Comb bound)

Typeset by Phoenix Photosetting, Chatham, UK
Printed and bound in Great Britain by Bookcraft, Bath, UK

02 03 04 05 06 07 BC 9 8 7 6 5 4 3 2 1

Contents

Contributors viii
Preface to the First Edition xiii
(*A colour plate section appears between pages 178 and 179*)

I Genome Sequence Analysis

1. Annotation of Microbial Genomes 3
   J Parkhill

2. From Sequence to Consequence: *In Silico* Hypothesis Generation and Testing 27
   M Pallen

3. The Atlas Visualization of Genomewide Information 49
   M Skovgaard, LJ Jensen, C Friis, HH Stærfeldt, P Worning, S Brunak and D Ussery

II Construction of DNA Microarrays

4. Microarray Design for Bacterial Genomes 67
   J Hinds, AA Witney and JK Vass

5. Glass Slide Microarrays for Bacterial Genomes 83
   J Hinds, KG Laing, JA Mangan and PD Butcher

III Comparative Nucleic Acid Analysis

6. Representational Difference Analysis of cDNA and Genome Comparisons 103
   DL Taylor, A Bart and LD Bowler

7. Application of DNA Microarrays for Comparative and Evolutionary Genomics 121
   N Dorrell, OL Champion and BW Wren

8. Gene Expression during Host–Pathogen Interactions: Approaches to Bacterial mRNA Extraction and Labelling for Microarray Analysis 137
   JA Mangan, IM Monahan and PD Butcher

9. High Throughput *In Vivo* Screens: Signature-tagged Mutagenesis 153
   N West, P Sansonetti and CM Tang

10. Further Strategies for Signature-tagged Mutagenesis and the Application of Oligonucleotide Microarrays for the Quantification of DNA-tagged Strains 167
    AV Karlyshev, N Dorrell, E Winzeler and BW Wren

IV Proteome Analysis

11. Advances in Bacterial Proteome Analysis 187
    SJ Cordwell

12. Discovery of Protein–Protein Interaction Using Two-hybrid Systems 209
    A Patel, KH Mellits and IF Connerton

V Applications of Microbial Genomics

13. Cloning the Metagenome: Culture-independent Access to the Diversity and Functions of the Uncultivated Microbial World 241
    J Handelsman, M Liles, D Mann, C Riesenfeld and RM Goodman

14. Reverse Vaccinology: from Genome to Vaccine 257
    JL Telford, M Pizza, G Grandi and R Rappuoli

15. Microbial Genomics for Antibiotic Target Discovery 271
    F Fan and D McDevitt

VI Case Studies – Bacteria

16. *Helicobacter pylori* Functional Genomics 291
    KJ Guillemin and NR Salama

17. *Streptomyces coelicolor* A3(2): from Genome Sequence to Function 321
    KF Chater, G Bucca, P Dyson, K Fowler, B Gust, P Herron, A Hesketh, G Hotchkiss, T Kieser, V Mersinias and CP Smith

18. Functional Analysis of the *Bacillus subtilis* Genome 337
    CR Harwood, A Wipat and Z Prágai

VII Case Studies – Parasites

19. *Plasmodium falciparum* DNA Microarrays and Interpretation of Data 371
    AA Witney, RM Anthony, TR Jones and DJ Carucci

20. Functional Analysis of the *Plasmodium falciparum* Genome Using Transfection 383
    AF Cowman, M Duraisingh, RA O'Donnell, T Triglia and BS Crabb

21. Chromosome Fragmentation as an Approach to Whole-genome Analysis in Trypanosomes 397
    JM Kelly and S Obado

Index 407

Series Advisors

Gordon Dougan, Director of The Centre for Molecular Microbiology and Infection, Department of Biological Sciences, Imperial College of Science, Technology and Medicine, London SW7 2AZ, UK

Graham J Boulnois, Schroder Ventures Life Sciences Advisers (UK) Limited, 71 Kingsway, London WC2B 6ST, UK

Jim Prosser, Department of Molecular and Cell Biology, University of Aberdeen, Institute of Medical Sciences, Foresterhill, Aberdeen AB25 2ZD, UK

Ian R Booth, Professor of Microbiology, Department of Molecular and Cell Biology, University of Aberdeen, Institute of Medical Sciences, Foresterhill, Aberdeen AB25 2ZD, UK

David A Hodgson, Reader in Microbiology, Department of Biological Sciences, University of Warwick, Coventry CV4 7AL, UK

David H Boxer, Department of Biochemistry, Medical Sciences Institute, Dundee DD1 4HN, UK

Contributors

Robert M Anthony Malaria Program, Naval Medical Research Center, Silver Spring, MD 20910, USA

Aldert Bart Department of Medical Microbiology, Academic Medical Center, University of Amsterdam, PO Box 22660, 1100 DD Amsterdam, The Netherlands

Lucas D Bowler Trafford Centre for Graduate Medical Education and Research, University of Sussex, Falmer, Brighton BN1 9RY, UK

Søren Brunak Center for Biological Sequence Analysis, BioCentrum-DTU, Building 208, The Technical University of Denmark, DK-2800 Kgs. Lyngsby, Denmark

Giselda Bucca Department of Biomolecular Sciences, UMIST, PO Box 88, Manchester M60 1QD, UK

Philip D Butcher Department of Medical Microbiology, St George's Hospital Medical School, Cranmer Terrace, London SW17 0RE, UK

Daniel J Carucci Malaria Program, Naval Medical Research Center, Silver Spring, MD 20910, USA

Olivia Champion Department of Infectious and Tropical Diseases, London School of Hygiene and Tropical Medicine, London WC1E 7HT, UK

Keith F Chater John Innes Centre, Norwich Research Park, Colney, Norwich NR4 7UH, UK

Ian F Connerton Division of Food Sciences, School of Biosciences, University of Nottingham, Sutton Bonington Campus, Loughborough LE12 5RD, UK

Stuart J Cordwell Australian Proteome Analysis Facility, Sydney, Australia 2109

Alan F Cowman The Walter and Eliza Hall Institute of Medical Research, Victoria 3050, Australia

Brendan S Crabb The Walter and Eliza Hall Institute of Medical Research, Victoria 3050, Australia

Nick Dorrell Department of Infectious and Tropical Diseases, London School of Hygiene and Tropical Medicine, London WC1E 7HT, UK

Manoj Duraisingh The Walter and Eliza Hall Institute of Medical Research, Victoria 3050, Australia

Paul Dyson School of Biological Sciences, University of Wales Swansea, Singleton Park, Swansea SA2 8PP, UK

Frank Fan Microbial, Musculoskeletal and Proliferative Diseases CEDD, GlaxoSmithKline, 1250 South Collegeville Road, Collegeville, PA 19426-0989, USA

Kay Fowler John Innes Centre, Norwich Research Park, Colney, Norwich NR4 7UH, UK

Carsten Friis Center for Biological Sequence Analysis, BioCentrum-DTU, Building 208, The Technical University of Denmark, DK-2800 Kgs. Lyngsby, Denmark

Robert M Goodman Department of Plant Pathology, University of Wisconsin, Madison, WI 53706, USA

Guido Grandi IRIS, Chiron S.p.A., Via Fiorentina 1, 53100 Siena, Italy

Karen J Guillemin Institute of Molecular Biology, University of Oregon, Eugene, OR 97403, USA

Bertolt Gust John Innes Centre, Norwich Research Park, Colney, Norwich NR4 7UH, UK

Jo Handelsman Department of Plant Pathology, University of Wisconsin, Madison, WI 53706, USA

Colin R Harwood Department of Microbiology and Immunology, Newcastle University, Newcastle upon Tyne NE2 4HH, UK

Paul Herron School of Biological Sciences, University of Wales Swansea, Singleton Park, Swansea SA2 8PP, UK

Andy Hesketh John Innes Centre, Norwich Research Park, Colney, Norwich NR4 7UH, UK

Jason Hinds Department of Medical Microbiology, St George's Hospital Medical School, Cranmer Terrace, London SW17 0RE, UK

Graham Hotchkiss Department of Biomolecular Sciences, UMIST, PO Box 88, Manchester M60 1QD, UK

Lars Juhl Jensen Center for Biological Sequence Analysis, BioCentrum-DTU, Building 208, The Technical University of Denmark, DK-2800 Kgs. Lyngsby, Denmark

Trevor R Jones Malaria Program, Naval Medical Research Center, Silver Spring, MD 20910, USA

Andrey V Karlyshev Department of Infectious and Tropical Diseases, London School of Hygiene and Tropical Medicine, London WC1E 7HT, UK

John M Kelly Department of Infectious and Tropical Diseases, London School of Hygiene and Tropical Medicine, London WC1E 7HT, UK

Tobias Kieser John Innes Centre, Norwich Research Park, Colney, Norwich NR4 7UH, UK

Kenneth G Laing Department of Medical Microbiology, St George's Hospital Medical School, Cranmer Terrace, London SW17 0RE, UK

Mark Liles Department of Plant Pathology, University of Wisconsin, Madison, WI 53706, USA

Joseph A Mangan Department of Medical Microbiology, St George's Hospital Medical School, Cranmer Terrace, London SW17 0RE, UK

David Mann Department of Plant Pathology, University of Wisconsin, Madison, WI 53706, USA

Damien McDevitt Microbial, Musculoskeletal and Proliferative Diseases CEDD, GlaxoSmithKline, 1250 South Collegeville Road, Collegeville, PA 19426-0989, USA

Kenneth H Mellits Division of Food Sciences, School of Biosciences, University of Nottingham, Sutton Bonington Campus, Loughborough LE12 5RD, UK

Vassilios Mersinias Department of Biomolecular Sciences, UMIST, PO Box 88, Manchester M60 1QD, UK

Irene M Monahan Department of Medical Microbiology, St George's Hospital Medical School, Cranmer Terrace, London SW17 0RE, UK

Samson Obado Department of Infectious and Tropical Diseases, London School of Hygiene and Tropical Medicine, London WC1E 7HT, UK

Rebecca A O'Donnell The Walter and Eliza Hall Institute of Medical Research, Victoria 3050, Australia

Mark Pallen Division of Immunity and Infection, Medical School, University of Birmingham, Birmingham B15 2TT, UK

Julian Parkhill The Sanger Institute, Wellcome Trust Genome Campus, Hinxton, Cambridge CB10 1SA, UK

Amit Patel Division of Food Sciences, School of Biosciences, University of Nottingham, Sutton Bonington Campus, Loughborough LE12 5RD, UK

Mariagrazia Pizza IRIS, Chiron S.p.A., Via Fiorentina 1, 53100 Siena, Italy

Zoltán Prágai Department of Microbiology and Immunology, Newcastle University, Newcastle upon Tyne NE2 4HH, UK

Rino Rappuoli IRIS, Chiron S.p.A., Via Fiorentina 1, 53100 Siena, Italy

Christian Riesenfeld Department of Plant Pathology, University of Wisconsin, Madison, WI 53706, USA

Nina Reda Salama Human Biology Division, Fred Hutchinson Cancer Research Center, 1100 Fairview Avenue North, PO Box 19024, Seattle, WA 98109, USA

Phillippe Sansonetti Unité de Pathologie Microbienne Moléculaire, Institut Pasteur, 28 rue du Dr Roux F-75724, Paris, Cédex 15, France

Marie Skovgaard Center for Biological Sequence Analysis, BioCentrum-DTU, Building 208, The Technical University of Denmark, DK-2800 Kgs. Lyngsby, Denmark

Colin P Smith Department of Biomolecular Sciences, UMIST, PO Box 88, Manchester M60 1QD, UK

Hans Henrik Stærfeldt Center for Biological Sequence Analysis, BioCentrum-DTU, Building 208, The Technical University of Denmark, DK-2800 Kgs. Lyngsby, Denmark

Christoph M Tang Centre for Molecular Microbiology and Infection, The Flowers Building, Imperial College of Science, Technology and Medicine, Armstrong Road, London SW7 2AZ, UK

Deborah L Taylor Trafford Centre for Graduate Medical Education and Research, University of Sussex, Falmer, Brighton BN1 9RY, UK

John L Telford IRIS, Chiron S.p.A., Via Fiorentina 1, 53100 Siena, Italy

Tony Triglia The Walter and Eliza Hall Institute of Medical Research, Victoria 3050, Australia

David Ussery Center for Biological Sequence Analysis, BioCentrum-DTU, Building 208, The Technical University of Denmark, DK-2800 Kgs. Lyngsby, Denmark

J Keith Vass Beatson Institute for Cancer Research, Garscube Estate, Glasgow G61 1BD, UK

Nicholas West Centre for Molecular Microbiology and Infection, The Flowers Building, Imperial College of Science, Technology and Medicine, Armstrong Road, London SW7 2AZ, UK

Elizabeth Winzeler Novartis Institute for Functional Genomics, 3115 Merryfield Row, Suite 200, San Diego, CA 92121-1125, USA

Anil Wipat Department of Computing Sciences, Newcastle University, Newcastle upon Tyne NE2 4HH, UK

Adam A Witney Department of Medical Microbiology, St George's Hospital Medical School, Cranmer Terrace, London SW17 0RE, UK

Peder Worning Center for Biological Sequence Analysis, BioCentrum-DTU, Building 208, The Technical University of Denmark, DK-2800 Kgs. Lyngsby, Denmark

Brendan W Wren Department of Infectious and Tropical Diseases, London School of Hygiene and Tropical Medicine, London WC1E 7HT, UK

Preface

The availability of whole genome sequences and allied high-throughput technologies means that we are in the midst of an unprecedented revolution in microbiology, eclipsing even the Van Leeuwenhook and Pasteur eras in terms of research opportunities. Such circumstances allow a holistic approach to be taken to the functional characterisation of microbes at the DNA and protein expression levels, which will fundamentally improve our understanding of what makes these organisms tick! In particular this should prove to be a golden era for the study of microbial pathogens, because despite progress over the last century, they still represent one of the most important threats to human health worldwide. Many infectious disease agents have never been controlled, or have re-emerged as global pathogens, while others pose a new threat. Media and scientific attention has focused on a range of problems, including the alarming spread of antibiotic resistance, the threat of bioterrorism, microbial contamination of the food chain, the global resurgence of tuberculosis and malaria, and other emerging and re-emerging infections triggered by life-style, political, economical and ecological changes. The need to gain an integrated and comprehensive understanding of the workings of our old microbial adversaries is greater than ever.

The aim of this book is to provide the researcher with an up-to-date collection of post genome technologies central to studying the function of microorganisms. This should facilitate the researcher in participating in the post genome bonanza in their given subject area. It has not been our intention to provide a laboratory-based manual, but rather to include a mixture of conceptual overviews, experimental approaches and referenced experimental procedures. Where new techniques or strategies are mentioned, more detailed protocols are given.

The book is divided into seven sections. The first section describes varied bioinformatic approaches that can be used to interrogate sequence data in a meaningful manner. However, such analysis only provides clues and does not establish gene function. The acquisition and analysis of genome sequence data is not an end in itself, but is a starting point for generating testable hypotheses to be conformed by 'wet laboratory' research. The avalanche of genome sequence data has coincided with important technological advances in several 'wet laboratory' research areas. Section two details the construction of amplified PCR product and oligonucleotide microarrays. Section three describes the application of this technology for comparative genome analysis, whole genome expression studies and high throughput *in vivo* screening of microbial genes. Section four describes proteome analysis which when coupled with microarray analysis liberates experimental approaches from the piece-meal study of individual genes or operons towards comprehensive analyses of the entire gene and protein complement of the microbial cell. Section five highlights wider applications of microbial genomics towards the identification of novel microbial species and the identification of new antimicrobial and vaccine candidates. Section six presents several case studies of individual pathogenic or non-pathogenic bacteria where functional genomic analysis have revolutionised our understanding of these model species. In the final section, extrapolating approaches used for bacteria, functional genomic approaches for the larger genome species of *Plasmodium* and *Trypanosomes* are described. The major omission from this volume is a description of the functional analysis of *Saccharomyces cerevisiae*. The *S. cerevisiae* research community have been at the forefront of functional genomic applications and a detailed description of yeast gene analysis is described in Volume 26 of this series.

Currently, we have only just began to scratch the surface in terms of the potential applications of functional microbial genomics to study microbes. The next few years promises to be a voyage of discovery in terms of developing our understanding of microbes, their phylogeny, ecology, physiology, pathogenesis and evolution, as well as their commercial exploitation. There has never been a better time to be a microbiologist!

In compiling this volume we are indebted to many contributors who are leading experts in their respective fields. Many have suppied up-to-date information and previously unpublished data. We thank them for the speed with which they have prepared their excellent manuscripts. Finally we thank the editorial staff at Elsevier, especially Tessa Picknett and Claire Minto.

Part I
Genome Sequence Analysis

1 Annotation of Microbial Genomes

Julian Parkhill
The Sanger Institute, Wellcome Trust Genome Campus, Hinxton, Cambridge CB10 1SA, UK

◆◆◆

CONTENTS

Introduction
Gene prediction
Stable RNA genes
Similarity searching
Domain/motif searching
Orthology, paralogy and synteny
Non-similarity-based methods
Horizontal gene transfer and other anomalies
Functional classification/Metabolic reconstruction
Concluding remarks

◆◆◆◆◆◆ INTRODUCTION

Annotation is the process by which useful information is added to a raw genomic DNA sequence, producing a framework for the interrogation and understanding of the sequence, and hopefully enhancing its utility for the downstream user. It must be emphasized from the outset that, while the DNA sequence itself can be objectively determined, and is likely to be highly accurate, all but the most simple annotation is an interpretation of the sequence, and is thus subject to error and misinterpretation. While genome annotators strive to be as objective and accurate as possible in their interpretation, a good scientist should always look carefully at others' interpretation of data, and ensure that they understand the methods by which it was reached.

The process of genome annotation falls into two broad sections: firstly, the prediction of the position and start site of all the genes; and, subsequently, the attempt to ascribe function to each of the gene products. Although these are to some extent linked and should always feed back to each other during the annotation process, we will deal with them separately here.

◆◆◆◆◆◆ GENE PREDICTION

Although gene prediction seems like a simple process and is often represented as a solved problem, especially in bacterial genomes, this is certainly not the case. A large proportion of the genes in any given genome can certainly be predicted with a good degree of accuracy and consistency by several different gene prediction programs (Table 1.1). However, for reasons that will be described in detail below, and without attempting to be overly critical of the programs themselves, these programs will always come across grey areas where gene prediction is difficult, and they will never reach 100% efficiency. Whilst for a statistician, >97% accuracy may be perfectly acceptable, as biologists we know that much of the interest in any system lies not in the well-understood core regions, but in the grey areas around the edges. For this reason, we must always look carefully at the output of these (or indeed any) bioinformatics tools and interpret the results they give according to our own knowledge.

The first and most obvious point to make about gene prediction is that it is not just a case of marking up open reading frames (ORFs). ORFs are simply defined as stretches of DNA larger than a given size, beginning with a start codon (ATG, GTG or TTG in most bacterial cases), and ending with an in-frame stop codon (TAA, TAG or TGA). These can be easily calculated from the DNA sequence by any number of different tools. The essence of gene prediction is to attempt to decide which of these ORFs are likely to be actually coding for a protein sequence. Although it should be fairly obvious that all ORFs are not coding sequences (CDSs), it can be easily shown by considering genomes of differing $G+C$ content. Bacterial genomes have been seen to have a fairly constant coding capacity of one gene per 1.1 kb, with few, if any, heavily overlapping genes. A simple prediction of ORFs >100 amino acids (aa) long in *Campylobacter jejuni* (Parkhill et al., 2000), which has a $G+C$ content of 30.5%, gives 1783 ORFs in 1641 kb (a density of 1.09 ORFs per kb). At the other extreme, *Streptomyces coelicolor* (http://www.sanger.ac.uk/Projects/S_coelicolor) has a $G+C$ content of 72.1%, contains 30 595 ORFs >100 aa in 8667 kb (a

Table 1.1 Commonly used microbial gene prediction programs

Program	Source	Reference
Glimmer	http://www.tigr.org/softlab/glimmer/glimmer.html	Delcher et al. (1999)
Orpheus	http://pedant.gsf.de/orpheus/	Frishman et al. (1998)
GeneMark GeneMark.hmm	http://opal.biology.gatech.edu/GeneMark/	Borodovsky and McIninch (1993), Lukashin and Borodovsky (1998)
CRITICA	http://geta.life.uiuc.edu/~gary/programs/CRITICA.html	Badger and Olsen (1999)

density of 3.53 ORFs per kb). The reason for this huge discrepancy is that the three stop codons are A + T-rich and are, therefore, rare in G + C-rich organisms, leading to a large excess of random non-coding ORFs, and the extension of true CDSs at the 5'-end. Clearly, we cannot rely on ORF prediction as a sensible measure and must, therefore, bring in other measures of coding likelihood to determine whether each ORF is a CDS or not, and for how much of its length.

In essence, virtually all methods for assessing the coding likelihood of a given ORF attempt to compare the properties of that ORF against the properties of known CDS within that organism. A set of genes known or likely to be coding is used to train the program to recognize other genes likely to be coding. Whilst this approach has many strengths, not least in the ability to be automated and applied very rapidly to entire genomes, it is important to understand exactly what the program is trying to do. The function demanded of the program is *not* 'find me all the genes in this genome', but 'find me all the CDSs *that look like the given training set* in this genome'.

Clearly, all the genes within any organism will not all be identical; the properties of different sets of genes within any given genome can vary for many reasons. Amongst these are coding constraints, such as genes encoding largely hydrophobic proteins, small basic ribosomal proteins, or repetitive proteins containing a preponderance of certain amino acids; many genes with these properties will not be seen by the gene prediction programs because they do not conform to the model of typical genes within that organism. Another reason for genes not appearing to match the typical properties of an organism is that they may have been recently acquired from another organism; such horizontally transferred genes may well be missed by programs looking for genes that match a given model. In addition to non-typical genes, genomes do contain non-protein coding sequences, such as RNA genes, DNA repeats and degenerate pseudogenes. Most gene prediction programs will attempt to predict coding sequences within these non-coding regions and these overpredictions must be weeded out manually, after looking at the putative genes in the context of other predicted sequence features.

For these reasons, and others, different gene-prediction programs will often come up with different sets of predicted genes and, even when they agree on the position of a gene, they may disagree on the chosen start site. It is, therefore, advisable always to run more than one gene-prediction program on any given genome and to integrate the results with a manual inspection of the sequence itself. Differing interpretations of which ORF is likely to be coding and where exactly the CDS starts can often be resolved by visual inspection. Genome-specific measures, such as codon preference and positional G + C content and genome-independent measures, such as amino-acid usage and BLASTX comparison against the protein databases can be used to determine which predicted gene is likely to be correct. Similar measures can be used to look for probable coding sequences in regions where none have been predicted by the programs used. Looking at where codon-usage and positional G + C plots begin to change, and integrating information such as ribosome-binding sites and searches against protein database sequences will assist in determining which of the alternative start sites is correct. A sequence visualization tool such as Artemis (Rutherford *et al.*, 2000) is ideal for this procedure (Figure 1.1).

◆◆◆◆◆◆ STABLE RNA GENES

Although the previous section has concentrated on protein coding sequences, genes can also encode stable RNA species, which can be structural, catalytic or regulatory. Prediction of stable RNA genes can be considerably more difficult than protein coding genes, as there is no generic signal, such as codon usage, third position GC bias, etc. for the programs to detect. Often, the conservation in stable RNA species may be of a particular structure rather than a specific sequence.

Genes for known RNA species can be found by simple DNA/DNA similarity searches. Obvious examples include rRNA genes and more specific genes, such as those for the RNA components of RNAse P (*rnpB*) and the protein secretion pathway (4.5S RNA) (see http://mbcr.bcm.tmc.edu/smallRNA/smallrna.html for further examples). Such genes often can only be reliably found in organisms that are closely related to those from which the original sequence was determined. The specialized RNA responsible for targeting the products of prematurely terminated transcripts for degradation (tmRNA) can also be found in this way, and a particularly good collection of known and predicted tmRNAs can be found on the tmRNA website (http://www.indiana.edu/~tmrna/; Williams, 2002). Such searches will, of course, only work when a known probe sequence is available; they cannot find novel RNA genes. One specific counterexample to this is the ability to search for tRNAs. The structure and function of tRNAs are fairly well understood – after all, they were the first nucleic acids to be completely sequenced (Holley, 1965) – and it is, therefore, possible to find the complete set of tRNAs in any organism with a computational approach. The standard tool for this is tRNA-SE (Lowe and Eddy, 1997).

Recent laboratory work has indicated that, perhaps unsurprisingly, many more stable RNA species exist in *Escherichia coli* than were previously known. These have been found by looking for matched promoters and terminators without an intervening coding sequence (Argaman *et al.*, 2001), and by microarray analysis of conserved regions (Wassarman *et al.*, 2001). Computational techniques are now becoming available to detect these previously unknown genes. One of the most promising is qrna (Rivas *et al.*, 2001), which uses the comparison between related genomes to detect conserved sequences having the property that the mutational differences between them indicate selection for conservation of secondary structure rather than coding capacity. One drawback to this technique is that it requires a pair of genomes with a particular degree of relatedness, a condition that is, however, likely to become more common in the future.

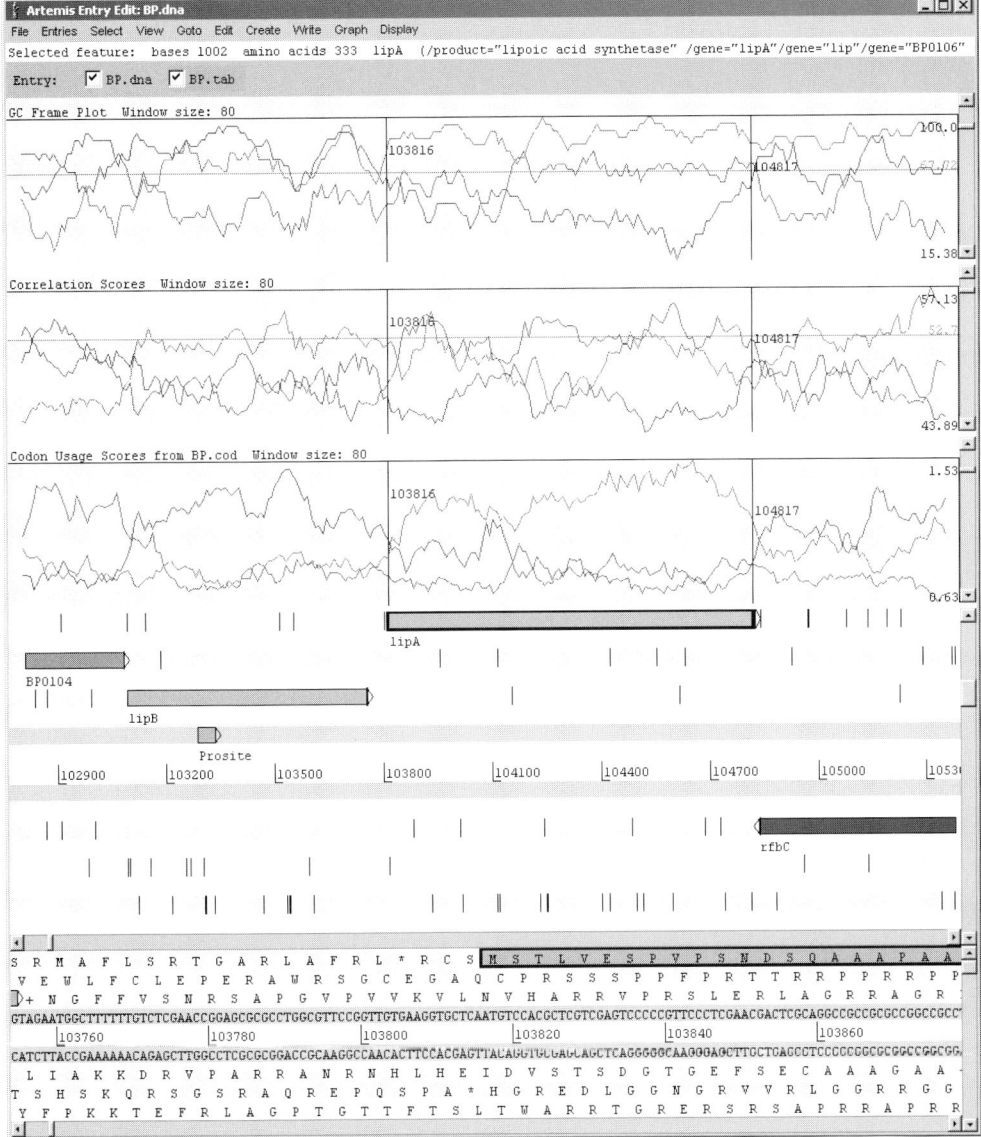

Figure 1.1. Illustration of coding parameters in Artemis. A short section of the genome of *Bordetella pertussis* is shown. In the bottom two panels is the six-frame translation of the sequence at two zoom levels, with stop codons represented as vertical bars in the zoomed-out upper panel, and the predicted genes marked as open boxes. The top graph (GC frameplot) shows a position-specific G + C content plot (FramePlot; Bibb et al., 1984). The coloured lines represent the G + C content of the first, second and third positions of the triplet code, relative to the start of the sequence. For example, red represents the third position for a CDS in frame 1, while green represents the third position for a CDS in frame 2, and blue the third position for a CDS in frame 3. In G + C-rich organisms, the third position of the codon tends to be more G + C-rich than the first and second. The second graph (correlation scores) indicates the results of a calculation of the correlation between the amino-acid usage of the coding frame within the specified window, and the general amino-acid usage calculated from genes in EMBL. The green line indicates the score for frame 1, blue for frame 2 and red for frame 3. Values close to the line at 52.7 indicate a strong likelihood of coding. This calculation is independent of codon usage and G + C content. The third plot indicates the codon usage scores compared with a codon usage table generated from *B. pertussis* genes in the public databases (http://www.kazusa.or.jp/codon/; Nakamura et al., 2000). In this case the red line indicates a good codon usage score in frame 1, green in frame 2 and blue in frame 3. Reproduced in colour between pages 178 and 179.

◆◆◆◆◆◆ SIMILARITY SEARCHING

Having a predicted gene set for the genome in hand, the next stage is to attempt to ascribe function to each of the putative genes. Fundamentally, this is done by some form of comparison between the novel sequence and sequences of previously determined function. This can be done directly, by sequence comparison with databases, or indirectly by comparing predicted proteins against models of known protein domains or motifs.

The most rigorous and sensitive methods of sequence comparison are those described by Smith and Waterman (1981) and Needleman and Wunsch (1970). Unfortunately, these algorithms are too computationally intensive to allow them to be used efficiently to compare sequences against databases of the sizes currently available, except with specialized hardware. Almost all current database searches, therefore, involve some form of tradeoff between sensitivity and speed. The two commonly used search programs FASTA (Pearson and Lipman, 1988) and BLAST (Altschul *et al.*, 1990) use different short cuts: an initial search for short perfect matches in the former case, and the use of hash tables in the second case. FASTA does perform a final Smith–Waterman alignment but only on the top-scoring hits. Given the different search mechanisms used, FASTA and BLAST will often arrive at slightly different lists of database hits, especially at the boundaries of detectable similarity. There is one

a

```
The best scores are:                                      initn init1  opt   z-sc E(535121)
TTK_ECOLI    TTK PROTEIN.                                  1085 1085 1085 1277.50
AAF64285     HYPOTHETICAL 24.2 KDA PROTEIN.                1061 1061 1066 1254.90
Q9KVD2       TRANSCRIPTIONAL REGULATOR, TETR FAMILY         856  856  868 1024.2  0
TTK_HAEIN    TTK PROTEIN HOMOLOG.                           623  623  662  783.9  0
O67927       TRANSCRIPTIONAL REGULATOR (TETR/ACRR F         125  125  172  213.6  0.00023
Q9X7X6       PUTATIVE REGULATORY PROTEIN.                   106  106  160  194.8  0.0026
Q9S3L4       AMTR PROTEIN.                                   96   96  147  183.5  0.011
Q59306       30S RIBOSOMAL PROTEIN S21.                      78   78  145  182.2  0.013
Q9KIL9       F58R (FRAGMENT).                                72   72  143  181.4  0.014
CAC01371     PUTATIVE TETR-FAMILY TRANSCRIPTIONAL R          72   72  143  179.1  0.019
Q9L078       PUTATIVE TETR-FAMILY REGULATORY PROTEI         128  101  142  178.1  0.022
Q9KXK1       PUTATIVE TETR-FAMILY REGULATORY PROTEI          78   51  142  176.9  0.026
CAC01492     PUTATIVE TRANSCRIPTIONAL REGULATORY PR          61   61  139  174.3  0.036
O67930       HYPOTHETICAL 22.1 KDA PROTEIN.                  86   86  137  172.8  0.044
AAG04792     PROBABLE TRANSCRIPTIONAL REGULATOR.             93   93  137  172.2  0.047
Q9WZT0       TRANSCRIPTIONAL REGULATOR, TETR FAMILY          57   57  135  170.5  0.058
Q9RY76       TRANSCRIPTIONAL REGULATOR, TETR FAMILY          45   45  134  169.1  0.07
★ Q9Z551     PUTATIVE TRANSCRIPTIONAL REGULATOR.            116  116  133  168.1  0.079
★ P96856     HYPOTHETICAL 21.9 KDA PROTEIN.                  75   75  133  167.9  0.082
CAC08387     PUTATIVE TETR-FAMILY TRANSCRIPTIONAL R          76   76  133  167.7  0.084
```

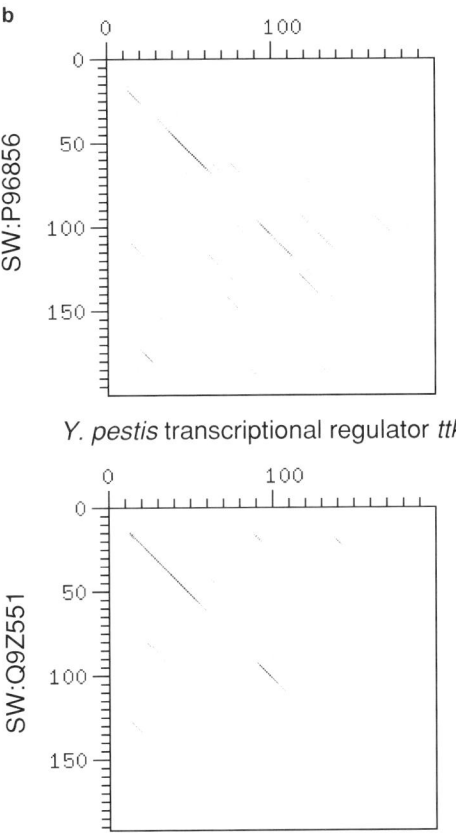

Figure 1.2. (a) FASTA results for a database search using the *Y. pestis* Ttk protein as a query. The statistically insignificant, but biologically significant, matches to the TetR-family proteins SW:Q9Z551 and SW:P96856 are marked with a star. (b) Dotplot comparisons of *Y. pestis* Ttk vs. SW:Q9Z551 and *Y. pestis* Ttk vs. SW:P96856 produced with Dotter (Sonnhammer and Durbin, 1995). The co-linear similarities can be seen on the main diagonal.

other important difference between the two programs: FASTA attempts to find the best *global* (i.e. end-to-end) alignment between the query and the database sequences, while BLAST will attempt to find the best *local* alignment, that is, the best segment of similarity irrespective of its length or position in each of the sequences. Each of these methods has its strengths, showing the extent of overall similarity in the first case, and giving the ability to find conserved domains only in the second. A careful investigator will, therefore, run and consider both types of searches in every case.

Given a list of database matches by either program, it is important to consider the significance of the match. 'Significance', in this case, can either mean statistical significance or biological significance. It is also important to remember that, while they are usually coincident, there are exceptions: statistical significance does not always imply biological significance, and biologically significant matches are not always statistically significant. Both BLAST and FASTA report the statistical significance of each match in the database, usually as an 'E' or 'P' value. Although the interpretation of these numbers is similar the derivation is different. 'E' represents the number of matches of that score that would be expected by chance, given query sequences and databases of the same size and content. 'P' represents the probability of the match occurring by chance, again given query sequences and databases of the same size and content. Obviously the lower the numbers the more significant the results; standard statistical measures suggest P-values of >0.05 are significant and >0.01 are very significant. Given the uncertainty around the 'twilight zone', researchers often quote values of $>10^{-10}$ as indicating a true match. However, it is important not to be seduced by the simplicity of absolute cutoff values; while easy to apply, they can be wrong, and the results should always be interpreted in the light of biological knowledge.

Three examples will serve to indicate the points made above that biological significance does not equal statistical significance and that absolute cutoff values are not always trustworthy. Measures of statistical significance can be confounded by distant but real relatives, by biased sequence composition, and by repetitive sequence.

Figure 1.2a shows the result of a FASTA search against the protein sequence database using the *Yersinia pestis* Ttk protein, a TetR-family

a

```
The best scores are:                                          initn init1  opt   z-sc  E(534512)
CAC14227    YAPH PROTEIN.                                     22769 22769 22769 21273.70
CAC14223    YAPD PROTEIN.                                      9153  9153  9153  8553.6    0
Q9JMS3      YCHA PROTEIN.                                      1831   773  1708  1595.5    0
AIDA_ECOLI  ADHESIN AIDA-I PRECURSOR.                          1013   340  1413  1320.2    0
Q9Z625      MISL.                                              1569  1020  1314  1229.6    0
P77286      HYPOTHETICAL 50.5 KDA PROTEIN.                     1358   704  1209  1136.0    0
Q9XCJ4      SHDA.                                               899   321  1129  1051.8    0
P75997      FROM BASES 1220357 TO 1232354                      1017   652  1034   974.5    0
Q52298      PLASMID PMYSH6000 VIRULENCE-ASSOCIATED             1156   428  1007   941.7    0
CAC14226    YAPG PROTEIN.                                      1208   771   986   922.8    0
CAC03614    HYPOTHETICAL 68.5 KDA PROTEIN.                      950   620   931   874.1    0
YDEK_ECOLI  HYPOTHETICAL 136.5 KDA LIPOPROTEIN IN               797   392   896   836.8    0
Q9JMS5      YCBB PROTEIN.                                       129    87   737   686.4  1.1e-30
★ O50379    PPE-FAMILY PROTEIN.                                 210    78   725   670.4  8.3e-30
★ O06304    HYPOTHETICAL 327.0 KDA PROTEIN PPE.                 121    71   721   667.4  1.2e-29
YFAL_ECOLI  HYPOTHETICAL 131.2 KDA PROTEIN IN UBIG              574   285   631   589.5  2.7e-25
YFAL_ECOLI  HYPOTHETICAL 131.2 KDA PROTEIN IN UBIG              574   285   631   589.5  2.7e-25
Q9XC47      OUTER MEMBRANE PROTEIN A.                           190    73   607   563.7  7.3e-24
★ O07231    RV0304C.                                            164   117   581   539.2  1.7e-22
★ YZ08_MYCTU HYPOTHETICAL PE-PGRS FAMILY PROTEIN RV             164    96   563   523.3  1.3e-21
```

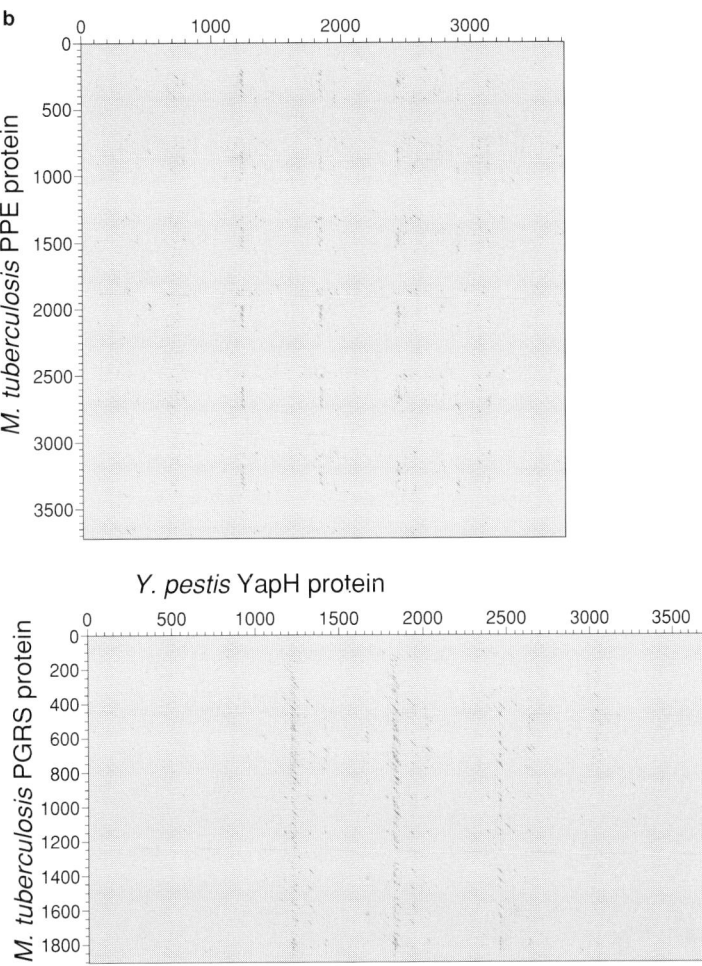

Figure 1.3. (a) FASTA results for a database search using the *Y. pestis* YapH protein as a query. The statistically significant, but biologically meaningless, matches to *M. tuberculosis* PPE and PGRS proteins are marked with a star. (b) Dotplot comparisons of *Y. pestis* YapH vs. *M. tuberculosis* PPE and PGRS proteins produced with Dotter. It can be seen that there is no single strong diagonal indicating co-linear similarity.

transcriptional regulator, as a query sequence. The top hits are unambiguous, matching equivalent proteins from related species with very low *E*-values. The matches further down the list are not statistically significant, with *E*-values above 0.05. However, these proteins are from the same family as Ttk; they are, or are predicted to be, TetR-family transcriptional regulators. Thus the match is biologically significant, despite not being statistically significant. If the top-scoring proteins had not been sequenced, leaving only the other matches, these could well have been dismissed. A dotplot comparison of Ttk with two of the lowest-scoring proteins (Figure 1.2b) underlines the point; there is clearly a co-linear similarity between the proteins.

Figure 1.3a shows the results of a similar FASTA search using the *Y. pestis* YapH protein, a putative extracellular adhesin molecule related to the *Escherichia coli* AIDA-1 adhesin (Benz and Schmidt, 1992). This is a very large protein (3705 aa) with a highly biased sequence; 49% of the amino-acid residues are Ala, Arg or Gly. After the real matches, with very low *E*-values, are a number of matches to proteins from *Mycobacterium tuberculosis* with apparently highly significant scores of 10^{-20} to 10^{-30}. These *M. tuberculosis* PPE and PGRS proteins are also large, and again have a preponderance of certain amino-acid residues, Ala and Gly among them (Cole *et al.*, 1998). However, despite this, there is no co-linear similarity between these proteins and YapH (Figure 1.3b) – the high-scoring matches are solely due to random matches between the biased amino acids, accentuated by the large size of both proteins.

The third example shows the effect of repetitive sequences on the apparent significance of database matches. Figure 1.4 shows the FASTA database search for the *Y. pestis* MukB protein. The equivalent protein in *E. coli* has been shown to be involved in chromosome partitioning during cell division (Niki *et al.*, 1991), and again this is a large protein, with a strong prediction for a coiled-coil secondary structure. As can be seen from the search results, there is a weak, but statistically significant, match to eukaryotic myosin. However, the three-dimensional structure of MukB has recently been solved (van den Ent *et al.*, 1999), and it has shown that, in fact, there is no structural similarity between myosin and MukB. The matches apparent in the database search are due entirely to the 7-amino-acid repeat present in both proteins. The overall significance of short, locally insignificant matches between the repeat units is artificially amplified by the frequent occurrence of the repeats themselves.

In addition to all of the above, it is important to consider all the matches in a database search, not just the best. The consistency of different matches with each other and with known motifs within the protein (see below) should be considered. Spurious, but relatively high-scoring matches can always occur by chance. Always consider the level of the match when assigning function; a weaker match to a characterized protein may indicate that the function of the protein is similar, but not identical. It may be possible, for instance, to specify a catalytic activity (e.g. 'oxidoreductase' or 'transporter') without being able to be specific about the actual substrate. The fact that a protein with a similar activity on a specific substrate is the best match does not necessarily indicate that the query sequence has exactly the same specificity.

Finally, when assigning function based on database matches, remember that the data within those databases are not infallible. Be especially wary of assigning function based on the assigned function from other investigators' annotation (they may be wrong), and always try to use the comparison with a biologically characterized sequence wherever possible. Failure to observe this rule can lead to transitive annotation, where chains of functional assignations are each based on a previous annotation. This can amplify the effects of an initial erroneous assignment, leading to apparently consistent, but wrong, annotation of many related proteins in the database.

```
The best scores are:                              initn init1 opt   z-sc E(543010)
MUKB_ECOLI  CELL DIVISION PROTEIN MUKB.           7800 7800 7818 7496.1    0
P71227      177 KDA PROTEIN INVOLVED IN CHROMOSOME 6530 6488 6948 6661.4   0
MUKB_HAEIN  CELL DIVISION PROTEIN MUKB HOMOLOG.   2361 2361 6195 5939.3    0
Q9KRC8      CELL DIVISION PROTEIN MUKB.           3643 3643 5794 5554.8    0
O68583      CELL DIVISION PROTEIN MUKB (FRAGMENT). 1060 1060 1060 1027.6   0
Q9NKT9      L6202.3.                               113   70  393  371.5 3.8e-13
Q9NKR1      L6202.3.                               113   70  360  339.9 2.2e-11
Q9JI55      PLECTIN (FRAGMENT).                    167   84  337  313.6 6.4e-10
Q16640      PLECTIN.                               217   79  321  298.1 4.6e-09
Q63731      NEURONAL MYOSIN HEAVY CHAIN.           164   84  311  293.9 7.9e-09
O87075      Z72F PROTEIN (FRAGMENT).               230  230  290  289.8 1.3e-08
Q15149      PLECTIN.                               211   79  311  288.3 1.6e-08
MYSN_CHICK  MYOSIN HEAVY CHAIN, NONMUSCLE (CELLULA 110   64  302  285.4 2.4e-08
CAB89415    DJ477O4.6 (CENTROSOMAL NEK2-ASSOCIATED 232   74  302  283.9 2.8e-08
O60588      CENTROSOMAL NEK2-ASSOCIATED PROTEIN 1. 232   74  302  283.9 2.8e-08
Q21000      SIMILARITY TO C. ELEGANS MYOSIN HEAVY  114   52  300  283.4  3e-08
AAG24132    F58G4.1 PROTEIN.                       114   52  300  283.4  3e-08
Q9NJ22      MYOSIN HEAVY CHAIN CATCH (SMOOTH) MUSC 135   77  294  280.7 4.3e-08
Q9NJ21      MYOSIN HEAVY CHAIN CARDIAC MUSCLE SPEC 135   77  294  280.7 4.3e-08
Q9NJ20      MYOSIN HEAVY CHAIN CARDIAC MUSCLE SPEC 135   77  294  280.6 4.4e-08
```

Figure 1.4. FASTA results for a database search using the *Y. pestis* MukB protein as a query. The matches against myosin at the end of the list are statistically significant, but caused solely by the repeat structure of the proteins.

◆◆◆◆◆◆ DOMAIN/MOTIF SEARCHING

One way of avoiding the problems of functional assignments based on simple similarity, outlined above, is to use protein family or motif searches. The essence of such techniques is to identify sequences or patterns that are common to all members of a particular protein family and use these to identify new members of that family. The strength of these techniques lies in increasing sensitivity while reducing false positives by considering only those attributes of the protein that are evolutionarily conserved and, therefore, functionally or structurally important. Such tools also allow proteins to be analysed in a modular or domain-based way. Proteins frequently have a modular architecture and this can often confuse simple similarity-based searches, where different domains within the protein can match different sets of database sequences.

The earliest system to use such techniques was PROSITE (Bairoch, 1991). This used a simple motif-based search, looking for specific patterns of particular residues within a protein. This method is ideal for finding short functional motifs, such as enzyme active sites or small-molecule binding sites that can be specified with a simple syntax. A good example is the ATP/ADP binding site [AG].{4}GK[ST] (A or G, four of any residue, G, K, S or T). This method, though highly specific, does lack the flexibility and sensitivity needed to identify larger and more diffuse features, such as entire protein domains – and, indeed, PROSITE has since been expanded to include profile information (Falquet et al., 2002).

A more sensitive method for looking at larger domains involves using profiles. Rather than looking for the presence or absence of specific residues at particular sites, these methods look at which residues are likely to be present at a particular site in all members of the family, and how frequently, and uses these to weight the occurrence of any of those residues at that site in the query sequence. These weights can be calculated over the length of the domain to give an overall score. Such methods allow for much more sensitivity in determining whether any particular query is part of a given family. A common algorithm used for generating and querying such profiles is the hidden Markov model (HMM), as implemented, for example, in the HMMER package (http://hmmer.wustl.edu/; Eddy, 1998). HMMs were originally developed for linguistic analysis, but have since proved themselves ideal for complex sequence analysis. Protein family/domain profiles of this form are, of course, only as good as the collected families that are used to generate the profiles. A large amount of effort has gone into collecting and curating families of proteins and domains, and a good example of this is the PFAM collection (http://www.sanger.ac.uk/Software/Pfam/; Bateman et al., 2002). PFAM has generally aimed for broad families of proteins representing functional domains with general catalytic or structural properties. This often increases sensitivity in finding diverse members of the family, although sometimes at the expense of losing more precise information such as substrate specificities. A more tightly focused approach has been taken by TIGRFAM (http://www.tigr.org/TIGRFAMs/), which uses closely defined families with specific substrate specificities and catalytic activities. This has the benefit of increasing the resolution of the technique, but at the cost of losing the sensitivity to identify more distant relatives.

Other protein family/domain investigation tools exist, such as SMART (http://smart.embl-heidelberg.de/; Letunic et al., 2002) and ProDom (http://prodes.toulouse.inra.fr/prodom/doc/prodom.html; Corpet et al., 2000). The majority of these systems has now been collected together under the Interpro umbrella (http://www.ebi.ac.uk/interpro/; Apweiler et al., 2001), facilitating multiple analyses, allowing comparison between different analysis packages, and giving a consistent overview of the types of domains and motifs present.

◆◆◆◆◆◆ ORTHOLOGY, PARALOGY AND SYNTENY

Further mechanisms for the refinement of functional assignment exist, one of the most useful of which is the concept of orthology and paralogy.

These concepts can be used as guides for functional transfer of information from one complete genome to another. The simplest definition of orthologues is that they are equivalent genes in a pair of organisms that are directly descended from the same gene in the last common ancestor of the two organisms (Figure 1.5a). Following on from this definition, paralogues are genes that are related to each other through a gene duplication event, either in the parent organism (Figure 1.5b), or in one of the descendants (Figure 1.5c). Orthologous genes are assumed to serve the same cellular or molecular function, whereas paralagous genes may either perform the same function (leading to functional redundancy) or may have diverged after duplication to perform a different function, or to act on different substrates. If, therefore, we can identify which genes are orthologues and which paralogues of a given functionally characterized gene, we can use this information to transfer the functional description of the characterized gene on to the novel gene.

Orthologues can be computationally identified in complete genomes by looking for the reciprocal best matches between the gene sets. A given gene in the first genome is compared against all of the genes in the second, the best match is extracted and then compared back against all the genes in the first genome. If the second gene identifies the original query as the best match in the first genome, then they are described as reciprocal best matches. The comparisons are best done with predicted protein sequences, ideally using global alignment tools such as FASTA, though BLASTP can also be used. Constraints on the relative lengths of the two sequences and on the minimum length of the match compared to the full sequences are also usually applied. Obviously these methods can only be used to identify orthologues where the complete genome of both organisms is known. This analysis has also been extended to cover many genomes, resulting in clusters of orthologous groups (COGs; http://www.ncbi.nlm.nih.gov/COG/; Tatusov et al., 2001), which can be very useful for correlating functional assignments across multiple genomes.

Unfortunately, the identification of orthologues by the computation of reciprocal best matches is not foolproof. Several biological processes can

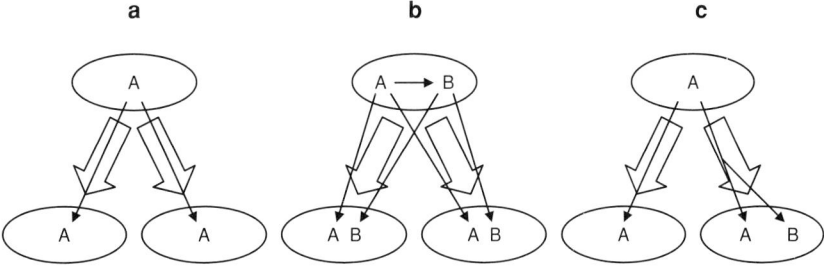

Figure 1.5. Illustration of orthology and paralogy. The upper ovals represent the ancestral species, and the lower ovals represent the descendant species; A and B represent genes. In each pair of descendant species A and A are orthologues, while A and B are paralogues. (a) represents a simple descent, (b) represents a gene duplication of A to B in the ancestral species, and (c) represents a gene duplication of A to B in one lineage only.

confound this process, leading to paralogous genes being identified as orthologues. One of these processes is that of gene loss. Consider a pair of duplicated genes (A and B) in an ancestral organism that have diverged and taken on different roles. If gene B is lost in one lineage, and gene A in another, the remaining genes will be identified as orthologous, when they are paralogues with different functions (Figure 1.6a). A similar scenario can be envisaged with gene acquisition. If gene A is lost in one descendant, which subsequently acquires a related gene B, A and B will be detected as orthologues when no such relationship exists (Figure 1.6b). In extreme cases, two individual lineages may independently acquire related genes that were not present in the ancestral organism. Again, these will be wrongly identified as orthologues. It should be stressed that these are not just hypothetical possibilities; although rare, examples can be seen in many pairs of related genomes.

The assignment of orthology can be checked in related genomes by looking at synteny, defined as conserved gene order and orientation. Where there has been some conservation of gene order, orthologues will generally be found in the same context as other ortholgous pairs between the genomes. Any large deviation from syntenic context, especially the presence of predicted orthologues within apparently unique gene islands, may indicate the need for caution in the assignment of orthology.

Even with the utmost care and attention in transferring annotation between genes, biology can still spring some surprises. It is possible for genes to be true orthologues in related organisms, and still have different functions. This is, of course, because genes can change their function within an organism, being recruited to serve new roles. A good example of this is an adhesin protein first identified in *Streptococcus sanguis*, where it was believed to be responsible for oral colonization by the bacterium (Fenno *et al.*, 1989). When related proteins were identified in other bacteria, they were also annotated as adhesins, and this description was applied to the family as a whole. As more sequences were generated, it became apparent that these genes were in fact part of a wider family of exported solute-binding proteins, and that this was the probable function of most of the members of this family. It appears that this transporter component has been recruited to act as an adhesin in this branch of the Streptococci, while retaining its original function in other related organisms.

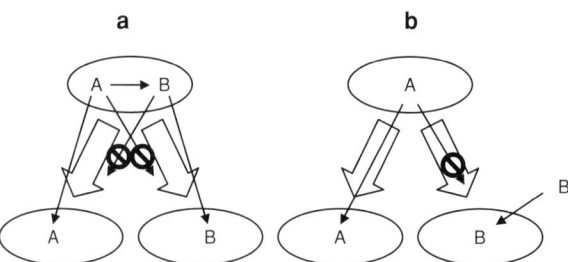

Figure 1.6. Illustration of problems with reciprocal best-match determination of orthology. In each case A and B would be detected as orthologues, although the true orthologues have been removed by gene loss, indicated by the barred circles.

An even more extreme example of protein recruitment was uncovered when the gene for the insect toxin of *Enterobacter aerogenes* was cloned and sequenced (Yoshida *et al.*, 2001). On analysis, it was discovered to be >98% identical to the *Escherichia coli* protein-folding chaperone GroEL, differing at only 11/548 aa. Furthermore, it could be shown, by site-directed mutagenesis, that changing any one of four variant amino acids in the *Escherichia coli* GroEL protein could confer insect-toxin activity on that protein. A sobering lesson for anybody attempting to predict function by similarity.

◆◆◆◆◆◆ NON-SIMILARITY-BASED METHODS

Whilst the only way of positively ascribing a specific function to a particular gene product remains laboratory experiment, the similarity-based methods outlined above may allow a specific function to be putatively attached with a reasonable degree of accuracy. Beyond this, it is possible to ascribe general functional classes or activities to unknown genes based on similarity to characterized proteins. However, there are still a large number of predicted proteins in microbial genomes with no apparent similarity to any other known protein, and very little functional information can be attached to these, except when general, non-specific, domains such as signal sequences or transmembrane domains can be identified. Between these two extremes there is a third class of predicted proteins: those that show similarity only to other predicted proteins for which no function has been suggested. These are generally termed 'conserved hypothetical proteins'. Obviously it is impossible to give any sort of functional prediction about such genes using similarity-based methods, at least until such time as one member of the family is characterized.

Other, non-similarity-based, techniques are now becoming available for investigating such proteins. While they are not intended to find specific functions for these proteins directly, they can indicate metabolic pathways or functional clusters in which they might be involved. At present most of these techniques are experimental and many of the programs used to implement them are proprietary. For these reasons, the techniques will only be outlined briefly in this chapter.

The basis of all of these techniques is to attempt to show an association between the conserved hypothetical genes and other genes of known function. There are three current methods for investigating this association. The first is to look at sequence clustering: when looking across genomes, does the unknown protein occur physically close to, or appear to be co-transcribed with, genes of known function? If this is the case when analysing many different genomes, there is a possibility that the unknown gene may be in the same metabolic pathway or functional group as the known genes, thus generating a hypothesis for laboratory testing (Overbeek *et al.*, 1999). The second technique is similar, but instead of looking at physical co-occurrence, it examines patterns of phylogenetic co-occurrence, i.e. does the unknown protein appear in particular phylogenetic subsets, and does that correlate with any proteins of known function? (Pellegrini et al., 1999). The third technique is based on the fact that multifunctional proteins exist, especially in eukaryotes, where two or more independent biochemical functions exist in different domains of the same protein. Frequently these independent biochemical functions form different parts, often sequential, of the same pathway. Equally importantly, these domains are often found in separate proteins in other organisms. The task then becomes that of looking for such domain fusions across diverse genomes and using them to associate unknown genes with genes of known function (Marcotte et al., 1999).

These techniques can be particularly useful when looking for genes to associate with biological or biochemical functions that have been demonstrated to be present, but for which the protein responsible is unknown. This can include 'missing' components in biochemical pathways, and several of these have now been discovered and subsequently confirmed experimentally using such methods (Daugherty et al., 2001).

◆◆◆◆◆◆ HORIZONTAL GENE TRANSFER AND OTHER ANOMALIES

The importance of horizontal gene transfer is still a source of much discussion, at least in terms of the transfer across large phylogenetic distances, or in eukaryotic organisms. Methods of detecting such transfers include looking for phylogenetic trees for individual genes that are incongruent with the accepted tree for the organism itself, and looking for the incongruent presence or absence of genes in different lineages. Such techniques are complex to interpret, and riddled with potential sources for error, including evolutionary rate variation of individual genes, limited sampling and differential lineage-specific genes loss. For these reasons, these techniques will not be considered in detail here; interested investigators would be wise to start with one of the many recent reviews of the subject (Eisen, 2000; Koonin et al., 2001).

In microbial genomes, the importance of horizontal gene transfer is accepted to a much greater degree, and is certainly easier to quantify, especially when looking at recent events (Lawrence and Ochman, 2002; Ochman et al., 2000). Especially in certain families of bacterial pathogens, such as the Enterobacteriaceae, many of the differences between strains and species can be attributed to fairly recent horizontal transfer (McClelland et al., 2001; Parkhill et al., 2001a; Perna et al., 2001). Such recent transfers can often be detected by looking at the variation in genomic signatures, including $G+C$ content, dinucleotide frequency and codon usage (Karlin, 2001) (Figure 1.7). The assumption behind such measurements is that individual genomes have characteristic levels or biases for each of these measurements. The simplest of these is $G+C$ content; it has been known for some time that different organisms contain different proportions of $G+C$ vs. $A+T$. However, with the advent of genomic sequencing it was discovered that this bias in $G+C$ content is fairly stable

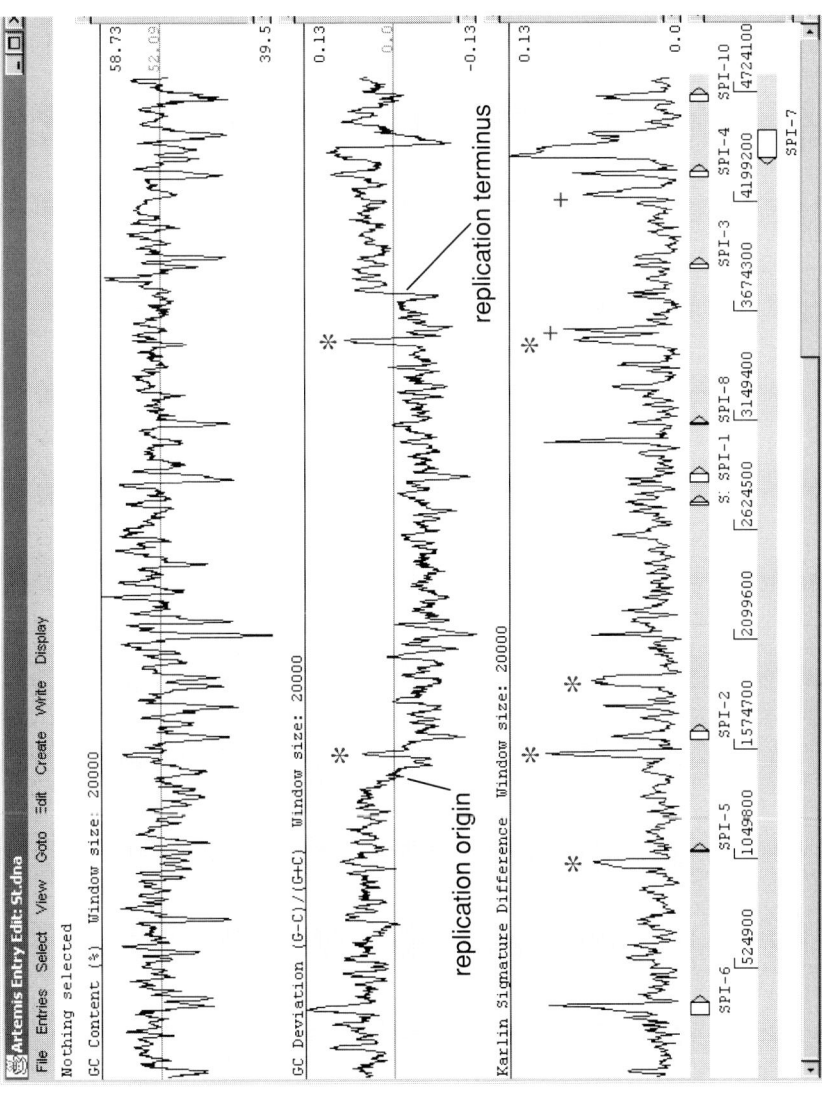

Figure 1.7. Anomalies in large-scale nucleotide plots. The complete genome of *Salmonella typhii* is illustrated with plots for G + C content, G/C bias and dinucleotide (Karlin) signature. It can be seen that the classical horizontally transferred *Salmonella* pathogenicity islands (SPIs), indicated by open boxes in the lower panel, show anomalies in one or more of the plots. Also indicated are prophages (*), which are horizontally acquired, and ribosomal protein operons (+), which are not.

across the whole genome. Thus, a region with a significant difference in G + C content from the genomic background may indicate that it has been recently acquired from an organism with a different genomic background. Further investigation revealed that the bias in usage of nucleotides extends further than just individual bases: different genomes also tend to have a characteristic dinucleotide content, with certain pairs of consecutive nucleotides being more common than others (Karlin and Burge, 1995). The deviation from this 'dinucleotide signature' can also be plotted across the genome, and again significant deviation from the genomic norm may indicate the presence of horizontally acquired genomic DNA. A further extension to this is the use of codon preference; different genomes have different patterns of preferred codons, and looking for deviations from the norm can be a profitable way of finding candidate horizontally transferred genes. Bacterial genomes exhibit another global bias, that of G/C skew. For some reason, probably a mutational asymmetry in the replication fork, there tend to be more G residues on the leading strand than the lagging strand (Lobry, 1996). Plotting $(G - C)/(G + C)$ over the genome will generally give a plot that crosses the axis twice, once at the origin of replication and once at the terminus, with the bulk of the plot being consistently above or below the axis (Figure 1.7, second panel). Significant deviations from this plot may either indicate recent insertions of DNA, or recent rearrangements, such as the large-scale insertion sequence (IS)-element mediated inversions seen in the genome of *Yersinia pestis* (Parkhill *et al.*, 2001b; Figure 1.8).

For each of these measurements, there are several provisos that must be considered. Obviously, they will not detect DNA that has been acquired from a very similar genomic background. In addition, each of these properties is imposed in some way on the genomic sequence (e.g G/C bias is thought to be imposed by asymmetries in the replication fork and differences in codon usage by the tRNA pool) and, therefore, the differences in externally acquired DNA will ameliorate over time until the signal is lost.

The final, and most important, proviso is that not all biases in genomic DNA are due to horizontal acquisition. Sequence constraints can also play a strong role in producing perturbations in any of these measurements. Such constrains include preferential coding for specific amino-acid subsets, such as those in large hydrophobic proteins, and small basic ribosomal proteins; indeed, the major ribosomal protein operon often shows the largest deviation in dinucleotide signature in many bacterial genomes. Large repetitive proteins containing a short or biased amino-acid repeat will often also give strong nucleotide and codon bias signals (the Gly-rich repetitive PGRS proteins of *Mycobacterium tuberculosis* are particularly noteworthy in this regard; Cole *et al.*, 1998).

Investigations of potential horizontally acquired islands in genomes should, therefore, be extended to include other possible indicators; large mobile islands, especially those carrying pathogenicity genes, may often carry mobility determinants, such as phage integrases or transposases. Those with phage-derived mobility genes may also be adjacent to tRNA genes, a common integration site for prophage. In closely related syntenic

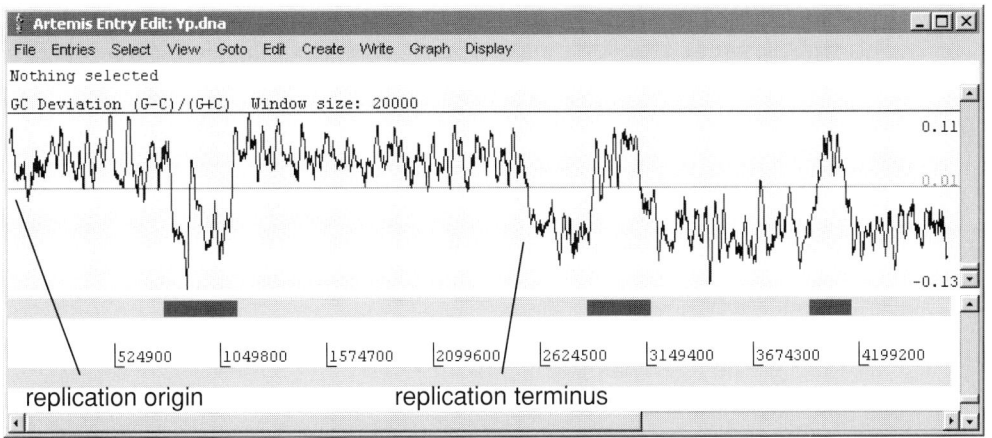

Figure 1.8. Anomalies in large-scale nucleotide plots. The G/C bias plot for the entire genome of *Yersinia pestis* is shown. The three large anomalies caused by genomic rearrangements are indicated with black bars. Compare this to the G/C bias plot of *Salmonella typhi* (Figure 1.7), where the majority of the plot is consistently above or below the axis, crossing at the origin and terminus of replication.

genomes, this analysis can also be augmented by looking directly at the genomic comparison for insertions and deletions, facilitated by a visualization tool, such as the Artemis Comparison Tool (ACT) (http://www.sanger.ac.uk/Software/ACT/).

◆◆◆◆◆◆ FUNCTIONAL CLASSIFICATION/METABOLIC RECONSTRUCTION

One of the final stages in any annotation is often an attempt to identify or reconstruct the metabolic capabilities of the organism under question. This analysis is usually facilitated by attempting to classify functionally all of the predicted gene products for which some kind of function can be inferred. The first attempt to produce a comprehensive hierarchical functional classification of the gene products of a single organism was for *E. coli* (Riley, 1993; Riley and Labedan, 1996), and the classification scheme developed for this project was adapted for many subsequent microbial genome projects. Classifying genes in this way allows the rapid identification of proteins required for known biochemical pathways, and the clustering of genes involved in specific cellular activities, such as host interaction. One-dimensional hierarchical classification schemes of this type do, however, suffer somewhat from the fact that gene products are only placed in one category, when they may participate in many different pathways or functions. Many analysis teams are now migrating to more complex classification schemes, the most common of which is the Gene Ontology (http://www.geneontology.org; Ashburner *et al.*, 2000), a collaborative attempt to develop a systematic descriptive vocabulary to describe and classify protein functions.

Metabolic reconstruction from genomic sequence can uncover unexpected capabilities and suggest new routes for subsequent research. A good example of this comes from the genome sequencing of *Mycobacterium tuberculosis* (Cole et al., 1998). *M. tuberculosis* was classically considered to be an obligate aerobe, so it was surprising when a number of genes potentially involved in anaerobic respiration were discovered in the genomic sequence. Subsequent laboratory investigation of these genes indicated that they were expressed under anaerobic conditions (Hutter and Dick, 1999), and might well be important for the virulence of the organism (Weber *et al.*, 2000).

Several excellent collections of metabolic pathway information are available to assist in metabolic reconstruction attempts, and good examples of these are the Kyoto Encyclopaedia of Genes and Genomes (KEGG; http://www.genome.ad.jp/kegg/), and EcoCyc (http://ecocyc.org/). Other tools exist, such as WIT (http://wit.mcs.anl.gov/WIT2/; Overbeek *et al.*, 2000), which can also be used to automate some of these procedures. Obviously the metabolism of any microbe can only be reconstructed as far as the biochemical knowledge of current model organisms allows. Genomic analysis cannot, on its own, predict completely novel metabolic capabilities, although it may be able to suggest routes for experimental investigation. It should always be remembered that the metabolism of any organism may be significantly different from current models, and any attempt to investigate novel systems by comparison with these may risk imposing an apparent uniformity that is not a true reflection of the capabilities of the organism under study. Sources such as the Minnesota Biocatalysis/Biodegradation database (http://umbbd.ahc.umn.edu/) give a useful insight into the breadth of biochemistry of which microorganisms are capable.

◆◆◆◆◆◆ CONCLUDING REMARKS

As can be concluded from many of the points made above, annotation of genomic information is an extremely powerful tool for the understanding of biology, but can be fraught with problems and pitfalls. Whilst the results of most programs and techniques used can be trusted most of the time, the annotator must always be aware of potential problems. It is important to be guided by biological knowledge and understanding, and not solely by bioinformatic analysis, in order to gain a true insight into the meaning of the genomic information available.

References

Altschul, S. F., Gish, W., Miller, W., Myers, E. W. and Lipman, D. J. (1990). Basic local alignment search tool. *J. Mol. Biol.* **215**, 403–410.

Apweiler, R., Attwood, T. K., Bairoch, A., Bateman, A., Birney, E., Biswas, M., Bucher, P., Cerutti, L., Corpet, F., Croning, M. D., Durbin, R., Falquet, L., Fleischmann, W., Gouzy, J., Hermjakob, H., Hulo, N., Jonassen, I., Kahn, D.,

Kanapin, A., Karavidopoulou, Y., Lopez, R., Marx, B., Mulder, N. J., Oinn, T. M., Pagni, M. and Servant, F. (2001). The InterPro database, an integrated documentation resource for protein families, domains and functional sites. *Nucleic Acids Res.* **29**, 37–40.

Argaman, L., Hershberg, R., Vogel, J., Bejerano, G., Wagner, E. G., Margalit, H. and Altuvia, S. (2001). Novel small RNA-encoding genes in the intergenic regions of *Escherichia coli*. *Curr. Biol.* **11**, 941–950.

Ashburner, M., Ball, C. A., Blake, J. A., Botstein, D., Butler, H., Cherry, J. M., Davis, A. P., Dolinski, K., Dwight, S. S., Eppig, J. T., Harris, M.A., Hill, D. P., Issel-Tarver, L., Kasarskis, A., Lewis, S., Matese, J. C., Richardson, J. E., Ringwald, M., Rubin, G. M. and Sherlock, G. (2000). Gene ontology: tool for the unification of biology. The Gene Ontology Consortium. *Nature Genet.* **25**, 25–29.

Badger, J. H. and Olsen, G. J. (1999). CRITICA: coding region identification tool invoking comparative analysis. *Mol. Biol. Evol.* **16**, 512–524.

Bairoch, A. (1991). PROSITE: a dictionary of sites and patterns in proteins. *Nucleic Acids Res.* **19**(Suppl), 2241–2245.

Bateman, A., Birney, E., Cerruti, L., Durbin, R., Etwiller, L., Eddy, S. R., Griffiths-Jones, S., Howe, K. L., Marshall, M. and Sonnhammer, E. L. (2002). The Pfam protein families database. *Nucleic Acids Res.* **30**, 276–280.

Benz, I. and Schmidt, M. A. (1992). AIDA-I, the adhesin involved in diffuse adherence of the diarrhoeagenic *Escherichia coli* strain 2787 (O126:H27), is synthesized via a precursor molecule. *Mol. Microbiol.* **6**, 1539–1546.

Bibb, M. J., Findlay, P. R. and Johnson, M. W. (1984). The relationship between base composition and codon usage in bacterial genes and its use for the simple and reliable identification of protein-coding sequences. *Gene* **30**, 157–166.

Borodovsky, M. and McIninch, J. (1993). GeneMark: parallel gene recognition for both DNA strands. *Computers Chemistry* **17**, 123–133.

Cole, S. T., Brosch, R., Parkhill, J., Garnier, T., Churcher, C., Harris, D., Gordon, S. V., Eiglmeier, K., Gas, S., Barry, C. E., Tekaia, F., Badcock, K., Basham, D., Brown, D., Chillingworth, T., Connor, R., Davies, R., Devlin, K., Feltwell, T., Gentles, S., Hamlin, N., Holroyd, S., Hornsby, T., Jagels, K., Krogh, A., McLean, J., Moule, S., Murphy, L., Oliver, K., Osborne, K., Quail, M. A., Rajandream, M. A., Rogers, J., Rutter, S., Seeger, K., Skelton, J., Squares, S., Squares, R., Sulston, J. E., Taylor, K., Whitehead, S. and Barrell, B. G. (1998). Deciphering the biology of *Mycobacterium tuberculosis* from the complete genome sequence. *Nature* **393**, 537–544.

Corpet, F., Servant, F., Gouzy, J. and Kahn, D. (2000). ProDom and ProDom-CG: tools for protein domain analysis and whole genome comparisons. *Nucleic Acids Res.* **28**, 267–269.

Daugherty, M., Vonstein, V., Overbeek, R. and Osterman, A. (2001). Archaeal shikimate kinase, a new member of the GHMP-kinase family. *J. Bacteriol.* **183**, 292–300.

Delcher, A. L., Harmon, D., Kasif, S., White, O. and Salzberg, S. L. (1999). Improved microbial gene identification with GLIMMER. *Nucleic Acids Res.* **27**, 4636–4641.

Eddy, S. R. (1998). Profile hidden Markov models. *Bioinformatics* **14**, 755–763.

Eisen, J. A. (2000). Horizontal gene transfer among microbial genomes: new insights from complete genome analysis. *Curr. Opin. Genet. Dev.* **10**, 606–611.

Falquet, L., Pagni, M., Bucher, P., Hulo, N., Sigrist, C. J., Hofmann, K. and Bairoch, A. (2002). The PROSITE database, its status in 2002. *Nucleic Acids Res.* **30**, 235–238.

Fenno, J. C., LeBlanc, D. J. and Fives-Taylor, P. (1989). Nucleotide sequence analysis of a type 1 fimbrial gene of *Streptococcus sanguis* FW213. *Infect. Immun.* **57**, 3527–3533.

Frishman, D., Mironov, A., Mewes, H. W. and Gelfand, M. (1998). Combining diverse evidence for gene recognition in completely sequenced bacterial genomes. *Nucleic Acids Res.* **26**, 2941–2947.

Holley, R. W. (1965). Structure of an alanine transfer ribonucleic acid. *JAMA* **194**, 868–871.

Hutter, B. and Dick, T. (1999). Up-regulation of narX, encoding a putative 'fused nitrate reductase' in anaerobic dormant *Mycobacterium bovis* BCG. *FEMS Microbiol. Lett.* **178**, 63–69.

Karlin, S. (2001). Detecting anomalous gene clusters and pathogenicity islands in diverse bacterial genomes. *Trends Microbiol.* **9**, 335–343.

Karlin, S. and Burge, C. (1995). Dinucleotide relative abundance extremes: a genomic signature. *Trends Genet.* **11**, 283–290.

Koonin, E. V., Makarova, K. S. and Aravind, L. (2001). Horizontal gene transfer in prokaryotes: quantification and classification. *Annu. Rev. Microbiol.* **55**, 709–742.

Lawrence, J. G. and Ochman, H. (2002). Reconciling the many faces of lateral gene transfer. *Trends Microbiol.* **10**, 1–4.

Letunic, I., Goodstadt, L., Dickens, N. J., Doerks, T., Schultz, J., Mott, R., Ciccarelli, F., Copley, R. R., Ponting, C. P. and Bork, P. (2002). Recent improvements to the SMART domain-based sequence annotation resource. *Nucleic Acids Res.* **30**, 242–244.

Lobry, J. R. (1996). Asymmetric substitution patterns in the two DNA strands of bacteria. *Mol. Biol. Evol.* **13**, 660–665.

Lowe, T. M. and Eddy, S. R. (1997). tRNAscan-SE: a program for improved detection of transfer RNA genes in genomic sequence. *Nucleic Acids Res.* **25**, 955–964.

Lukashin, A. V. and Borodovsky, M. (1998). GeneMark.hmm: new solutions for gene finding. *Nucleic Acids Res.* **26**, 1107–1115.

Marcotte, E. M., Pellegrini, M., Ng, H. L., Rice, D. W., Yeates, T. O. and Eisenberg, D. (1999). Detecting protein function and protein–protein interactions from genome sequences. *Science* **285**, 751–753.

McClelland, M., Sanderson, K. E., Spieth, J., Clifton, S. W., Latreille, P., Courtney, L., Porwollik, S., Ali, J., Dante, M., Du, F., Hou, S., Layman, D., Leonard, S., Nguyen, C., Scott, K., Holmes, A., Grewal, N., Mulvaney, E., Ryan, E., Sun, H., Florea, L., Miller, W., Stoneking, T., Nhan, M., Waterston, R. and Wilson, R. K. (2001). Complete genome sequence of *Salmonella enterica* serovar Typhimurium LT2. *Nature* **413**, 852–856.

Nakamura, Y., Gojobori, T. and Ikemura, T. (2000). Codon usage tabulated from international DNA sequence databases: status for the year 2000. *Nucleic Acids Res.* **28**, 292.

Needleman, S. B. and Wunsch, C. D. (1970). A general method applicable to the search for similarities in the amino acid sequence of two proteins. *J. Mol. Biol.* **48**, 443–453.

Niki, H., Jaffe, A., Imamura, R., Ogura, T. and Hiraga, S. (1991). The new gene *mukB* codes for a 177 kd protein with coiled-coil domains involved in chromosome partitioning of *E. coli*. *EMBO J.* **10**, 183–193.

Ochman, H., Lawrence, J. G. and Groisman, E.A. (2000). Lateral gene transfer and the nature of bacterial innovation. *Nature* **405**, 299–304.

Overbeek, R., Fonstein, M., D'Souza, M., Pusch, G.D. and Maltsev, N. (1999). Use of contiguity on the chromosome to predict functional coupling. *In Silico Biol.* **1**, 93–108.

Overbeek, R., Larsen, N., Pusch, G. D., D'Souza, M., Selkov, E., Jr, Kyrpides, N., Fonstein, M., Maltsev, N. and Selkov, E. (2000). WIT: integrated system for high-throughput genome sequence analysis and metabolic reconstruction. *Nucleic Acids Res.* **28**, 123–125.

Parkhill, J., Wren, B. W., Mungall, K., Ketley, J. M., Churcher, C., Basham, D., Chillingworth, T., Davies, R. M., Feltwell, T., Holroyd, S., Jagels, K., Karlyshev, A. V., Moule, S., Pallen, M. J., Penn, C. W., Quail, M. A., Rajandream, M. A., Rutherford, K. M., van Vliet, A. H., Whitehead, S. and Barrell, B. G. (2000). The genome sequence of the food-borne pathogen *Campylobacter jejuni* reveals hypervariable sequences. *Nature* **403**, 665–668.

Parkhill, J., Dougan, G., James, K. D., Thomson, N. R., Pickard, D., Wain, J., Churcher, C., Mungall, K. L., Bentley, S. D., Holden, M. T., Sebaihia, M., Baker, S., Basham, D., Brooks, K., Chillingworth, T., Connerton, P., Cronin, A., Davis, P., Davies, R. M., Dowd, L., White, N., Farrar, J., Feltwell, T., Hamlin, N., Haque, A., Hien, T. T., Holroyd, S., Jagels, K., Krogh, A., Larsen, T. S., Leather, S., Moule, S., O'Gaora, P., Parry, C., Quail, M., Rutherford, K., Simmonds, M., Skelton, J., Stevens, K., Whitehead, S. and Barrell, B. G. (2001a). Complete genome sequence of a multiple drug resistant *Salmonella enterica* serovar Typhi CT18. *Nature* **413**, 848–852.

Parkhill, J., Wren, B. W., Thomson, N. R., Titball, R. W., Holden, M. T., Prentice, M. B., Sebaihia, M., James, K. D., Churcher, C., Mungall, K. L., Baker, S., Basham, D., Bentley, S. D., Brooks, K., Cerdeno-Tarraga, A. M., Chillingworth, T., Cronin, A., Davies, R. M., Davis, P., Dougan, G., Feltwell, T., Hamlin, N., Holroyd, S., Jagels, K., Karlyshev, A. V., Leather, S., Moule, S., Oyston, P. C., Quail, M., Rutherford, K., Simmonds, M., Skelton, J., Stevens, K., Whitehead, S. and Barrell, B. G. (2001b). Genome sequence of *Yersinia pestis*, the causative agent of plague. *Nature* **413**, 523–527.

Pearson, W. R. and Lipman, D. J. (1988). Improved tools for biological sequence comparison. *Proc. Natl Acad. Sci. USA* **85**, 2444–2448.

Pellegrini, M., Marcotte, E. M., Thompson, M. J., Eisenberg, D. and Yeates, T. O. (1999). Assigning protein functions by comparative genome analysis: protein phylogenetic profiles. *Proc. Natl Acad. Sci. USA* **96**, 4285–4288.

Perna, N. T., Plunkett, G., 3rd, Burland, V., Mau, B., Glasner, J. D., Rose, D. J., Mayhew, G. F., Evans, P. S., Gregor, J., Kirkpatrick, H. A., Posfai, G., Hackett, J., Klink, S., Boutin, A., Shao, Y., Miller, L., Grotbeck, E. J., Davis, N. W., Lim, A., Dimalanta, E. T., Potamousis, K. D., Apodaca, J., Anantharaman, T. S., Lin, J., Yen, G., Schwartz, D. C., Welch, R. A. and Blattner, F. R. (2001). Genome sequence of enterohaemorrhagic *Escherichia coli* O157:H7. *Nature* **409**, 529–533.

Riley, M. (1993). Functions of the gene products of *Escherichia coli*. *Microbiol. Rev.* **57**, 862–952.

Riley, M. and Labedan, B. (1996). *Escherichia coli* gene products: physiological functions and common ancestries. In Escherichia coli *and* Salmonella: *Cellular and Molecular Biology* (Neidhardt, F. C. and Curtiss, R., eds), pp. 2118–2202. Washington, DC: ASM Press.

Rivas, E., Klein, R. J., Jones, T. A. and Eddy, S. R. (2001). Computational identification of noncoding RNAs in E. coli by comparative genomics. *Curr. Biol.* **11**, 1369–1373.

Rutherford, K., Parkhill, J., Crook, J., Horsnell, T., Rice, P., Rajandream, M. A. and Barrell, B. (2000). Artemis: sequence visualization and annotation. *Bioinformatics* **16**, 944–945.

Smith, T. F. and Waterman, M. S. (1981). Identification of common molecular subsequences. *J. Mol. Biol.* **147**, 195–197.

Sonnhammer, E. L. and Durbin, R. (1995). A dot-matrix program with dynamic threshold control suited for genomic DNA and protein sequence analysis. *Gene* **167**, GC1–10.

Tatusov, R. L., Natale, D. A., Garkavtsev, I. V., Tatusova, T. A., Shankavaram, U. T., Rao, B. S., Kiryutin, B., Galperin, M. Y., Fedorova, N. D. and Koonin, E. V. (2001). The COG database: new developments in phylogenetic classification of proteins from complete genomes. *Nucleic Acids Res.* **29**, 22–28.

van den Ent, F., Lockhart, A., Kendrick-Jones, J. and Lowe, J. (1999). Crystal structure of the N-terminal domain of MukB: a protein involved in chromosome partitioning. *Structure Fold Des.* **7**, 1181–1187.

Wassarman, K. M., Repoila, F., Rosenow, C., Storz, G. and Gottesman, S. (2001). Identification of novel small RNAs using comparative genomics and microarrays. *Genes Dev.* **15**, 1637–1651.

Weber, I., Fritz, C., Ruttkowski, S., Kreft, A. and Bange, F. C. (2000). Anaerobic nitrate reductase (narGHJI) activity of *Mycobacterium bovis* BCG in vitro and its contribution to virulence in immunodeficient mice. *Mol. Microbiol.* **35**, 1017–1025.

Williams, K. P. (2002). The tmRNA Website: invasion by an intron. *Nucleic Acids Res.* **30**, 179–182.

Yoshida, N., Oeda, K., Watanabe, E., Mikami, T., Fukita, Y., Nishimura, K., Komai, K. and Matsuda, K. (2001). Protein function. Chaperonin turned insect toxin. *Nature* **411**, 44.

2 From Sequence to Consequence: *In Silico* Hypothesis Generation and Testing

Mark Pallen

Division of Immunity and Infection, Medical School, University of Birmingham, Birmingham B15 2TT, UK

◆◆

CONTENTS

Introduction
Methods
Parsing proteins into domains
Genomic context
Hypothesis formulation
Worked examples
From *in silico* to *in vitro* or *in vivo*
Conclusions

◆◆◆◆◆◆ INTRODUCTION

Bacteriologists are facing a flood of sequence data. At the time of writing, 57 complete bacterial genome sequences have been published, together with 13 archaeal sequences, but a further 265 prokaryotic genome-sequencing projects are under way (Genomes OnLine Database; Bernal *et al.*, 2001). This means that within a few years we will have over 300 prokaryotic genome sequences, releasing roughly 600 000 protein-coding gene sequences into the public databases. Extrapolating from the finished genomes, we can expect around 25–33% of these protein-coding sequences, i.e. 150 000–200 000 to be of completely unknown function!

How are we to make sense of all this information? One way is to exploit the range of high-throughput functional genomics approaches described elsewhere in this volume. Another is to try to use the sequence data themselves to devise hypotheses that can be tested in the laboratory and that can inform our view of the biology of the organism(s) from which they came. This kind of *in silico* hypothesis generation predates the genomic era, but becomes much more effective when one can take a broad view across the whole of biology and its attendant 'sequence space'. It then often becomes clear that assumptions about what is typical of a bacterial system, based on observations in a single model organism, are simplistic or even erroneous when placed in the context of an analysis drawing on dozens or hundreds of genome sequences, for example, the archetypal sortase gene in *Staphylococcus aureus* is rather atypical in not clustering with genes of its substrates (Pallen *et al.*, 2001a).

Several impediments limit our ability to draw the clearest and best-informed conclusions from the sequence data. Annotated completed genome sequences and their attendant protein sequences are available in the National Center for Biotechnology Information (NCBI; Benson *et al.*, 2002), European Molecular Biology Laboratory (EMBL; Stoesser *et al.*, 2002) and Japanese (Tateno *et al.*, 2002) public databases, but sequences from projects in progress, although usually in the public domain, are scattered across on-line sites around the world and are usually unannotated. Any attempt to formulate novel hypotheses should aim to draw on this best, most current data, but searching all the unfinished genome projects individually is very time consuming. With that in mind, several attempts have been made to gather up these sequences into a single searchable database (Tateno *et al.*, 2002; The Institute for Genomic Research), including my own ViruloGenome site (http//www.vge.ac.uk). These not only save time by allowing many in-progress sequences to be searched simultaneously and by facilitating follow-up of hits by easy sequence retrieval but, in the case of ViruloGenome, also allow more sensitive searches through, for example, position-specific iterative basic local alignment sequence tool (PSI-BLAST) searches over more comprehensive databases.

Use of unfinished sequence data can bring other problems. Broad-based analyses of protein families can often add value to a genome sequence, for example, our analysis of sortase-like proteins and their substrates has been cited in some definitive genome sequence papers (Hoskins *et al.*, 2001; Pallen *et al.*, 2001; Tettelin *et al.*, 2001). However, publication of analyses of extended regions of a genome could jeopardize the publication of a definitive description of the sequence by those who did the hard work in completing it (Marshall, 2002). For this reason, careful attention should be paid to acceptable use policies associated with the data and, when in doubt, it is advisable to contact the owners of the sequence.

Also, one must always remember that any analyses on incomplete genome sequences and their predicted proteins must be considered tentative until the genome is complete. Although an incomplete genome sequence is better than none, Parkhill (2000) has stressed the importance of finishing genome sequences as crucial genes may lurk in the gaps.

Even finished genome sequence data present problems in that annotation is usually superficial, is often misleading and is sometimes plain wrong (Bork and Koonin, 1998; Galperin and Koonin, 1998; Brenner, 1999; Dandekar *et al.*, 2000; Devos and Valencia, 2001; Hegyi and Gerstein, 2001; Karp *et al.*, 2001; Suzek *et al.*, 2001; Wang and Zhang, 2001; Bacs *et al.*, 2002; Ouzounis and Karp, 2002). This is not surprising, given that sequence annotation is a labour-intensive process that represents a major bottleneck in genome-sequencing projects. All too often the annotation simply reflects the title line of the highest scoring BLAST hit, prefixed with the term 'putative' – an annoying habit, dubbed 'putativism' by Wassenaar and Gaastra (2001). This can mean that a function of the database protein can be assigned to the search protein in error, because the function resides in a part of the database protein sequence not represented in the BLAST result (Hegyi and Gerstein, 2001). Alternatively, the careless annotator may blur the distinction between the molecular action of a protein or domain (e.g. iron-binding protein) and its physiological role or location (e.g. periplasmic) and inappropriately transfer functional assignments to a distant homologue embedded in an entirely different physiological context (e.g. in an organism lacking a periplasm – see Entrez protein entry NP_269229 or try searching the Entrez protein database with the terms 'periplasmic' and 'firmicutes').

Once made, such errors can be propagated through the databases (see worked example 1 below). For this reason, one should never accept any functional assignment without drilling back through the annotation and literature trail to the original evidence that linked function to sequence, and then checking that this evidence is indeed pertinent to the sequence in front of you. Sometimes the trail runs cold where an annotated sequence has been deposited in the database without a link to a publication. In such cases, it is often worth searching PubMed (http://www.pubmed.gov) with the names of those who deposited the sequence (there usually is an associated publication) and/or contacting the individuals concerned by e-mail (search for their contact details using Google (http://www.google.com)). Searching for cross-references to carefully annotated resources such as SwissProt (http://www.expasy.ch/sprot/) and InterPro (http://www.ebi.ac.uk/interpro/) can also often save time.

However, once you have drilled down to the paper describing the experiments, take time to appraise the experimental evidence and how it fits in with the rest of the literature and with the *in silico* data. Sometimes you can confidently dismiss even experimental data as flawed, particularly when armed with a compelling bioinformatics analysis (see worked example 1 below).

A final major impediment to the proper use of sequence data in hypothesis formulation is the careless attitude of many scientists. All too often excellent laboratory science is accompanied by superficial and erroneous sequence analysis (Pallen *et al.*, 1999; Reid *et al.*, 2001). At best this simply clutters up the literature with rubbish, at worst it means that time is wasted at the bench trying to investigate unsupportable and ill-conceived hypotheses. To minimize the risk of errors, it is essential to adopt the same healthy scepticism and self-doubt that one would apply to laboratory work. Always ask the questions: 'Am I fooling myself? What other sequence information could prove this idea right or wrong?' Remember that almost all evidence from sequence analyses is circumstantial (which is why the words 'putative' and 'probable' feature so heavily in such analyses); direct evidence comes only from laboratory-based experiments. However, before leaping to confirm *in silico*

Table 2.1. Selected hypotheses on bacterial protein function generated from sequence analysis

Hypotheses	References	In silico evidence	Supporting evidence	Laboratory confirmed?	References
1 Coiled-coil domains mediate protein–protein interactions in type III secretion systems	Pallen et al. (1997)	~50% of type III-secreted proteins have coiled coils	Only ~5% of all E. coli proteins have coiled coils; EspA pilus resembles flagellum, secreted through homologue system; flagellum assembly relies on coiled-coil interactions	Yes	Delahay et al. (1999), Day and Plano (2000), Daniell et al. (2001)
2a PDZ domains in periplasmic proteases mediate C-terminal substrate selection 2b They bind to ssrA-encoded tag	Pallen and Ponting (1997), Pallen and Wren (1997)	PDZ domains found by homology searches	PDZ domains in eukaryotic proteins usually recognize C-terminal sequences; periplasmic proteases stress-induced; ssrA-tagging targets misfolded proteins	Yes	Beebe et al. (2000), Spiers et al. (2002)
3a Sortase-mediated protein sorting to cell surface occurs in many Gram-positives 3b Sortases form covalent links between fimbrial monomers	Pallen et al. (2001a)	Sortases found by VGE-PSI-BLAST	Genes for sortase-substrates cluster with sortase genes Sortase-like proteins needed for fimbrial assembly; fimbrial filaments covalently linked; sortases are trans-peptidases	Not yet	—
4a Novel ADP-ribosyltransferases occur in many bacteria 4b Novel toxins in Salmonella typhi and Streptococcus pyogenes	Pallen et al. (2001b)	Found by VGE-PSI-BLAST	Adjacent gene for B subunit in S. typhi; S. pyogenes gene close relative of known S. aureus toxin	Not yet	—
5a TolB contains a beta-propeller 5b This mediates protein–protein interactions	Ponting and Pallen (1999a)	Found by PSI-BLAST	Beta-propellers involved in protein–protein interactions; TolB interacts with other proteins	Yes	Abergel et al. (1999), Carr et al. (2000), Ray et al. (2000)

Table 2.1. Continued

Hypotheses	References	In silico evidence	Supporting evidence	Laboratory confirmed?	References
6a Tricorn proteases contains PDZ and beta-propeller domains	Ponting and Pallen (1999b), Pallen et al. (2001c)	Found by PSI-BLAST	Beta-propeller mediates substrate selection in prolyl oligopeptidase; PDZs mediate substrate selection in (2) above	Yes	Brandstetter et al. (2001)
6b These mediate substrate selection				Not yet	
7a Homologues of M. tuberculosis ESAT-6 exist in other bacteria	Pallen (2002)	ESAT-6/YukE link found by VGE-PSI-BLAST	In all cases where ESAT-6 homologue gene found, YukA-like membrane-bound ATPase gene found nearby; ESAT-6 secreted into supernatant; YukA-like proteins contain FtsK/SpoIIIE domains, involved in pumping DNA in other proteins; Gram-negative type IV secretion systems export proteins and DNA; ESAT-6 is similar in size to fimbrial proteins	Not yet	—
7b ESAT-6 homologues exported to supernatant					
7c YukAB homologues are membrane-bound ATPases involved in exporting YukE/ESAT-6-like proteins					
7d ESAT-6 part of a multi-protein complex involved in protein export				Preliminary evidence for YukE–YukE interaction	
8 FHA domains in bacteria mediate pSer/pThr-dependent protein interactions	—	Bacterial FHA domains found by PSI-BLAST (manuscript in preparation)	Tend to lie in proteins encoded by gene clusters containing Ser/Thr kinases	Not yet	—

hypotheses in the laboratory, one should seek to confirm findings from an initial computer-based analysis with clues from an alternative *in silico* approach (Table 2.1). As can be seen in the case studies below and in Table 2.1, the most convincing and satisfying analyses draw on interlocking clues, as in a crossword puzzle, so that the resulting evidence is compelling, even if circumstantial. Just as with laboratory-based work, *in silico* initially speculative hypotheses are strengthened and refined by multiple rounds of analyses.

◆◆◆◆◆◆ METHODS

My aim in this chapter is to introduce the reader to methods and databases that are freely available, either through the web or in the form of free software that can be downloaded and installed on your desktop computer. More sophisticated analyses can be done by bioinformaticians skilled in the use of the Unix operating system and scripting or programming languages. However, most busy bench-based microbiologists do not have the time to acquire these skills and so must rely on the methods described here or alternatively should consider recruiting a bioinformatician into their research group. In addition, space does not permit full explanations of how to use every program described – readers are encouraged to follow up references and look at on-line documentation. Instead, I will concentrate on illustrating the kind of reasoning required when using sequence analyses to formulate hypotheses.

◆◆◆◆◆◆ PARSING PROTEINS INTO DOMAINS

It is well accepted that many human proteins consist of more than one domain, where a domain is defined as a structurally discrete unit that can fold independently of the rest of the protein and that can be recognized as a stretch of sequence homology that can be found in different contexts in different proteins. It is becoming increasing clear that this is also true of many bacterial proteins (Pallen and Ponting, 1997; Pallen *et al.*, 1997, 2001c; Ponting, 1997, 1999; Aravind and Ponting, 1999; May and Ponting, 1999; Ponting and Pallen, 1999a, b; Ponting *et al.*, 1999; Copley *et al.*, 2001; Daniell *et al.*, 2001). This is of key importance in sequence analysis, where domains act as the minimum unit of sequence similarity, so that two proteins will often show homology to one another in just one domain (Hegyi and Gerstein, 2001). As noted above, ignoring this fact can lead to the propagation of annotation errors. It can also make database searches more difficult, as hits in the output to a common domain can swamp weak hits to an adjacent less common domain. For this reason, it is important to try to break down – or 'parse' – protein sequences into their constituent domains during an analysis (Ponting and Birney, 2000; Ponting, 2001; Ponting and Dickens, 2001).

Initial quality control

Almost all protein sequences in the databases are actually predictions derived from DNA sequences. The quality of these predictions varies. The simplest way of identifying protein-coding sequences is to rely on predictions of open-reading frames longer than a certain length, say 100 codons (this is the method used in Artemis). However, this approach substantially overpredicts proteins, particularly in GC-rich genomes, as the three stop codons (TAA, TAG, TGA) are AT-rich. More sophisticated methods, such as Glimmer (Salzberg *et al.*, 1998; Delcher *et al.*, 1999), rely on Markov models, but are not accessible to the average microbiologist (although Glimmer-predicted proteins for many incomplete genomes can be obtained from ViruloGenome). However, even Glimmer's performance varies according to the quality of the data used to train it, so predictions from small data sets should be treated with caution. An alternative approach, given that protein sequences are better conserved than DNA sequences, is to rely on homology searches to identify protein-coding regions. When used on whole genomes, as in Orpheus (Frishman *et al.*, 1998), this is highly processor intensive, but for manual analysis of short stretches of sequence is straightforward and informative.

Two problems dog the prediction of bacterial protein sequences. Firstly, it is often hard to be certain which is the initiation codon (Suzek *et al.*, 2001). Aligning the protein with its homologues may provide clues, as can the discovery of a potential signal peptide. However, it is important to keep an open mind on this until direct experimental evidence is obtained. Secondly, protein-coding regions may contain frameshifts, which can be identified when homology to a related protein sequence spans more than one reading frame (Brown *et al.*, 1998; Medigue *et al.*, 1999). Important clues to the presence of frameshifts, that should ring alarm bells during a sequence analysis, are when a predicted protein is shorter or longer than its closest homologues, or contains an anomalous segment that is conserved in all other sequences in a multiple alignment. Frameshifts can be artefacts of poor sequencing or, in good-quality sequence, can indicate gene inactivation. Frameshift errors in sequences (and mispredictions of start sites) can ruin a program of experimental work, for example, when one is attempting to tag or express a protein, or investigate a protein–protein interaction. From my own experience, I can confirm that months of laboratory time can be wasted working on truncated products that do not represent a real full-length protein. The take-home message is that you should always resequence the DNA encoding of any proteins you want to work on and never trust anyone else's sequence!

Not all frameshifts are sequence artefacts – in some cases, particularly if there is more than one frameshift, one can assume that the full-length protein is not expressed and that one is dealing with a pseudogene. [These are particularly common in organisms that have recently changed their niche (Andersson and Andersson, 1999a,b, 2001; Cole *et al.*, 2001; Eiglmeier *et al.*, 2001; McClelland *et al.*, 2001; Parkhill *et al.*, 2001).]. However, programmed frameshifting is a well-recognized method of bacterial gene regulation, so a frameshifted gene may well be intact and expressing a full-length protein in a proportion of the bacterial population, particularly if the frameshift occurs in a DNA region that is liable to slipped-strand mis-pairing, that is, a short sequence repeat, a homopolymeric tract (van Belkum *et al.*, 1998; Henderson *et al.*, 1999; Parkhill *et al.*, 2000; Baranov *et al.*, 2001; Linton *et al.*, 2001).

Structural features

The first step in domain parsing is to look for structural features that can be predicted from the sequence. Probably the most important of these are transmembrane (TM) domains, which can be recognized by a variety of programs (Sonnhammer *et al.*, 1998; Krogh *et al.*, 2001; Moller *et al.*, 2001; Ikeda *et al.*, 2002). As any extensive sequences flanking a TM domain can be assumed to lie on either side of the membrane, one can also assume that they form separate domains. A special case is the signal peptide, which is recognized by TM prediction programs as a transmembrane domain next to the N-terminus, but can be specifically identified using other programs such as SignalP (Nielsen *et al.*, 1997, 1999; Menne *et al.*, 2000). It can be hard to distinguish signal peptides from other TM domains where the N-terminus has been incorrectly predicted. In such cases, manually trimming back the suspicious N-terminal extension then running the protein through a signal-peptide detection program can help. Structural features such as signal peptides and TM domains can be used to predict the localization of a protein from its sequence. Several such predictive features have been usefully incorporated in the program PSORT (Nakai and Horton, 1999).

Two other features that can be identified in protein sequences and often overlap are sequence repeats and low-complexity sequences. Low-complexity sequences consist of stretches of sequence where, instead of calling on the full range of amino acids, the sequence is heavily biased towards just a handful of residues. These stretches of low-complexity sequence can be detected by a range of programs and often lie between domains or form structural elements as simple repeats, e.g. in surface proteins (Andrade *et al.*, 2001; Kajava, 2001). Larger, more complex repeats can form components of domains, e.g. WD repeats in beta-propellers (Ponting and Pallen, 1999a,b; Li and Roberts, 2001), or can represent domains in their own right, e.g. the two PDZ domains at the end of HtrA-like proteases (Pallen and Ponting, 1997; Pallen and Wren, 1997). Both low-complexity sequences and complex repeats can be visualized using dot plots, e.g. with the excellent Dotlet (Junier and Pagni, 2000).

The coiled-coil domain represents a special type of structural feature built from low-complexity sequence repeats that can be predicted from the sequence, using programs such as COILS and Multicoil (Lupas, 1997; Wolf *et al.*, 1997; Newman *et al.*, 2000). Various methods can be used to improve confidence in the prediction (see COILS documentation; http://www.ch.embnet.org/software/coils/COILS_doc.html), including comparing the results of both programs (Sonhammer and Wootton, 2001). An additional marker of significance, for both TM and coiled-coil predictions is that similar predictions hold for homologous regions from a range of related proteins.

Homology searches

Space does not permit a fuller discussion of homology, which was covered in the previous chapter, here – the reader is also referred to reviews elsewhere (Doyle and Gaut, 2000; Thornton and DeSalle, 2000; Ponting, 2001). However, it is important to stress the distinction between 'sequence homology' and 'sequence similarity'. The term 'homology' should be used only to describe sequences that arise through descent with modification from a common ancestor. Sequences can be similar but not homologous, e.g. if of low complexity (see above). Although strong unequivocal full-length homology can be picked up with a simple BLAST search, more care is needed to detect weak homologies and/or homologies that apply to only one region of a protein. Several painstaking manual runs of trial-and-error using programs that search with multiple sequences are required in all but the simplest investigations (Park *et al.*, 1998).

Two methods are commonly used to find these weak homologies – PSI-BLAST searches (Altschul *et al.*, 1997; Altschul and Koonin, 1998; Aravind and Koonin, 1999; Muller *et al.*, 1999; Taylor and Brown, 1999; Friedberg *et al.*, 2000; Schaffer *et al.*, 2001) and searches with hidden Markov models (HMMs; Eddy, 1998). The NCBI's PSI-BLAST is a modification of standard BLASTP (BLAST for proteins) that involves iterative rounds of searching with a steadily more refined positive-specific similarity matrix. PSI-BLAST ranks as one of the most sensitive methods for finding weak homologies. However, it should be used with caution. In particular, be wary of incorporation into the model of low-complexity sequences (including coiled coils) or sequences from families with large numbers of members (e.g. cytochromes, ABC transporters), as these will often contaminate otherwise useful search results with spurious hits (Aravind and Koonin, 1998; Jones and Swindells, 2002).

One of the disadvantages of the NCBI's PSI-BLAST is that it works only against the so-called non-redundant protein database, which does not include protein sequences deduced from incomplete genome sequences. We have overcome this problem in ViruloGenome (http://www.vge.ac.uk), building a library that includes both the non-redundant (NR) database and a database of Glimmer-predicted proteins from unfinished genome sequences. Our ViruloGenome PSI-BLAST thus represents probably the world's most powerful publicly available resource for finding distant homologies among bacterial proteins.

An alternative to PSI-BLAST is to search a query sequence against HMMs built from multiple alignments of known domains. Although you can build your own HMMs and search with them using, e.g. HMMer (Eddy, 1998; http://hmmer.wustl.edu/), it is usually easier to search publicly available collections, such as the PFAM (Bateman *et al.*, 2002) and SMART (Schultz *et al.*, 1998; Letunic *et al.*, 2002) databases. These sites also incorporate predictions of structural domains, such as signal peptides, and so can provide a handy one-stop shop for parsing your protein into its constituent domains.

What if only part of the protein sequence is recognized in homology searches? One can extract the protein sequences that give hits and perform a multiple alignment using CLUSTALW and JalView (http://circinus.ebi.ac.uk:6543/cgi-bin/clustalw.cgi), extending the regions detected in the homology searches by trial and error to see if the homology covers larger regions. If it is clear that the homology does indeed stop at a certain point in the sequence, then it is likely that you have hit a domain boundary. In such cases, try repeating the PSI-BLAST searches on the bits of the sequence outside the region you have already identified, as weak hits to these regions may have been masked in the initial searches by hits to the already identified region.

If you are interested in determining the extent of a protein or domain family, rather than investigating a single protein, iterative rounds of searching and multiple alignment can be useful. Sequences with unimpressive BLAST e-values may clearly belong in your domain family, when viewed in a multiple alignment. Performing searches with the most distant relatives of your starting sequence, although time consuming, might identify even more distant homologues (Park *et al.*, 1997, 1998).

◆◆◆◆◆◆ GENOMIC CONTEXT

Several whole-genome analyses have shown that proteins encoded by genes in conserved clusters tend to interact and/or have similar physiological or structural roles (Dandekar *et al.*, 1998; Enright *et al.*, 1999; Marcotte *et al.*, 1999; Overbeek *et al.*, 1999a,b; Huynen *et al.*, 2000a,b; Lathe *et al.*, 2000; Snel *et al.*, 2000). Thus, if your gene of unknown function clusters with genes of known function, especially if they are in the same operon, then one can hypothesize that the unknown and known genes share similar functions. This hypothesis will be strengthened if homologues of your gene are embedded in similar gene clusters in a range of microorganisms.

◆◆◆◆◆◆ HYPOTHESIS FORMULATION

Once you have completed a round of bioinformatics analysis, it is time to start formulating hypotheses. This process can range from straightforward predictions to more complex speculations (see Table 2.1). The key point is that any speculations should lead to predictions that can then be tested in the laboratory. Like many sequence analyses, this will be an iterative process, with each cycle of speculation, prediction and testing leading to refinements in the hypothesis. As the pace and cost of laboratory work is much slower than bioinformatics analyses, it pays to refine and test the hypothesis as far as possible *in silico* before starting a program of work into the laboratory, although sometimes predictions can be right for the wrong reasons! (Iyer *et al.*, 2001). Also, one must weigh up the strength of the *in silico* prediction against the potential rewards and the difficulty of testing the hypothesis in the laboratory. A highly speculative prediction on an important problem may be worth pursuing even if technically demanding, whereas for less important problems, predictions may be worth following up only if technically straightforward or strongly supported.

◆◆◆◆◆◆ WORKED EXAMPLES

Here I provide two worked examples of how the above approaches can be used on bacterial proteins. Alternative similar worked examples are available from other sources (http://www.ocms.ox.ac.uk/~ponting/methmb/example.html; http://www.ncbi.nlm.nih.gov/Education/BLASTinfo/information3.html).

Drilling back to the source of a misleading annotation

In a recent paper (Pallen, 2002), I hypothesized that the YukA family of membrane-associated ATPases, which occur in a range of Gram-positive bacteria, played a role in the export of a family of small proteins homologous to ESAT-6 from *M. tuberculosis* and YukE from *B. subtilis*. Working through the YukA homologues in the public databases, I was surprised to see the YukA homologue from *Staphylococcus aureus* N315 annotated as 'conserved hypothetical protein, similar to diarrheal toxin' (Entrez UID 15925989). It is hard to see how a membrane-associated protein export pump could also moonlight as a 'diarrheal toxin'. The source of this annotation soon became apparent – a protein sequence (BAA04134) for a so-called diarrhoeal toxin, BceT, from *Bacillus cereus*. However, this protein (366 amino acids) is much smaller than other YukA homologues (typically around 1500 residues) and the homology it shows spans a region in other YukA-like occupied by one and a half ATPase domains. This suggests it is an artefactual protein fragment, rather than a full-length sequence. A BlastX search with the associated nucleotide sequence finds regions of homology at the protein level between sequences from upstream of the supposed start codon and the YukA-like proteins, suggesting that the start of the BceT sequence has been misidentified. BLAST searches were then performed against sequences from the incomplete *B. cereus* genome project. A BLASTX search revealed that the BceT protein sequence was, as I expected, an internal fragment of a much larger protein (1344 amino acids), while a BLASTN search found no hits in the *B. cereus* sequence to the region upstream of the coding sequence, suggesting that the sequence was a chimera between *B. cereus* DNA and DNA from some other source.

The next step was a review of the literature on BceT. This confirmed that the experimental basis for believing that BceT was a toxin was weak – just one unrepeated study (Agata *et al.*, 1995) which showed that a crude cell lysate from *E. coli* containing the relevant DNA fragment appeared to exhibit toxicity to Vero cells and rabbit ileal loops. Furthermore, Granum and Lund (1997) found that the *bceT* DNA sequence was detectable in only 40% of 95 strains of *B. cereus* screened by the polymerase chain reaction (PCR), and only two of five food poisoning strains, and a recent review article concluded '... no evidence exists that the 41 kDa enterotoxin T has been involved in any cases of *B. cereus* food poisoning... The exact role enterotoxin T gene plays in *B. cereus* food poisoning remains unclear'. Thus, in conclusion, BceT would appear to be 'the toxin that never was' and it is all too easy to imagine students wasting months or years trying to characterize the putative 'diarrheal toxin' of *S. aureus*.

Finding distant homologues: the 1318 family

Sequencing the *Campylobacter jejuni* genome revealed the existence of around two dozen potentially phase-variable genes, where short repeats or homopolymeric tracts provided a likely mechanism for phase variation (Parkhill *et al.*, 2000; Linton *et al.*, 2001). These genes tended to lie in gene clusters implicated in the synthesis of surface structures. Two of the potentially phase-variable genes (Cj1318, Cj1335/6) are in a family of seven paralogous genes of unknown function, named the 1318 family. These genes cluster in flagellar operons in *C. jejuni*. A preliminary analysis of these genes was carried out for the genome sequence paper in the summer of 1999, which involved numerous labour-intensive searches using scattered resources. With the setting up of our ViruloGenome site and the passage of over 2 years since that first analysis, a fresh analysis of the 1318 family will provide a suitable worked example of sequence analysis in the postgenomic era.

Our analysis begins with scrutiny of the genomic context of the seven 1318 family genes from *C. jejuni* (Cj1318, Cj1333, Cj1334, Cj1335/6, Cj1337, Cj1341c), using the genome sequence visualization and annotation tool, Artemis (Rutherford *et al.*, 2000). Cj1318 itself lies adjacent to *neuB3*, a N-acetylneuraminic acid synthetase essential for the biosynthesis of the flagellum (Karlyshev *et al.*, 2000). The other family members are clustered around the genes encoding the flagellar structural subunits – four on one side, two on the other. It thus seems reasonable to assume that the physiological role of the Cj1318 family proteins is some aspect of flagellar biosynthesis in *C. jejuni*. By analogy with *neuB3*, one might posit an enzymatic activity involved in post-translational modification of flagellar subunits, which are known to be sialylated in *C. jejuni* (Karlyshev *et al.*, 2002).

We retrieve the proteins in FASTA format, and join Cj1335 and Cj1336 into one sequence, using Artemis to identify where the join should take place. Next, we perform a multiple alignment of the proteins with Jalview (http://circinus.ebi.ac.uk:6543/cgi-bin/clustalw.cgi), which shows that they can all be aligned along almost their full lengths, suggesting they all have the same domain structure (data not shown). We then attempt to parse them into their constituent domains using PFAM and SMART searches (Schultz *et al.*, 1998; Bateman *et al.*, 2002). The SMART searches predicted a coiled-coil domain in one of the proteins (residues 534–569 in

Cj1318), with patches of low-complexity sequence at the N- and C-terminal extremities of most sequences. PFAM searches find no hits, except a weak partial match of ~50 residues (*e*-value 0.0036) in Cj1340c and CJ1341c to DUF115 (PFAM entry PF01973), a ~200-residue domain of unknown function found in several archaeal proteins. Searches with Multicoil (Wolf *et al.*, 1997) predict a single coiled-coil domain close to the C-terminus in all the 1318 family proteins, with scores ranging from 0.2 to 0.6 (1.0 is certainty). Although these are not conclusive scores, one could speculate that this coiled-coil domain mediates protein–protein interactions (either self–self or self–other).

We now perform a PSI-BLAST search on the ViruloGenome site (http://www.vge.ac.uk), starting with Cj1318. Mindful of the presence of regions of low-complexity sequence and coiled coils, we leave the filter on. We allow the search to repeat to convergence at iteration 7, with no manual intervention. A stunning array of sequences from bacteria and archaea is identified as members of the Cj1318 superfamily (data not shown). Looking at the most distantly but confidently related sequences, it is clear the similarity is focused on the region spanning residues ~240 to ~380 (data not shown). Unfortunately, all are annotated as purely hypothetical proteins, with the exception of three lipopolysaccharide (LPS) core biosynthesis proteins – RfaZ from *E. coli* K12, RfaZ from *Salmonella typhimurium* LT2, and WaaZ from *E. coli* F632 (Figure 2.1).

An attempt to perform the PSI-BLAST search in reverse by starting with RFAZ_SALTY stutters to a halt after three iterations using the default *e*-value for automatic inclusion of sequences in a subsequent round. However, after manual inclusion of a sequence from *C. acetobutylicum* (with a reasonable *e*-value of 0.054) at iteration 3, the search finds the 1318 family by iteration 5.

```
Cj1318  220  VYNLPQMITHPSYKELLSKRKNLSDTAIIVSTGP-SLTKQLPLLKKYASKATIFCADSSY  278
              + N+ +        + SK    SD II +GP S    L +L+       +   + S
WaaZ      1  MKNIRYIDKKDVENLIESK---TSDDVIIFLSGPTSQKTPLSVLQTRD----VIAVNGSA   53

Cj1318  279  PILAKHGIKPDYVCMLE------RTELTAEFFNHDF------------GEFDKDIVF--I  318
              L  H I P     + +      R +  +F                  E DK  +
WaaZ     54  QYLLSHNIIPYIYVLTDVRFLHQRRDDFYKFSQRSRYTIVNVDVYEHASEEDKRYILQNC  113

Cj1318  319  CAGVVHPK-----AIEYLKGRNLVITQKVLAFPYYINLKDFSYAAVGL------------  361
                   +        I+ +K    ++++        +          VG
WaaZ    114  LVLRSFYRREKGGLIKKIKFN--ILSRIHKELLISVPFSKKGRL-VGFCKDINLGYCSCH  170

Cj1318  362  SVAHTLSYLATYLSHKNIIFIGQDLAYAENGNSHPDDYQNSANYESQMYEHILTEAYGGN  421
              +VA    +A L +  II  G DL +          + +          M    + + +
WaaZ    171  TVAFAAIQIAYSLKYARIICSGLDLTGS------CSRFYDEDKNP--MPSELTRDLF---  219

Cj1318  422  GKVETHSIWLLFKNWFENEMIPNTRKMGITTYNCTE  457
                    I   F+              I  YN ++
WaaZ    220  ------KILPFFR-------FMRENIEDINIYNLSD  242
```

Figure 2.1. PSI-BLAST alignment between Cj1318 (Entrez UID 15792641) and WaaZ (Entrez UID 3132869).

Unfortunately, scrutiny of the papers associated with the *E. coli* and *Salmonella* sequences reveals that, although their genes cluster with LPS core biosynthesis genes, the proteins are of unknown function (MacLachlan *et al.*, 1991; Klena *et al.*, 1992; Heinrichs *et al.*, 1998a,b; Regue *et al.*, 2001). However, Heinrichs *et al.* (1998a) conclude, by a process of elimination, that WaaZ is a glycosyltransferase responsible for adding a third 3-deoxy-D-*manno*-oct-2-ulosonic acid (KdoIII) moiety to the LPS core. The discovery that a distant homologue of Cj1318 is probably a glycosyltransferase involved in modification of a surface structure fits in nicely with our existing ideas about the 1318 family. We can now hypothesize that the 1318 family proteins are glycosyltransferases (some phase variable) involved in flagellar modification and that residues conserved between Cj1318 and WaaZ (Figure 2.1) form crucial components of the active site. We look forward to experimental verification of these suggestions!

◆◆◆◆◆◆ FROM *IN SILICO* TO *IN VITRO* OR *IN VIVO*

The first question one has to address when taking sequence-based predictions into the laboratory is 'are the gene and its associated protein expressed?. If you are lucky, you may be able to glean this information from pre-existing transcriptional fusion, microarray or two-dimensional gel electrophoresis data, either published for the organism in question or available through a collaborator. If this is not possible, you may consider performing reverse transcription-PCR or northern blots, or making your own transcriptional fusions. However, as many important virulence genes are not expressed under laboratory conditions, a negative result here does not mean you should dismiss the gene as a pseudogene. One approach that can be used to prove that the gene and its protein are expressed – and are even synthesized *in vivo* – is to express and purify a soluble fragment from the protein in *E. coli* (e.g. using Invitrogen's gateway system), then perform Western blots against the protein with sera from infected humans or animals. Although there are some obvious caveats (cross-reactions may occur with other proteins from other pathogens or commensals or from vaccine preparations), this approach allows one to show that a gene and its protein is expressed *in vivo*, even when you do not know the conditions regulating its expression.

Once you are convinced a protein is expressed, you are likely to want to examine its localization, processing, etc. The traditional way to do this is to raise an antibody to the protein expressed and purified in *E. coli*, or, if the protein is toxic and/or insoluble, to a synthetic peptide patterned on the protein sequence. However, this can be expensive, time consuming and wasteful of rabbits. An alternative approach for organisms that can be manipulated genetically is to label the protein with an epitope tag and then use a commercially available antitag antibody to detect and characterize the protein (Jarvik and Telmer, 1998; Fritze and Anderson, 2000; Uzzau *et al.*, 2001).

Next, you will want to check that the protein does what you think it should. For example, you may wish to test its enzymatic activity using biochemical approaches. This can sometimes be done even if the target of the enzyme is not known. For example, it is possible to find the intracellular targets of ADP-ribosylating toxins by mixing the toxin with eukaryotic cell lysates (Otto *et al.*, 2000). Alternatively, you may wish to test predictions on protein–protein interactions, using a variety of approaches: yeast or bacterial two-hybrid systems (see Chapter 13; Karimova *et al.*, 1998), immunoprecipitation, far-Western blots, etc.). If you find an interaction, you can then begin to dissect out the determinants responsible by site-directed mutagenesis, alanine-scanning, domain swapping or truncation.

Another approach to testing your predictions is to investigate the effects of gene deletion or mutation. If you are working with a model organism, such as *Bacillus subtilis*, you may even be able to obtain a deletion mutant 'off the shelf' from a functional analysis consortium (Biaudet *et al.*, 1997; see Chapter 18). For many organisms, including now *E. coli*, it is possible to make mutants quickly and easily using PCR-amplified selectable marker genes tagged through the primers with target-specific sequences. In many cases, your *in silico* analyses will provide clues as to what phenotypes to test for in the mutant strain. In the worst case, where you have no clues, you can still test the mutant for a range of stress-induced and other phenotypes.

◆◆◆◆◆◆ CONCLUSIONS

This sequence-rich postgenomic world of microbiology is full of opportunity, with plenty of low-hanging fruit, ripe for the picking. To work at the bench in this brave new world without being able to mine sequence data is like trying to sail the ocean without charts. But to analyse sequences without testing predictions at the bench is like gazing at charts without ever leaving home! With hundreds of thousands of new gene sequences set to arrive on our desktops in the next few years, the postgenomic revolution will keep this and future generations of microbiologists busy for years to come. The challenge is to adopt the same rigour *in silico* as we do in the laboratory to produce many thousands of challenging and insightful new sequence-fuelled hypotheses.

Acknowledgements

I acknowledge the BBSRC for funding ViruloGenome, and Nick Loman, Alex Lam and Arshad Khan for their work on it. I thank the Sanger Wellcome Trust Institute, TIGR, University of Oklahoma, University of Washington and other centres for making incomplete genome data publicly available. Thanks are due to Chris Ponting for stimulating my interest in this field.

References

Abergel, C., Bouveret, E., Claverie, J. M., Brown, K., Rigal, A., Lazdunski, C. and Benedetti, H. (1999). Structure of the *Escherichia coli* TolB protein determined by MAD methods at 1.95 Å resolution. *Structure Fold Des.* **7**, 1291–1300.

Agata, N., Ohta, M., Arakawa, Y. and Mori, M. (1995). The *bceT* gene of *Bacillus cereus* encodes an enterotoxic protein. *Microbiology* **141**, 983–988.

Altschul, S. F. and Koonin, E. V. (1998). Iterated profile searches with PSI-BLAST – a tool for discovery in protein databases. *Trends Biochem. Sci.* **23**, 444–447.

Altschul, S. F., Madden, T. L., Schaffer, A. A., Zhang, J., Zhang, Z., Miller, W. and Lipman, D. J. (1997). Gapped BLAST and PSI-BLAST: a new generation of protein database search programs. *Nucleic Acids Res.* **25**, 3389–3402.

Andersson, J. O. and Andersson, S. G. (1999a). Genome degradation is an ongoing process in *Rickettsia*. *Mol. Biol. Evol.* **16**, 1178–1191.

Andersson, J. O. and Andersson, S. G. (1999b). Insights into the evolutionary process of genome degradation. *Curr. Opin. Genet. Dev.* **9**, 664–671.

Andersson, J. O. and Andersson, S. G. (2001). Pseudogenes, junk DNA, and the dynamics of *Rickettsia* genomes. *Mol. Biol. Evol.* **18**, 829–839.

Andrade, M. A., Perez-Iratxeta, C. and Ponting, C. P. (2001). Protein repeats: structures, functions, and evolution. *J. Struct. Biol.* **134**, 117–131.

Aravind, L. and Koonin, E. V. (1999). Gleaning non-trivial structural, functional and evolutionary information about proteins by iterative database searches. *J. Mol. Biol.* **287**, 1023–1040.

Aravind, L. and Ponting, C. P. (1999). The cytoplasmic helical linker domain of receptor histidine kinase and methyl-accepting proteins is common to many prokaryotic signalling proteins. *FEMS Microbiol. Lett.* **176**, 111–116.

Baranov, P. V., Gurvich, O. L., Fayet, O., Prere, M. F., Miller, W. A., Gesteland, R. F., Atkins, J. F. and Giddings, M. C. (2001). RECODE: a database of frameshifting, bypassing and codon redefinition utilized for gene expression. *Nucleic Acids Res.* **29**, 264–267.

Bateman, A., Birney, E., Cerruti, L., Durbin, R., Etwiller, L., Eddy, S. R., Griffiths-Jones, S., Howe, K. L., Marshall, M. and Sonnhammer, E. L. (2002). The Pfam protein families database. *Nucleic Acids Res.* **30**, 276–280.

Beebe, K. D., Shin, J., Peng, J., Chaudhury, C., Khera, J. and Pei, D. (2000). Substrate recognition through a PDZ domain in tail-specific protease. *Biochemistry* **39**, 3149–3155.

Benson, D. A., Karsch-Mizrachi, I., Lipman, D. J., Ostell, J., Rapp, B. A. and Wheeler, D. L. (2002). GenBank. *Nucleic Acids Res.* **30**, 17–20.

Bernal, A., Ear, U. and Kyrpides, N. (2001). Genomes OnLine Database (GOLD): a monitor of genome projects world-wide. *Nucleic Acids Res.* **29**, 126–127.

Biaudet, V., Samson, F. and Bessieres, P. (1997). Micado – a network-oriented database for microbial genomes. *Comput. Appl. Biosci.* **13**, 431–438.

Bocs, S., Danchin, A. and Medigue, C. (2002). Re-annotation of genome microbial CoDing-Sequences: finding new genes and inaccurately annotated genes. *BMC Bioinformatics* **3**, 5.

Bork, P. and Koonin, E. V. (1998). Predicting functions from protein sequences – where are the bottlenecks? *Nature Genet.* **18**, 313–318.

Brandstetter, H., Kim, J. S., Groll, M. and Huber, R. (2001). Crystal structure of the tricorn protease reveals a protein disassembly line. *Nature* **414**, 466–470.

Brenner, S. E. (1999). Errors in genome annotation. *Trends Genet.* **15**, 132–133.

Brown, N. P., Sander, C. and Bork, P. (1998). Frame: detection of genomic sequencing errors. *Bioinformatics* **14**, 367–371.

Carr, S., Penfold, C. N., Bamford, V., James, R. and Hemmings, A. M. (2000). The structure of TolB, an essential component of the tol-dependent translocation system, and its protein–protein interaction with the translocation domain of colicin E9. *Structure Fold Des.* **8**, 57–66.

Cole, S. T., Eiglmeier, K., Parkhill, J., James, K. D., Thomson, N. R., Wheeler, P. R., Honore, N., Garnier, T., Churcher, C., Harris, D., Mungall, K., Basham, D., Brown, D., Chillingworth, T., Connor, R., Davies, R. M., Devlin, K., Duthoy, S., Feltwell, T., Fraser, A., Hamlin, N., Holroyd, S., Hornsby, T., Jagels, K., Lacroix, C., Maclean, J., Moule, S., Murphy, L., Oliver, K., Quail, M. A., Rajandream, M. A., Rutherford, K. M., Rutter, S., Seeger, K., Simon, S., Simmonds, M., Skelton, J., Squares, R., Squares, S., Stevens, K., Taylor, K., Whitehead, S., Woodward, J. R. and Barrell, B. G. (2001). Massive gene decay in the leprosy bacillus. *Nature* **409**, 1007–1011.

Copley, R. R., Russell, R. B. and Ponting, C. P. (2001). Sialidase-like Asp-boxes: sequence-similar structures within different protein folds. *Protein Sci.* **10**, 285–292.

Dandekar, T., Snel, B., Huynen, M. and Bork, P. (1998). Conservation of gene order: a fingerprint of proteins that physically interact. *Trends Biochem. Sci.* **23**, 324–328.

Dandekar, T., Huynen, M., Regula, J. T., Ueberle, B., Zimmermann, C. U., Andrade, M. A., Doerks, T., Sanchez-Pulido, L., Snel, B., Suyama, M., Yuan, Y. P., Herrmann, R. and Bork, P. (2000). Re-annotating the *Mycoplasma pneumoniae* genome sequence: adding value, function and reading frames. *Nucleic Acids Res.* **28**, 3278–3288.

Daniell, S. J., Delahay, R. M., Shaw, R. K., Hartland, E. L., Pallen, M. J., Booy, F., Ebel, F., Knutton, S. and Frankel, G. (2001). Coiled-coil domain of enteropathogenic *Escherichia coli* type III secreted protein EspD is involved in EspA filament-mediated cell attachment and hemolysis. *Infect. Immun.* **69**, 4055–4064.

Day, J. B. and Plano, G. V. (2000). The *Yersinia pestis* YscY protein directly binds YscX, a secreted component of the type III secretion machinery. *J. Bacteriol.* **182**, 1834–1843.

Delahay, R. M., Knutton, S., Shaw, R. K., Hartland, E. L., Pallen, M. J. and Frankel, G. (1999). The coiled-coil domain of EspA is essential for the assembly of the type III secretion translocon on the surface of enteropathogenic *Escherichia coli*. *J. Biol. Chem.* **274**, 35969–35974.

Delcher, A. L., Harmon, D., Kasif, S., White, O. and Salzberg, S. L. (1999). Improved microbial gene identification with GLIMMER. *Nucleic Acids Res.* **27**, 4636–4641.

Devos, D. and Valencia, A. (2001). Intrinsic errors in genome annotation. *Trends Genet.* **17**, 429–431.

Doyle, J. J. and Gaut, B. S. (2000). Evolution of genes and taxa: a primer. *Plant Mol. Biol.* **42**, 1–23.

Eddy, S. R. (1998). Profile hidden Markov models. *Bioinformatics* **14**, 755–763.

Eiglmeier, K., Parkhill, J., Honore, N., Garnier, T., Tekaia, F., Telenti, A., Klatser, P., James, K. D., Thomson, N. R., Wheeler, P. R., Churcher, C., Harris, D., Mungall, K., Barrell, B. G. and Cole, S. T. (2001). The decaying genome of *Mycobacterium leprae*. *Lepr. Rev.* **72**, 387–398.

Enright, A. J., Iliopoulos, I., Kyrpides, N. C. and Ouzounis, C. A. (1999). Protein interaction maps for complete genomes based on gene fusion events. *Nature* **402**, 86–90.

Friedberg, I., Kaplan, T. and Margalit, H. (2000). Evaluation of PSI-BLAST alignment accuracy in comparison to structural alignments. *Protein Sci.* **9**, 2278–2284.

Frishman, D., Mironov, A., Mewes, H. W. and Gelfand, M. (1998). Combining diverse evidence for gene recognition in completely sequenced bacterial genomes. *Nucleic Acids Res.* **26**, 2941–2947.

Fritze, C. E. and Anderson, T. R. (2000). Epitope tagging: general method for tracking recombinant proteins. *Meth. Enzymol.* **327**, 3–16.

Galperin, M. Y. and Koonin, E. V. (1998). Sources of systematic error in functional annotation of genomes: domain rearrangement, non-orthologous gene displacement and operon disruption. In Silico *Biol.* **1**, 55–67.

Granum, P. E. and Lund, T. (1997). *Bacillus cereus* and its food poisoning toxins. *FEMS Microbiol. Lett.* **157**, 223–228.

Hegyi, H. and Gerstein, M. (2001). Annotation transfer for genomics: measuring functional divergence in multi-domain proteins. *Genome Res.* **11**, 1632–1640.

Heinrichs, D. E., Monteiro, M. A., Perry, M. B. and Whitfield, C. (1998a). The assembly system for the lipopolysaccharide R2 core-type of *Escherichia coli* is a hybrid of those found in *Escherichia coli* K-12 and *Salmonella enterica*. Structure and function of the R2 WaaK and WaaL homologs. *J. Biol. Chem.* **273**, 8849–8859.

Heinrichs, D. E., Yethon, J. A. and Whitfield, C. (1998b). Molecular basis for structural diversity in the core regions of the lipopolysaccharides of *Escherichia coli* and *Salmonella enterica*. *Mol. Microbiol.* **30**, 221–232.

Henderson, I. R., Owen, P. and Nataro, J. P. (1999). Molecular switches – the ON and OFF of bacterial phase variation. *Mol. Microbiol.* **33**, 919–932.

Hoskins, J., Alborn, W. E., Jr, Arnold, J., Blaszczak, L. C., Burgett, S., DeHoff, B. S., Estrem, S. T., Fritz, L., Fu, D. J., Fuller, W., Geringer, C., Gilmour, R., Glass, J. S., Khoja, H., Kraft, A. R., Lagace, R. E., LeBlanc, D. J., Lee, L. N., Lefkowitz, E. J., Lu, J., Matsushima, P., McAhren, S. M., McHenney, M., McLeaster, K., Mundy, C. W., Nicas, T. I., Norris, F. H., O'Gara, M., Peery, R. B., Robertson, G. T., Rockey, P., Sun, P. M., Winkler, M. E., Yang, Y., Young-Bellido, M., Zhao, G., Zook, C. A., Baltz, R. H., Jaskunas, S. R., Rosteck, P. R., Jr, Skatrud, P. L. and Glass, J. I. (2001). Genome of the bacterium *Streptococcus pneumoniae* strain R6. *J. Bacteriol.* **183**, 5709–5717.

Huynen, M., Snel, B., Lathe, W. and Bork, P. (2000a). Exploitation of gene context. *Curr. Opin. Struct. Biol.* **10**, 366–370.

Huynen, M., Snel, B., Lathe, W., 3rd and Bork, P. (2000b). Predicting protein function by genomic context: quantitative evaluation and qualitative inferences. *Genome Res.* **10**, 1204–1210.

Ikeda, M., Arai, M., Lao, D. M. and Shimizu, T. (2002). Transmembrane topology prediction methods: a re-assessment and improvement by a consensus method using a dataset of experimentally-characterized transmembrane topologies. In Silico *Biol.* **2**, 19–33.

Iyer, L. M., Aravind, L., Bork, P., Hofmann, K., Mushegian, A. R., Zhulin, I. B. and Koonin, E. V. (2001). Quod erat demonstrandum? The mystery of experimental validation of apparently erroneous computational analyses of protein sequences. *Genome Biol.* **2**, RESEARCH0051.

Jarvik, J. W. and Telmer, C. A. (1998). Epitope tagging. *Annu. Rev. Genet.* **32**, 601–618.

Jones, D. T. and Swindells, M. B. (2002). Getting the most from PSI-BLAST. *Trends Biochem. Sci.* **27**, 161–164.

Junier, T. and Pagni, M. (2000). Dotlet: diagonal plots in a web browser. *Bioinformatics* **16**, 178–179.

Kajava, A. V. (2001). Review: proteins with repeated sequence – structural prediction and modeling. *J. Struct. Biol.* **134**, 132–144.

Karimova, G., Pidoux, J., Ullmann, A. and Ladant, D. (1998). A bacterial two-hybrid system based on a reconstituted signal transduction pathway. *Proc. Natl. Acad. Sci. USA* **95**, 5752–5756.

Karlyshev, A. V., Linton, D., Gregson, N. A. and Wren, B. W. (2002). A novel paralogous gene family involved in phase-variable flagella-mediated motility in *Campylobacter jejuni*. *Microbiology* **148**, 473–480.

Karp, P. D., Paley, S. and Zhu, J. (2001). Database verification studies of SWISS-PROT and GenBank. *Bioinformatics* **17**, 526–532; discussion 533–534.

Klena, J. D., Pradel, E. and Schnaitman, C. A. (1992). Comparison of lipopolysaccharide biosynthesis genes *rfaK*, *rfaL*, *rfaY*, and *rfaZ* of *Escherichia coli* K-12 and *Salmonella typhimurium*. *J. Bacteriol.* **174**, 4746–4752.

Krogh, A., Larsson, B., von Heijne, G. and Sonnhammer, E. L. (2001). Predicting transmembrane protein topology with a hidden Markov model: application to complete genomes. *J. Mol. Biol.* **305**, 567–580.

Lathe, W. C., 3rd, Snel, B. and Bork, P. (2000). Gene context conservation of a higher order than operons. *Trends Biochem. Sci.* **25**, 474–479.

Letunic, I., Goodstadt, L., Dickens, N. J., Doerks, T., Schultz, J., Mott, R., Ciccarelli, F., Copley, R. R., Ponting, C. P. and Bork, P. (2002). Recent improvements to the SMART domain-based sequence annotation resource. *Nucleic Acids Res.* **30**, 242–244.

Li, D. and Roberts, R. (2001). WD-repeat proteins: structure characteristics, biological function, and their involvement in human diseases. *Cell Mol. Life Sci.* **58**, 2085–2097.

Linton, D., Karlyshev, A. V. and Wren, B. W. (2001). Deciphering *Campylobacter jejuni* cell surface interactions from the genome sequence. *Curr. Opin. Microbiol.* **4**, 35–40.

Lupas, A. (1997). Predicting coiled-coil regions in proteins. *Curr. Opin. Struct. Biol.* **7**, 388–393.

MacLachlan, P. R., Kadam, S. K. and Sanderson, K. E. (1991). Cloning, characterization, and DNA sequence of the rfaLK region for lipopolysaccharide synthesis in *Salmonella typhimurium* LT2. *J. Bacteriol.* **173**, 7151–7163.

Marcotte, E. M., Pellegrini, M., Ng, H. L., Rice, D. W., Yeates, T. O. and Eisenberg, D. (1999). Detecting protein function and protein–protein interactions from genome sequences. *Science* **285**, 751–753.

Marshall, E. (2002). Data sharing. DNA sequencer protests being scooped with his own data. *Science* **295**, 1206–1207.

May, A. P. and Ponting, C. P. (1999). Integrin alpha- and beta 4-subunit-domain homologues in cyanobacterial proteins. *Trends Biochem. Sci.* **24**, 12–13.

McClelland, M., Sanderson, K. E., Spieth, J., Clifton, S. W., Latreille, P., Courtney, L., Porwollik, S., Ali, J., Dante, M., Du, F., Hou, S., Layman, D., Leonard, S., Nguyen, C., Scott, K., Holmes, A., Grewal, N., Mulvaney, E., Ryan, E., Sun, H., Florea, L., Miller, W., Stoneking, T., Nhan, M., Waterston, R. and Wilson, R. K. (2001). Complete genome sequence of *Salmonella enterica* serovar Typhimurium LT2. *Nature* **413**, 852–856.

Medigue, C., Rose, M., Viari, A. and Danchin, A. (1999). Detecting and analyzing DNA sequencing errors: toward a higher quality of the *Bacillus subtilis* genome sequence. *Genome Res.* **9**, 1116–1127.

Menne, K. M., Hermjakob, H. and Apweiler, R. (2000). A comparison of signal sequence prediction methods using a test set of signal peptides. *Bioinformatics* **16**, 741–742.

Moller, S., Croning, M. D. and Apweiler, R. (2001). Evaluation of methods for the prediction of membrane spanning regions. *Bioinformatics* **17**, 646–653.

Muller, A., MacCallum, R. M. and Sternberg, M. J. (1999). Benchmarking PSI-BLAST in genome annotation. *J. Mol. Biol.* **293**, 1257–1271.

Nakai, K. and Horton, P. (1999). PSORT: a program for detecting sorting signals in proteins and predicting their subcellular localization. *Trends Biochem. Sci.* **24**, 34–36.

Newman, J. R., Wolf, E. and Kim, P. S. (2000). A computationally directed screen identifying interacting coiled coils from *Saccharomyces cerevisiae*. *Proc. Natl Acad. Sci. USA* **97**, 13203–13208.

Nielsen, H., Engelbrecht, J., Brunak, S. and von Heijne, G. (1997). A neural network method for identification of prokaryotic and eukaryotic signal peptides and prediction of their cleavage sites. *Int. J. Neural Syst.* **8**, 581–599.

Nielsen, H., Brunak, S. and von Heijne, G. (1999). Machine learning approaches for the prediction of signal peptides and other protein sorting signals. *Protein Eng.* **12**, 3–9.

Otto, H., Tezcan-Merdol, D., Girisch, R., Haag, F., Rhen, M. and Koch-Nolte, F. (2000). The *spvB* gene-product of the *Salmonella enterica* virulence plasmid is a mono(ADP-ribosyl)transferase. *Mol. Microbiol.* **37**, 1106–1115.

Ouzounis, C. A. and Karp, P. D. (2002). The past, present and future of genome-wide re-annotation. *Genome Biol.* **3**, COMMENT2001.

Overbeek, R., Fonstein, M., D'Souza, M., Pusch, G. D. and Maltsev, N. (1999a). Use of contiguity on the chromosome to predict functional coupling. In Silico *Biol.* **1**, 93–108.

Overbeek, R., Fonstein, M., D'Souza, M., Pusch, G. D. and Maltsev, N. (1999b). The use of gene clusters to infer functional coupling. *Proc. Natl Acad. Sci. USA* **96**, 2896–2901.

Pallen, M. J. (2002). The ESAT-6/WXG100 superfamily and a new Gram-positive secretion system? *Trends Microbiol.* **10**, 209–212.

Pallen, M. J. and Ponting, C. P. (1997). PDZ domains in bacterial proteins. *Mol. Microbiol.* **26**, 411–413.

Pallen, M. J. and Wren, B. W. (1997). The HtrA family of serine proteases. *Mol. Microbiol.* **26**, 209–221.

Pallen, M. J., Dougan, G. and Frankel, G. (1997). Coiled-coil domains in proteins secreted by type III secretion systems. *Mol. Microbiol.* **25**, 423–425.

Pallen, M., Wren, B. and Parkhill, J. (1999). 'Going wrong with confidence': misleading sequence analyses of CiaB and clpX. *Mol. Microbiol.* **34**, 195.

Pallen, M. J., Lam, A. C., Antonio, M. and Dunbar, K. (2001a). An embarrassment of sortases – a richness of substrates? *Trends Microbiol.* **9**, 97–102.

Pallen, M. J., Lam, A. C., Loman, N. J. and McBride, A. (2001b). An abundance of bacterial ADP-ribosyltransferases – implications for the origin of exotoxins and their human homologues. *Trends Microbiol.* **9**, 302–307; discussion 308.

Pallen, M. J., Lam, A. C. and Loman, N. (2001c). Tricorn-like proteases in bacteria. *Trends Microbiol.* **9**, 518–521.

Park, J., Teichmann, S. A., Hubbard, T. and Chothia, C. (1997). Intermediate sequences increase the detection of homology between sequences. *J. Mol. Biol.* **273**, 349–354.

Park, J., Karplus, K., Barrett, C., Hughey, R., Haussler, D., Hubbard, T. and Chothia, C. (1998). Sequence comparisons using multiple sequences detect three times as many remote homologues as pairwise methods. *J. Mol. Biol.* **284**, 1201–1210.

Parkhill, J. (2000). In defense of complete genomes. *Nature Biotechnol.* **18**, 493–494.

Parkhill, J., Wren, B. W., Mungall, K., Ketley, J. M., Churcher, C., Basham, D., Chillingworth, T., Davies, R. M., Feltwell, T., Holroyd, S., Jagels, K., Karlyshev, A. V., Moule, S., Pallen, M. J., Penn, C. W., Quail, M. A., Rajandream, M. A., Rutherford, K. M., van Vliet, A. H., Whitehead, S. and Barrell, B. G. (2000). The genome sequence of the food-borne pathogen *Campylobacter jejuni* reveals hypervariable sequences. *Nature* **403**, 665–668.

Parkhill, J., Wren, B. W., Thomson, N. R., Titball, R. W., Holden, M. T., Prentice, M. B., Sebaihia, M., James, K. D., Churcher, C., Mungall, K. L., Baker, S., Basham,

D., Bentley, S. D., Brooks, K., Cerdeno-Tarraga, A. M., Chillingworth, T., Cronin, A., Davies, R. M., Davis, P., Dougan, G., Feltwell, T., Hamlin, N., Holroyd, S., Jagels, K., Karlyshev, A. V., Leather, S., Moule, S., Oyston, P. C., Quail, M., Rutherford, K., Simmonds, M., Skelton, J., Stevens, K., Whitehead, S. and Barrell, B. G. (2001). Genome sequence of *Yersinia pestis*, the causative agent of plague. *Nature* **413**, 523–527.

Ponting, C. P. (1997). Evidence for PDZ domains in bacteria, yeast, and plants. *Protein Sci.* **6**, 464–468.

Ponting, C. P. (1999). Chlamydial homologues of the MACPF (MAC/perforin) domain. *Curr. Biol.* **9**, R911–913.

Ponting, C. P. (2001). Issues in predicting protein function from sequence. *Brief Bioinform.* **2**, 19–29.

Ponting, C. P. and Birney, E. (2000). Identification of domains from protein sequences. *Meth. Mol. Biol.* **143**, 53–69.

Ponting, C. P. and Dickens, N. J. (2001). Genome cartography through domain annotation. *Genome Biol.* **2**, Comment 2006.

Ponting, C. P. and Pallen, M. J. (1999a). A beta-propeller domain within TolB. *Mol. Microbiol.* **31**, 739–740.

Ponting, C. P. and Pallen, M. J. (1999b). Beta-propeller repeats and a PDZ domain in the tricorn protease: predicted self-compartmentalisation and C-terminal polypeptide-binding strategies of substrate selection. *FEMS Microbiol. Lett.* **179**, 447–451.

Ponting, C. P., Aravind, L., Schultz, J., Bork, P. and Koonin, E. V. (1999). Eukaryotic signalling domain homologues in archaea and bacteria. Ancient ancestry and horizontal gene transfer. *J. Mol. Biol.* **289**, 729–745.

Ray, M. C., Germon, P., Vianney, A., Portalier, R. and Lazzaroni, J. C. (2000). Identification by genetic suppression of *Escherichia coli* TolB residues important for TolB–Pal interaction. *J. Bacteriol.* **182**, 821–824.

Regue, M., Climent, N., Abitiu, N., Coderch, N., Merino, S., Izquierdo, L., Altarriba, M. and Tomas, J. M. (2001). Genetic characterization of the *Klebsiella pneumoniae* waa gene cluster, involved in core lipopolysaccharide biosynthesis. *J. Bacteriol.* **183**, 3564–3573.

Reid, S. D., Green, N. M., Buss, J. K., Lei, B. and Musser, J. M. (2001). Multilocus analysis of extracellular putative virulence proteins made by group A *Streptococcus*: population genetics, human serologic response, and gene transcription. *Proc. Natl Acad. Sci. USA* **98**, 7552–7557.

Rutherford, K., Parkhill, J., Crook, J., Horsnell, T., Rice, P., Rajandream, M. A. and Barrell, B. (2000). Artemis: sequence visualization and annotation. *Bioinformatics* **16**, 944–945.

Salzberg, S. L., Delcher, A. L., Kasif, S. and White, O. (1998). Microbial gene identification using interpolated Markov models. *Nucleic Acids Res.* **26**, 544–548.

Schaffer, A. A., Aravind, L., Madden, T. L., Shavirin, S., Spouge, J. L., Wolf, Y. I., Koonin, E. V. and Altschul, S. F. (2001). Improving the accuracy of PSI-BLAST protein database searches with composition-based statistics and other refinements. *Nucleic Acids Res.* **29**, 2994–3005.

Schultz, J., Milpetz, F., Bork, P. and Ponting, C. P. (1998). SMART, a simple modular architecture research tool: identification of signaling domains. *Proc. Natl Acad. Sci. USA* **95**, 5857–5864.

Snel, B., Bork, P. and Huynen, M. (2000). Genome evolution. Gene fusion versus gene fission. *Trends Genet.* **16**, 9–11.

Sonnhammer, E. L. and Wootton, J. C. (2001). Integrated graphical analysis of protein sequence features predicted from sequence composition. *Proteins* **45**, 262–273.

Sonnhammer, E. L., von Heijne, G. and Krogh, A. (1998). A hidden Markov model for predicting transmembrane helices in protein sequences. *Proc. Int. Conf. Intell. Syst. Mol. Biol.* **6**, 175–182.

Spiers, A., Lamb, H. K., Cocklin, S., Wheeler, K. A., Budworth, J., Dodds, A. L., Pallen, M. J., Maskell, D. J., Charles, I. G. and Hawkins, A. R. (2002). Differential proteolytic responses of HtrA and Tsp to recognition of the ssrA-encoded peptide. (Submitted).

Stoesser, G., Baker, W., van den Broek, A., Camon, E., Garcia-Pastor, M., Kanz, C., Kulikova, T., Leinonen, R., Lin, Q., Lombard, V., Lopez, R., Redaschi, N., Stoehr, P., Tuli, M. A., Tzouvara, K. and Vaughan, R. (2002). The EMBL nucleotide sequence database. *Nucleic Acids Res.* **30**, 21–26.

Suzek, B. E., Ermolaeva, M. D., Schreiber, M. and Salzberg, S. L. (2001). A probabilistic method for identifying start codons in bacterial genomes. *Bioinformatics* **17**, 1123–1130.

Tateno, Y., Imanishi, T., Miyazaki, S., Fukami-Kobayashi, K., Saitou, N., Sugawara, H. and Gojobori, T. (2002). DNA Data Bank of Japan (DDBJ) for genome scale research in life science. *Nucleic Acids Res.* **30**, 27–30.

Taylor, W. R. and Brown, N. P. (1999). Iterated sequence databank search methods. *Comput. Chem.* **23**, 365–385.

Tettelin, H., Nelson, K. E., Paulsen, I. T., Eisen, J. A., Read, T. D., Peterson, S., Heidelberg, J., DeBoy, R. T., Haft, D. H., Dodson, R. J., Durkin, A. S., Gwinn, M., Kolonay, J. F., Nelson, W. C., Peterson, J. D., Umayam, L. A., White, O., Salzberg, S. L., Lewis, M. R., Radune, D., Holtzapple, E., Khouri, H., Wolf, A. M., Utterback, T. R., Hansen, C. L., McDonald, L. A., Feldblyum, T. V., Angiuoli, S., Dickinson, T., Hickey, E. K., Holt, I. E., Loftus, B. J., Yang, F., Smith, H. O., Venter, J. C., Dougherty, B. A., Morrison, D. A., Hollingshead, S. K. and Fraser, C. M. (2001). Complete genome sequence of a virulent isolate of *Streptococcus pneumoniae*. *Science* **293**, 498–506.

Thornton, J. W. and DeSalle, R. (2000). Gene family evolution and homology: genomics meets phylogenetics. *Annu. Rev. Genomics Human Genet.* **1**, 41–73.

Uzzau, S., Figueroa-Bossi, N., Rubino, S. and Bossi, L. (2001). Epitope tagging of chromosomal genes in *Salmonella*. *Proc. Natl Acad. Sci. USA* **98**, 15264–15269.

van Belkum, A., Scherer, S., van Alphen, L. and Verbrugh, H. (1998). Short-sequence DNA repeats in prokaryotic genomes. *Microbiol. Mol. Biol. Rev.* **62**, 275–293.

Wang, J. and Zhang, C. T. (2001). Identification of protein-coding genes in the genome of *Vibrio cholerae* with more than 98% accuracy using occurrence frequencies of single nucleotides. *Eur. J. Biochem.* **268**, 4261–4268.

Wassenaar, T. M. and Gaastra, W. (2001). Bacterial virulence: can we draw the line? *FEMS Microbiol. Lett.* **201**, 1–7.

Wolf, E., Kim, P. S. and Berger, B. (1997). MultiCoil: a program for predicting two- and three-stranded coiled coils. *Protein Sci.* **6**, 179–189.

3 The Atlas Visualization of Genomewide Information

Marie Skovgaard, Lars Juhl Jensen, Carsten Friis, Hans Henrik Stærfeldt, Peder Worning, Søren Brunak and David Ussery*
Center for Biological Sequence Analysis, BioCentrum-DTU, Building 208, The Technical University of Denmark, DK-2800 Kgs. Lyngby, Denmark

CONTENTS

Introduction
Construction of the visualization software
Use of atlases to visualize DNA information
Atlases for visualizing genomewide RNA expression
Atlases for visualizing global prediction of protein function
Concluding remarks

◆◆◆◆◆◆ INTRODUCTION

The wealth of information contained in a microbial genome is not easy to comprehend at all scales. Even after the genome of an organism has been sequenced, the problem of gaining an overview of the newly acquired data still remains. One way to obtain an overview is to visualize positional features at the chromosome level: we have developed a method, Atlases, for showing correlations between position-dependent information in sequenced chromosomes.

The DNA sequence is not only hard to comprehend because of its size but also because the genomic sequence is not linked in a simple way to the biology of the organism. For example, in examining the AT content of genomes, a property often reported in genome-sequencing papers, considerable variation is observed. Figure 3.1 shows the percentage AT of 25 different proteobacterial genomes, ranging from 33% to 74% AT. The AT content does not appear to correlate with the proteobacterial subdivisions.

---

* To whom correspondence should be addressed. Tel: (+45) 45 25 24 88; Fax: (+45) 45 93 15 85; email: dave@cbs.dtu.dk

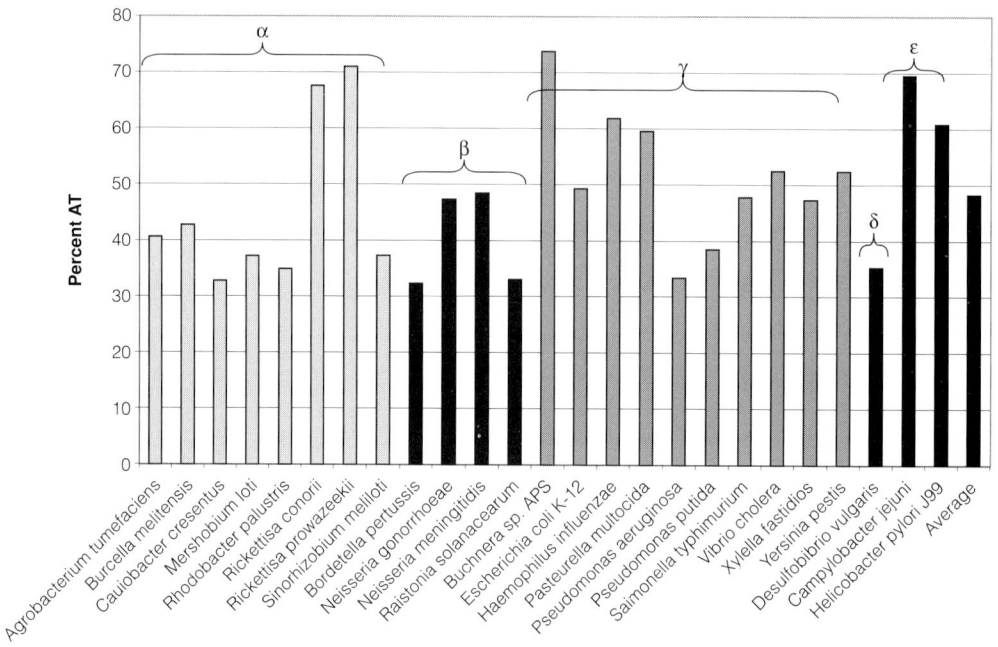

Figure 3.1. Percentage AT in 25 proteobacter genomes. The genomes are grouped into subdivisions and the last bar is the average of all 25 genomes.

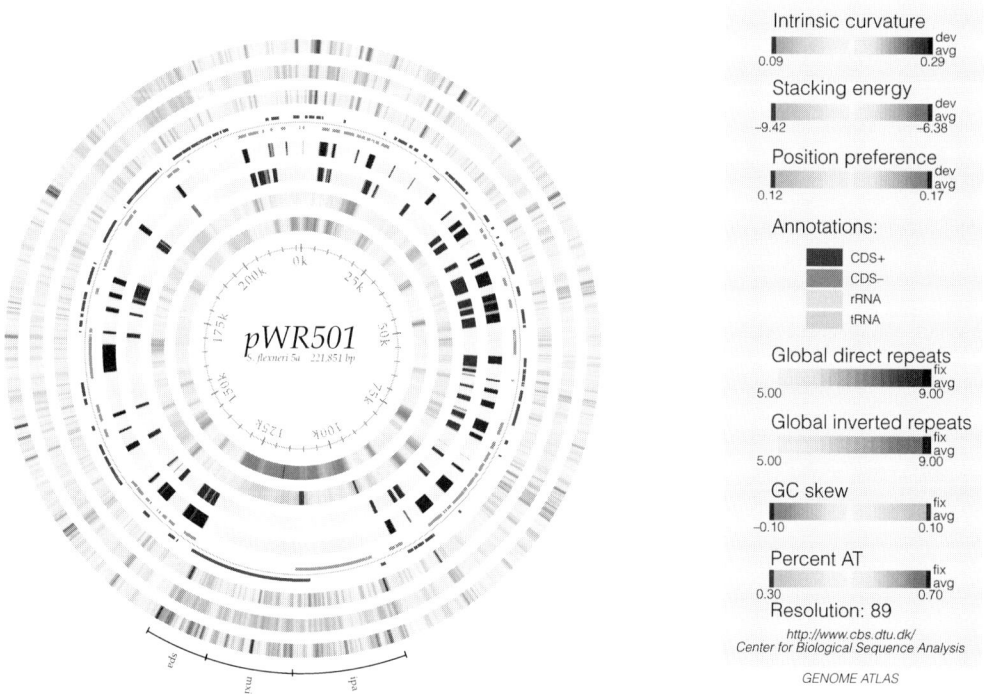

Figure 3.2. GenomeAtlas for *Shigella flexneri* 5a virulence plasmid pWR501. The marked regions contain three loci which code for a total of 34 virulence-related genes.

The percentage AT of a chromosome reflects only an *average* property of the chromosome. However, the AT content is not homogeneously distributed throughout the DNA. Often there are clusters of AT-rich and AT-poor regions. For example, most promoter regions are more AT-rich than the average coding sequences (Ozoline *et al.*, 1999; Pedersen *et al.*, 2000). In many cases the variations between regions will tell more than the average value, as exemplified in Figure 3.2, where an AT-rich region is found to contain genes involved in pathogenesis.

The AT content within a region of a chromosome is a very simple property to calculate from the nucleotide sequence. More complex features like DNA curvature or major groove compressibility, which reflect structural properties of a given region, can be estimated directly from the sequence and give biological insight. Additional information can be accessed by looking at the genes encoded by the chromosome. Once the location of the genes is known, it is possible to visualize both experimentally determined expression levels and RNA sequence features. By translating the RNA sequence to protein sequence, it is also possible to visualize properties of the proteome, such as protein function.

◆◆◆◆◆◆ CONSTRUCTION OF THE VISUALIZATION SOFTWARE

In order to be able to visualize such diverse data, a flexible software tool is needed. We have developed a computer program, GeneWiz, which enables us to visualize a complete chromosome compactly. The program creates either a circular or linear graphical representation of the entire chromosome or of a specified subsection. Sufficiently large regions that display significant variation from the rest of the chromosome can be readily found. In order to be able to see deviations of smaller regions, a zoomed Atlas must be made.

Each feature, such as AT content or gene expression level, is represented as a separate lane in the atlas. The value is at each position colour coded according to a user-specified colour scale. We generally use colour scales where regions of extreme values are highlighted (these can be one-ended or two-ended) whereas typical values are grey. If required, the plot can be smoothed by a running average.

The properties to be visualized must be present in the form of one value per basepair (bp) in the chromosome. For simple sequence features like the AT content, this is the natural format, whereas for data such as gene expression levels, the value for each gene must be mapped on to the corresponding range of basepairs. In addition to the data series, the annotations from a GenBank file can be displayed using a series of icons with user-defined colours. This allows for the identification of short or long annotated regions of interest.

GeneWiz is solely a visualization program and is not capable of calculating the data used in the different atlases. All data must be calculated and properly formatted. While this obviously adds to the work of creating an atlas, it gives great flexibility, as these data can come from any source. In this paper we use simple measures generated from lookup tables, publicly available programs like BLAST (Altschul *et al.*, 1997), methods developed in-house like ProtFun (Jensen *et al.*, 2002) as well as experimentally determined expression data.

◆◆◆◆◆◆ USE OF ATLASES TO VISUALIZE DNA INFORMATION

GenomeAtlases

The GenomeAtlas is a general atlas made for all the fully sequenced microbial chromosomes found in public databases (Jensen *et al.*, 1999; Pedersen *et al.*, 2000). The GenomeAtlas is a combination of some generally informative parameters and can be used as an offset for identifying unique regions or special features for the given chromosome. The GenomeAtlas for all publicly available sequenced chromosomes can be found at http://www.cbs.dtu.dk/services/GenomeAtlas/.

Introducing the parameters

To generate GenomeAtlas plots, a number of parameters are calculated for the DNA double helix based on the nucleotide sequence. These parameters belong to three categories: repeats, structural parameters, and parameters directly related to the base composition. These three categories are combined into a common atlas where the parameters are visualized, giving the values of the parameters as the intensity of the colour (Jensen *et al.*, 1999).

Structural parameters

A number of measures for the local structure of DNA have been devised, most of which are based on dinucleotide or trinucleotide models that have been obtained by fitting either experimental results or theoretical estimates (Pedersen *et al.*, 1998, 2000).

Intrinsic curvature is a property of DNA that is closely related to anomalous gel mobility, as DNA fragments with high intrinsic curvature will migrate slower on polyacrylamide gels than markers with the same length. In this work we have used the CURVATURE program (Shpigelman *et al.*, 1993), which is based on a wedge model (Trifonov and Sussman, 1980; Ulanovsky *et al.*, 1986), for prediction of intrinsic curvature. From a set of dinucleotide values for the twist, wedge and direction angles, the three-dimensional path of a 21 bp fragment is calculated. Curvature profiles for longer sequences can thus be calculated using a 21 bp running window. Curves are often encountered upstream of highly expressed genes (Bracco *et al.*, 1989).

Stacking energy relates to the interaction energy between adjacent basepairs in the DNA double helix. The total stacking energy of a DNA segment can be estimated from the set of dinucleotide values determined by quantum mechanical calculations on crystal structures (Ornstein et al., 1978). All stacking energies are negative since base stacking is an energetically favourable interaction that serves to stabilize the double helix. This means that regions with large stacking energies are strongly stabilized and, therefore, less likely to destack or melt than regions with less negative stacking energies.

The position preference is a measure of helix flexibility based on a set of 32 trinucleotide values giving the log-odds of the minor groove facing outwards when wrapped around a histone octamer (Satchwell et al., 1986). On this scale, a value of zero represents no preference of the trinucleotide for specific positions in the nucleosomes, while large absolute values mean that the trinucleotide has strong preference. Because large absolute values thereby imply that the sequence is inflexible, a measure of flexibility is obtained by removing the sign from the original trinucleotide values (Pedersen et al., 1998). On that scale, low values correspond to high bendability.

Base composition

The trivial way to parameterize the base composition is to simply use the G-, A-, T- and C-contents. A drawback of this representation is that the four parameters are mutually correlated as they sum to one. An alternative parameterization for the base composition is $A + T$ and $G - C$. In addition to being mutually independent measures, they also have the advantage of being easier to interpret in a biological context.

The AT content is strongly correlated to the structural parameters described above, especially the stacking energy. AT-rich regions usually destack more readily, have a higher intrinsic curvature and are less flexible. The parameter $G - C$, similar to the GC skew (McLean et al., 1998) reflects a general bias of purines towards the leading strand of DNA replication (Tillier and Collins, 2000). Since the GC skew has almost no correlation to the structural properties of DNA, the AT content contains nearly all the structural information arising from the mononucleotide composition.

Repeat elements

Repeats are multiple copies of the same sequence at different locations on a piece of DNA. The repeats can be found either by a very accurate method using a basic algorithm that finds the highest degree of homology for an R bp long repeat within a window of length W (Jensen et al., 1999), or by cutting the sequence up in fragments and using the heuristic alignment algorithm BLAST (Altschul et al., 1997) to find the homologous regions with the length R. The basic algorithm is more accurate than BLAST but it is also computationally demanding, therefore, BLAST is used on large sequences. There are two kinds of repeats. A direct repeat is a sequence that is present in at least two copies on the same strand, whilst two copies located on opposite strands will give rise to an inverted repeat.

GenomeAtlases of pathogenicity plasmids

A GenomeAtlas can give a quick overview of a given chromosome, and thereby be the reason for further analysis of a given organism, or a more specific search for a given feature can be made by looking through a collection of atlases. The latter was the case when a study of pathogenicity islands in bacterial plasmids was based on the knowledge of the correlation between pathogenicity islands and variation in AT content, such as the toxin genes in plasmid pO157 from pathogenic *E. coli* strains (Friis *et al.*, 2000). Another example of the correlation between pathogenicity islands and changes in AT content can also be found in the large virulence plasmid of *Shigella flexneri* (GenBank accession number AF348706) (Venkatesan *et al.*, 2001).

The atlas of the *Shigella flexneri* 5a virulence plasmid pWR501 (Figure 3.2) reveals an AT-rich area, which is strongly curved, will destack or melt more readily than the rest of the plasmid, and is more rigid. This region encodes a locus of genes (*ipa–mxi–spa*) involved in the pathogenic invasion of mammalian cells, and includes a type III secretion pathway (Schuch *et al.*, 1999; Page *et al.*, 2001).

Variations in AT content are obviously not always correlated with the presence of toxic genes; another indication of potential pathogenic regions can be the localization of multiple repeats, especially insertion sequence (IS) elements (Hacker *et al.*, 1997; Hacker and Kaper, 2000). Large numbers of direct and inverted repeats can be seen in Figure 3.2. Typically, global direct repeats account for around 3% or less of most bacterial chromosomes (data not shown, but 'GenomeAtlases' for all sequenced genomes can be found on our web page). Many of the repeats (especially the global inverted repeats) are reflective of IS elements. Note that the AT-rich *ipa–mxi–spa* region is the largest region free of repeats in the plasmid.

Another example of a plasmid with many repeats is the plasmid pBtoxis* from the spore-forming bacteria *Bacillus thuringiensis* subsp. *israelensis*. Like pWR501, the repeats in pBtoxis are scattered all over the plasmid (Figure 3.3). A search was made for genes from transposable elements like transposases and integrases and, by doing a simple BLAST search against SWISSPROT, these genes were located and they were indeed found to be associated with the repeats. In the case of this plasmid, the presence of transposable elements was known long before the plasmid was sequenced (Mahillon *et al.*, 1994), but the GenomeAtlas can be used as an easy method for localization of transposons and IS elements.

B. thuringiensis is used in agriculture as an alternative to synthetic chemical pesticides. It produces parasporal crystals that have insecticidal activity, and the genes that are believed to be responsible for this activity are marked in Figure 3.3 (Schnepf *et al.*, 1998). The transposable elements in pBtoxis are at least partly responsible for the high degree of genetic plasticity that makes *B. thuringiensis* adaptable to a variety of environments. However, it should also lead to caution in the use, since *B. thuringiensis*,

---

* These sequence data were produced by the Microbial Genomes Sequencing Group at the Sanger Institute and can be obtained from ftp://ftp.sanger.ac.uk/pub/pathogens/bti/.

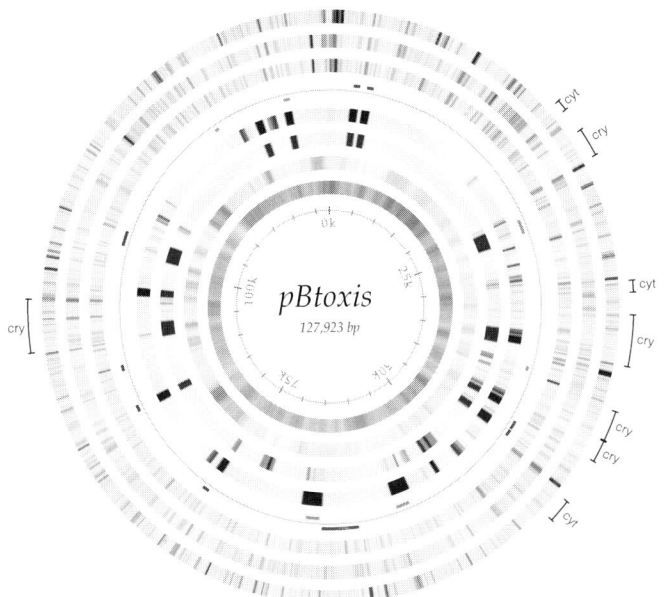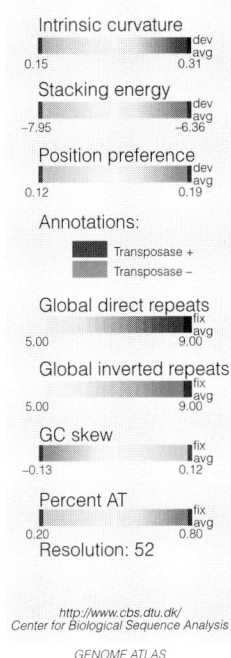

Figure 3.3. GenomeAtlas for *Bacillus thuringiensis* pBtoxis. The insecticide activity comes from the marked *cry* and *cyt* genes.

based on genetic evidence, is from the same species as *Bacillus anthracis* and *Bacillus cereus*, both human pathogens (Helgason *et al.*, 2000).

Some of the repeats that are not associated with genes from transposable elements seem to be copies of *cry*, the gene for the pesticide crystal protein. The three *cyt* genes produce cytolytic delta-endotoxins; the absence of repeats in this area indicates that they are not similar to each other at the nucleotide level.

In the case of the two plasmids presented here, the atlas was used as a method to screen large plasmids for signs that indicated the presence of toxic genes. Many pathogenic regions within plasmids might not be found in this way, but the atlas serves as a very strong method for initial examination of the sequences (Friis *et al.*, 2000).

A custom-made DNA atlas

The GenomeAtlas is our 'standard' atlas, which can capture interesting features of a chromosome. As an example, consider chromosome 1 from the protozoan *Leishmania major*, an intracellular pathogen of the immune system. This chromosome has an unusual organization of its genes, with the 79 protein coding genes being in two large clusters. The first 29 genes are coded on one strand, whilst the last 50 genes are on the other strand (Myler *et al.*, 1999). From the GenomeAtlas* a correlation between

---

* The atlas can be seen at http://www.cbs.dtu.dk/services/GenomeAtlas/Eukaryotes/Leishmania/major/.

intergenic regions and global repeats can be observed. In order to further investigate the possible relationship between other structural parameters and intergenic regions, we constructed a custom atlas (see Figure 3.4).

Several properties of the chromosome are revealed by the base composition parameters (AT content and GC skew). The telomeres and a region around 80 kbp have a much higher AT content than the rest of the chromosome. Also a shift in the GC skew is observed around 80 kbp, correlating with the unusual gene organization. This is in agreement with the region being proposed as the origin of replication (McDonagh et al., 2000).

More direct than inverted repeats are found in this chromosome. Some of these arise from gene duplications, the most obvious example being the two genes around position 240 kbp. Even though gene duplications are observed, the direct repeats still exhibit a slight preference for noncoding regions. This preference is much stronger for inverted repeats, which occur exclusively in intergenic regions, as shown in Table 3.1.

The exclusive localization of inverted repeats in intergenic regions prompted an interest in whether other DNA structural elements might also be preferentially positioned within non-coding regions. Runs of purines (or pyrimidines) as well as alternating pyrimidine/purine

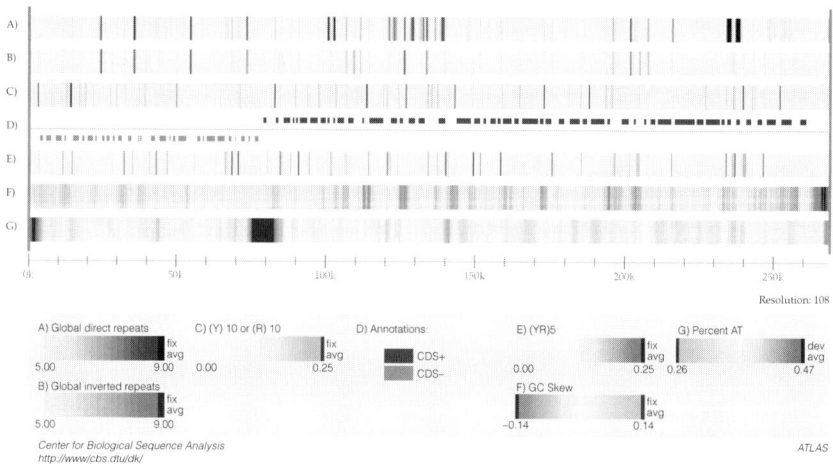

Figure 3.4 Specialized atlas for *Leishmania major* chromosome 1 (Freidlin chromosome 1, 268 984 bp).

Table 3.1 Characteristics of coding and non-coding sequences in *Leishmania major* chromosome 1

DNA property	Length (bp)	% direct	% inverted	% (Y)$_{10}$	% (YR)$_5$	% AT
Coding	140 229	4.5	0.0	1.2	2.6	34.6
Non-coding	128 755	6.0	4.6	11.2	6.9	39.4
Whole chromosome	268 984	5.2	2.2	6.0	4.7	36.9

Y_{10}, pyrimidine stretches of length 10; YR_5 alternating pyrimidine/purine stretches of length 10.

stretches occur more often than would be expected from the base composition of *Leishmania major* (Ussery *et al.*, 2002). The location of these regions was visualized by plotting the location of all such stretches of at least 10 bp. Many purine stretches can adopt an A-DNA conformation, whereas pyrimidine/purine stretches that are GC-rich can adopt a Z-DNA conformation. There is a strong preference (about tenfold; see '%(Y)$_{10}$' column in Table 3.1) for purine stretches in the intergenic regions, while the pyr/pur regions are less strongly correlated with the non-coding DNA.

◆◆◆◆◆◆ ATLASES FOR VISUALIZING GENOMEWIDE RNA EXPRESSION

It has often been said that even non-coding DNA is far from a random string of bases. Since helix structure is a function of that same string of bases, this statement must apply to structural features as well. These structural features are suspected of affecting not just the termination of transcription as mentioned earlier, but also the rate of transcription itself.

Almost unheard of 5 years ago, genomewide mRNA expression analysis has become mainstream in most major microbiological laboratories. With this technology, it is feasible to examine the transcription levels for an entire microbial genome under a broad range of different circumstances, and to some degree reverse engineer regulatory pathways (Spellman *et al.*, 1998).

By visualizing measured levels of transcription in an atlas, it becomes possible to examine whether correlations exist between the mRNA expression levels and DNA structural properties or base composition. Such correlations could be expected due to chromatin packing (Ussery *et al.*, 2001). Also the relationship between the level of transcription and chromosomal location may reveal interesting aspects (Hughes *et al.*, 2000b).

ExpressionAtlas

The strength of genome-wide RNA expression analysis lies in the ability to measure the expression levels of an entire genome simultaneously. When analysing several arrays with thousands of genes, one is faced with much the same problem as when analysing whole-genome sequences: the sheer amount of data makes it hard to obtain an overview. The ExpressionAtlas is a way of visualizing expression experiments taking into account chromosomal position and other factors suspected of being involved in transcription, such as DNA structure and repeats.

In the example shown, the average intensities from cDNA arrays (Cho *et al.*, 1998) were used as an estimate of the constitutive expression levels of genes in *Saccharomyces cerevisiae*. Alternatively log-fold changes could be plotted to highlight regulated genes. We chose average intensities to ensure comparability to the predicted expression levels also displayed on the atlas.

Neural networks trained on average expression values from *E. coli* microarray experiments predicted the expression level of each gene. The predicted levels of expression were normalized to a range from 0 to 1. As input to the neural networks, the trinucleotide frequencies of the coding regions were used. These 64 frequencies were calculated without taking the reading frame into account. This representation was chosen because the majority of DNA structural properties can be captured at the trinucleotide level. In this way we can capture possible correlation between the structural properties of the coding DNA and the expression levels.

Both the experimentally measured and the predicted expression levels are displayed in the atlas in Figure 3.5, together with position preference, global repeats and AT content. The AT content and the global repeats were included to give a general view of the composition of the chromosome, whereas the position preference, being a measure of the flexibility of the double helix, is expected to be correlated with the expression levels (Pedersen *et al.*, 2000). The inverted repeats clearly mark the telomeric regions.

Comparing the actual expression levels with the levels predicted by neural networks reveals a strong correlation between the two. A similar, albeit weaker, correlation is observed between expression levels and the position preference measure. The fact that neural networks trained on the prokaryote *E. coli* data can predict highly expressed genes in a eukaryote implies the existence of universal DNA properties that influence transcription. The correlation with the position preference measure suggests that helix flexibility plays a part in this. Speculations on such generic features of expressed genes have been proposed before (Sharp and Li, 1987).

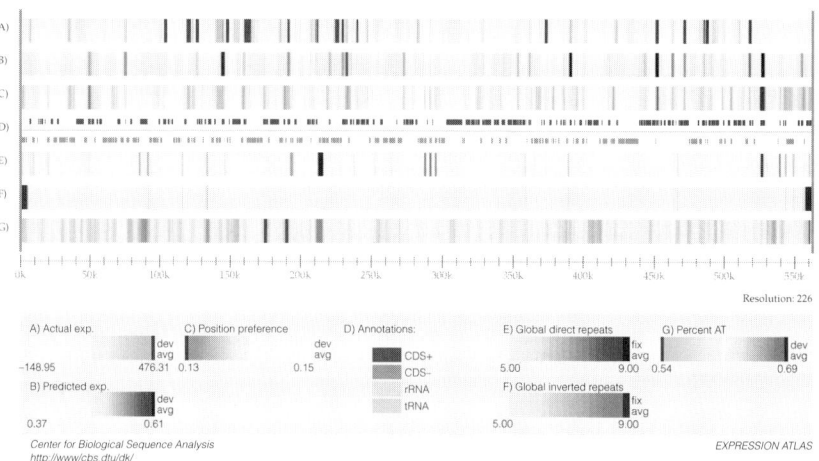

Figure 3.5. The ExpressionAtlas of *Saccharomyces cerevisiae* chromosome VIII (562 639 bp total). Lane A shows the average intensities from cDNA arrays, indicating the constitutively expressed genes, whilst lane B is the predicted expression.

◆◆◆◆◆◆ ATLASES FOR VISUALIZING GLOBAL PREDICTION OF PROTEIN FUNCTION

By looking at the ExpressionAtlas, it is possible to identify regions with genes that are highly expressed (and possibly regulated) under one or more experimental conditions. It is at this point obvious to ask what the function of these genes might be.

Unfortunately, the function of a large fraction of genes remains unknown in most fully sequenced chromosomes. Of the 30 000–50 000 genes believed to be present in the human genome, no more than 40–60% can be assigned a functional role based on homology to known proteins. Even though the situation is a bit more favourable when looking at simpler model organisms like *S. cerevisiae* and *C. elegans*, the function of more than 30% of the predicted protein sequences still remains unknown.

In newly sequenced chromosomes, most of the functional annotation of genes is based on homology inference. Using methods such as BLAST (Altschul *et al.*, 1997), homologous proteins are identified by sequence similarity and the function is inferred from the knowledge about the homologues. However, it is usually the case that somewhere between 30% and 50% of the proteins give no matches to proteins of known function. These are known as 'orphan' proteins.

Traditionally, protein function has been viewed as something directly related to the conformation of the polypeptide chain. However, as the three-dimensional structure currently is quite hard to calculate from the sequence (Lesk *et al.*, 2001), a computational strategy for the elucidation of orphan protein function may benefit also from the prediction of functional attributes, which are more directly related to the linear sequence of amino acids.

Our approach to function prediction is based on the fact that a protein is not alone when performing its biological task. As it will have to operate using the same cellular machinery for modification and sorting as all the other proteins do, one can expect some conservation of essential types of post-translational modifications (PTMs). Because reasonably precise methods for prediction of PTMs from sequence exist today, our prediction method, which integrates such relevant features to assign orphan protein to functional class, can be applied to all proteins where the sequence is known (Gupta *et al.*, 2002; Jensen *et al.*, 2002). This is in contrast to methods that rely on clustering of co-expressed genes (Eisen *et al.*, 1998), prediction of gene fusions and/or phylogenetic profiles (Marcotte *et al.*, 1999a,b; Pellegrini *et al.*, 1999; Hughes *et al.*, 2000a).

For any function prediction method, the ability to assign the relationship correctly depends strongly on the function classification scheme used. We predict a scheme of 12 cellular functions that is closely related to the 14-class classification originally proposed by Riley for the *E. coli* genome (Riley, 1998). The system consists of an ensemble of neural networks for each functional category, each neural network having a different combination predicted protein features as its input. The networks were trained exclusively on human protein sequences, but perform well on a wide selection eukaryotes (including *S. cerevisiae*). For each protein sequence the outputs of these neural networks are subsequently combined into a probability for each category.

We have applied this software to all predicted protein sequences from *S. cerevisiae* chromosome VIII. Based on our performance estimates of the method on *S. cerevisiae* sequences, we have selected a subset of eight categories out of the original 12-category system. The probabilistic scores of each protein sequence were mapped on to the position in the chromosome of the corresponding gene. Figure 3.6 shows the resulting FunctionAtlas along with the actual expression levels also shown in the ExpressionAtlas.

One feature that is visible from a FunctionAtlas is clusters of genes with related functions. Examples of this include the regions 10–50 k and 250–260 k that contain very large numbers of predicted transport and binding proteins. The regions 50–60 k, 335–350 k and 410–420 k are predicted to contain a large number of genes involved in replication or transcription, several of which are likely to serve a regulatory role according to our predictions. Since genes of related function are known to often cluster (although the extent varies from organism to organism) predicted functional clusters can be trusted more than individual predictions. If the function of some of the genes within a cluster is known and in agreement with the prediction, as is the case for several of the clusters mentioned, this obviously adds to the evidence.

Another possibility is to correlate the predicted protein function to expression data. Close inspection of Figure 3.6 reveals that many of the constitutively highly expressed genes are predicted to be involved in energy metabolism, although overall this is a quite rare category. A hypergeometric test of the underlying data verifies that this correlation is indeed significant at a 95% confidence level. A large number of highly

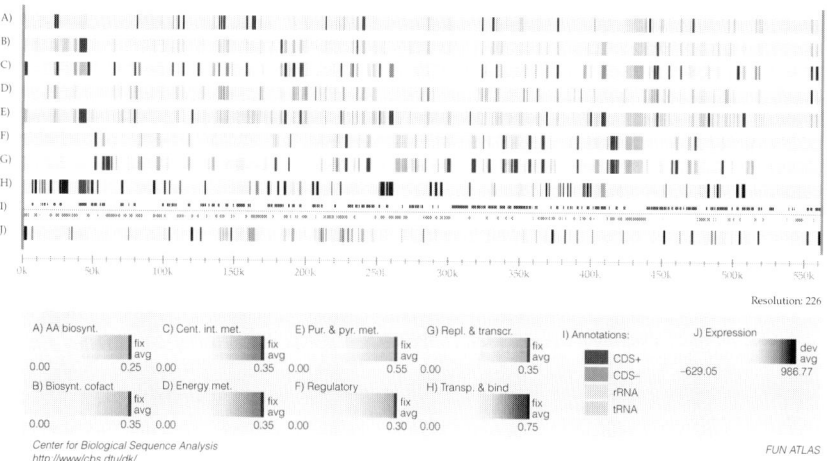

Figure 3.6. The FunctionAtlas of *Saccharomyces cerevisiae* chromosome VIII (562 638 bp). Lane A, amino acid biosynthesis; lane B, biosynthesis of co-factors; lane C, central intermediary metabolism; lane D, energy metabolism; lane E, purine and pyrimidine metabolism; lane F, regulatory function; lane G, replication and transcription; lane H, transport and binding; lane J, average intensity from cDNA experiments (the same as in Figure 3.5).

expressed transcripts for proteins involved in replication and transcription can also be identified, although no correlation between function and expression level is found in this case.

◆◆◆◆◆◆ CONCLUDING REMARKS

In summary, we have shown several different applications of Atlases for visualizing different types of information, within the context of the whole plasmid or chromosome. Essentially, any type of information concerning the DNA, RNA or protein can be plotted along the chromosome, allowing for rapid analysis of global properties in a serendipitous manner. The atlas gives the researcher the option to view the calculated or experimentally measured data in a position-dependent way, and thereby see correlations between a feature and its position or variation in a feature within the chromosome.

The atlas can be used to spot variation in different features within a region but not all the information can be viewed at the same time. For a bacterial chromosome of the same size as *E. coli*, only features with approximately the same size as a gene or larger can be observed. This means that some of the features shown for *L. major* with repeats in intergenic regions would not be visible in the *E. coli* genome. If variation in a smaller scale is to be seen for a large chromosome, a shorter region should be visualized.

Acknowledgements

The authors would like to acknowledge and thank the people at CBS for their help, in particular Ramneek Gupta, Steen Knudsen, Anders Krogh and Anders Gorm Pedersen. This work was supported by a grant from the Danish National Research Foundation.

References

Altschul, S., Madden, T., Schaffer, A., Zhang, J., Zhang, Z., Miller, W. and Lipman, D. (1997). Gapped BLAST and PSI-BLAST: a new generation of protein database search programs. *Nucleic Acids Res.* **25**, 3389–3402.

Bracco, L., Kotlarz, D., Kolb, A., Diekmann, S. and Buc, H. (1989). Synthetic curved DNA sequences can act as transcriptional activators in *Escherichia coli*. *EMBO J.* **8**, 4289–4296.

Cho, R., Campbell, M., Winzeler, E., Steinmetz, L., Conway, A., Wodicka, L., Wolfsberg, T., Gabrielian, A., Landsman, D., Lockhart, D. and Davis, R. W. (1998). A genome-wide transcriptional analysis of the mitotic cell cycle. *Mol. Cell* **2**, 65–73.

Eisen, M., Spellman, P., Brown, P. and Botstein, D. (1998). Cluster analysis and display of genome-wide expression patterns. *Proc. Natl Acad. Sci. USA* **95**, 14863–14868.

Friis, C., Jensen, L. and Ussery, D. (2000). Visualization of pathogenicity regions in bacteria. *Genetica* **108**, 47–51.

Gupta, R., Jensen, L. and Brunak, S. (2002). Orphan protein function and its relation to glycosylation. In *Ernst Schering Research Foundation Proceedings* (Mewes, H., Weiss, B. and Seidel, H., eds) Ch 13, pp. 275–294. Berlin: Springer-Verlag.

Hacker, J. and Kaper, J. (2000). Pathogenicity islands and the evolution of microbes. *Annu. Rev. Microbiol.* **54**, 641–679.

Hacker, J., Blum-Oehler, G., Muhldorfer, I. and Tschape, H. (1997). Pathogenicity islands of virulent bacteria: structure, function and impact on microbial evolution. *Mol. Microbiol.* **23**, 1089–1097.

Helgason, E., Økstad, O., Caugant, D., Johansen, H., Fouet, A., Mock, M., Hegna, I. and Kolstø, A. (2000). *Bacillus anthracis, Bacillus cereus*, and *Bacillus thuringiensis* – one species on the basis of genetic evidence. *Appl. Environ. Microbiol.* **66**, 2627–2630.

Hughes, T., Marton, M., Jones, A., Roberts, C., Stoughton, R., Armour, C., Bennett, H., Coffey, E., Dai, H., He, Y., Kidd, K., King, A., Meyer, M., Slade, D., Lum, P., Stepaniants, S., Shoemaker, D., Gachotte, D., Chakraburtty, K., Simon, J., Bard, M. and Friend, S. (2000a). Functional discovery via a compendium of expression profiles. *Cell* **102**, 109–126.

Hughes, T., Roberts, C., Dai, H., Jones, A., Meyer, M., Slade, D., Burchard, J., Dow, S., Ward, T., Kidd, M., Friend, S. and Marton, M. (2000b). Widespread aneuploidy revealed by DNA microarray expression profiling. *Nat. Genet.* **25**, 333–337.

Jensen, L., Friis, C. and Ussery, D. (1999). Three views of the *E. coli* genome. *Res. Microbiol.* **150**, 773–777.

Jensen, L., Gupta, R., Blom, N., Devos, D., Tamames, J., Kesmir, C., Nielsen, H., Stærfeldt, H., Rapacki, K., Workman, C., Andersen, C., Knudsen, S., Krogh, A., Valencia, A. and Brunak, S. (2002). *Ab initio* prediction of human orphan protein function from post-translational modifications and localization features. *J. Mol. Biol.* **319**, 1257–1265.

Lesk, A. M., Lo Conte, L. and Hubbard, T. (2001). Assessment of novel fold targets in CASP4: predictions of three-dimensional structures, secondary structures and interresidue contacts. *Proteins* **45** (Suppl. 5), 98–118.

Mahillon, J., Rezsohazy, R., Hallet, B. and Delcour, J. (1994). IS231 and other *Bacillus thuringiensis* transposable elements: a review. *Genetica* **93**, 13–26.

Marcotte, E., Pellegrini, M., Ng, H., Rice, D. W., Yeates, T. and Eisenberg, E. (1999a). Detecting protein function and protein–protein interactions from genome sequences. *Science* **285**, 751–753.

Marcotte, E., Pellegrini, M., Thompson, M., Yeates, T. and Eisenberg, D. (1999b). A combined algorithm for genome-wide prediction of protein function. *Nature* **402**, 83–86.

McDonagh, P., Myler, P. and Stuart, K. (2000). The unusual gene organization of *Leishmania major* friedlin chromosome 1 may reflect novel transcription processes. *Nucleic Acids Res.* **28**, 2800–2803.

McLean, M., Wolfe, K. and Devine, K. (1998). Base composition skews, replication orientation, and gene orientation in 12 prokaryotic genomes. *J. Mol. Evol.* **47**, 691–696.

Myler, P., Audleman, L., de Vos, T., Hixson, G., Kiser, P., Magness, C., Rickel, E., Sisk, E., Sunkin, S., Swartzell, S., Westlake, T., Bastein, P., Fu, G., Ivens, A. and Stuart, K. (1999). *Leishmania major* friedlin chromosome 1 has an unusual distribution of protein-coding genes. *Proc. Natl Acad. Sci. USA* **96**, 2902–2906.

Ornstein, R., Rein, R., Breen, D. and MacElroy, R. (1978). An optimized potential function for the calculation of nucleic acid interaction energies. I. Base stacking. *Biopolymers* **17**, 2341–2360.

Ozoline, O., Deev, A., Arkhipova, M., Chasov, V. and Travers, A. (1999). Proximal transcribed regions of bacterial promoters have a non-random distribution of A/T tracts. *Nucleic Acids Res.* **27**, 4768–4774.

Page, A., Fromont-Racine, M., Sansonetti, P., Legrain, P. and Parsot, C. (2001). Characterization of the interaction partners of secreted proteins and chaperones of *Shigella flexneri*. *Mol. Microbiol.* **42**, 1133–1145.

Pedersen, A., Baldi, P., Chauvin, Y. and Brunak, S. (1998). DNA structure in human RNA polymerase II promoters. *J. Mol. Biol.* **281**, 663–673.

Pedersen, A., Jensen, L., Stærfeldt, H., Brunak, S. and Ussery, D. (2000). A DNA structural atlas of *E. coli*. *J. Mol. Biol.* **299**, 907–930.

Pellegrini, M., Marcotte, E., Thompson, M., Eisenberg, D. and Yeates, T. (1999). Assigning protein functions by comparative genome analysis: protein phylogenetic profiles. *Proc. Natl Acad. Sci. USA* **96**, 4285–4288.

Riley, M. (1998). Systems for categorizing functions of gene products. *Curr. Opin. Struct. Biol.* **8**, 388–392.

Satchwell, S., Drew, H. and Travers, A. (1986). Sequence periodicities in chicken nucleosome core DNA. *J. Mol. Biol.* **191**, 659–675.

Schnepf, E., Crickmore, N., Rie, J. V., Lereclus, D., Baum, J., Feitelson, J., Zeigler, D. and Dean, D. (1998). *Bacillus thuringiensis* and its pesticidal crystal proteins. *Microbiol. Mol. Biol. Rev.* **62**, 775–806.

Schuch, R., Sandlin, R. and Maurelli, A. (1999). A system for identifying post-invasion functions of invasion genes: requirements for the *Mxi-Spa* type III secretion pathway of *Shigella flexneri* in intercellular dissemination. *Mol. Microbiol.* **34**, 675–689.

Sharp, P. M. and Li, W. H. (1987). The codon adaption index – a measure of directional synonymous codon usage bias, and its potential applications. *Nucleic Acids Res.* **15**, 1281–1295.

Shpigelman, E., Trifonov, E. and Bolshoy, A. (1993). CURVATURE: software for the analysis of curved DNA. *CABIOS* **9**, 435–444.

Spellman, P., Sherlock, G., Zhang, M., Iyer, V., Anders, K., Eisen, M., Brown, P., Botstein, D. and Futcher, B. (1998). Comprehensive identification of cell cycle-regulated genes of the yeast *S. cerevisiae* by microarray hybridization. *Mol. Biol. Cell* **9**, 3273–3297.

Tillier, E. and Collins, R. (2000). The contributions of replication orientation, gene direction, and signal sequences to base-composition asymmetries in bacterial genomes. *J. Mol. Evol.* **50**, 249–257.

Trifonov, E. and Sussman, J. (1980). The pitch of chromatin DNA is reflected in its nucleotide sequence. *Proc. Natl Acad. Sci. USA* **77**, 3816–3820.

Ulanovsky, L., Bodner, M. and Trifonov, E. (1986). Curved DNA: design, synthesis, and circularization. *Proc. Natl Acad. Sci. USA* **83**, 862–866.

Ussery, D. W., Larsen, T., Wilkes, K., Friis, C., Worning, P., Krogh, A. and Brunak, S. (2001). Genome organisation and chromatin structure in *Escherichia coli*. *Biochimie* **83**, 201–212.

Ussery, D., Soumpasis, D., Brunak, S., Stærfeldt, H., Worning, P. and Krogh, A. (2002). Bias of purine stretches in sequenced genomes. *Computers Chem.* **26**, 531–541.

Venkatesan, M., Goldberg, M., Rose, D., Grotbeck, E., Burland, V. and Blattner, F. (2001). Complete DNA sequence and analysis of the large virulence plasmid of *Shigella flexneri*. *Infect. Immun.* **69**, 3271–3285.

Part II
Construction of DNA Microarrays

4 Microarray Design for Bacterial Genomes

Jason Hinds,[1] Adam A Witney[1] and J Keith Vass[2]

[1]Department of Medical Microbiology, St George's Hospital Medical School, Cranmer Terrace, London SW17 0RE, UK
[2]Beatson Institute for Cancer Research, Garscube Estate, Glasgow G61 1BD, UK

CONTENTS

Introduction
General design considerations
Design process for a PCR product-based microarray
Summary
Relevant web resources

◆◆◆◆◆◆ INTRODUCTION

Microarrays provide a parallel analysis of complex mixtures of nucleic acid molecules by detecting the hybridization of nucleic acids in solution with sequence-specific 'probe elements' immobilized on the microarray surface. This process is imperfectly understood but is assumed to be subject to the same physicochemical parameters that influence solution hybridization. Microarrays are mainly used to detect mRNA or cDNA to study gene expression, or to detect DNA in the analysis of gene complement. The design of microarrays is a critical factor in the quality and usefulness of the data generated. Considerations for whole genome array design not only depend on simple gene representation for a given genome, but also on parameters that affect nucleic acid hybridization kinetics and specificity, such as G + C content, melting temperature (T_m) values, primer specificity and secondary structure. Furthermore, array design for comparative genomics of multiple strains requires study of genome information from a range of sequenced genomes. Bioinformatics is, therefore, an essential component of array design.

For bacterial pathogens, comparative and functional genomics using microarrays enable the study of virulence and pathogenicity, strain typing, molecular epidemiology, gene expression and gene regulation, often through the use of defined mutants or clinical isolates in comparison with sequenced strain types. Over 60 bacterial genomes have been completely sequenced (Doolittle, 2002) and many species-specific features of each genome are apparent. Such features include gene families, repetitive elements in both intragenic and intergenic regions, insertion sequences, phage sequences, as well as the expected domain homologies in functionally related genes.

The approach to array design described in this chapter is that used for spotted polymerase chain reaction (PCR) product microarrays (see Chapter 5). Therefore, the general aim of the design process is to select a pair of PCR primers that will amplify a single PCR product for use as an element on the array. This PCR product will ideally represent a single gene in the genome with maximum specificity. This chapter represents the collective experience of array element design accrued from the construction of five bacterial whole genome PCR product-based arrays: *Mycobacterium tuberculosis*, *Haemophilus influenzae*, *Yersinia pestis*, *Campylobacter jejuni* and *Streptococcus pneumoniae*. The design process is continually evolving to overcome the challenges that each genome presents in light of rapidly emerging and changing genome-sequencing information.

◆◆◆◆◆◆ GENERAL DESIGN CONSIDERATIONS

There is no single approach to the design of microarrays that addresses all the issues that arise owing to the complexities of genome structure, PCR primer design and DNA hybridization parameters. This section highlights the key issues that must be addressed when planning the design of a microarray and discusses the relevant points.

Array technology

There are several types of DNA microarray platforms available and the most suitable format will be governed in part by the accessibility, cost and intended application. The formats successfully used for bacteria include both oligonucleotide and PCR product-based arrays, produced by either robotically spotting the presynthesized oligonucleotide/PCR product element on to the substrate or, as in the case of the Affymetrix GeneChip™, *in situ* synthesis of oligonucleotide array elements on the substrate. Whilst commercial products provide an attractive, off-the-shelf solution with full technical support, this comes at a cost in terms of expense, lack of flexibility and the limited availability of arrays for particular bacteria.

The spotted microarray platform is the most commonly used approach at present that permits both custom design and in-house production for suitably equipped laboratories. The spotted oligonucleotide array is perhaps the most attractive as, once the oligonucleotides have been designed and synthesized, there is then simply the matter of printing them on the array substrate. For a PCR product array, following the design and synthesis of the PCR primer oligonucleotides, the considerable task of generating and verifying PCR products must first be undertaken before printing can occur. In terms of the ease of manufacture, oligonucleotide arrays present an obvious advantage over PCR product arrays. However, when design and predicted performance of array elements are also considered, the choice is less clear cut.

Oligonucleotides for spotted microarrays generally range from 40 to 70 nucleotides in length and, therefore, offer increased sequence specificity when compared to longer PCR products. Whilst we have no direct experience of oligonucleotide array design, intuitive criteria for oligonucleotides would include: (1) length, which influences specificity and sensitivity; (2) absence of secondary structure, self-annealing and repeats; (3) fixed G + C% range; and (4) melting temperature. However, the design of oligonucleotides for hybridization analysis is based on imperfect algorithms. This is due to a poor understanding of hybridization properties of small oligonucleotides, partly due to tertiary structure and the availability of only parts of the molecule (Southern *et al.*, 1999). Straightforward *in silico* design must, therefore, be supported by an empirical evaluation of the performance of the oligonucleotide in hybridization, subsequently followed by redesign, if required, which can become costly. It is not clear whether the proprietary methods used to design commercial oligonucleotide arrays include an empirical selection based on hybridization performance. However, development of a greater understanding of the design process is leading to improved algorithms (Rouillard *et al.*, 2002).

In contrast, the design of PCR products for spotted microarrays is a much more developed and established technique. The algorithms and software for primer design for PCR amplification have been more widely applied and understood than those for oligonucleotide design, simply because this approach has been in existence for longer. Our own experience with a number of whole genome bacterial microarrays has shown the spotted PCR product approach to be robust and reliable. Therefore, this chapter will focus on the design for spotted PCR product arrays, although much of the discussion and issues raised are relevant to all types of array platforms.

Genome sequence and annotation

The starting premise for microarray design and construction has been the availability of a fully annotated and published genome sequence. This criterion presents a defined reference that the microarray can be based upon, helping to standardize gene predictions and annotation. Clearly, the published genome sequence and annotation is only the best prediction at the point of submission, and so this initial genome annotation may be revised as additional biological information and improved sequence analysis tools become available. Following the initial array design there is, therefore, an ongoing process whereby new information with regards to gene prediction and sequence annotation must be included within the physical microarray and downstream analysis.

Another source of sequence information that may require inclusion in the microarray design, either at the outset or at a subsequent stage, may be the publication of genome sequences of additional strains or of related bacterial species. Whilst a core set of genes will be common to all strains of a species or all species of a genus to a lesser extent, there will be a number of strain/species-specific genes not present on the microarray. Inclusion of these additional genes enables the microarray to have wider and more comprehensive applications and introduces the concept of pan-species rather than strain-specific arrays. Axiomatically, a microarray cannot detect any sequence, the complement of which is not printed on the array and thus the presence of novel genes within a particular new strain or species will always be undetectable by this approach.

For both gene expression and genomic analyses, genome sequence divergence (i.e. sequence polymorphism) is a potential problem. This problem is arguably most acute with the short array elements, such as oligonucleotides, where a few nucleotides difference may substantially destabilize heteroduplex hybridization. This is unlikely to be a problem with longer PCR products unless substantial deletions or rearrangements occur. In such a case we would hope that hybridization would be disrupted to inform us that some change had taken place between bacterial strains. This re-emphasizes the point that the array design is influenced by the intended application. For example, an advantage of a short oligonucleotide array, such as the Affymetrix GeneChip™, is that it is able to detect single nucleotide polymorphisms. In addition, the design of the Affymetrix GeneChip™ consists of numerous oligonucleotide elements within each gene and intergenic region, which will be representative of more of the genome in terms of physical position, thereby providing a greater resolution to the analysis when compared to a single element for each gene.

Genomic features

If the intended use of the microarray is to achieve global genome-wide comparison of gene expression or gene complement then clearly each gene in the genome should be represented on the array. Ensuring that every gene is represented on the array is, in part, dependent on the accurate prediction and annotation of genes from the sequence analysis. For bacterial genomes, in which the number of genes in the genome will be in the order of one to several thousand, the gene numbers are not restrictive to the synthesis of large numbers of PCR products. This is especially true when the PCR process can be undertaken in a high-throughput, 96-well format utilizing liquid handling robotics. Elements on the array do not have to be solely restricted to the genes, but can also include genomic features such as pseudogenes, potential intergenic regulatory regions, ribosomal and other RNA species, insertion sequences, and phage.

Multiple genomic copies of genes present another issue for consideration. There are two possible approaches to deal with this, either have one element on the array to represent all identical copies of the gene or reflect the genomic copy number in the number of elements on the array. A good example of the problem is the presence of multiple copies of genes associated with insertion sequences in bacterial genomes. The approach that we have taken is to generate a PCR product for each copy of the insertion sequence gene, ensuring that the number of elements on the array equals the number of genes in the genome. Although identical in sequence, these PCR products are generated in distinct PCR reactions amplified with oligonucleotide primers that were synthesized independently. This provides a useful internal control not only for the PCR amplification but also for the downstream data analysis. One thought was that having the same number of elements on the array as genomic copy number may help to reduce biased signal intensities through changed hybridization kinetics resulting from essentially increased copy number. The concept of one element for one gene helps simplify the initial analysis process, and access to the predicted cross-hybridization information between genes and array element allows such effects to be accounted for.

Oligonucleotide primers for PCR

Having accounted for the genomic and gene features that must be incorporated into the array design, there is now the need to consider the practicalities of amplifying a PCR product to make each element on the array. The primary concern is designing the oligonucleotide primers required for the PCR amplification. Careful selection of PCR primer pairs is necessary to ensure specificity of amplification to give a single product and also performance in hybridization analysis. The primer pairs for each element of the array must also have similar properties so that the PCR process can be automated using the same PCR cycling parameters for all reactions simultaneously. Therefore, the most important characteristic for any primer is the melting temperature (T_m). Not only should the T_m for each primer be closely matched to the other primer in the pair but also to the primer pairs for all the other genes. Standard criteria for primer design apply, such as the avoidance of self-annealing or formation of hairpin fold-back structures, and the use of software such as Primer 3 (Rozen and Skaletsky, 1998) gives excellent results with the correct parameter choice. The ability to batch process the design for all elements simultaneously is invaluable, avoiding the otherwise unwieldy and tedious gene by gene approach. Thus, matching PCR primer pair requirements to achieve maximum and specific amplification of single amplicons under standard conditions for PCR cycling is important and critical for the straightforward construction of the array.

PCR product selection

The potential use of the array, the desired specificity of the array elements and the practicalities of PCR amplification determine the decision as to which part of the gene is represented on the array. In selecting the PCR products to form the array element there remain the standard set of requirements for multiple probes aimed at making the hybridization process uniform, including GC content, T_m and length of the PCR product. The entire coding sequence of a gene may be used, as was prepared for an *Escherichia coli* array (Richmond *et al.*, 1999), or alternatively a fragment of the gene may be used. In practical terms, the importance of such consideration is exemplified in the case of the *M. tuberculosis* genome. A substantial number of genes in the *M. tuberculosis* genome are in excess of 5 kb, with a high G + C content making any standardized format for long-range PCR difficult, and so for this reason shorter PCR products are preferred. Furthermore, in situations where a gene has a high degree of homology with other genes in the genome, either because of domain similarity, related functionality or because it is a member of a multigene family, then it would be advantageous to choose a region of a gene with minimal intragenomic homology.

Addressing these issues so as to minimize non-specific hybridization or cross-hybridization has been a main consideration in our array design process. This is termed a minimal cross-hybridizing array (see Figure 4.1) and involves the generation of PCR products of gene fragments. In general terms, when considering a PCR product to represent *geneA* on the microarray, regions within *geneA* that share sequence similarity with *geneB* and *geneC* should be avoided. PCR product 1 would clearly be the optimal choice as this is amplified from a region of *geneA* that shares no significant sequence similarity with other genes. The benefit of this choice would be that the signal obtained from that element on the array would be gene-specific for *geneA*. PCR products 2 and 3 would be less desirable as they share sequence similarity with *geneB* and/or *geneC* and, therefore, any signal obtained from those elements may arise not only from the target *geneA* but also through cross-hybridization with *geneB* and/or *geneC*. The MCH approach should ideally incorporate a unique PCR product for

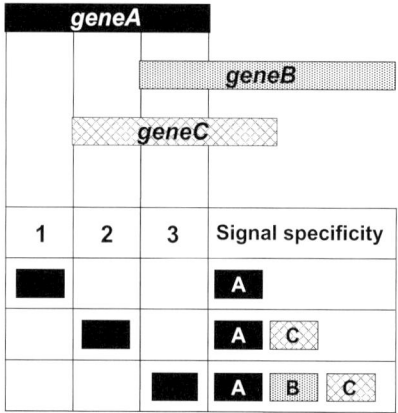

Figure 4.1. Illustration of the principle of minimal cross-hybridization array design. Genes A, B and C share significant regions of homology, indicated by positional overlap. PCR products designed from regions 1, 2 or 3 of gene A will have signal specificity as indicated for gene A, genes A and C or genes A, B and C, respectively.

each gene to preserve hybridization data clarity through gene-specific signals. However, if a unique PCR product is not possible, then the PCR product with the least amount of similarity to non-target genes needs to be selected for minimal cross-hybridization.

The *M. tuberculosis* array provides a very good example of the benefits of this minimal cross-hybridization (MCH) design. The genome sequence of *M. tuberculosis* revealed a high proportion of repetitive DNA, most notably in two large multigene families, the PE and PPE genes (Cole et al., 1998). Using this MCH approach, the number of gene-specific elements on the array was increased by more than 500 genes when compared to the alternative approach of using whole open reading frame (ORF) PCR products (see Table 4.1). This ensured that gene-specific signal for the maximum number of genes was obtained. The problematic (in terms of cross-hybridization) PE and PPE multigene families are highlighted to demonstrate the reduction in cross-hybridization achieved for these highly homologous genes. About 10% of the coding capacity of the genome is devoted to these two unrelated families of acidic, glycine-rich proteins and so clearly it is important to be able to differentiate between the individual family members using the array. The PE and PPE genes contain multiple copies of the polymorphic repetitive sequences referred to as PGRSs and major polymorphic tandem repeats (MPTRs), respectively. The 99 members of the PE family all have a highly conserved N-terminal domain of ~100 amino acids. Similarly, the PPE genes have conserved N-terminal domains with C-terminal segments that vary markedly in sequence and length. Clearly, exclusion of these repetitive conserved areas from the PCR product is important. The MCH design permitted 70% of PE genes and 68% of PPE genes to be differentiated by hybridization compared with 21% and 6%, respectively, if whole ORFs were used. Similar advantages of MCH design were gained with the *M. tuberculosis* gene families involved in fatty acid degradation and polyketide synthesis.

The length of the PCR products that make up the array elements is another important consideration. Short PCR fragments give weaker signals than longer ones (Stillman and Tonkinson, 2001). Although this is largely compensated for by ratio analysis in two-dye microarrays, low-abundance mRNA species with small probe array elements may produce

Table 4.1. An example of how the minimal cross-hybridization (MCH) array design approach for the *M. tuberculosis* genome improved the number of gene-specific elements on a microarray. The figures indicate the number of genes (percentage) for which a unique PCR product was designed that had no significant predicted cross-hybridization by BLASTN to other genes in the genome

Genes	MCH design	Whole ORF	MCH design gains
All genes	3475/3924 (89%)	2967/3924 (76%)	508
PPE	46/68 (68%)	4/68 (6%)	42
PE-PGRS	9/61 (15%)	0/61 (0%)	9
PE	21/38 (55%)	8/38 (21%)	13

signal intensity ratios that are less statistically robust in the data distribution and may be filtered from the analysis as unreliable data. If there are no constraints with regard to cross-hybridization, then PCR products of greater length are preferable to increase the signal obtained in hybridization.

Another choice with regard to positional location is the 5' to 3' bias, which has been suggested as a criterion that might result in more sensitive and representational signal from transcribed mRNA. For eukaryotes, a 3' bias would be advantageous when driving cDNA synthesis from the 3' end of the mRNA with oligo(dT) as the resultant cDNA will also be biased towards the 3' end of every gene. Since bacterial mRNA does not have polyA tails (see Chapter 8) and cDNA synthesis is primed randomly, there would be little benefit from a 3' bias. However, when considering mRNA transcription, sequences are transcribed from the 5' to 3' end. Thus, a 5' bias in probe elements on an array may provide a better reflection of *de novo* transcriptional initiation. In our own design approach there has been no positional biasing of PCR products within the genes, the overriding criterion that influenced any positional effect was the reduction of cross-hybridization.

Summary

The general rules that emerge from these considerations are simple to state but difficult to achieve. The effort invested at each of the stages outlined, from genome sequence to PCR product, has great influence over the quality of both the array and data generated.

◆◆◆◆◆◆ DESIGN PROCESS FOR A PCR PRODUCT-BASED MICROARRAY

The previous section reviewed some of the important aspects that influence the design of microarrays. In the following section the practicalities and approach that we have taken in this process are outlined. The process of designing primer pairs and PCR products for each gene in a genome is conceptually straightforward. However, from the bioinformatics perspective it is instructive to break the process into steps: (1) acquisition of systematic names and nucleotide sequence for each of the genes to be present on the array; (2) design multiple PCR products for each gene; and (3) selection of the optimum PCR product for each gene.

Sequence acquisition of genomic features

The general starting point for the design process is an EMBL or GenBank file for the complete genome sequence. This includes the co-ordinates of each predicted genomic feature, annotation associated with each of these and the nucleotide sequence. An essential for genomic projects is a systematic naming convention and many annotations have a systematic nomenclature as well as common gene names. These systematic names provide a unique identifier for each of the genes and give some indication as to relative chromosome position and direction of transcription.

The first step in the process is to extract the relevant information from the EMBL or GenBank files, a process not always straightforward due to subtle variation in these 'standard' formats. The systematic gene name, genome co-ordinates and strand of the coding sequence are extracted from the EMBL or GenBank files, and this information is used to extract the nucleotide sequence for each gene. A series of custom Perl scripts are utilized to achieve this task and results in the generation of a multiple FASTA file: essentially a text file containing a list of all the gene sequences with an identifier associated with each of them. The FASTA file represents the core set of genes that will be represented on the array and is used in the subsequent primer design process. Therefore, any additional genes for inclusion on the array can be added to this file. Maximum genome coverage is aided by including all genes, annotated as both coding sequences (CDS) and pseudogenes (commonly, a gene identified as a pseudogene in one strain/species may be present as a CDS in another strain or species making their inclusion worthwhile).

Whilst we have tended to design arrays based upon completed and annotated genome sequences, it is possible to work with unfinished sequences, but these are more problematic. The design process simply requires a list of gene sequences to be included on the array. Therefore, the first task with an unfinished sequence is to identify the genes and extract the nucleotide sequences. Gene prediction is not an automatic and systematic process, although programs such as Glimmer (Delcher et al., 1999) have been developed to aid this task. If the sequence exists as a set of contigs that may include overlaps or gaps, then this also presents an added level of complexity. Whilst gene sequences are all that are required physically to design the array, some annotation for each of the genes would be valuable to gain maximum benefit from the array and downstream analysis. Performing genome-wide similarity searches requires good comparative genome knowledge and considerable post-computer-prediction scrutiny. Furthermore, when the sequence is completed and annotated, there is the need to match the published information with that derived independently.

Multiple strain coverage

The increased number of bacterial genome sequences completed has meant that, in some instances, there is sequence available for a number of strains of the same bacterial species or closely related species. The impact of comparative functional genomics grows and so there is a real need to cover multiple strains or species on a single composite array. Accordingly, pan-species or pan-strain arrays can be made rather than simply the strain-specific arrays based on a single genome sequence.

There are two types of differences between genome sequences that must be accounted for when designing arrays to cover multiple strains. On one level there are annotation differences whereby the DNA sequences are identical but there is variation in the gene predictions and, on another level, there are true differences in the DNA sequence due to the presence of divergent or novel genes. Generally, within a bacterial species, the vast majority of genes are common to all strains but there will be a variable number of additional strain-specific genes that should also be included on the array. For example, the two sequenced strains of *S. pneumoniae*, TIGR4 (Tettelin *et al.*, 2001) and R6 (Hoskins *et al.*, 2001), have been compared at the DNA level (see Figure 4.2) using ACT (Artemis Comparison Tool, Sanger Institute). The genome comparison reveals that the majority of genes are present in both strains and there are a small number of genes unique to each strain. Therefore, compiling a list of all the gene sequences from one strain plus the novel gene sequences from the extra strain creates a virtual genome that is inclusive of both strains and can feed into the primer design process. As noted in the previous section, the first step in the primer design process is sequence acquisition to generate a multiple FASTA file of all the gene sequences that require representation on the array. Therefore, when designing multiple-strain composite arrays, we include an extra procedure to generate this file.

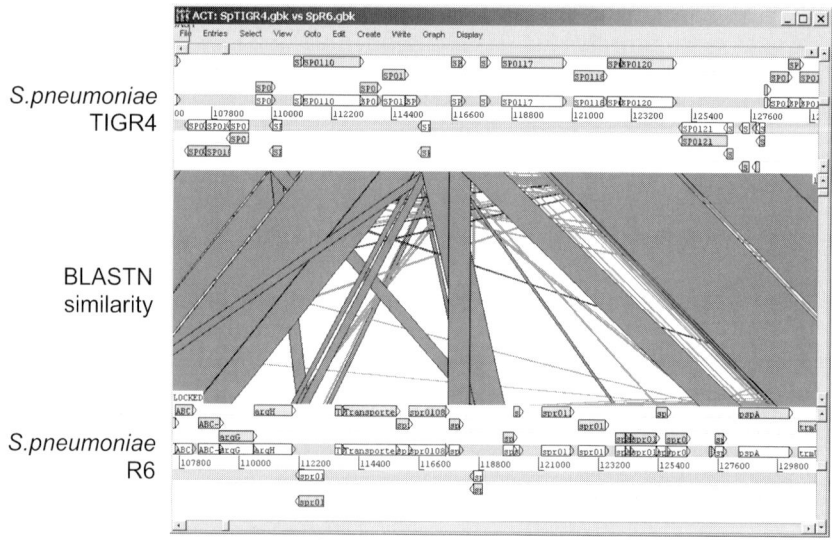

Figure 4.2. Genome comparison of two *S. pneumoniae* strains using the Artemis comparison tool. When comparing the R6 strain with the TIGR4 strain of *S. pneumoniae*, novel regions of DNA present in the R6 strain but absent in the TIGR4 strain are revealed. These are shown above as lacking any BLASTN similarity matches and are displayed as gaps in alignment between the two genomes. The visualization tool may also reveal annotation differences between strains whereby the DNA sequence is identical but additional genes have been predicted. Inclusion of these strain-specific genes on the array, in addition to all genes present in the reference strain, produces a composite array that covers multiple strains of a particular bacterial species or genus. (Artemis Comparison Tool, Sanger Institute).

The multi-strain comparison follows a stepwise progression to determine the strain-specific genes not present in any of the strains previously analysed in the process. The process starts with the gene sequences from the first strain, termed the reference strain, and compares each of the additional strains in turn to identify and select the strain-specific genes not found in strains analysed prior to this. The process is diagrammatically represented in Figure 4.3 and described as follows.

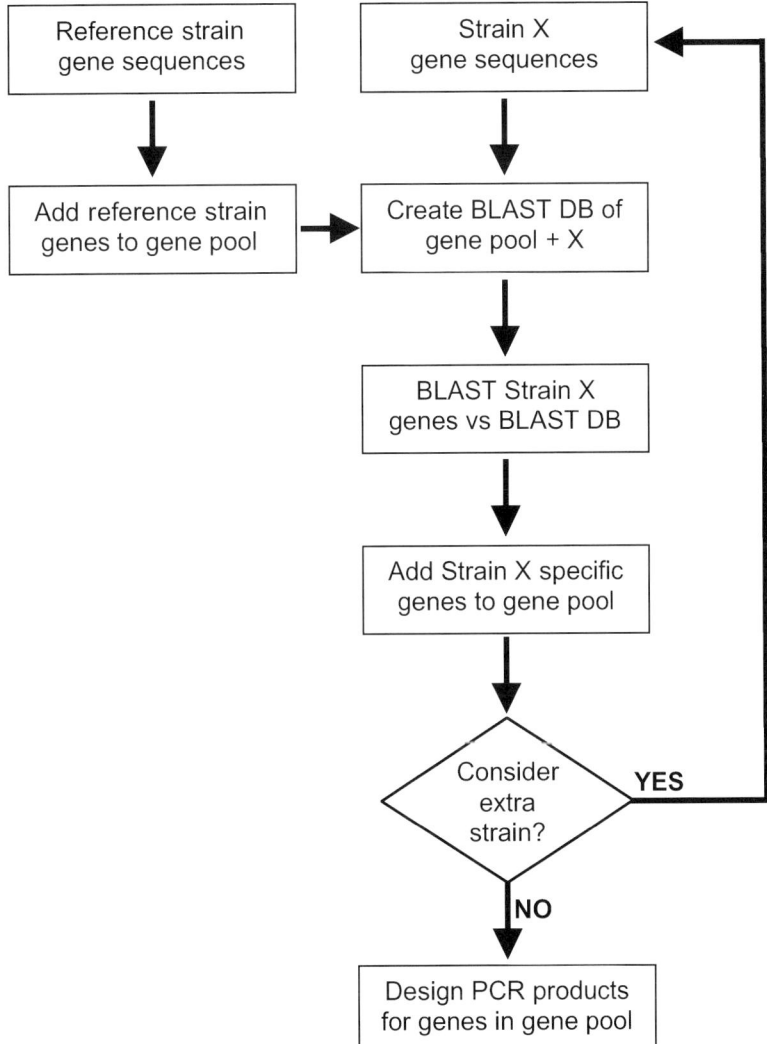

Figure 4.3. Schematic diagram to illustrate the approach taken to create a composite array representative of several bacterial strains. The purpose is to build up a pool of all genes present in the reference strain plus genes specific to any additional strains analysed. As each subsequent strain is compared, any genes found to be absent from the pool and thus specific to that strain are in turn added to the pool. The final pool of gene sequences created contains a representative sequence for all genes in all strains and is used in the primer design process.

1. Acquire sequences of all the genes/pseudogenes in the reference strain.
2. Add reference strain gene sequences to the gene pool.
3. Create a BLAST database of the reference strain gene sequences.
4. Acquire the gene sequences of the strain to be compared (Strain X).
5. Add the Strain X gene sequences to the BLAST database.
6. BLAST the Strain X gene sequences against the BLAST database.
7. Select Strain X genes that only have significant BLAST hits against Strain X genes and no other genes in the database. (Note: Significant if the BLAST bit score is more than double the bit score of the next best hit.)
8. Add Strain X specific genes to gene pool.
9. Create a BLAST database of all genes in the gene pool.
10. Repeat process from step 4 for each additional strain.

This process results in a growing list of genes in the gene pool that contains all of the genes present in the first reference strain plus the genes found to be novel in each of the subsequent strains. Creating the multiple FASTA file of the gene sequences for all the genes in the final gene pool permits PCR products to be designed, resulting in the desired multistrain or pan-species array.

Primer design

Having successfully generated the multiple FASTA file of gene sequences in the previous step in the process either from a single genome sequence or the multiple-strain comparison, the next step in the process is to design oligonucleotide primers to PCR amplify fragments for each of the genes. The aim is to evaluate ten potential PCR products for each gene but this number may be reduced due to restrictions imposed on the primer design software by gene features such as short length, a high degree of repetitive DNA or an unusual G + C content.

We have successfully used Primer3 software (Rozen and Skaletsky, 1998) to design the PCR primers for a number of whole genome arrays. The general parameters that we use in the design software are a primer optimum size of 22 bases (range 20–25), an optimum T_m of 60°C (range of 50–80°C, maximum difference of 7°C) and an optimum size of 600 bases (range 50–800) with all other parameters set as default. We have found these to work well for a number of genomes with varying G + C content to produce primers that perform successfully in PCR amplification and PCR products that work reliably in array hybridizations. The main aim is to achieve closely matched T_m values for the two primers in a pair and for all pairs relative to each other, thus enabling the high-throughput PCR required downstream. It is also important to try and select products as similar in size and G + C content as possible, although this comes secondary to T_m matching and the avoidance of cross-hybridization, which is described in the next section on PCR product selection. Achieving complete genome coverage in one design run is often the most difficult part of the design process. We occasionally have to resort to a more manual design for a handful of genes for which the software failed to design primers due to some of the reasons indicated above. For example, this final manual approach may require additional intergenic sequence to be included to increase the length of DNA available for very short genes or modification of the general parameters to account for extremes in G + C content.

PCR product selection

Having designed ten potential PCR products for each gene, there is then the need to select the optimal PCR product to represent each gene on the array. The over-riding aim is that the longest PCR product with the minimal amount of cross-hybridization is selected. Therefore, the ranked PCR product selection criteria applied to achieve this are that the PCR product must be unique to the target gene; if not, then the PCR product should have minimal cross-hybridization by non-target genes and, lastly, that the PCR product of the greatest size is selected. As before, all file handling and manipulations are completed using a series of custom Perl scripts that generate input files, run the necessary programs, parse output files and make selections.

The selection process utilizes a BLAST (Altschul *et al.*, 1997) comparison of the potential PCR product sequences against the gene sequences. BLAST is an extremely useful tool as it measures the number of basepairs in the predicted heteroduplex, thus allowing evaluation of the likelihood of cross-hybridization. A multiple FASTA file of all the potential PCR product nucleotide sequences is generated and queried against a BLAST database of the gene sequences. The BLAST output is filtered to only consider hits below a certain probability ($e < 0.001$) and from this output the optimal PCR product for each gene is selected by the ranked criteria.

Ideally products with only a single good-quality hit against the corresponding gene sequence would be considered. To select the optimal PCR product from a number of PCR products, which all have a single good-quality hit, the BLAST bit score is used for selection. The bit score not only reflects the expectation of the hit (all good-quality hits may have an equal expectation of zero) but also the length of the match. Therefore, selecting the PCR products with the highest bit score selects the PCR product of the greatest length. If it was not possible to select a PCR product with a single good-quality hit, then the product with the minimal number of good-quality hits was selected; again the bit score is used to differentiate between products with an identical number of hits to maximize the length of the PCR product.

Having designed and selected the optimal PCR product for the array, the final output from the process is a table of primer sequences and attributes plus a summary of the predicted cross-hybridization for each PCR product.

Quality control of primer prediction

It is important to check the output of the design process independently by some form of manual method and also a computational checking program that is independent of the design software. This is essential considering the subsequent large investment in money when ordering the primers as well as the time involved constructing and using the array.

A manual assessment can be achieved using a genome visualization tool such as Artemis (Rutherford et al., 2000). By this method, around 10% of the primer sequences can reasonably be checked to confirm that the primer pairs match the appropriate sequence position within a gene, that the 5′–3′ orientation of the primers matches the gene orientation and also highlight any unusual PCR products. For example, detection of a small PCR product designed for a large gene would initiate a follow-up check to ensure that this is influenced by cross-hybridization and thus results from the design process.

Clearly, there is great value in a manual approach but a more systematic computational method is required to be able to check all the designed primers. For this, a distinct set of Perl scripts is used to check the primer sequences against the complete genome sequence. The systematic checks performed ensure that: (1) the primers are within the target gene sequence; (2) the primers within a pair match opposite DNA strands in the correct 5′–3′ orientation to ensure amplification; (3) the expected size of PCR product from the primer design matches that predicted from the genome sequence; and also (4) there is no mispriming that may affect the specificity of the PCR. The systematic check of all primers against the genome sequence permits the production of a file that defines the position of each PCR product within the genome sequence. This information can be overlayed on the genome sequence in Artemis to provide visualization of all the PCR products on the genome.

◆◆◆◆◆◆ SUMMARY

The relative biological simplicity of bacterial genomes makes microarray design a straightforward process. Compared to complex eukaryotic genomes, bacteria do not have introns or alternative splicing, have seldom been annotated to have overlapping genes on opposite strands and also have high gene density. The starting point for constructing whole genome microarrays is the fully sequenced and annotated genome, and bioinformatics has a major role to play in the design process. This is particularly true when designing arrays that cover the genomes of multiple strains of a single species, whereby there may be additional genes from multiple strains compared to a sequenced reference strain. Although a truism, it is safe to say that better design aids better data analysis, and this is most important when cross-hybridization between several genes is present. Deconvolution of complex data sets to account for the effect of multiple cross-hybridization on signal intensity will not be straightforward. Thus, appropriate use of predictive informatics in array design to minimize cross-hybridization is necessary. It is also evident from experience with designing arrays for a range of bacterial species that each genome presents distinct considerations that require biological knowledge and interpretation over and above the automatic predictive algorithms that computer-based design can provide. We have described the principles of batched PCR primer design and hybridization probe design and have used this successfully in the construction of bacterial pathogen spotted glass slide microarrays (see Chapter 5).

◆◆◆◆◆◆ RELEVANT WEB RESOURCES

ACT	http://www.sanger.ac.uk/Software/ACT/
Affymetrix	http://www.affymetrix.com
Artemis	http://www.sanger.ac.uk/Software/Artemis/
Glimmer	http://www.tigr.org/softlab/glimmer/glimmer.html
NCBI BLAST	ftp://ncbi.nlm.nih.gov/blast/
OligoArray	http://berry.engin.umich.edu/oligoarray/
Operon	http://www.operon.com
Perl	http://www.perl.com/
PremierBiosoft	http://www.premierbiosoft.com/
Primer3	http://www-genome.wi.mit.edu/genome_software/other/primer3.html
WU-BLAST	http://blast.wustl.edu/

References

Altschul, S. F., Madden, T. L., Schaffer, A. A., Zhang, J., Zhang, Z., Miller, W. and Lipman D. J. (1997). Gapped BLAST and PSI-BLAST: a new generation of protein database search programs. *Nucleic Acids Res.* **25**, 3389–3402.

Cole, S. T., Brosch, R., Parkhill, J., Garnier, T., Churcher, C., Harris, D., Gordon, S. V., Eiglmeier, K., Gas, S, Barry C. E., 3rd, Tekaia, F., Badcock, K., Basham, D., Brown, D., Chillingworth, T., Connor, R., Davies, R., Devlin, K., Feltwell, T., Gentles, S., Hamlin, N., Holroyd, S., Hornsby, T., Jagels, K. and Barrell, B. G. (1998). Deciphering the biology of *Mycobacterium tuberculosis* from the complete genome sequence. *Nature* **393**, 537–544.

Delcher, A. L, Harmon, D., Kasif, S., White, O. and Salzberg, S. L. (1999). Improved microbial gene identification with GLIMMER. *Nucleic Acids Res.* **27**, 4636–4641.

Doolittle, R. F. (2002). Biodiversity: microbial genomes multiply. *Nature* **416**, 697–700.

Hoskins, J., Alborn, W. E., Jr, Arnold, J., Blaszczak, L. C., Burgett, S., DeHoff, B. S., Estrem, S. T., Fritz, L., Fu, D. J., Fuller, W., Geringer, C., Gilmour, R., Glass, J. S., Khoja, H., Kraft, A. R., Lagace, R. E., LeBlanc, D. J., Lee, L. N., Lefkowitz, E. J., Lu, J., Matsushima, P., McAhren, S. M., McHenney, M., McLeaster, K., Mundy, C. W., Nicas, T. I., Norris, F. H., O'Gara, M., Peery, R. B., Robertson, G. T., Rockey, P., Sun, P. M., Winkler, M. E., Yang, Y., Young-Bellido, M., Zhao, G., Zook, C. A., Baltz, R. H., Jaskunas, S. R., Rosteck, P. R., Jr, Skatrud, P. L. and Glass, J. (2001). Genome of the bacterium *Streptococcus pneumoniae* strain R6. *J. Bacteriol.* **183**, 5709–5717.

Richmond, C. S., Glaser, J. D., Mau, R., Jin, H. and Blattner, F. R. (1999) Genome-wide expression profiling in *Escherichia coli* K-12. *Nucleic Acids Res.* **27**, 3821–3835.

Rouillard, J. M., Herbert, C. J. and Zuker, M. (2002). OligoArray: genome-scale oligonucleotide design for microarrays. *Bioinformatics* **18**, 486–487.

Rozen, S. and Skaletsky, H. J. (1998). Primer3. http://www-genome.wi.mit.edu/genome_software/other/primer3.html

Rutherford, K., Parkhill, J., Crook, J., Horsnell, T., Rice, P., Rajandream, M-A. and Barrell, B. (2000). Artemis: sequence visualization and annotation. *Bioinformatics* **16**, 944–945.

Southern, E. M., Mir, K. and Shchepinov, M. (1999). Molecular interactions on microarrays. *Nature Genet.* **21(S)**, 5–9.

Stillman, B. A. and Tonkinson, J. L. (2001). Expression microarray hybridization kinetics depend on length of the immobilized DNA but are independent of immobilization substrate. *Analyt. Biochem.* **295,** 149–157.

Tettelin, H., Nelson, K. E., Paulsen, I. T., Eisen, J. A., Read, T. D., Peterson, S., Heidelberg, J., DeBoy, R. T., Haft, D. H., Dodson, R. J., Durkin, A. S., Gwinn, M., Kolonay, J. F., Nelson, W. C., Peterson, J. D., Umayam, L. A., White, O., Salzberg, S. L., Lewis, M. R., Radune, D., Holtzapple, E., Khouri, H., Wolf, A. M., Utterback, T. R., Hansen, C. L., McDonald, L. A., Feldblyum, T. V., Angiuoli, S., Dickinson, T., Hickey, E. K., Holt, I. E., Loftus, B. J., Yang, F., Smith, H. O., Venter, J. C., Dougherty, B. A., Morrison, D. A., Hollingshead, S. K. and Fraser, C. M. (2001). Complete genome sequence of a virulent isolate of *Streptococcus pneumoniae*. *Science* **293**, 498–506.

5 Glass Slide Microarrays for Bacterial Genomes

Jason Hinds, Kenneth G Laing, Joseph A Mangan and Philip D Butcher
Department of Medical Microbiology, St George's Hospital Medical School, Cranmer Terrace, London SW17 0RE, UK

◆◆

CONTENTS

Introduction
Construction of spotted PCR product microarrays
Analysis of DNA and RNA samples by hybridization
Summary

◆◆◆◆◆◆ INTRODUCTION

The rapidly expanding availability of bacterial whole genome sequences (Doolittle, 2002; http://www.tigr.org/tdb/mdb/; http://www.sanger.ac.uk/Projects/Microbes/) has coincided with the development of a range of technologies to study genes at the genome-wide scale. These tools include microarrays, which allow the simultaneous analysis of every gene in the genome, either at the DNA or mRNA level (Ye et al., 2001). This permits investigation into both microbial evolution due to natural selection of genetic content as well as adaptive responses mediated by changes in gene expression.

Bacterial microarrays have been used successfully for comparative analysis of the genomes of related strains by DNA hybridization, whereby deletions or sequence divergence in one strain can be identified by comparison to another reference strain. This approach has been used with whole genome microarrays for various bacteria such as *Campylobacter jejuni* (Dorrell et al., 2001), *Mycobacterium tuberculosis* (Behr et al., 1999; Kato-Meeda et al., 2001) and *Helicobacter pylori* (Salama et al., 2000). As reviewed in Chapter 7 by Dorrell et al., comparative genomic studies have provided key insights into genotypic complements that correlate with virulence and inform genome evolution studies.

The ability to analyse simultaneously every mRNA transcript in a genome, termed the transcriptome, also provides a very valuable tool.

Examination of global changes in gene expression, for example following drug treatment or in defined mutants, reveals the linkage between genome and functional pathways. Furthermore, the elucidation of gene expression patterns of pathogenic bacteria both under conditions that mimic the host environment *in vitro* and during *in vivo* models of infection will provide a valuable insight into the mechanism of bacterial pathogenesis and host–pathogen interactions. Whole genome expression monitoring using microarray hybridization provides a powerful approach to achieving these goals and should provide useful targets for rational design of new drugs and vaccine candidates for bacterial pathogens.

Application of microarrays in microbiology

Microarrays have advanced sufficiently to become a high-throughput and robust technology in bacterial functional genomics and have, therefore, been utilized for numerous and widespread applications within microbiology. These include the following.

- Differential gene expression: comparison of mRNA extracted under varying environmental conditions, following drug treatment or *in vivo* infection (Richmond et al., 1999; Wilson et al., 1999; Ye et al., 2000).
- Gene regulation: expression profile analysis of defined regulatory mutants in comparision to wild-type organisms (de Saizieu et al., 2000).
- Comparative genomics: assessing the DNA similarity of bacterial strains, species or clinical isolates and determining phylogenetic relatedness (Behr et al., 1999; Salama et al., 2000; Dorrell et al., 2001; Kato-Meeda et al., 2001).
- Genome plasticity: monitoring DNA variation within an individual strain or species.
- Molecular epidemiology: fingerprinting, gene deletion mapping, insertion sequence (IS) location.
- Diagnosis: pan-pathogen arrays with multiplex polymerase chain reaction (PCR) or probes.

Array technologies and terminologies

The term 'microarray' has been applied to technologies as varied as high-density arrangements of elements on nitrocellulose or nylon membranes, glass slides, silica wafers and acrylamide pads. To add to this confusion, 'DNA chip' has also been used interchangeably with 'microarray' so that the many different technology platforms available have become considered as one. Furthermore, terms used to describe components of the whole microarray process have also been used rather loosely so that standard identifiers have had to be defined. This next section will highlight and differentiate the common technologies available and also try to clarify some of the confusing terminology used.

Membrane arrays

These are generally referred to as 'macroarrays', owing to the larger format used and the lower density (~100/cm^2) of elements on the array. Elements printed on nylon membranes may include sequence defined DNA as PCR products or clones, random unknown sequences derived from cDNA libraries or even bacterial colonies. The membranes are hybridized to labelled samples in a similar manner to the method for Southern or Northern blots, with the sample labelled either radioactively or using other non-radioactive methods. This type of technology may be referred to as 'one-colour', as only a single channel of information can be gained from a membrane at any one time. Therefore, to achieve the comparison of samples, the hybridization signals from one membrane must be compared to signals from another membrane hybridized with the comparator sample.

Affymetrix GeneChips™

These consist of oligonucleotides synthesized *in situ* on silica wafers at very high density (~200 000/cm^2) through a process of nucleotide photochemistry and photolithography. Each array element is built up in a combinatorial fashion such that a nucleotide is sequentially added, up to a total of around 25 bases, to the extending oligonucleotide element on the substrate (http://www.affymetrix.com). This type of technology may be referred to as 'DNA chips' rather than 'microarrays', owing to the similarity of the production processes used in semiconductor manufacture, the term reflecting the *in situ* synthesis of the array elements that differs from other array platforms. For a genome, the Affymetrix GeneChip™ consists of multiple oligonucleotides for each gene plus oligonucleotides within the intergenic regions. For each sequence-matched oligonucleotide there is also a single base mismatched oligonucleotide on the chip, present as a control to determine the cutoff for true hybridization. In theory, every base in a bacterial genome and every mutation in every base may be represented on this type of platform (Gingeras and Rosenow, 2000) and could thus be used to probe not only for the expression and presence of genes, but also for studying gene homology and gene sequence. Similar to membrane arrays, this technology is 'one-colour', as only one channel of information is gained from a single array. Therefore, comparisons are made by the comparative analysis of individual chips that have been hybridized with different samples. These arrays are very powerful tools, gradually becoming more readily available to the research community. Their unique utility lies in the analysis of complex genomes requiring a very high density of elements, but for bacterial genomes with a maximum of only several thousands of genes, this is less restrictive and so the cost–benefit analysis is reduced.

Spotted arrays

The availability of ever-cheaper oligonucleotides, high-throughput PCR technology and the development of arraying robotics have provided the means to produce customized arrays in-house. This technology relies on first synthesizing the array elements and subsequently spotting them on to the array substrate, thus are referred to as spotted arrays. The approach is similar to that used for membrane arrays, the fact that the elements are spotted on to solid-phase substrates such as coated glass microscope slides at higher density (~2000/cm^2) differentiates them as 'microarrays'. The elements spotted on to the arrays may take the form of double-stranded DNA or oligonucleotides. Double-stranded DNA elements are often in the form of PCR products, generated from gene-specific PCR or amplified clone inserts of DNA/cDNA libraries, whereas oligonucleotide elements are simply pre-synthesized single-stranded DNA in the size range of 40–70 nucleotides. Complete genome sequence data enable the design of whole genome arrays, whereby at least one element is produced for every gene in the genome, but may also include intergenic regions, any other additional genomic features or genes from other organisms. One of the advantages of the spotted arrays is the flexibility provided that allows the arrays to be easily customized to suit particular applications. Another benefit of the spotted arrays is that they may be used as a 'two-colour' technology so that the direct comparison of two samples can be achieved on a single array. Described in more detail later in this chapter and outlined in Figure 5.1, this direct comparison is achieved by fluorescently labelling the samples with fluorophores that have distinct excitation and emission wavelengths permitting two channels of information to be acquired from each array.

Terminology

Conventional molecular biology terminology has used the term 'probe' to describe the labelled sample in solution that would hybridize to a bound

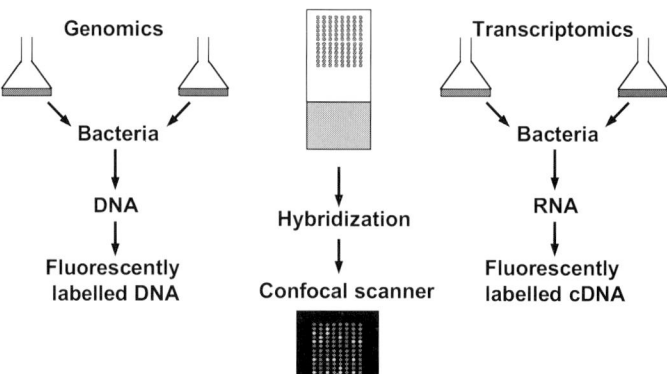

Figure 5.1. Schematic of experimental approach for the use of a bacterial whole-genome microarray to study comparative genomics and gene expression profiles. The two DNA or RNA preparations to be compared on the microarray are extracted from the bacteria and each of the samples is labelled separately with one or other of the two fluorescent dyes, commonly Cy3 and Cy5. The labelled samples are mixed and co-hybridized under a coverslip on the array. When scanned using a dual laser scanner the relative intensities of the two fluors hybridized to each reporter indicate the relative abundance of the particular DNA or mRNA species in the two samples. Reproduced in colour between pages 178 and 179.

target. However, with microarrays the term 'probe' has commonly been used to describe the element attached to the substrate and not the labelled sample hybridized to the array. Clearly, this has led to considerable confusion and the need for standard terminology which has been implemented through development of MAGE-ML (MicroArray Gene Expression Markup Language; http://www.mged.org). MAGE-ML proposes to use the term 'reporter' to refer to the elements bound to the microarray surface and 'sample' as the labelled nucleic acid in the hybridization solution that is derived from the DNA or RNA preparation, thus avoiding use of the term 'probe' altogether.

Outline of approach to array construction

Our group have used the spotted PCR product approach successfully to produce whole genome DNA microarrays for the bacterial pathogens *Campylobacter jejuni*, *Mycobacterium tuberculosis*, *Haemophilus influenzae*, *Yersinia pestis* and *Streptococcus pneumoniae*. The next section reviews our approach to array construction and highlights some of the issues for consideration and alternative approaches.

❖❖❖❖❖❖ CONSTRUCTION OF SPOTTED PCR PRODUCT MICROARRAYS

The approach that we have adopted to construct spotted PCR product microarrays is based on the pioneering work of Pat Brown's laboratory at Stanford University (http://cmgm.stanford.edu/pbrown/) and many of the protocols described are based on those developed and published by his and other groups (Hegde *et al.*, 2000; Diehl *et al.*, 2001). The methods used provide a simple yet effective method for generating whole-genome DNA microarrays that are within the resources of academic research groups to produce custom arrays in-house. The past few years have seen an explosion of new developments and technologies that are available commercially such that the choices available at each stage in the process are overwhelming and somewhat daunting to anyone starting out. The aim of this section is to provide a sound basis for array construction that is cost-effective and works reliably, highlighting some of the options and enhancements possible at each stage. There are two approaches to generating spotted PCR product microarrays that we have undertaken, based on the starting material available and the financial restraints imposed on a project.

Sequence-defined clone array: an alternative low-cost approach

Owing to the ongoing expansion of bacterial genomics, ordered sequence-defined genomic libraries are normally generated as a resource in sequencing projects. As a consequence, a subset of minimally overlapping clones is available with the potential to represent each individual gene. Such a resource has utility in the generation of low-cost whole-genome bacterial arrays such as that generated for *Campylobacter jejuni* (Dorrell et al., 2001). This approach may be achieved via the amplification of individual cloned inserts following the selection of ordered clones containing minimal overlap. Since a cloned library is being used for amplification, a single set of primers specific for flanking regions in the cloning vector may be utilized to generate a library of sequence-defined PCR products for arraying.

The limitation of using such an approach lies in the potential for cross-hybridization, since some elements on the array may be representative of multiple genes. Therefore, the identification of a signal of interest from a hybridization under such circumstances may require some downstream analysis to elucidate the composition of a specific complex signal and is likely to involve further experimentation. However, this limitation is balanced against the biological information that can be gained from such microarrays and the relative cost effectiveness of such an approach.

Gene-specific approach

Chapter 4 by Hinds *et al.* describes the methods that we have used to design whole genome PCR product-based microarrays in more detail. In general, gene-specific oligonucleotide primers are designed to enable PCR amplification of gene fragments to form the elements on the array. The PCR products for each gene are designed to have minimal cross-hybridization to other genes in the genome. Primers ranging from 22 to 25 bases in size are designed to have closely matched melting temperature to aid high-throughput PCR amplification of products ranging in size from 70 to 800 bases.

The primer sequences used are simply based on the gene sequence, no additional tags are included in the primer sequence primarily due to the extra costs that would be incurred. Clearly, adding extra bases to form the universal tag on each of the primer sequences significantly increases costs. A disadvantage of this is that any subsequent PCR amplifications required to regenerate PCR products have to be performed using the original gene-specific primers, involving all the associated variation in yield and success of amplification with the different primer pairs. However, we estimate that from 100 µl PCR reactions we are able to produce approximately 5000 microarrays to give some context to this potential problem.

An adaptation of this approach using gene-specific primers is to incorporate flanking sequences into the primers used. For this purpose, T3 and T7 phage promoter sequences are frequently adopted. The principal advantage accrued from the extension of primers with tag sequences is that, after an initial amplification with gene-specific tagged primers, a second round of PCR is carried out using the primers specific to the tags alone. This has several advantages: (1) it enables the generation of numerous copies of the PCR product with a universal set of primers at reduced cost and without any future need for gene-specific primers; (2) it permits a second standard round of PCR amplification following the first gene-specific step to reduce potential variation in the amount of PCR product printed; and also (3) it provides a means to determine the relative quantity of each PCR product present on the microarray by hybridization with an oligonucleotide against the flanking sequences and enable correction for spot-to-spot variation in loading.

PCR amplification of array elements

Amplification of the PCR products for the elements on the array is undertaken in a high-throughput, 96-well format using fully automated liquid handling and PCR cycling robotics provided by the RoboAmp 4200 (MWG Biotech Ltd; http://www.mwg-biotech.com). An advantage of this particular robot platform is the incorporation of technology to use the non-cross-contamination (NCC) 96-well plates. This NCC design ensures that when using the 96-well plate only a single well is open at any one time, greatly reducing the possibility of cross-contamination between wells which would impact on the specificity of PCR products.

To aid the high-throughput process, oligonucleotide primers for PCR are provided by the supplier (MWG Biotech Ltd; http://www.mwg-biotech.com) at a standard concentration in 96-well plates suited to the robotic platform. The primers are supplied as 96-well plates of 50 µM forward primers, 50 µM reverse primers and an additional plate containing an equimolar mix of the forward and reverse primers. The mixed primer plates are used directly for the PCR amplification by simply adding an aliquot of the mixed primers to a mastermix of all the other reaction components, namely enzyme, buffer, nucleotides and genomic DNA template.

When starting out on the PCR process for a new genome, optimization of the PCR conditions using a single 96-well plate is an essential first step before embarking on amplification of the whole genome. This ensures that the general PCR parameters used will result in a high success rate to return a good yield of specific PCR products. Through modification of either the annealing temperature or the PCR mastermix, ideally aim to achieve a success rate of generating greater than 90% of the PCR products in the plate. The success of PCR reactions is determined by agarose gel electrophoresis and ethidium bromide staining of the DNA. A representative gel of 2 µl of the PCR reaction from the first round of PCR amplification is provided in Figure 5.2 to indicate the typical success rate obtained in terms of numbers, product yield and specificity. We have found that for primers designed with a melting temperature (T_m) of 60°C an annealing temperature of 52–55°C generally gives an excellent success rate. For PCR amplification on this genome-wide scale must maintain a balance between avoiding non-specific multiple PCR products at lower annealing temperatures and reducing the success rate at higher annealing temperatures. The standard composition of a 100 µl PCR reaction used is 5 U HotStar *Taq* DNA polymerase (Qiagen; http://www.qiagen.com), 1×PCR buffer including 1.5 mM Mg^{2+}, 0.5 µM of each primer, 200 µM dNTPs and 10 ng genomic DNA template. Slight modification of this basic

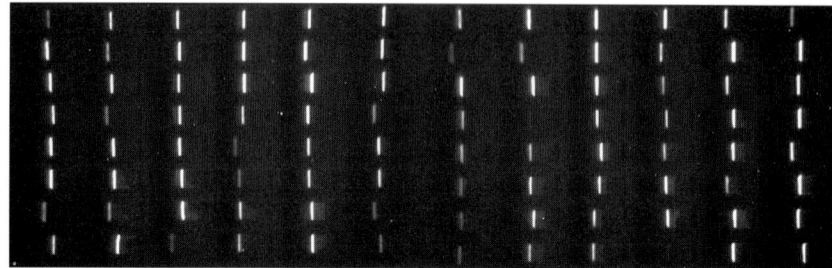

Figure 5.2. Agarose gel electrophoresis and ethidium bromide staining of PCR products generated in 96-well plates. Representative gel to demonstrate that for each PCR product able to validate single band specificity, suitable amount of product and correct size relative to neighbours.

PCR mastermix may be required and was found to be necessary for amplification of PCR products for the *M. tuberculosis* microarray, presumably due to the higher G + C content of the DNA, requiring inclusion of the manufacturer's PCR additive Solution Q and 3% dimethylsulphoxide (DMSO). Numerous enzymes and suppliers were initially evaluated to achieve optimal PCR amplification. The use of a hot start *Taq* polymerase is essential when using the robotic systems to ensure that all reactions are activated at the same time as reaction setup times may vary considerably within a plate. For this same reason, proof-reading enzymes with single-strand exonuclease activity should also be avoided.

The need to optimize the first attempt PCR and gain the maximal success rate becomes evident when you consider that, for those PCR reactions that did not result in a single, good-quality PCR product, there must now be a follow-up process to complete all the missing or multiple PCR products required for whole genome coverage. Figure 5.3 illustrates the steps that are taken to obtain the complete set of PCR products for an array. The experience of going through the process for several bacterial arrays has enabled us to establish which methods work and refine the process. The aim with each subsequent stage is to successfully complete the PCR amplification for as many of the products as possible. The final stage in attempting to complete the array for only a handful of genes can become a very time-consuming process. Table 5.1 summarizes the process of PCR amplifica-

Table 5.1. Summary of the percentage of the total PCR products successfully obtained at each stage in the initial and follow-up stages of PCR product amplification for four bacterial whole-genome arrays

Array	Initial PCR (%)	Repeat gel (%)	Repeat PCR (%)	Modify annealing (%)	Change additive (%)	Reorder oligonucleotides (%)
M. tuberculosis	89.8	4.2	0.7	0.9	3.6	2.1
H. influenzae	94.7	2.9	0.9	1.5	0.3	0.6
Y. pestis	96.5	1.1	1.4	0.7	0.1	1.0
C. jejuni	92.9	2.5	3.3	0.9	0	2.5

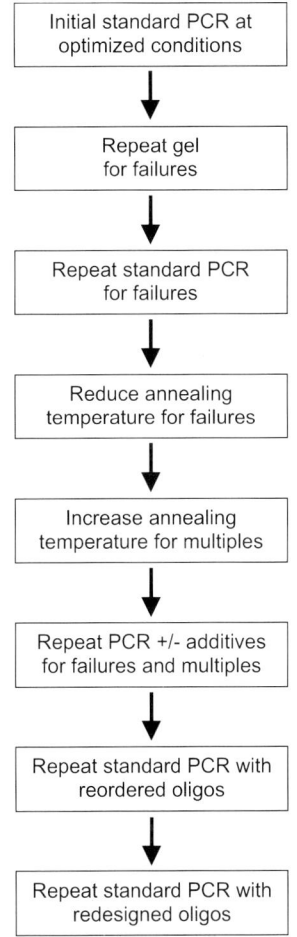

Figure 5.3. Schematic diagram of approach to complete PCR amplification of array elements. Following the initial PCR a series of repeats and modifications to the PCR cycling parameters and PCR mastermix composition are undertaken, achieving the successful production of a high yield of specific PCR product to represent each gene in the genome on the array.

tion for four different whole-genome arrays to indicate the percentage of the total genes amplified at each stage to achieve 100% completion.

From the figures in Table 5.1 it is obvious that the vast majority of PCR products are generated at the first attempt. It is certainly worthwhile repeating both the gel visualization and subsequent repeat of the standard PCR of failures due to possible liquid handling errors by the robots that have missed out essential components. Although the percentages involved in the follow-up stages appear low, there can still be a considerable number of genes involved that require completion. Methods to increase or decrease the annealing temperature and add or remove any PCR additives help to pick up a good number of the missing products, but a point is soon reached at which any further modifications, such as altering Mg^{2+} concentration, seem to have no benefit. A final stage to obtain the remaining PCR products has been to reorder exactly the same primers from the supplier. This has tended to work first time using standard PCR conditions, presumably due to the accepted low failure rate in the original oligonucleotide synthesis and purification process, as it only requires one defective primer in the pair to be problematic. If reordering the same primer fails then redesigning the primer sequences is undertaken, although this is rarely necessary.

Validation of PCR

A major task in validating the PCR products generated prior to arraying is to assess each PCR product by agarose gel electrophoresis to ensure single band identity of the expected size (Figure 5.2). In addition, for 5% of the PCR products DNA sequence verification is also obtained to confirm the correct identity, aiming to sequence four or five products from each of the 96-well plates. An aliquot of each PCR reaction is also stored in the freezer so that any interesting array data can initially be verified by sequencing the PCR product on the array, thus adding to the sequence validation of the array.

Preparation of PCR products for printing

Having completed and validated all the PCR reactions the next step is to prepare them for printing, which involves steps aimed at purification, concentration and resuspension in the print solution. Ideally at this point we may consider trying to quantify and equalize PCR product concentrations prior to printing, although any real benefits gained through achieving this are largely outweighed by the practicalities. As a general guide, if 2 µl of the PCR reaction is observed on an ethidium bromide-stained agarose gel, then this is considered at a sufficient concentration for the element to give signal from a hybridization. The slides only have a limited binding capacity for the DNA elements so, as long as DNA concentrations are above a threshold of around 100 ng/µl, it is likely that excess DNA will be spotted and the binding sites will be saturated.

The PCR reactions require some form of purification before they can be spotted and we use a simple precipitation method. A total of 50 µl of the 100 µl PCR reaction is precipitated, the other 50 µl is stored in the freezer as a backup and for further sequence verification of array elements if required. Standard precipitation in 96-well plates with propan-2-ol, sodium acetate and 5 µg of glycogen as an inert carrier are followed by a 70% ethanol wash, aspirated and air dried. The precipitated PCR products are resuspended in 15 µl 50% DMSO and transferred to 384-well plates for printing. The choice of 50% DMSO is beneficial for printing for two reasons: it greatly reduces the amount of evaporation of the samples, and also helps denature the double-stranded DNA to improve signal from the array in hybridizations. Other spotting solutions that can be considered include 3 × salt sodium citrate (SSC), phosphate buffer, glycerol or betaine plus a number of commercial solutions. An alternative to precipitation is to use the 96-well format membrane purification systems available that may be undertaken robotically, these include the Multiscreen96 (Millipore; http://www.millipore.com) and the QIAquick96 (Qiagen; http://www.qiagen.com) PCR purification systems.

Printing arrays

PCR products are deposited on to coated glass microscope slides at high density using a BioRobotics MicroGridII arrayer (Apogent Discoveries). This arrayer uses split-pin technology that permits spotting to over 100 slides from a single visit to the source plates, giving uniform spot size and consistency. Currently the standard approach is to use home-made poly-L-lysine or commercial amino-silane coated slides such as GAPSII™ slides (Corning). Our own experience with home-made poly-L-lysine slides, following essentially the original Stanford protocols, is that they produce arrays of high quality with good signal and low background. When making a cost comparison to commercial supplies it is difficult to justify the often considerable additional expense based on the results obtained.

Each PCR product is printed on the array, ideally as replicates, with the positioning of the products ideally in a random manner to avoid any spatial effects. It is also advisable to have replicate spots spaced at distance so that any local artefacts do not affect all of the replicate spots arising to misleading data. A number of controls are also printed to monitor efficiencies in cDNA labelling, hybridization and scanning sensitivities, and provide quantitative correlates of standardized signals and acceptable experimental data. Cy3 and Cy5 modified oligonucleotides are spotted at the corners of each subgrid as orientation markers or 'landing lights'. Positive controls include dilution series of ribosomal RNA gene sequences. Negative controls are made up using spotting solution and carry-over controls as well as non-homologous human genes such as β-actin and glycerol phosphate dehydrogenase (GAPDH). These non-homologous genes may also act as useful controls to enable spiking of the cDNA labelling reactions using *in vitro* runoff transcripts to monitor the labelling and hybridization efficiencies more effectively (Figure 5.4).

The area of attachment chemistries is one where there is considerable commercial development and improvements that will lead to increased sensitivity and reproducibility. There are numerous other types of surface chemistries available, few of which are touched upon here. These include three-dimensional (3D)-link activated glass slides for covalently attaching DNA (SurModics). These slides immobilize amine-terminated PCR products and oligonucleotides through end-point attachment. This orients the DNA while the polymeric coating holds it away from the surface of the glass slide. Scheicher & Schuell FAST™ slides are coated with a proprietary microporous polymer, which binds DNA in a non-covalent but irreversible manner that does not require modification to the nucleic acids being arrayed. Scheicher & Schuell also make CAST™ Slides

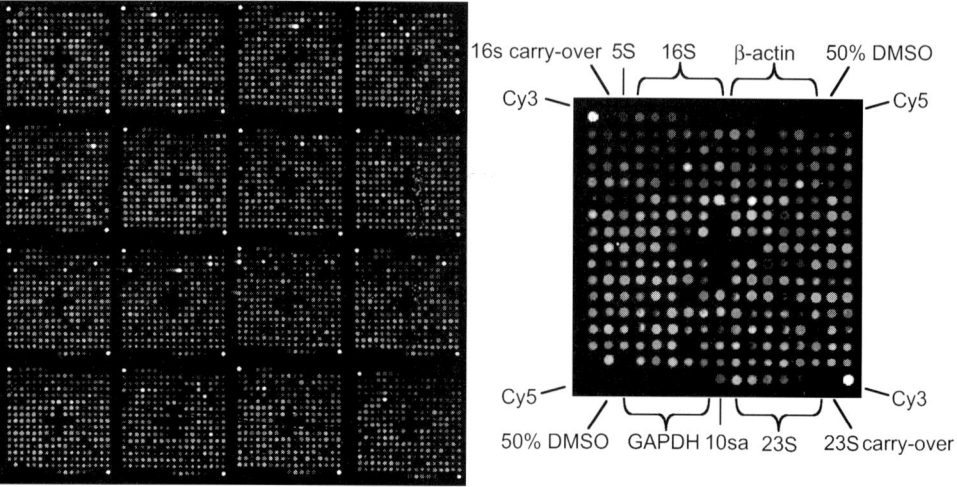

Figure 5.4. Microarray layout for *M. tuberculosis* whole-genome array of 3924 genes, indicating positions of orientation markers, positive and negative controls. The array consists of 16 subgrids of 17 × 17 reporter elements at a density of ~2000/cm^2, spot-to-spot pitch of 240 µm and typical spot diameter of 150–180 µm so that a single copy of the genome fits under a 22 × 22 mm coverslip. Controls included in each subgrid include Cy3/Cy5 landing lights, dilution series of 16S and 23S ribosomal RNA genes, dilution series of human β-actin and GAPDH genes, as well as print solution and carryover controls.

with a layer of Nytran® SuPerCharge positively charged nylon membrane affixed to the slide for high DNA binding capacity.

It is also possible to generate PCR products that are 5′-amino modified to increase the retention of arrayed products on to the solid support. The reason for covalent attachment or high-affinity binding is to maximize the concentration of immobilized DNA resulting in an enhanced signal to noise ratio giving rise to increased sensitivity and reproducibility of detection without loss of target during hybridization. Such covalent attachment or high-affinity association may also produce an array that has the potential to be used multiple times, with rounds of stripping and reprobing. Furthermore, 5′ modification of each primer ensures binding to the slide of both strands (5′–3′ and 3′–5′) of the amplicons, thereby permitting the essential hybridization analysis of both sense and antisense gene transcripts.

Post-print processing

Following printing there is the need to process the arrays before use in hybridization. There are many reasons for this process, which includes the need to attach the DNA to the coated glass slide, remove any excess unbound DNA, block potential non-specific binding sites prior to hybridization to reduce background and also denature the double-stranded DNA.

The arrays are initially rehydrated briefly for 5 s in the steam over boiling water, snap heated and ultraviolet (UV) cross-linked at 200 mJ to fix the DNA to the slide. They are then washed vigorously in 1 × SSC 0.05%

sodium dodecyl sulphate (SDS) for 1 min followed by a wash in 0.06×SSC for 1 min to remove excess DNA. The next blocking step uses freshly prepared 6 g succinic anhydride dissolved in 335 ml 1-methyl-2-pyrrolidinone with 15 ml of sodium borate pH 8.0 added just prior to use and the arrays are washed vigorously for 2 min followed by more gentle agitation for a further 15–20 min. The purpose of this step is to block any non-specific binding sites available on the array surface by acylating any unreacted amine groups not used for DNA binding between the spots, thus helping to reduce background due to the binding of the labelled sample to the array surface. Vigorous washing is required for all these steps to ensure that excess DNA is washed clear from the array to prevent it from binding to adjacent areas on the array surface that causes smearing or 'comet tails'. Following the blocking step the arrays are immersed and washed for 2 min in boiling water and then washed in ice-cold 95% ethanol, serving to both wash off the blocking solution and also denature the double-stranded DNA. The postprint processing step is important as it determines the quality of the data generated from the arrays.

Biological validation of arrays

Once the arrays have been printed, it is then useful to be able to validate biologically that they are working correctly. There are a number of approaches towards this end that essentially perform tests to see if the expected results are obtained. One such approach is the hybridization of independent sequence-verified DNA libraries to observe if the elements relating to the DNA within a chosen clone or BAC are detected. This helps to confirm that the tracking process from design, through PCR and into printing has enabled the correct identification of array elements. Other possibilities are to perform experiments for which certain results are already known, such as defined deletions in a comparative genomics study or the overexpression of particular genes in response to a stress such as heat shock. This validation adds to the various quality control steps throughout the process to increase confidence in the finished array and the results generated with it.

◆◆◆◆◆◆ ANALYSIS OF DNA AND RNA SAMPLES BY HYBRIDIZATION

The general approach to analysing DNA and RNA samples using the microarray is summarized in Figure 5.1. Using the spotted microarray platform outlined above, fluorescent labelling of the DNA or RNA preparations is the first step in the process before the labelled samples are hybridized on the array. The ability to selectively incorporate nucleotides tagged with fluorescent dyes that have different excitation and emission wavelengths means that it is possible to label differentially the samples to be compared. Therefore, samples that have been labelled with different dyes can be co-hybridized to the array in a direct competitive hybridization, the labelled DNA or cDNA competing with each other for hybridization to the specific reporter elements immobilized on the array. The amount of hybridization to each element of DNA or cDNA within each of the samples reflects their relative abundance in the two samples. The hybridization is quantified by measuring the light emitted from each of the elements on the array when excited by lasers of different wavelengths. Commonly Cyanine3 and Cyanine5 dyes are used to enable a two-way comparison using a dual-laser scanner, although a number of dyes are now available with distinct excitation and emission wavelengths that provide the potential for more than a two-way comparison when using scanners with multiple lasers.

The methodologies used for hybridization analysis of DNA and RNA are extremely similar, the only differences being due to how the preparations are initially labelled with the fluorescent dyes. The basic methods for labelling DNA preparations use Klenow DNA polymerase in combination with random hexamers to incorporate labelled nucleotides into the DNA synthesized by random priming, whereas RNA preparations are labelled using reverse transcriptase and random hexamers to incorporate labelled nucleotides during first-strand cDNA synthesis. The quality of the DNA and RNA preparations used in the labelling process directly influence the quality of the data obtained from the arrays and more detailed considerations as to RNA extraction and strategies to drive cDNA synthesis are extensively discussed in Chapter 8 by Mangan *et al.*

DNA labelling

A total of 2 μg DNA are mixed with 3 μg random primers (Invitrogen) and water to 41.5 μl, denatured at 95°C for 5 min and snap cooled on ice. Added to this are 5 μl 10×enzyme buffer, 1 μl dNTP mix (5 mM dA/G/TTP, 2 mM dCTP), 1.5 μl Cy3 or Cy5 dCTP (Amersham Biosciences) and 1 μl Klenow DNA polymerase (5 U/μl; Invitrogen) followed by incubation at 37°C for 90 min in the dark.

RNA labelling

A total of 5–10 μg of total RNA are mixed with 3 μg random primers (Invitrogen) and water to 11 μl, denatured at 95°C for 5 min and snap cooled on ice. Added to this are 5 μl 5× first strand buffer, 2.5 μl 100 mM DTT, 2.3 μl dNTP mix (5 mM dA/G/TTP, 2 mM dCTP), 1.7 μl Cy3 or Cy5 dCTP (Amersham Biosciences) and 2.5 μl SuperScriptII (200 U/μl; Invitrogen) followed by incubation at 25°C for 10 min then 42°C for 90 min in the dark.

Purification of labelled sample

The Cy3 and Cy5 labelled samples to be co-hybridized are mixed together and purified using a Qiagen MinElute PCR Purification kit according to the manufacturer's instructions. This purification method has been found to be effective and has the added advantage that the purified samples can be eluted in the small volume of water (10.5 µl) required for hybridization.

Hybridization

Arrays to be hybridized are first prehybridized at 65°C for 20 min in a solution of 3.5 × SSC, 0.1% SDS and 10 mg/ml bovine serum albumin (Sigma). Following prehybridization the arrays are washed in water for 1 min, propan-2-ol for a further 1 min and spun dry in a centrifuge. For hybridization, 3.2 µl 20 × SSC and 2.3 µl 2% SDS are added to the 10.5 µl of purified sample to form 16 µl of hybridization mixture composed of 4 × SSC 0.3% SDS. This hybridization solution is denatured at 95°C for 2 min, allowed to cool slightly at room temperature, placed on the array and a 22 × 22 mm coverslip lowered carefully over the array to trap the hybridization solution underneath. The arrays are then placed in a hybridization chamber and incubated submerged in a water bath at 65°C for 16–20 h in the dark.

Washing

After hybridization the arrays are first washed in 400 ml 4 × SSC 0.3% SDS at 65°C for 2 min, taking care to float the coverslip off the array before commencing more vigorous washing. Two additional washes in 400 ml 0.06 × SSC at room temperature for 2 min are then undertaken before the slides are dried by centrifugation.

Image acquisition and analysis

The final stage is to acquire the data from the array using appropriate microarray scanning equipment. The equipment that we currently use is an Affymetrix 428 dual-laser confocal microarray scanner (http://www.affymetrix.com) equipped with lasers matched to the wavelengths of Cy3 and Cy5. Scanning is performed according to the manufacturer's instructions and, at this point, it is important to ensure that the best dynamic range of the data is achieved by adjusting the gain settings so that the highest signal from a non-control element is just under saturation. The image acquired from the array forms the raw data so any errors in the capture of this will follow through into downstream data analysis.

The images captured from the two channels, corresponding to the two samples compared on the array, are then quantified using commercial software such as ImaGene (BioDiscovery Inc; http://www.biodiscovery.com). This software and similar packages go through processes to locate and identify the elements on the array and for each of the two channels measure pixel intensities of signal from within the element and background around the elements. This software may also perform some quality control measures that can be flagged in the output files so that downstream analysis that occurs independently of the raw array image data may have some indication as to the quality of the raw data.

The challenge following the extraction of the raw data is to apply methods to analyse and understand the data with regard to the biological context. This is a major undertaking and goes beyond the remit that we have tried to address in this chapter by outlining methods to construct and use microarrays. There are many numerous approaches possible to analyse the data, ranging from simple spreadsheet calculations to high-powered software using complex algorithms and, as with all aspects of microarray technology, there are a plethora of options available. GeneSpring (Silicon Genetics; http://www.sigenetics.com) is a commercial software package that we routinely use to visualize and analyse the data.

◆◆◆◆◆◆ SUMMARY

The acquisition of complete bacterial genome sequence has coincided with important advances in the development of high-density nucleic acid arrays and hybridization technology. Although 'chip' technology holds great promise for the future, at the present time spotted arrays represent an accessible format with sufficient reporter element density to interrogate bacterial genomes by comparative genomics and gene expression profiling. A rapidly growing literature using microarrays in functional genomics studies on bacteria confirms the reality and robust nature of spotted DNA microarrays for small genome organisms and also supports the use of total RNA samples for transcriptome analysis. However, microarray experimental systems are an emerging technology and optimization of nearly every step is an ongoing process. Thus, current methods are subject to development and have limitations and pitfalls. This should not be an argument against their widespread use, since the power of arrays as they are now is of undisputed biological significance and may be regarded as a proven technology. The future challenges for microarrays lies with mathematical and statistical analysis of high-dimensional data sets that are produced even in simple biological experiments.

Acknowledgements

The authors wish to acknowledge the financial support of The Wellcome Trust for establishing a multicollaborative microbial pathogen microarray facility under the Functional Genomics Resources Initiative.

References

Behr, M. A., Wilson, M. A., Gill, W. P., Salamon, H., Schoolnik, G. K., Rane, S. and Small P. M. (1999). Comparative genomics of BCG vaccines by whole-genome DNA microarray. *Science* **284**, 1520–1523.

de Saizieu, A., Gardes, C., Flint, N., Wagner, C., Kamber, M., Mitchell, T. J., Keck, W., Amrein, K. E. and Lange, R. (2000). Microarray-based identification of a novel *Streptococcus pneumoniae* regulon controlled by an autoinduced peptide. *J. Bacteriol.* **182**, 4696–4703.

Diehl, F., Grahlmann, S., Beier, M. and Hoheisel, J. (2001). Manufacturing DNA microarrays of high spot homogeneity and reduced background signal. *Nucleic Acids Res.* **29**, E38.

Doolittle, R. F. (2002). Biodiversity: microbial genomes multiply. *Nature* **416**, 697–700.

Dorrell, N., Mangan, J. A., Laing, K. G., Hinds, J., Linton, D., Al-Ghusein, H., Barrell, B. G., Parkhill, J., Stoker, N. G., Karlyshev, A. V., Butcher, P. D. and Wren, B. W. (2001). Whole genome comparison of *Campylobacter jejuni* human isolates using a low-cost microarray reveals extensive genetic diversity. *Genome Res.* **11**, 1706–1715.

Gingeras, T. R. and Rosenow, C. (2000). Studying microbial genomes with high-density oligonucleotide arrays. *ASM News* **66**, 463–468.

Hegde, P., Qi, R., Abernathy, K., Gay, C., Dharap, S., Gaspard, R., Earle-Hughes, J., Snesrud, E., Lee, N. and Quackenbush, J. (2000). A concise guide to cDNA microarray analysis. *Biotechniques* **29**, 548–562.

Kato-Maeda, M., Rhee, J. T., Gingeras, T. R., Salamon, H., Drenkow, J., Smittipat, N. and Small, P. M. (2001). Comparing genomes within the species *Mycobacterium tuberculosis*. *Genome Res.* **11**, 547–554.

Richmond, C. S., Glasner, J. D., Mau, R., Jin, H. and Blattner, F. R. (1999). Genome-wide expression profiling in *Escherichia coli* K-12. *Nucleic Acids Res.* **27**, 3821–3835.

Salama, N., Guillemin, K., McDaniel, T. K., Sherlock, G., Tompkins, L. and Falkow, S. (2000). A whole-genome microarray reveals genetic diversity among *Helicobacter pylori* strains. *Proc. Natl Acad. Sci. USA* **97**, 14668–14673.

Wilson, M., DeRisi, J., Kristensen, H. H., Imboden, P., Rane, S., Brown, P. O. and Schoolnik, G. K. (1999). Exploring drug-induced alterations in gene expression in *Mycobacterium tuberculosis* by microarray hybridization. *Proc. Natl Acad. Sci. USA* **96**, 12833–12838.

Ye, R. W., Tao, W., Bedzyk, L., Young, T., Chen, M. and Li, L. (2000). Global gene expression profiles of *Bacillus subtilis* grown under anaerobic conditions. *J. Bacteriol.* **182**, 4458–4465.

Ye, R. W., Wang, T., Bedzyk, L. and Croker, K. M. (2001). Applications of DNA microarrays in microbial systems. *J. Microbiol. Meth.* **47**, 257–272.

Part III
Comparative Nucleic Acid Analysis

6 Representational Difference Analysis of cDNA and Genome Comparisons

Deborah L Taylor[1], Aldert Bart[2], and Lucas D Bowler[1]

[1] Trafford Centre for Graduate Medical Education and Research, University of Sussex, Falmer, Brighton, BN1 9RY, UK
[2] Department of Medical Microbiology, Academic Medical Center, University of Amsterdam, PO Box 22660, 1100 DD Amsterdam, The Netherlands

CONTENTS

Introduction
Representational difference analysis
cDNA representational difference analysis
Future prospects

◆◆◆◆◆◆ INTRODUCTION

Representational difference analysis (RDA) is a subtractive hybridization technique, which combines PCR amplification with kinetic enrichment to isolate DNA sequences present in one population, but not in another. In its original form, RDA was a procedure for the identification of differences between two complex genomes (Lisitsyn et al., 1993), but the approach has since been modified for use with bacteria (Tinsley and Nassif, 1996; Bart et al., 2000), and for use in the identification of differences in gene expression (Hubank and Schatz, 1994; Bowler et al., 1999). RDA is a flexible, relatively inexpensive and highly effective technique. Furthermore, it has the advantage that it does not require that sequence data be available for the species or strain under investigation.

For comparisons between bacterial genomes ('genomic RDA'), populations of chromosomal DNA restriction fragments (called 'representations') are used as the starting material. Defined oligonucleotide adapters are then ligated on to the 5'-ends of the DNA molecules comprising one population (known as the 'tester') but not the other (the 'driver'). Reiterative hybridization/selection steps then follow. The tester population is hybridized against excess driver, and the resulting mixture treated with Taq DNA polymerase, thereby adding the complement of the defined oligonucleotide adapters to the 3'-ends of the tester fragments. Accordingly, only tester molecules that reanneal to other tester-originating sequences, will yield molecules with adapter sequences at both the 5'- and 3'-ends of the resulting double-stranded sequences, and as a consequence, only these molecules (representing the differences, or 'targets') will be exponentially amplifiable by the polymerase chain reaction (PCR). Selection for target sequences is increased by kinetic enrichment. This process takes advantage of the second-order kinetics of DNA reannealing (i.e. that abundant sequences will reanneal faster than less abundant ones). Accordingly, more abundant DNA species in a mixture of fragments are further enriched by reannealing to low C_0t values (the product of initial concentration and time), and subsequent purification of the resulting double-stranded molecules. In RDA, this is achieved by degradation (using mung bean nuclease) of any remaining single-stranded molecules after an initial PCR amplification step. The target

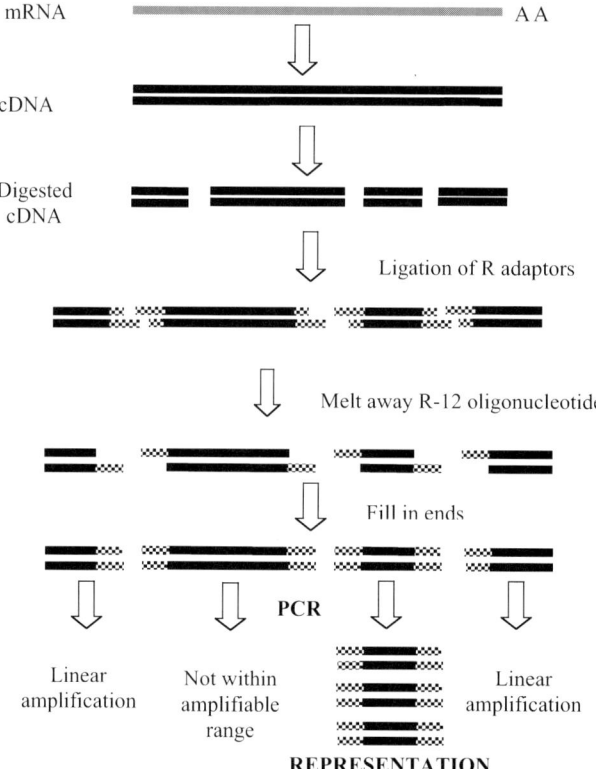

Figure 6.1. Generation of bacterial cDNA representations. cDNA (black) is made from total RNA (grey) using random hexamers. The cDNA is restricted to completion with *Dpn*II, and R-Bgl 12/24 adapters ligated to it (checked shading). The R-Bgl-12mer is melted away, the gap filled by DNA *Taq* polymerase and the R-Bgl-24mer used as an oligonucleotide primer to amplify the cDNA in a PCR reaction. Only fragments of sizes within the amplifiable range of the polymerase will be amplified. This is the representation.

sequences will be amplified exponentially by this PCR and will thus finish at a much higher concentration than non-target tester sequences (thereby forming relatively more double-stranded molecules). By degrading the single-stranded DNA, only the double-stranded sequences remain (enriched for the target sequences). Successive iterations of the PCR-coupled subtractive process result in the production of PCR products (the enriched target), which represent the differences between the representations. These can then be cloned and characterized.

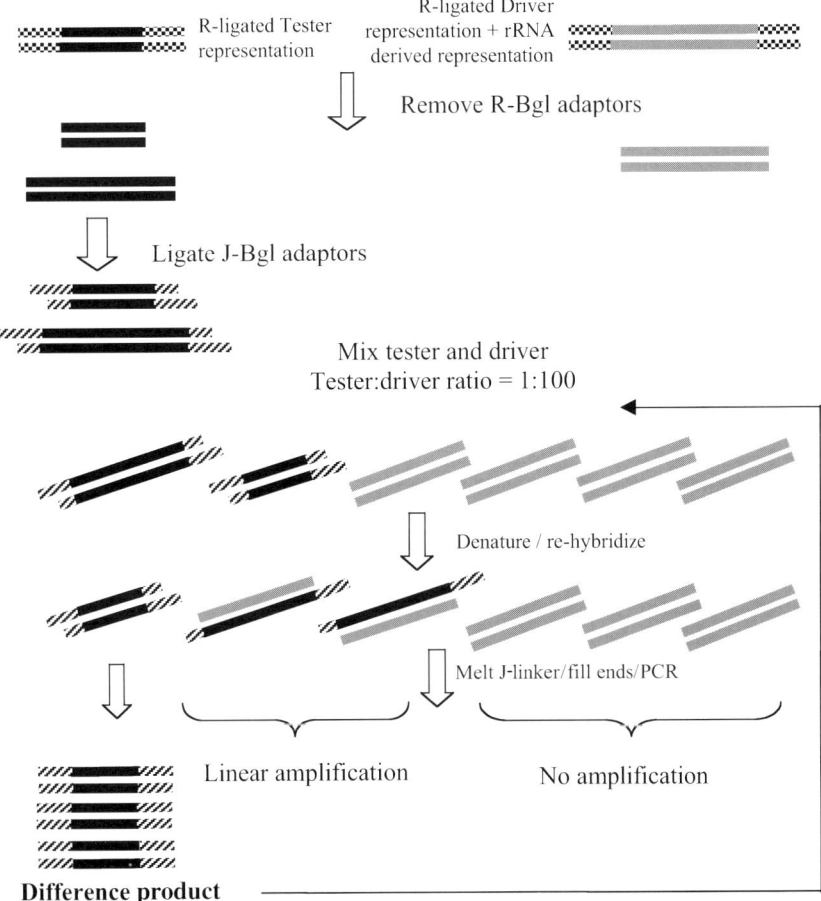

Figure 6.2. Overview of hybridization/amplification steps in RDA. In cDNA RDA, the R-Bgl adaptors (checked lines) are removed from the tester (black) and driver (grey) representations by restriction with *Dpn*II. J-Bgl-12/24 adaptors (diagonal shading) are then ligated to the tester representation. The driver and tester populations are mixed in a 100 : 1 ratio (50 driver : 50 rRNA-derived driver supplement : 1 tester), denatured and allowed to re-anneal. The J-Bgl-24mer is then used as a primer in a PCR amplification reaction. Only tester–tester fragments will be amplified. These are the difference products (DP1). The process is then repeated by replacing the J-Bgl adaptors from the DP1 with N-Bgl adaptors and increasing the driver : tester ratio. For genomic RDA, a digest of chromosomal DNA is used without generating representative amplicons. A set of adaptors are then ligated to the tester only and the process carried out as for cDNA RDA.

For the analysis of differences in gene expression, a modification of RDA, cDNA RDA, has been described (Hubank and Schatz, 1994; Bowler *et al.*, 1999, 2001) in which the starting material is cDNA derived from RNA preparations, rather than genomic DNA. cDNA RDA accordingly targets only genes that are expressed at the time the RNA is isolated.

As with RDA, cDNA RDA can be divided into a number of phases: the generation of representations (the PCR amplicons representing the RNAs isolated from given bacterial populations); the PCR-coupled subtractive hybridization of the different representations; and the cloning and screening of the resultant products (which represent the differences between the two populations that were compared). Here we provide detailed protocols for both RDA and cDNA RDA.

Figures 6.1 and 6.2 illustrate the principle of the RDA/cDNA RDA process schematically.

◆◆◆◆◆◆ REPRESENTATIONAL DIFFERENCE ANALYSIS

Bacterial strains and plasmids

Bacterial strains to be compared are grown under conditions relevant to their individual requirements. *Escherichia coli* Top10F' (Invitrogen) was used for propagating clones in the vector pCR2.1 (Invitrogen). Plasmid-carrying *E. coli* strains are routinely grown on Luria–Bertani (LB) plates at 37°C, supplemented with appropriate antibiotics, X-gal (5-bromo-4-chloro-3-indolyl-ß-D-galactopyranoside) and IPTG (isopropyl-ß-D-thiogalactopyranoside) for selection purposes where relevant, according to the manufacturers' protocols.

Isolation chromosomal DNA

High-quality chromosomal DNA from both bacterial strains to be compared (i.e. 'tester' and 'driver' strains) is isolated as described by Akopyanz *et al.* (1992). Alternatively, chromosomal DNA can be isolated using various commercially available kits. Both the Genomic-Tip 100/G system (Qiagen) and the PUREGENE DNA Isolation Kit (Gentra systems) work well in our experience.

Production of input material

With most eukaryotes, genomic RDA requires an initial reduction in the complexity of the input material. This is necessary because the sequence complexity of eukaryotic chromosomal DNA is too high to permit complete hybridization during the subtraction stages of the procedure (Hubank and Schatz, 1999). This 'simplification' of the genome is achieved by digesting total genomic DNA with a six-base cutting restriction endonuclease and subsequent amplification of the resulting DNA fragments by PCR. By using a six-base cutter a high proportion of the digested fragments do not fall into the amplifiable range (of approximately 100 bp–1.2 kb), thereby reducing the complexity of the amplicon, so that the resulting product contains no more than about 10% of the original total genome (Lisitsyn et al., 1993). In contrast, given their generally smaller genome size, this is not necessary with bacterial RDA, and, accordingly, unamplified restriction fragments, produced using a four-base cutting enzyme, are used as the input material.

High-performance liquid chromatography (HPLC)-purified adaptor oligonucleotides were obtained from Perkin Elmer, The Netherlands, and suspended in ultra-high purity (UHP) water as 0.2 mM stocks. The oligonucleotide adaptor sequences used are: J-Bgl-12 (5'-GATCTGTTCATG-3'), and J-Bgl-24 (5'-ACCGACGTCGACTATCCATGAACA-3'). Input material for the subtractive hybridization was produced by restriction of chromosomal DNA and ligation of the adaptor oligonucleotides to the tester DNA as shown in Protocol 1.

Protocol 1. Construction of input material

1. Digest 100 μg each of both tester and driver chromosomal DNA to completion with Sau3A. Extract this with phenol–chloroform and ethanol precipitate.
2. Resuspend driver DNA in 30 μl sterile ultra-high purity (UHP) water (to a final concentration of approx. 2.5 μg/μl). Similarly, resuspend tester DNA in 30 μl sterile UHP water, and dilute 10 μl of this with 40 μl UHP water (to a final concentration of approx. 500 ng/μl). The remaining 20 μl of the undiluted material can be used in a tester–tester control.
3. Ligate the J-Bgl-adaptors, J-Bgl-12 (5'-GATCTGTTCATG-3') and J-Bgl-24 (5'-ACCGACGTCGACTATCCATGAACA-3') to the tester chromosomal DNA by taking 2 μl (approx. 1 μg) diluted tester digest, adding 2.5 μl 0.2 mM J-Bgl-12, 2.5 μl 0.2 mM J-Bgl-24, 3 μl 10 × ligase buffer, and 14 μl sterile UHP water. Anneal the oligonucleotide adaptors using a PCR thermocycler, by heating the mixture to 50°C for 2 min, followed by cooling to 10°C at no more than 1°C/min.
4. Add 1 μl T4 DNA ligase, mix and incubate for 18 h at 16°C.
5. Purify the ligation by phenol–chloroform extraction. Precipitate the DNA by adding 0.1 vol. 3 M sodium acetate, pH 5.3, and 2.5 vols 100% ethanol. Centrifuge at 16 000g, for 30 min. Wash with 70% EtOH, resuspend in 30 μl sterile UHP water.

Note. It can be useful to include both positive and negative controls in the experiments. For example, as a positive control, tester DNA may be supplemented with an appropriate target (e.g. Bacteriophage lambda DNA or a known plasmid). As a negative control, original tester DNA can be used as both control-tester and control-driver. The positive control should give rise to target-derived amplicons, the negative control should not yield any amplicon.

Subtractive hybridization (see Figure 6.2)

The input material derived from the population containing the putative target genes is designated the tester, while the control material is known as the driver, since it is used to 'drive out' fragments common to both samples. The tester population is generated by ligation of a defined adaptor consisting of unphosphorylated 12- and 24-base oligonucleotides. The 12-mer provides the appropriate end structure to permit ligation of the 24-mer to the digested DNA, and is not itself covalently linked to the DNA. The driver and tester are then mixed, denatured and allowed to anneal. From this process, three types of double-stranded molecules can be formed (see Figure 6.1): driver : driver hybrids form most frequently, but these lack the adaptor sequences, cannot generate a primer binding site during the initial *Taq* DNA polymerase fill-in reaction and are, therefore, not amplified in the subsequent PCR. Driver : tester hybrids are the next most common product, and represent sequences common to both tester and driver populations. Because the driver strand is unable to generate a primer-binding site, these fragments can only 'amplify' linearly. If, on the other hand, a target fragment in the tester has no complement in the driver (i.e. represents a sequence unique to the tester population), then it can only anneal to a complementary strand originating from the tester itself, thereby generating fragments with the 24-mer oligonucleotides ligated at each end. The 12-mers are then melted away, and a fill in reaction is performed to generate the complementary primer binding site for the subsequent PCR, using the 24-mer itself as primer. Because only the tester : tester hybrids have the primer-binding site at both ends, they alone will amplify exponentially. Protocol 2 describes how the subtractive hybridization is performed.

Protocol 2. Subtractive hybridization

1. Combine 4 µl J-ligated tester (± 0.133 ng) and 4 µl driver (± 10 µg), and 25 µl 100% ethanol. Mix well to precipitate the DNA, centrifuge at 16 000g, for 30 min. Carefully wash with 70% ethanol. Air-dry pellet, resuspend thoroughly in 4 µl 3×EE buffer (30 mM EPPS (*N*-[2-hydroxylethyl]piperazine-*N*′-[3-propanesulfonic acid]; HEPPS), 3 mM EDTA, pH 8.0).
2. Denature DNA at 96°C for 5 min in a PCR thermocycler.
3. Add 1 µl 5 M NaCl (prewarmed).
4. Hybridize for 20 h at 67°C.
5. To 0.5 µl of the above hybridization mixture, add 40 µl 10×PCR buffer, 12 µl 10 mM dNTP, 3 µl Taq polymerase (5 U/µl), and 343.5 µl sterile UHP water.
6. Incubate in PCR thermocycler for 5 min at 72°C.
7. Add 2 µl 0.2 mM J-Bgl-24 primer.
8. Perform 10 cycles of 1 min at 95°C, 3 min at 70°C.
9. Precipitate PCR mix with 2.5 vols 100% ethanol, centrifuge at 16 000g, for 30 min. Wash with 70% ethanol.
10. Resuspend pellet in 18 µl sterile UHP water, add 20 µl 2×mung bean nuclease reaction buffer, add 2 µl (10 U/µl) mung bean nuclease (MBN), mix and incubate for 1 h at 30°C.

11. Inactivate mung bean nuclease by adding 160 μl 50 mM Tris–HCl pH 8.9, incubate for 5 min at 98°C.
12. Use 40 μl of the above mixture as a template in a PCR reaction with 40 μl 10×PCR buffer, 12 μl 10 mM dNTP, 3 μl Taq polymerase (15 U), 2 μl 0.2 mM J-Bgl-24 primer, 303 μl sterile UHP water (final vol. is 400 μl). Perform 20 cycles of 1 min at 95°C, 3 min at 70°C, followed by a final extension for 10 min at 72°C.
13. Precipitate DNA by adding 0.1 vol. 3 M sodium acetate, pH 5.3, and 2.5 vols 100% ethanol, centrifuge at 16 000g, for 30 min. Wash with 70% EtOH, resuspend in 50 μl sterile UHP water.
14. Run out a 5 μl sample on a 2% non-denaturing agarose gel. If tester and driver contain relatively few differences, discrete amplicons will be visible.

Cloning of difference products

The purified amplicons (the difference products) are cloned into a relevant vector system as described in Protocol 3. The number of amplicons seen following electrophoresis will depend on the number of differences that exist between tester and driver DNA populations. A relatively low number of differences will normally yield discrete amplicons after only one round of subtractive hybridization. However, the absence of discrete banding usually indicates a greater number of differences between the two DNA populations being compared, and/or relatively inefficient subtractive hybridization. In this instance, a second round of subtractive hybridization should be carried out in the same manner, but using a second, different, pair of oligonucleotide adaptors (e.g. N-Bgl-12 (5'-GATCTTCCCTCG-3'), and N-Bgl-24 (5'-AGGCAACTGTGCTATCCGAGGGAA-3')), ligated to the tester. In some instances where weak bands are seen, the optional reamplification step prior to cloning described in Protocol 3 can be of value.

Protocol 3. Cloning of difference products

1. Run remainder of each sample on a 2% non-denaturing agarose gel in Tris–borate–EDTA buffer, pH 8.3 (TBE). Carefully excise each band from the agarose gel, using a sharp razor blade.
2. Purify amplicons using the QIAquick gel extraction system (QIAGEN Ltd) or similar, according to manufacturers' instructions.
3. Optional: reamplify the amplicon. Use 1 μl of the purified amplicon as template in a PCR reaction with 5 μl 10×PCR buffer, 1 μl 10 mM dNTP, 1 μl Taq polymerase (5 U), 0.5 μl 0.2 mM J-Bgl-24 primer, 42.5 μl sterile UHP water to a final volume of 50 μl. Perform 30 cycles of 1 min at 95°C, 3 min at 70°C, followed by a final extension for 10 min at 72°C.
4. Clone the amplicon in the pCR2.1 vector, according to manufacturers' instructions. Recombinant plasmid DNA can then be isolated (using standard methods) and sequenced, for identification of the amplicons.

Note. Multiple transformants should be tested for each amplicon, as amplicons tend to be heterogeneous, i.e. different DNA species can be present with the same electrophoretic mobility.

◆◆◆◆◆◆ cDNA REPRESENTATIONAL DIFFERENCE ANALYSIS

cDNA RDA has been used for the identification of differences in gene expression in a variety of different bacterial species, both Gram negative and Gram positive (Bowler *et al.*, 1999; Becker *et al.*, 2001). The following section describes the methodology using an analysis of differential gene expression in the animal pathogen *Streptococcus uberis*, during growth in fresh cow's milk, as an example.

Bacterial strains and plasmids

Streptococcus uberis strain 0140J, used in this study was kindly provided by Drs J. Leigh and P. Ward (Institute of Animal Health, Compton, UK). *S. uberis* is routinely grown in Todd-Hewitt (TH) broth (DIFCO) at 37°C or on TH agar plates supplemented with 5% sheep blood.

For cDNA RDA analysis, *S. uberis* in TH broth was grown overnight, washed in HEPES (N-[2-hydroxyethyl]piperazine-N'-[2-ethanesulphonic acid]) buffered saline, and used to inoculate either fresh TH broth (to provide the driver component) or sterile skimmed milk (to provide tester). Liquid cultures were grown for approximately 4–6 h (to approximately mid-exponential phase, as determined by previous growth rate analyses).

Escherichia coli Top10F' (Invitrogen), was used for propagating clones in the vector pCR2.1 (Invitrogen). Plasmid-carrying *E. coli* strains were routinely grown on Luria–Bertani plates at 37°C, supplemented with appropriate antibiotics, X-gal (5-bromo-4-chloro-3-indolyl-ß-D-galactopyranoside), and IPTG (isopropyl-ß-D-thiogalactopyranoside) for selection purposes where relevant, according to the manufacturers' protocols.

Isolation of total RNA

Since bacterial mRNAs possess no, or relatively short, poly(A) tails (O'Hara *et al.*, 1995), poly(T)-affinity purification techniques cannot be used to isolate them from total RNA populations. Accordingly, when applying cDNA RDA to the analysis of differential gene expression in bacteria, total RNA rather than purified message must be used as the starting material.

cDNA RDA requires the isolation of high-quality representative RNA preparations (described in Protocol 4). RNA species are sensitive to degradation by ribonucleases, therefore, the isolation of good-quality RNA requires that all solutions and equipment used should be RNase-free. The use of sterile, certified RNase-free, disposable plasticware is recommended, and where glassware must be used, this should be foil sealed and baked for 4 h at 200°C before use. The skin can be a major source of nuclease contamination, and powder-free gloves should be worn at all times. UHP water is used in preference to diethyl pyrocarbonate (DEPC)-treated water, as residual traces of DEPC can inhibit some enzymatic processes. All solutions should be sterilized by filtration (using RNAse-free filters) where required.

Protocol 4. Isolation of total RNA

1. Harvest cells by centrifugation at 8000 rpm for 5 min at 4°C in a microfuge. Resuspend the cell pellet in 200 µl of RNase-free water and add to a RNase-free 2 ml centrifuge tube containing acid-washed glass beads.
2. Add 500 µl of acid phenol, 100 µl of chloroform and 500 µl of detergent solution (9.6 ml Decon 90, 24 ml 500 mM sodium acetate pH 4.0, 66.4 ml RNase-free water).
3. Process the tube immediately in a Fastprep (Q.BIOgene) for 3×20 s at full speed and cool on ice for 10 min.
4. Centrifuge the sample for 10 min at 16 000g in a microfuge at 4°C and take the aqueous phase to a fresh Eppendorf tube containing 500 µl of absolute ethanol. Precipitate the samples at −70°C for 1 h and harvest the RNA by centrifugation at 16 000g for 30 min at 4°C.
5. Wash the RNA pellet in 70% ethanol, dry and resuspend in 100 µl RNase-free 1 mM sodium citrate containing 100 U Prime RNase inhibitor (Eppendorf Ltd), or similar.
6. Quantification and crude quality assessment is done by examination on a 1% non-denaturing agarose gel in TBE buffer stained with ethidium bromide.
7. Aliquots of total RNA should be stored at −80°C until use.
8. Contaminating DNA must be removed from RNA samples by treatment with DNase I according to manufacturers' instructions before cDNA synthesis.

Note. In cDNA RDA it is important that RNA preparations are free of DNA contamination before cDNA synthesis is carried out. Accordingly it is advisable to confirm this by PCR before proceeding further. The addition of MnCl to the DNase I buffer to a final concentration of 1 mM can improve the efficiency of the DNase I digestion.

---

cDNA synthesis

cDNA synthesis is performed by random priming of total RNA as indicated in Protocol 5. We have found that a modified protocol using the Amersham Pharmacia Timer Saver cDNA synthesis kit gives good results, although other methods may be used.

> **Protocol 5. cDNA synthesis**
>
> 1. Synthesize cDNA, e.g. using a commercial cDNA synthesis kit, such as the Pharmacia TimeSaver cDNA synthesis kit according to the manufacturers' instructions. Use approx. 5 µg total RNA as the template, and random hexamers (0.037 µg/reaction), to prime.
> 2. After 15 min incubation of the first strand reaction at 37°C, add 200 U of RNase H⁻ reverse transcriptase (e.g. Superscript II, Invitrogen), to the reaction mix and continue incubation for a further 1 h.
> 3. Carry out the second-strand incubation at 12°C for 1 h, then 22°C for a further 2 h.
> 4. Purify the cDNA using a spin column (e.g. Amersham Pharmacia S-300 columns), according to manufacturers' instructions. Make up the final volume to 142 µl with TE (10 mM Tris, 1 mM EDTA pH 8.0).

Isolation of S. uberis 16S and 23S rRNA genes

Given the abundance of ribosomal RNA (rRNA) in total RNA preparations, we have found it useful to supplement the normal driver component with additional rRNA-derived material to increase selection against these common sequences during the subsequent subtractive hybridization steps. The rRNA genes are isolated by PCR as shown in Protocol 6. The resulting PCR products are then used to generate rRNA-specific representations to add to the normal driver component. The sequences of the 16S and 23S ribosomal gene primers used will, of course, depend upon the bacterial species concerned. However, the 16S and 23S primer pairs indicated in Table 6.1 are routinely used by us to isolate streptococcal rRNA genes.

> **Protocol 6. Isolation of S. uberis 16S and 23S rRNA genes**
>
> 1. To 0.2 µg of chromosomal DNA template, add 20 µl 5 × PCR buffer (335 mM Tris–HCl pH 8.9, 20 mM $MgCl_2$, 80 mM $(NH_4)_2SO_4$, 166 µg/ml bovine serum albumen), 8 µl dNTP mixture and 20 µl 5 M betaine solution. Add 100 pmol each of the relevant primers (see Protocol 4) and make up to 99 µl with sterile UHP water. Denature DNA in a thermal cycler by heating to 96°C for 5 min. Add 2.5 U Taq polymerase, cycle reactions: 96°C for 1 min, 58°C for 1 min and 72°C for 3 min, for 25 cycles, with a final extension at 72°C for 10 min.
> 2. Examine reactions on a 1% non-denaturing agarose gel. The primer pairs used should give rise to products of approx. 1450 bp and 2850 bp, respectively.
> 3. Resolve remainder of reaction on a 1% non-denaturing agarose gel. Excise the bands, and purify using a DNA purification kit according to manufacturers' instructions. Dilute products to 0.5 mg/ml. These purified products are the starting material for the generation of the rRNA gene-derived representations.
>
> *Note.* The addition of betaine to the PCR reaction mix is not essential, although it does enhance amplification of the rRNA sequences.

Table 6.1. cDNA RDA primers/adaptors

R-Bgl-12	5'-GAT CTG CGG TGA-3'
R-Bgl-24	5'-AGC ACT CTC CAG CCT CTC ACC GCA-3'
J-Bgl-12	5'-GAT CTG TTC ATG-3'
J-Bgl-24	5'-ACC GAC GTC GAC TAT CCA TGA ACA-3'
N-Bgl-12	5'-GAT CTT CCC TCG-3'
N-Bgl-24	5'-AGG CAA CTG TGC TAT CCG AGG GAA-3'
16Sup	5'-CTTGTTACGACTTCACCCCA-3'
16Sdn	5'-TGGCTCAGGACGAACGCT-3'
23Sup	5'-CCTGATCATCTCTCAGGGCT-3'
23Sdn	5'-CCTTGGCACTAGAAGCCGA-3'

Generation of representations (see Figure 6.1)

Ideally, a cDNA RDA representation would contain at least one fragment from every gene that was expressed at the time the RNA was isolated. Accordingly, to achieve the highest degree of representation of the derived cDNA, whilst ensuring as many of the resulting restriction fragments as possible are amplifiable, the cDNA is digested with a four-cutter restriction enzyme (e.g. *Dpn*II), which maximizes the generation of amplifiable fragments (Hubank and Schatz, 1999). Following restriction, oligonucleotide adaptors (the R-Bgl-adaptors, see Table 6.1) are ligated on to the cut cDNA as described in Protocol 7, and PCR carried out as shown in Protocol 8. The resulting product is, however, limited in its complexity by the ability of each component to amplify within the mixture. Templates that are either too large or too small do not amplify efficiently under what are effectively competitive conditions, and the product, therefore, only truly 'represents' the amplifiable proportion of the digest (under the conditions of PCR used in cDNA RDA, amplifiable fragments tend to lie in the range of approx. 100 bp–1.2 kb). It is this amplified representation that then serves as the starting material for successive rounds of PCR-coupled subtractive hybridization.

Protocol 7. Ligation of R-Bgl-12/24 adaptors

1. Add 16 μl of 10 × *Dpn*II buffer (as supplied with enzyme) to the 142 μl cDNA preparation in TE. Add 2 μl *Dpn*II and incubate for 3 h at 37°C.
2. For the rRNA-derived amplicons, digest 1–2 μg of each purified PCR product by adding 1 μl of *Dpn*II, 10 μl 10 × *Dpn*II buffer, and sterile water to a total volume of 100 μl. Incubate for 3 h at 37°C.
3. Phenol-extract the reactions, add 1 μl (15 μg) glycogen carrier, 30 μl 10 M ammonium acetate, 600 μl of cold 100% ethanol, and incubate at −20°C for 30 min.
4. Precipitate the DNA at 16 000*g* for 30 min at 4°C. Wash pellet with 70% ethanol. Air-dry and resuspend in 20 μl TE, transfer to a 0.5 ml PCR tube.

5. Add 24 µl sterile water, 6 µl ligase buffer (as supplied with T4 DNA ligase) and 4 µl each of 0.25 mM R-Bgl-12 and 0.5 mM R-Bgl-24 adaptor/primers (see Protocol 4).
6. Anneal oligonucleotide adaptors in a PCR machine, by heating the reaction to 50°C for 2 min, then cool to 10°C at 1°C/min.
7. Add 2 µl T4 DNA ligase, mix well, and incubate for 18 h at 14°C.

Protocol 8. Generation of representations

1. Dilute the ligation by adding 120 µl TE.
2. For a pilot reaction, use 3 µl diluted ligation, 139 µl sterile UHP water, 40 µl 5 × PCR buffer, 16 µl 4 mM dNTP mix, and 1 µl 0.5 mM R-Bgl-24 primer. Place in a 0.5 ml PCR tube.
3. Incubate reactions at 72°C for 3 min in a thermal cycler. Add 5 U Amplitaq DNA polymerase, and continue incubation for a further 5 min. Cycle reactions at 95°C for 1 min and then 72°C for 3 min for 24 cycles.
4. Remove 10 µl aliquots at intervals from about cycle 16 onwards (the optimum number of cycles for a representation usually lies between 17 and 24 cycles).
5. Resolve the 10 µl samples with size and concentration standards by electrophoresis on a 1.5% non-denaturing agarose gel.
6. From the gel, select the number of cycles that generates suitable representations. Set up 9 × 200 µl PCR reactions for each sample intended for use as driver, and 3 × 200 µl reactions for each sample intended for use as tester.
7. Cycle reactions for the determined number of cycles, finishing with a 10 min extension at 72°C.
8. Phenol-extract the reactions, add 0.1 vol. of 3 M sodium acetate (pH 5.3), an equal volume of 2-propanol, and precipitate the DNA on ice for 30 min.
9. Pellet the DNA by centrifugation at 16 000g for 30 min at 4°C. Wash the pellet with 70% ethanol, air dry and resuspend in TE to give a concentration of approx. 0.5 mg/ml.
10. Determine the quality and concentration of DNA samples by running 1 µl on a 1.5% non-denaturing agarose gel.

It is crucial that amplification of representations is kept within the linear range if the relative proportions of species are to be maintained with respect to the starting RNA populations. Accordingly, pilot reactions should be carried out for each representation to be generated to establish the number of PCR cycles required to produce a suitable representation. For a 'suitable' representation, a 10 µl sample resolved by electrophoresis on a 1.5% agarose gel should give a smear ranging in size from approximately 100 bp–1.2 kb and contain approximately 0.5 µg DNA. Too few cycles yields insufficient material for subtraction, and overamplification can bias the populations. For the rRNA-derived representation a pilot reaction is not required, as we use the same number of cycles determined as optimum for the other and driver component.

As long as pilot reactions are performed to determine the most appropriate number of cycles, representations appear to reflect the overall RNA composition remarkably well. This is crucial if you intend to use cDNA RDA to detect relative, as well as absolute differences. It is important to remember that, while cDNA RDA is highly reliable for the discovery of target genes whose abundance differs markedly between the representations, the detection of relative differences of a lower magnitude, although possible, is not very efficient. cDNA RDA is most effective in detecting transcripts when the differences in the level of expression between two populations is relatively high; however, the methodology can be modified to facilitate the identification of transcripts with lower levels of differential expression using a technique known as 'melt depletion'. This process effectively depletes representations of low copy sequences. Melt depletion of linker-ligated cDNA (representation) is performed by denaturing (melting) unmixed driver samples at 98°C, then allowing them to reanneal for 1 h, a duration determined empirically to allow much more efficient reannealing for more abundant species than rarer species. Subsequent amplification of the annealed product by PCR generates a driver population with considerable bias against low abundance sequences, as only annealed fragments are capable of amplification (Hubank and Schatz, 1994).

Preparation of driver and tester components

To produce the driver and tester components, representations are digested with *Dpn*II to remove the R-Bgl- adaptors, and new oligonucleotide adaptors (the J-Bgl adaptor set, see Table 6.1) ligated on as indicated in Protocol 9. Other restriction enzymes may be substituted for *Dpn*II, but this will necessitate redesign of the adaptors.

Protocol 9. Preparation of driver and tester components

1. Digest 100 µg each of driver and rRNA-derived representations with *Dpn*II at 37°C as described previously, in a total volume of 100 µl.
2. Phenol-extract the reactions and add 0.1 vol. of 3 M sodium acetate (pH 5.3), and an equal volume of 2-propanol to each digest, and precipitate on ice for 30 min.
3. Pellet DNA at 16 000g for 30 min at 4°C. Wash with 70% ethanol, air-dry and resuspend in 150 µl TE. Combine 16S and 23S representations.
4. Determine concentration of driver and rRNA-derived representations. Adjust concentration to approx. 0.5 mg/ml with TE. These represent the driver components.
5. Digest 10 µg of tester DNA with 1 µl *Dpn*II in a final volume of 60 µl. Incubate for 3 h at 37°C.

> 6. Phenol-extract the reactions, add 0.1 vol. of 3 M sodium acetate (pH 5.3), and 3 vol. of cold ethanol to each digest, and precipitate at −20°C for 30 min.
> 7. Harvest the DNA at 16 000g for 30 min at 4°C. Wash with 70% ethanol, air-dry and resuspend in 20 µl TE.
> 8. Remove digested R-Bgl-adaptors using a spin column purification system (e.g. Pharmacia S-300 columns), according to manufacturers' instructions, and estimate DNA concentration.
> 9. Take 1 µg of purified DNA and attach J-Bgl-adaptors following Protocol 4.
> 10. Dilute ligation to 10 ng/µl with TE. This preparation is the J-ligated tester.

Subtractive hybridization and amplification of differences
(see Figure 6.2)

cDNA RDA is a flexible and highly effective technique in which target cDNA fragments are sequentially enriched by favourable hybridization kinetics and subsequently amplified by PCR. cDNA RDA differs from display-based techniques in that the positive selection of differences and removal of sequences common to both groups simplifies the interpretation of results and greatly facilitates identification of the differentially expressed genes. Furthermore, the exponential degree of enrichment achieved by the use of PCR in cDNA RDA enables the detection of very rare transcripts and allows application of the technique to very low amounts of starting material. The PCR-coupled subtractive hybridization procedure is described in Protocol 10. Because of random annealing events, many amplified molecules present at the DP1 stage will not represent genuine differences. A second round of subtractive hybridization and PCR is, therefore, usually required. However, because of the partial enrichment that has already been carried out, this can be performed at higher stringencies (driver : tester ratios). A high degree of background smearing at this stage can indicate incomplete denaturation of driver and tester components prior to hybridization (or failure of the MBN digestion). Protocol 11 shows the procedure for the generation of the second difference product (DP2).

> **Protocol 10. Subtractive hybridization and amplification of differences**
>
> (See Figure 6.2 for a schematic representation of the procedure.)
> 1. Combine 5 µg of digested driver, 5 µg digested rRNA-derived representation and 0.1 µg of J-ligated tester in a 0.5 ml microcentrifuge tube. Make up to 100 µl with sterile UHP water. This gives a driver : tester ratio of 100 : 1 (50 : 50 : 1).
> 2. Phenol-extract reaction, add 0.1 vol. of 3 M sodium acetate (pH 5.3), and 3 vols of cold ethanol and precipitate at −70°C for 30 min.
> 3. Incubate tube containing precipitate at 37°C for 1 min, then spin at

16 000*g* for 20 min at 4°C to collect DNA. Very carefully wash the pellet with 70% ethanol, air-dry and resuspend very thoroughly in 4 µl 3 × EE buffer (30 mM EPPS, 3 mM EDTA pH 8.0) by pipetting.
4. Incubate at 37°C for 5 min, vortex vigorously, and briefly spin to collect DNA solution at the bottom of the tube.
5. Overlay solution with a few drops of mineral oil (even with heated lid machines), and in a thermal cycler denature DNA for 5 min at 98°C, cool to 67°C and incubate for 24 h to allow complete annealing.
6. Dilute hybridization mix with 196 µl TE, mix well by pipetting, remove to a fresh tube and vortex thoroughly. This diluted, hybridized DNA is then used to generate the first difference product.
7. For each subtraction set up two PCR reactions, comprising: 122 µl sterile UHP water, 40 µl 5 × PCR buffer, and 16 µl of 4 mM dNTP mix, and 20 µl diluted hybridization mix.
8. In a PCR machine, incubate reactions at 72°C for 3 min, add 1 µl (5 U) Amplitaq DNA polymerase, and continue incubation at 72°C for a further 5 min.
9. Add 1 µl of 0.5 mM J-Bgl-24 primer and cycle at 95°C for 1 min and 70°C for 3 min for 11 cycles, with a final extension at 72°C for 10 min.
10. Combine the two reactions and phenol-extract. Add 100 µg tRNA (Sigma), 0.1 vols. 3 M sodium acetate (pH 5.3) and an equal volume of 2-propanol. Precipitate on ice for 30 min.
11. Spin down precipitate at 16 000*g* for 20 min at 4°C. Wash carefully with 70% ethanol and resuspend in 20 µl TE.
12. Add 4 µl mung bean nuclease buffer, 2 µl mung bean nuclease (MBN), and make up volume to 40 µl with sterile UHP water. Incubate at 30°C for 45 min.
13. Terminate reaction by adding 160 µl of 50 mM Tris–HCl (pH 8.9) and heating to 98°C for 5 min. Cool the reaction on ice.
14. On ice, set up a PCR reaction comprising: 122 µl sterile UHP water, 40 µl 5 × PCR buffer, 16 µl 4 mM dNTP mix, and 1 µl 0.5 mM J-Bgl-24 oligo. Add 20 µl of the MBN-treated DNA. Incubate the reaction in a PCR machine at 95°C for 1 min, add 1 µl (5 U) Amplitaq DNA polymerase, cycle 95°C for 1 min and 70°C for 3 min, for 18 cycles, with a final extension at 72°C for 10 min.
15. Estimate DNA concentration by gel electrophoresis of a 10 µl sample on a 1.5% non-denaturing agarose gel, with standards.
16. Phenol-extract the reactions, add 0.1 vols 3 M sodium acetate (pH 5.3), an equal volume of 2-propanol, and precipitate on ice for 30 min.
17. Centrifuge at 16 000*g* for 30 min at 4°C. Wash with 70% ethanol, air-dry, and resuspend in TE to 0.5 µg/µl.
18. This is the first difference product (DP1).

Note. The initial 11 cycle PCR (step 9), helps prevent the loss of genuine but rare difference products during the precipitation steps prior to the MBN digestion. It is advisable to check the efficiency of the MBN digestion by comparison of treated and untreated aliquots on a 1.5% non-denaturing agarose gel. In our experience, the use of fresh enzyme is vital. MBN appears to work poorly if more than a few months old.

Protocol 11. Generation of a second difference product

1. Digest 2 µg of DP1 with *Dpn*II in a total volume of 100 µl. Incubate for 3 h at 37°C.
2. Phenol-extract the reactions, add 1 µl (15 µg) glycogen carrier, 0.1 vol. sodium acetate (pH 5.3) and 3 vols cold ethanol. Precipitate DNA at −20°C for 30 min.
3. Collect precipitates at 16 000g for 30 min at 4°C. Wash with 70% ethanol, air-dry, and resuspend in 20 µl TE (approx. 100 ng/µl).
4. Take 2 µl of the restricted DP1, add 3 µl 10 × ligase buffer, 2 µl each of 0.25 mM N-Bgl-12 and 0.5 mM N-Bgl-24 adaptors (see Table 6.1), and make up to a final volume of 29 µl with water.
5. Anneal oligonucleotides in a PCR machine, by heating to 50°C for 2 min, then cool to 10°C at 1°C/min. Add 1 µl T4 DNA ligase, mix well, and incubate for 18 h at 14°C.
6. Dilute ligation by adding 130 µl TE (approx. 1.25 ng/µl).
7. Combine 5 µg of digested driver, 5 µg digested rRNA-derived representation, with 10 µl of N-Bgl-ligated DP1. This gives a driver–tester ratio of 800 : 1 (400 : 400 : 1).
8. Repeat the hybridization and amplification procedure as described in Protocol 10 (items 2–17); remembering to use the N-Bgl-24 primer during the PCR steps.
9. Resuspend the pellet in TE to 0.5 µg/µl. This is the second difference product (DP2). Resolve difference products on a 2% agarose-1000 gel (Invitrogen) with standards. At this stage, the bands can be excised and cloned, if they appear well defined, or used for a third round of subtractive hybridization and amplification by removing the N-Bgl-adaptors and substituting with J-Bgl-adaptors.

Generation of further difference products

There are both advantages and disadvantages in deciding whether to continue to a third round of subtraction. DP2 may contain poorly defined bands and a third round of subtractive hybridization and amplification can increase the definition of the products. However, it can also result in the loss of some difference products, in particular those from transcripts that are expressed at relatively low levels.

For the generation of DP3, the procedure remains the same as in Protocol 11 with the following modifications: you should digest DP2 with *Dpn*II to remove the N-Bgl adaptors, and replace those with J-Bgl adaptors. Dilute the J-Bgl-ligated DP2 to 1 ng/µl with TE. Set up hybridizations using driver to tester ratios of between 5000 and 20 000 : 1. You then generate DP3 according to Protocol 11, setting up four PCR reactions for each subtraction. Finally, resuspend the pellet from the four combined reactions to a final concentration of approximately 0.5 µg/ml in TE. This is DP3.

Cloning of difference products

The purified amplicons (difference products) are cloned into a relevant vector system essentially as described in Protocol 3, except the optional step (step 3) is omitted. As with RDA, the amplicons are heterogeneous and accordingly multiple transformants should be screened for each.

Since the identified differences should only be either present or upregulated in the tester, each putative difference should be checked for validity by Southern hybridization against the original representations, and ideally also by reverse transcriptase-PCR, using an independently isolated RNA preparation as a template (cloning of high numbers of 'false-positives' usually indicates that the driver:tester ratio was too low).

The application of cDNA RDA to *S. uberis* has so far resulted in the identification of 11 transcripts that are differentially expressed in milk. Of the sequences isolated to date, most show significant homology to known proteins of other streptococcal species. Many of the transcripts identified are glucose repressible, and are involved in the biosynthesis or metabolism of macromolecules. Although these transcripts are not classical virulence determinants, they almost certainly contribute to the overall pathogenicity of *S. uberis* during infection of the bovine udder. In addition we have identified a number of membrane-associated proteins involved in the uptake of nutrients or trace elements under conditions of limitation. Figure 6.3 shows the result of a typical cDNA RDA experiment carried out with *S. uberis*. This work is ongoing.

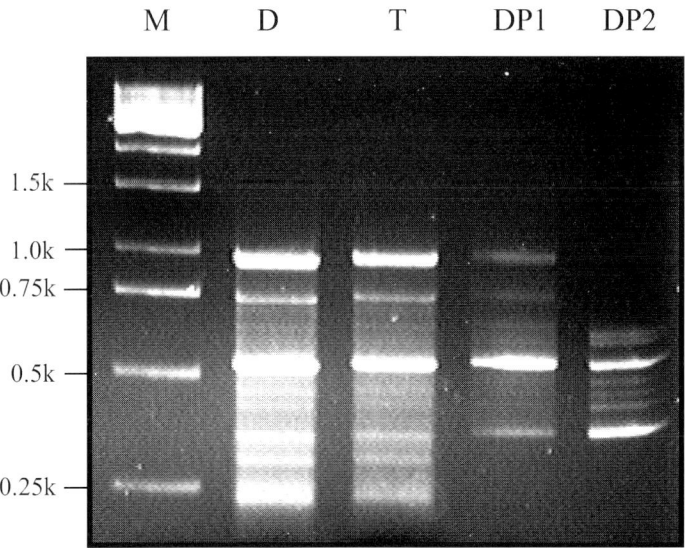

Figure 6.3. Difference products generated during analysis of differential gene expression in *S. uberis* during growth in milk. M, molecular mass markers; D, derived driver population; T, tester; DP1, difference products after one round of PCR-coupled subtractive hybridization; DP2, after two rounds. The discrete bands seen in DP2 are cloned and sequenced.

◆◆◆◆◆◆ FUTURE PROSPECTS

It has recently been shown by Kim et al. (2001), that the use of cDNA RDA in combination with microarrays may provide the most effective way to identify differentially expressed genes. The advantage of this hybrid approach is likely to be even greater with prokaryotes, particularly for the examination of host–pathogen interactions or where starting material is very limited, with an amplification and enrichment stage, such as provided by cDNA RDA, that is likely to prove essential.

Acknowledgements

Drs L. D. Bowler and D. L. Taylor would like to acknowledge financial support for their research from the BBSRC and The Meningitis Research Foundation.

References

Akopyanz, N., Bukanov, N. O., Westblom, T. U., Kresovich, S. and Berg, D. E. (1992). DNA diversity among clinical isolates of *Helicobacter pylori* detected by PCR-based RAPD fingerprinting. *Nucleic Acids Res.* **20**, 5137–5142.

Bart, A., Dankert, J. and Van der Ende, A. (2000). Representational difference analysis of *Neisseria meningitidis* identifies sequences that are specific for the hyper-virulent lineage III clone. *FEMS Microbiol. Lett.* **188**, 111–114.

Becker, P., Hufnagle, W., Peters, G. and Herrmann, M. (2001). Detection of differential gene expression in biofilm-forming versus planktonic populations of *Staphylococcus aureus* using micro-representational-difference analysis. *Appl. Environ. Microbiol.* **67**, 2958–2965.

Bowler, L. D., Hubank, H. and Spratt, B. G. (1999). Representational difference analysis of cDNA for the detection of differential gene expression in bacteria: development using a model of iron-regulated gene expression in *Neisseria meningitidis*. *Microbiology* **145**, 3529–3537.

Bowler, L. D., Bart, A. and van der Ende, A. (2001). *Meningococcal Disease*. Totowa, NJ: Humana Press.

Hubank, M. and Schatz, D. G. (1994). Identifying differences in mRNA expression by representational difference analysis of cDNA. *Nucleic Acids Res.* **22**, 5640–5648.

Hubank, M. and Schatz, D. G. (1999). cDNA representational difference analysis: a sensitive and flexible method of identification of differentially expressed genes. *Math. Enzymol.* **303**, 325–349.

Kim, S., Zeller, K., Dang, C. V., Sandgren, E. P. and Lee, L. A. (2001). A strategy to identify differentially expressed genes using representational difference analysis and cDNA arrays. *Analyt. Biochem.* **288**, 141–148.

Lisitsyn, N., Lisitsyn, N. and Wigler, M. (1993). Cloning the differences between two complex genomes. *Science* **259**, 946–951.

O'Hara, E. B., Chekanova, J. A., Ingle, C. A., Kushner, Z. R., Peters, E. and Kushner, S. R. (1995). Polyadenylation helps regulate mRNA decay in *Escherichia coli*. *Proc. Natl Acad. Sci. USA* **92**, 1807–1811.

Tinsley, C. R. and Nassif, X. (1996). Analysis of genetic differences between *Neisseria meningitidis* and *Neisseria gonorrhoeae*: two closely related bacteria expressing two different pathogenicities. *Proc. Natl Acad. Sci. USA* **93**, 11109–11114.

7 Application of DNA Microarrays for Comparative and Evolutionary Genomics

Nick Dorrell, Olivia L Champion and Brendan W Wren
Department of Infectious and Tropical Diseases, London School of Hygiene and Tropical Medicine, London WC1E 7HT, UK

◆◆

CONTENTS

Introduction
Which type of DNA microarray?
Selected examples of the application of DNA microarrays for comparative genomics
Limitations of DNA microarrays
Future perspectives

◆◆◆◆◆◆ INTRODUCTION

Traditional phylogenetic classification of bacteria to study evolutionary relatedness is based on the characterization of a limited number of genes, rRNA/rDNA or signature sequences. However, owing to the acquisition of DNA through lateral gene transfer, the differences between closely related bacterial strains can be significant. By contrast, whole-genome sequencing comparisons allow a multitude of genes to be compared. Already several bacterial species have had more than a single representative sequenced (e.g. *Escherichia coli*, *Helicobacter pylori*, *Neisseria meningitidis*, *Staphylococcus aureus* and several *Chlamydia* species). Nevertheless, genome sequencing still remains an expensive pursuit and such comparisons are limited to only a few strains. DNA microarrays represent an efficient technology for whole-genome comparisons, allowing a bird's eye view of the absence and presence of genes in a given genome compared to the reference genome on the microarray. This technology coupled with the continued availability of genome sequence data is set to revolutionize our ability to distinguish microbes. DNA microarray analyses have revealed a vast diversity both between genera and within species. Uncovering the mechanisms behind this variability will enable us to better understand the phylogeny, physiology, ecology and evolution of microbes. Genome comparisons between pathogens and non-pathogens within a species are particularly useful for identifying determinants important in virulence, transmission and host specificity. This chapter will summarize the application of DNA microarrays to investigate the genome content of several bacterial species, and in particular how this knowledge has led to a better understanding of the evolution of microbial virulence.

◆◆◆◆◆◆ WHICH TYPE OF DNA MICROARRAY?

PCR product-based DNA microarrays

DNA microarrays generally consist of amplified gene fragments representing individual genes (see Chapter 5). Currently, up to 50 000 gene fragments can be spotted on to a single microscope slide using robotic technology. The advantages of this technology are: (1) flexibility in the design of the array; (2) the relative ease of production; and (3) the relative low cost. Multiple identical microarrays can now be robotically printed in batches of over a hundred in a single run. Most of the cost in printing such arrays is due to the synthesis of oligonucleotide primer pairs required for the amplification of target gene fragments. This cost can be dramatically reduced by utilizing clone libraries as the template for amplifying the PCR products instead. For example, ordered pUC clone libraries that are a byproduct of most microbial genome sequencing projects can be used by selecting an optimum clone to represent every gene in the genome (see Figure 7.1) (Dorrell *et al.*, 2001). In such cases a single pair of primers based on the vector sequence can be used to amplify gene fragments to produce a low-cost representative genome microarray. However, many of the optimum clones chosen will not be gene specific and may contain adjacent genes or gene fragments. This can result in cross-hybridization, where the gene probe sequence will be present in more than one PCR product on the array, resulting in non-representative changes in hybridization signal intensities.

Even with gene-specific microarrays, for some microbial species, potential problems will still occur with cross-hybridization between gene fragments representing different but very similar genes, meaning that genetic rearrangements, insertions, inversions and duplications are difficult to detect using spotted PCR products (Brosch *et al.*, 2001). Therefore, the design of the oligonucleotide primer pairs to amplify target gene fragments is critical in DNA microarray design (see Chapter 4). Additionally, point mutations will not be identified using PCR product-based DNA microarrays.

Oligonucleotide-based DNA microarrays

Oligonucleotide arrays do not rely upon PCR amplification of gene targets and areas of specificity within any gene sequence can be chosen for

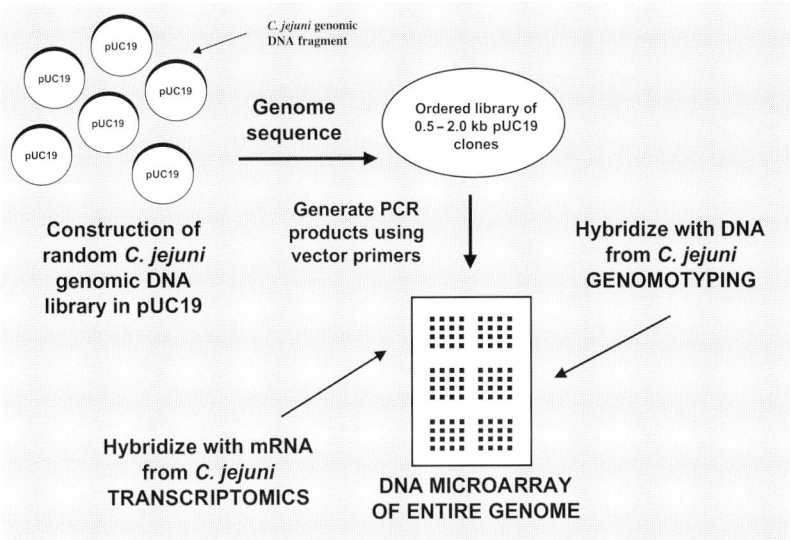

Figure 7.1. Strategy for the construction of low-cost DNA microarrays and functional analysis of a sequenced genome. The random shotgun library used as a starting point for most bacterial genome sequencing projects forms an ordered plasmid library upon annotation of the complete genome sequence. An optimum clone for each gene in the *Campylobacter jejuni* NCTC 11168 genome was selected and PCR products amplified using a single pair of pUC19 vector primers. These PCR products were printed on to glass slides to create a low-cost DNA microarray representative of the whole genome.

hybridization analysis. However, the sequence of target DNA must be known prior to oligonucleotide synthesis. The two most frequently used formats are the Affymetrix and the Qiagen Operon systems. In the Affymetrix format (www.affymetrix.com), oligonucleotides are synthesized *in situ* on a derivatized glass surface using a combination of photolithography and combinatorial chemistry (Pease *et al*, 1994). The synthesized oligonucleotides are usually 20–25 bases in length. The GeneChip® *E. coli* Genome Array was the first Affymetrix microarray product for the analysis of gene expression in a prokaryotic organism. The Qiagen Operon format (www.operon.com) uses an optimized 70-mer oligonucleotide to represent each gene in a genome. Each 70-mer is designed to have optimal specificity for its target gene and is melting-temperature normalized. An Array-Ready Oligo Set™ for the *Mycobacterium tuberculosis* genome consists of 70-mer probes representing 4269 *M. tuberculosis* genes. The probes were designed from the predicted sequence of the *M. tuberculosis* H37RV strain and an additional 371 sequences from the CDC-1551 strain.

Oligonucleotide-based microarrays offer many advantages over PCR product-based microarrays, including a reduction in cross-hybridization and an increase in the differentiation of overlapping genes or highly homologous regions. For example, mutant alleles or single nucleotide polymorphisms (SNPs) can be detected using such oligonucleotide microarrays owing to the shorter probe size compared to PCR product microarrays. A single nucleotide difference between the target and probe can prevent hybridization. Using a GeneChip® Yeast Genome S98 Array, 3000 polymorphisms between two strains of *Saccharomyces cerevisiae* were identified (Winzeler *et al.*, 1998). Similarly, another GeneChip was used to screen for deletion polymorphisms across whole genomes of different *M. tuberculosis* clones (Kato-Maeda *et al.*, 2001). By using oligonucleotide probes on the array that contain either small sequence variations or perfect match sequence, the signal intensity recorded following hybridization with labelled target should reveal which probe is perfectly complementary to the target sequence. However, this approach has not always proved successful, especially for the detection of small frameshift mutations. It is also difficult to assay simultaneously sequence tracts with localized regions of high G-C and A-T content when directly hybridizing mutation-containing target sequences to probes on an array (Hacia, 1999). Use of a multiplex assay using a modified PCR to amplify each amplicon evenly followed by a ligase detection reaction has allowed detection of small insertions and deletions through screening reaction products on a DNA microarray (Favis *et al.*, 2000).

Oligonucleotide arrays remain very expensive and, in the case of the Affymetrix system, the technology itself is proprietary. Currently, there are only a limited number of whole microbial genome arrays in this format. As the number of oligonucleotide arrays for different microorganisms increases and competition reduces the cost of these arrays, this technology platform should predominate.

Microelectronic chip arrays

Microarray-based assays have several limitations, including the requirement of a high concentration of target DNA for efficient rapid, passive hybridization and problems with achieving uniform stringency conditions during hybridization. Active microelectronic chip arrays extend the power of microarrays through the use of electronics (Edman *et al.*, 1997; Sosnowski *et al.*, 1997). Electronic addressing and/or hybridization is achieved by selective application of a direct current positive bias to individual microelectrodes beneath the selected test sites. This causes rapid transport and concentration of negatively charged nucleic acid molecules over selected locations on the microelectronic array. The nucleic acid may then be immobilized by direct attachment to a permeation layer overlying the microelectrode, or by hybridization to a previously addressed and attached nucleic acid. This electronic biasing allows improvements in both time and efficiency of hybridization reactions in a site-specific manner. Reversal of the electric potential then allows rapid removal of unhybridized molecules as well as a continuous adjustment of stringency.

Most biological molecules, including DNA, RNA and oligonucleotides, are charged and can be moved and concentrated electronically. The Nanogen NanoChip™ system (www.nanogen.com) takes advantage of this by manipulating the charge at each test site; nucleic acids can be rapidly moved and concentrated enhancing the flexibility and accuracy as compared to other array systems. Hybridization of complementary single-stranded DNA is rapid (in seconds rather than in hours) and electronic stringency removes the unbound and non-specifically bound DNA after hybridization. Multiple test sites can be analysed simultaneously from a single sample. Currently the NanoChip™ system is expensive and the number of discriminatory hybridization sites is limited.

The NanoChip™ system has been used to simplify the process of bacterial identification (Westin *et al.*, 2001). Species-specific bacterial identification of clinical specimens is often limited to a few species, owing to the difficulties of performing multiplex reactions. In addition, discrimination of amplicons is time consuming and laborious. A microelectronic chip array was used simultaneously to amplify and discriminate six genes, which are representative of different bacterial identification assays: *E. coli gyrA*, *Salmonella gyrA*, *Campylobacter gyrA*, *E. coli parC*, *Staphylococcus mecA* and *Chlamydia* cryptic plasmid. The NanoChip™ system can detect both plasmid and transposon genes and can also discriminate strains carrying antibiotic resistance SNPs. Application of this type of system should greatly simplify and speed up the identification of microbes and should have particular application in diagnostic laboratories.

◆◆◆◆◆◆ SELECTED EXAMPLES OF THE APPLICATION OF DNA MICROARRAYS FOR COMPARATIVE GENOMICS

DNA microarrays to investigate genome diversity

It is proposed that bacterial genomes consist of a core of genetic material that is conserved in most strains, a minimal gene set shared by the vast majority of bacteria and a flexible pool of strain-specific genes allowing the organism to adapt to its environment (Dobrindt and Hacker, 2001). These principles are borne out in work carried out on the intestinal pathogen *Campylobacter jejuni* (Dorrell *et al.*, 2001). Through microarray analysis of 11 *C. jejuni* strains, 1100 core genes (out of 1654 annotated genes in the sequenced strain) that contribute mainly to metabolic, biosynthetic, cellular and regulatory processes, but also many putative virulence determinants, were defined (see Table 7.1). The presence of virulence determinants within the species-specific genes was not surprising as all of the strains studied were isolates from patients with gastroenteritis. A subset of genes consisting of approximately 21% of the gene content in the sequenced strain *C. jejuni* NCTC 11168 was highly divergent in the 11 isolates tested, representing strain-specific genes. Many of these divergent genes are located in regions which are associated with the biosynthesis of surface structures including the flagellar, the lipo-oligosaccharide (LOS) and the newly identified capsule. These genes appear to be dispensable to the *Campylobacter* species, as they are absent or highly divergent in one or more isolates. The *C. jejuni* capsule biosynthesis locus is characterized by a central region containing genes responsible for the synthesis of the polysaccharide repeat units, which make up the capsule, flanked by the genes (*kpsSC* and *kpsFDETM*; see Figure 7.2) involved in the translocation of these repeat units to the cell surface (Karlyshev *et al.*, 2000). It was suggested that the capsule was the serodeterminant of the Penner serotyping system for *C. jejuni*, and not LOS as previously thought (Karlyshev *et al.*, 2000). Further analysis of the microarray data relating to the capsule biosynthesis locus revealed conservation of all the genes in this region in strains with the same Penner serotype as *C. jejuni* NCTC 11168 (serotype O:2). By contrast, strains of a different serotype to the sequenced strain lacked between 5 and 12 *C. jejuni* NCTC 11168 genes in the central biosynthetic region, whilst conserving the *kps* genes, providing further evidence that the capsule locus accounts for Penner serotype specificity in *C. jejuni* (see Figure 7.2).

A *Helicobacter pylori* composite DNA microarray has been constructed containing a superset of open reading frames (ORFs) from both sequenced strains (NCTC 26695 and J99) (Salama *et al.*, 2000). The results of genomic comparisons of 15 clinical isolates revealed a minimal functional core of

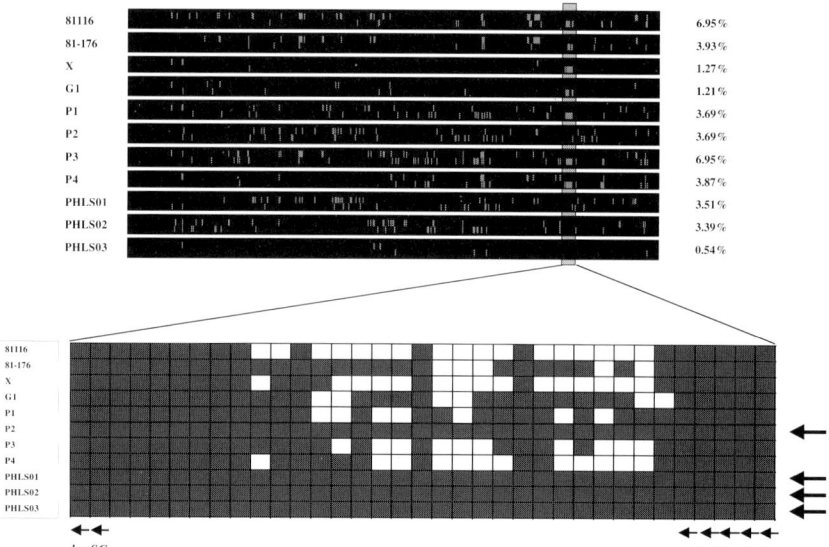

Figure 7.2. *Campylobacter jejuni* strain comparisons highlighting the capsule biosynthesis locus responsible for Penner serotype specificity. The top panel highlights genes identified as absent or highly divergent by microarray analysis in 11 human isolates compared against the sequenced strain NCTC 11168, which are indicated as vertical lines depicting the position of the gene in a linear representation of the 1.64 Mb genome from Cj0001 to Cj1654. Lines above the horizontal line indicate genes on the forward strand and lines below indicate genes on the reverse strand. The strain analysed is indicated to the left of the top panel and the percentage of genes that are absent or highly divergent compared to NCTC 11168 is indicated to the right. The bottom panel highlights the 42.6 kb capsule biosynthetic locus (Cj1413 to Cj1448) showing the genes identified as absent as a shaded box. The black arrows to the right indicate strains with the same Penner serotype as the sequenced strain.

1281 genes, common to all strains tested (see Table 7.1). These core genes encode most cellular and metabolic processes while the strain-specific genes encode cell surface proteins, transposases and restriction modification proteins. In this study a clustering algorithm was used for the analysis of strain-specific genes and the identification of potentially co-inherited genes. Predicted amino acid similarities indicated functions that may aid the bacterium under certain environmental stresses, including three strain-specific genes with similarity to *virB4* and *traG*, both ATPases for type IV secretion systems (see Chapter 16).

DNA microarray analysis of 36 *Staphylococcus aureus* strains of divergent clonal lineages, including methicillin-resistant strains and organisms causing toxic shock syndrome, revealed 78% of genes were common to all strains examined with 22% proving to be strain specific (see Table 7.1) (Fitzgerald *et al.*, 2001). Detailed analysis of these data identified 18 chromosomal regions consisting of more than three contiguous ORFs that were absent in one or more strains compared with the reference strain COL. The size of these chromosomal regions of difference (RD) varied from 3 to 50 kb. Genes encoding virulence determinants or proteins mediating antibiotic resistance are associated with ten of these RDs. It was apparent that multiple deletion, integration and recombination events have contributed to the variation in the RDs and the *S. aureus* genome overall. At least 19 genes that would encode proteins with homology to mediators of lateral gene transfer, such as integrases and transposases, are present in the reference strain COL and 14 of these are located in the RDs. Also most RDs were widely distributed among strains of divergent clonal lineages that have not shared a recent ancestor. Together this suggests that horizontal gene transfer has played a fundamental role in the evolution of pathogenic *S. aureus*, especially by assortive recombination of RDs containing virulence and antibiotic-resistance genes. For example, it was shown that the *mec* gene had been horizontally transferred into distinct *S. aureus* chromosomal backgrounds at least five times, indicating that methicillin-resistant strains have evolved independently on multiple occasions.

Similarly, using PCR products representing each ORF spotted on nylon membranes, the evolutionary history of the entire *E. coli* chromosome was traced by examining the distribution of the 4290 ORFs from the *E. coli* sequenced strain MG1655 among strains of known genealogical relationship (Ochman and Jones, 2000). These data were used to deduce the incidence of gene transfer and gene loss. A total of 30 deletion and 37 insertion events were required to account for the distribution of all genes now present in the MG1655 chromosome. Of the 4290 ORFs present in *E. coli* strain MG1655, 3782 were common to all the other strains examined (see Table 7.1). However, given the variation in gene content and chromosome size, *E. coli* strains can contain over a megabase of unique DNA, conferring traits that distinguish them from other members of the species. Indeed strains vary widely in the frequencies of deletions, which probably accounts for the variation in genome size within the species. The diversity of *E. coli* strains is also borne out in the sequence comparison between the non-pathogenic laboratory strain *E. coli* MG1655 and the

Table 7.1. Examples of comparative genomics with a single bacterial species using DNA microarrays

Bacterial strain(s) present on the microarray	Number of strains analysed	% of genome which appears strain-specific	Reference
Escherichia coli K-12 MG1655	4	10	Ochman and Jones (2000)
Mycobacterium tuberculosis H37Rv	19	0.3	Kato-Maeda *et al.* (2001)
Helicobacter pylori J99 and NCTC 26695	15	22	Salama *et al.* (2000)
Campylobacter jejuni NCTC 11168	11	21	Dorrell *et al.* (2001)
Staphylococcus aureus COL	36	22	Fitzgerald *et al.* (2001)
El Tor 01 *Vibrio cholerae* N16961	9	1	Dziejman *et al.* (2002)

human pathogenic strain *E. coli* O157:H7, where the latter was found to contain 1387 new genes encoded in strain-specific clusters of diverse sizes (Perna *et al.*, 2001).

DNA microarrays to investigate genome plasticity

Whole-genome microarray experiments have been applied to assess the genome plasticity of the laboratory *Mycobacterium tuberculosis* strain H37Rv with the closely related species *Mycobacterium bovis* and several strains of a serially passaged *M. bovis* BCG vaccine strain (Behr *et al.*, 1999). The original BCG vaccine strain was passaged *in vitro* 230 times, resulting in a strain with maintained immunogenicity but reduced virulence. Since then, this strain has continued to be passaged, resulting in a collection of daughter strains with different phenotypes. The genome composition of *M. bovis*, *M. tuberculosis* and the BCG daughter strains were compared in order to provide an explanation for the variable BCG efficacy in human trials. The differences observed indicate that, since original derivation, the BCG strains have evolved, which may have caused loss of protective efficacy. Compared to *M. tuberculosis*, 11 RDs were identified containing 91 ORFs, which were deleted from some or all of the pathogenic *M. bovis* strains tested. For the attenuated BCG vaccine strains, up to an additional five RDs containing 38 ORFs had been lost. The most common class of genes lost from BCG strains were transcriptional regulators. These results show that microarray analysis can be used to reconstruct the genealogy of related strains at the genome level.

The study of genetic variability within natural populations of pathogenic bacteria may provide insight into their evolution and pathogenesis. An Affymetrix GeneChip® representing all 3924 ORFs and 738 intergenic regions of *M. tuberculosis* strain H37Rv was used to detect small-scale genomic deletions among 19 clinically and epidemiologically well-characterized isolates of *M. tuberculosis* (Kato-Maeda *et al.*, 2001). A clone was defined as a group of isolates that shared sufficient properties such that they are likely to represent progeny of the same progenitor (Orskov and Orskov, 1983). Polymorphisms were not detected between clonal isolates H37Ra and H37Rv or among three isolates from individuals involved in a chain of disease transmission. The patterns of deleted sequences in the remaining 14 clones were all different, suggesting this is a suitable genotyping system for epidemiological studies.

A DNA microarray based on the recently sequenced El Tor O1 *Vibrio cholerae* strain N16961 was used to investigate the genetic similarity among *V. cholerae* strains isolated from diverse geographical regions over most of the 20th century (Dziejman *et al.*, 2002). This study was used to address the issue whether strains responsible for the current seventh cholera pandemic have genes encoding gain-of-function traits that may have displaced pre-existing classical *V. cholerae* strains. A microarray consisting of >93% of the genes of the El Tor O1 N16961 strain was used to analyse a collection of nine strains of diverse global origin isolated between 1910 and 1992. In contrast to *C. jejuni*, *H. pylori*, *S. aureus* and *E. coli*, the maximum number of genes missing from any of the test strains was 49 representing about 1% of the genome (see Table 7.1). This suggests a remarkable conservation of genomic information among *V. cholerae* strains, despite their isolation over the past century. Of the gene differences, some could differentiate classical biotype strains from El Tor biotype strains. Two putative chromosomal islands (VSP-I and VSP-II) with a deviant G + C content were identified. Dziejman *et al.* propose that the genes associated with the VSP-I and VSP-II islands may encode key properties that led to the global success of the seventh pandemic strain as agents of endemic and pandemic cholera. These genes may encode adaptive properties that allow these strains to withstand nutrient deprivation or stresses, and thus survive more efficiently in aquatic environments than strains before the seventh pandemic. These environmental adaptations may allow more efficient colonization of the human or non-human hosts (e.g. algae and crustaceans). The genes identified by microarray analysis can now be deleted systematically to determine their potential role in human infection and in promoting the fitness of *V. cholerae* in environmental ecosystems.

Isolates of the gastric pathogen *H. pylori* harvested from different individuals are highly polymorphic. Strain variation has also been observed between isolates from the same infected individual. Isolates of the sequenced *H. pylori* strain J99 were obtained from the source patient after a 6-year interval and analysed using the *H. pylori* microarray described above (Salama *et al.*, 2000). Microarray analysis revealed differences in the genetic content among all the isolates, and patterns of genetic diversity for these new isolates were distinct both when compared with each other and to the original J99 isolate (Israel *et al.*, 2001b). Previous analysis by randomly amplified polymorphic DNA PCR and sequence analysis had indicated that these new isolates were closely related to the original J99 strain.

This indicates that, within an apparently homogeneous population, remarkable genetic differences exist among single-colony *H. pylori* isolates, which is consistent with a model of continuous microevolution of *H. pylori* within its cognate human host (see Chapter 16).

DNA microarrays to investigate the pathological outcomes of infection

Whole-genome DNA microarray analysis was applied to the study of two *H. pylori* isolates that induce distinct pathological outcomes *in vivo*: gastric ulcer strain B128, and duodenal ulcer strain G1.1 (Israel *et al.*, 2001a). The DNA from these strains was hybridized to a whole-genome *H. pylori* microarray to identify putative virulence determinants responsible for the different clinical outcomes. This study showed that the regulation of epithelial cell responses related to inflammation depends on the presence of an intact 40 kb *cag* pathogenicity island (see Chapter 16). Thus the microarray results suggest that the less pro-inflammatory strain had undergone a large deletion of the *cag* pathogenicity island, providing a genetic explanation for the different disease outcomes.

One of the conclusions from the microarray analysis of *S. aureus* strains was that the toxic shock syndrome (TSS) epidemic of the 1970s and 1980s was caused by host factors (the use of a new super absorbent tampon) rather than the rapid global dissemination of a hyperinvasive strain (Fitzgerald *et al.*, 2001). Of the *S. aureus* strains isolated from women with urogenital-associated TSS, about 90% have a distinct multilocus enzyme genotype, designated ET 234 (Musser *et al.*, 1990). Although the DNA microarray data confirmed that ET 234 strains are genetically related and have shared a common ancestor, it also revealed that these strains are not genetically identical and that the last ancestor has not been very recent in evolutionary time.

The study of genetic variability among 19 clinically and epidemiologically well-characterized isolates of *M. tuberculosis* has also provided insight into their pathogenesis (Kato-Maeda *et al.*, 2001). The Affymetrix *M. tuberculosis* strain H37Rv GeneChip® was used to detect small-scale genomic deletions in these isolates and compared to the documented morbidity of the patient from whom the strain was isolated. As the number of deletions increased, the likelihood that the bacteria would cause pulmonary cavitation decreased, suggesting that the accumulation of mutations tends to diminish the pathogenicity of *M. tuberculosis*.

DNA microarrays to investigate genomes from related species

Microarray studies have also been conducted where the bacterium of interest has not been sequenced, but has a close relative with a fully sequenced genome. For example, *Wigglesworthia glossinidia*, a member of the enterobacteriaceae, is an obligate endosymbiont of the tsetse fly, which relies on the bacterium for fertility and nutrition. Symbiotic associations with microorganisms are pivotal in many insects, but the functional role of obligate symbionts can be difficult to study, owing to the problem of growing these organisms *in vitro*. The *W. glossinidia* genome is less than 770 kb, about one-sixth that of the related free-living bacteria *E. coli* (4.6 Mb). In order to gain an insight into the composition of the genome, *W. glossinidia* genomic DNA was hybridized to an *E. coli* DNA microarray, revealing 650 orthologous genes, corresponding to approximately 85% of the genome (Akman and Aksoy, 2001). Many of the genes retained in the *W. glossinidia* genome are involved in cell processes, DNA replication, transcription and translation. However, genes encoding transport proteins, chaperones, biosynthesis of co-factors, and some amino acids, were also identified in significant numbers, suggesting an important role for these proteins in the bacterium's symbiotic lifestyle. This is a good example of how a bacterial microarray can be used to obtain broad genome information for a closely related organism in the absence of complete genome sequence data.

Similarly, the diazotrophic maize endophyte *Klebsiella pneumoniae* 342 is closely related to *E. coli* K-12. Genomic DNA from *K. pneumoniae* 342 was hybridized to a microarray containing 96% of the annotated *E. coli* K-12 ORFs (Dong *et al.*, 2001). A set of 83 genes that were known to be common between *E. coli* and *K. pneumoniae* were used to divide the fluorescence ratios into categories of genes with: (1) >75% identity; (2) >55% and <75% identity; and (3) <55% identity between the two organisms at the nucleotide level. Using this criterion of 55% identity or greater, 70% of the *E. coli* K-12 genes were found in the *K. pneumoniae* 342 genome. This analysis provided information on genes with: (1) high identity between the two organisms; (2) functions that were less highly conserved (such as carbon compound metabolism); and (3) *E. coli* K-12 ORFs with little or no identity in *K. pneumoniae* 342. This application of microarrays enables the identification of many genes in previously uncharacterized bacterial species without the requirement for genome sequencing. However, genes unique to the test bacterium (*K. pneumoniae* 342 in this case) are not identified by this methodology.

Microorganisms play an integral and unique role in ecosystem function and sustainability. Understanding the structure and composition of microbial communities, and their responses and adaptations to environmental changes, is of great importance. Characterization and detection of microbial populations in natural environments is a complex task, particularly because of the number of different organisms involved. In contrast to studies using pure cultures, microarray-based analysis of environmental DNA isolated from mixed microbial populations presents a number of technical challenges (Wu *et al.*, 2001). A prototype microarray was constructed for understanding the specificity, sensitivity and quantification of microarray hybridization within the context of complex environmental microbial samples. The microarray contained approximately 100 functional genes encoding nitrite reductases, ammonia mono-oxygenases and methane mono-oxygenases, which are key enzymes in the ecosystem processes of denitrification, nitrification and methane oxidation, respectively. This microarray was used to assess functional gene diversity and distribution among different environmental isolates (Wu *et al.*, 2001). A linear relationship was found between signal intensity and DNA concentration for pure and mixed cultures. However, the specificity of the probe–target hybridization was found to depend on the sequence divergence in target genes. This affects the quantitative capacity of the microarray for measuring the relative abundance of targeted genes in complex environmental samples. Genes possessing less than 80–85% sequence identity were differentiated under hybridization conditions of high stringency. Altering the stringency of the reaction resulted in variation in hybridization, thus enabling detection of highly divergent and specific microbial populations. Thus it would appear that microarray hybridization has potential for revealing functional gene composition in natural microbial communities, but that further experimental optimization work is required.

◆◆◆◆◆◆ LIMITATIONS OF DNA MICROARRAYS

The major limitation with the application of DNA microarray-based approaches to whole genome studies is that the data obtained are restricted to the genes present in the reference strain(s) that are included on the microarray. Horizontally transferred DNA is largely responsible for the dissemination of virulence traits amongst bacteria. Rapid identification of acquired DNA in outbreak strains remains difficult as currently whole-genome sequencing of microbes is expensive and impractical. Although microarray-based approaches can identify genes absent or highly divergent in an interesting but unsequenced test strain, no information can be obtained on genes that are present in the test strain, but absent in the reference strain on the array. Therefore, the use of other whole-genome-based techniques is important for increasing our knowledge of novel strain-specific genes within a bacterial species.

Subtractive hybridization allows strain-specific DNAs to be selected directly and is an attractive methodology as it is unnecessary to score any particular phenotype or to do extensive mapping or sequencing at the outset (Straus and Ausubel, 1990) (see Chapter 6). An efficient and sensitive method for identifying bacterial strain-specific DNAs has been developed based on suppression subtractive hybridization, a method used to study gene expression in eukaryotes (Akopyants *et al.*, 1998). Pools of genomic DNA fragments from the strain of interest are depleted, by hybridization and PCR, of sequences that are also in a reference strain. The remaining DNA fragments highly enriched for sequences unique to the strain of interest can then be cloned for further analysis. Using this methodology it was possible to identify 18 genes specific to a monkey-colonizing *H. pylori* strain compared to the unrelated sequenced *H. pylori* strain 26695 (Akopyants *et al.*, 1998). Suppression subtractive hybridization was also used to identify genomic differences between the uropathogenic *E. coli* strain 536 and the non-pathogenic *E. coli* K-12 strain MG1655 (Janke *et al.*, 2001). In total, 22 DNA fragments were isolated which were specific for strain 536. Five of these fragments showed homology to known virulence determinants and four fragments matched genes for lipopolysaccharide or capsule biosynthesis and a siderophore receptor. Whereas two fragments were highly specific for uropathogenic *E. coli*, the other fragments could also be detected among all the other tested wild-type strains.

Another more recently developed technique is bacterial comparative genomic hybridization, which couples two-dimensional DNA electrophoresis, allowing separation of complex mixtures of DNA fragments based on size and sequence composition, and comparative genomic hybridization, which reveals regions of gain (or loss) of DNA sequences through differential hybridization of probes generated from test and reference genomic DNA. This technique has been used to investigate isogenic variants of the pathogen *Pseudomonas aeruginosa*, by detecting a single-copy gene insertion responsible for gentamicin resistance (Malloff et al., 2001).

◆◆◆◆◆◆ FUTURE PERSPECTIVES

Microarray analysis is a powerful enabling technology that allows global comparative analysis of the gene content between different strains in a given species. Comparative genomics using microarrays from a range of bacterial pathogens such as *H. pylori*, *C. jejuni*, *E. coli* and *S. aureus*, clearly demonstrates the diversity and adaptability of these specialized groups of organisms. Studies have revealed much evidence of lateral gene transfer and recombination. This supports an evolutionary scenario involving vertical diversification by mutagenesis, punctuated by frequent lateral gene transfer, resulting in a global mosaic genome structure. By contrast, *M. tuberculosis* and *V. cholerae* appear to exhibit very limited genome diversity.

All microarray analyses are limited by the genetic information on the array. However, as more genomes are sequenced and the capacity of arrays is increased further, information on arrays can be used to interrogate a given bacterial genome or mixture of genomes. Because of the range of G + C content in bacterial genes and genomes, hybridization conditions have to be carefully considered when using mixed bacterial species for microarray analysis. Another stumbling block to reaping the benefits of DNA microarrays is insufficient bioinformatic tools to analyse data. In the future, sustained improvements in software, computing speed and information storage will dramatically increase the scale of problems we tackle to understand the basic biology and evolution of the microbes. The development of a database of nucleotide differences among strains should allow the design of a universal microbial pathogen microarray, which would have wide applications in studying the epidemiology, population genetics, molecular phylogeny and evolution of microbial pathogens, as well as diagnostic applications. A 'lateral gene transfer' microarray consisting of genes from mobile elements, such as pathogenicity islands, phage and plasmid sequences, may have multiple applications. For example, it could be used in active microbial surveillance as an early warning system to alert public health officials to the presence or potential emergence of a more virulent pathogen. To date, most applications of DNA microarrays have been directed towards bacterial pathogens. No doubt the knowledge garnered from these studies will be applied to well-designed intervention strategies to reduce the burden of infectious disease. These types of microarray studies will be the forerunner for microbial pathogens with larger genomes and for symbiont and commensal microorganisms. Currently, we have only just began to scratch the surface in terms of the potential applications of DNA microarrays. The next few years will be a voyage of discovery in terms of developing our understanding of the microbial world.

Acknowledgements

We acknowledge financial support for our research from the BBSRC and MRC. We thank the Sanger Centre for the provision of clones to generate a *C. jejuni* microarray. We thank Jason Hinds and Stewart Hinchliffe for critical review of the manuscript.

References

Akman, L. and Aksoy, S. (2001). A novel application of gene arrays: *Escherichia coli* array provides insight into the biology of the obligate endosymbiont of tsetse flies. *Proc. Natl Acad. Sci. USA* **98**, 7546–7551.

Akopyants, N. S., Fradkov, A., Diatchenko, L., Hill, J. E., Siebert, P. D., Lukyanov, S. A., Sverdlov, E. D. and Berg, D. E. (1998). PCR-based subtractive hybridization and differences in gene content among strains of *Helicobacter pylori*. *Proc. Natl Acad. Sci. USA* **95**, 13108–13113.

Behr, M. A., Wilson, M. A., Gill, W. P., Salamon, H., Schoolnik, G. K., Rane, S. and Small, P. M. (1999). Comparative genomics of BCG vaccines by whole-genome DNA microarray. *Science* **284**, 1520–1523.

Brosch, R., Pym, A. S., Gordon, S. V. and Cole, S. T. (2001). The evolution of mycobacterial pathogenicity: clues from comparative genomics. *Trends Microbiol.* **9**, 452–458.

Dobrindt, U. and Hacker, J. (2001). Whole genome plasticity in pathogenic bacteria. *Curr. Opin. Microbiol.* **4**, 550–557.

Dong, Y., Glasner, J. D., Blattner, F. R. and Triplett, E. W. (2001). Genomic interspecies microarray hybridization: rapid discovery of three thousand genes in the maize endophyte, *Klebsiella pneumoniae* 342, by microarray hybridization with *Escherichia coli* K-12 open reading frames. *Appl. Environ. Microbiol.* **67**, 1911–1921.

Dorrell, N., Mangan, J. A., Laing, K. G., Hinds, J., Linton, D., Al-Ghusein, H., Barrell, B. G., Parkhill, J., Stoker, N. G., Karlyshev, A. V., Butcher, P. D. and Wren, B. W. (2001). Whole genome comparison of *Campylobacter jejuni* human isolates using a low-cost microarray reveals extensive genetic diversity. *Genome Res.* **11**, 1706–1715.

Dziejman, M., Balon, E., Boyd, D., Fraser, C. M., Heidelberg, J. F. and Mekalanos, J. J. (2002). Comparative genomic analysis of *Vibrio cholerae*: genes that correlate with cholera endemic and pandemic disease. *Proc. Natl Acad. Sci. USA* **99**, 1556–1561.

Edman, C. F., Raymond, D. E., Wu, D. J., Tu, E., Sosnowski, R. G., Butler, W. F., Nerenberg, M. and Heller, M. J. (1997). Electric field directed nucleic acid hybridization on microchips. *Nucleic Acids Res.* **25**, 4907–4914.

Favis, R., Day, J. P., Gerry, N. P., Phelan, C., Narod, S. and Barany, F. (2000). Universal DNA array detection of small insertions and deletions in BRCA1 and BRCA2. *Nature Biotechnol.* **18**, 561–564.

Fitzgerald, J. R., Sturdevant, D. E., Mackie, S. M., Gill, S. R. and Musser, J. M. (2001). Evolutionary genomics of *Staphylococcus aureus*: insights into the origin of methicillin-resistant strains and the toxic shock syndrome epidemic. *Proc. Natl Acad. Sci. USA* **98**, 8821–8826.

Hacia, J. G. (1999). Resequencing and mutational analysis using oligonucleotide microarrays. *Nature Genet.* **21**, 42–47.

Israel, D. A., Salama, N., Arnold, C. N., Moss, S. F., Ando, T., Wirth, H. P., Tham, K. T., Camorlinga, M., Biaser, M. J., Falkow, S. and Peek, R. M., Jr (2001a). *Helicobacter pylori* strain-specific differences in genetic content, identified by microarray, influence host inflammatory responses. *J. Clin. Invest.* **107**, 611–620.

Israel, D. A., Salama, N., Krishna, U., Rieger, U. M., Atherton, J. C., Falkow, S. and Peek, R. M., Jr (2001b). *Helicobacter pylori* genetic diversity within the gastric niche of a single human host. *Proc. Natl Acad. Sci. USA* **98**, 14625–14630.

Janke, B., Dobrindt, U., Hacker, J. and Blum-Oehler, G. (2001). A subtractive hybridisation analysis of genomic differences between the uropathogenic *E. coli* strain 536 and the *E. coli* K-12 strain MG1655. *FEMS Microbiol. Lett.* **199**, 61–66.

Karlyshev, A. V., Linton, D., Gregson, N. A., Lastovica, A. J. and Wren, B. W. (2000). Genetic and biochemical evidence of a *Campylobacter jejuni* capsular polysaccharide that accounts for Penner serotype specificity. *Mol. Microbiol.* **35**, 529–541.

Kato-Maeda, M., Rhee, J. T., Gingeras, T. R., Salamon, H., Drenkow, J., Smittipat, N. and Small, P. M. (2001). Comparing genomes within the species *Mycobacterium tuberculosis*. *Genome Res.* **11**, 547–554.

Malloff, C. A., Fernandez, R. C. and Lam, W. L. (2001). Bacterial comparative genomic hybridization: a method for directly identifying lateral gene transfer. *J. Mol. Biol.* **312**, 1–5.

Musser, J. M., Schlievert, P. M., Chow, A. W., Ewan, P., Kreiswirth, B. N., Rosdahl, V. T., Naidu, A. S., Witte, W. and Selander, R. K. (1990). A single clone of *Staphylococcus aureus* causes the majority of cases of toxic shock syndrome. *Proc. Natl Acad. Sci. USA* **87**, 225–229.

Ochman, H. and Jones, I. B. (2000). Evolutionary dynamics of full genome content in *Escherichia coli*. *EMBO J.* **19**, 6637–6643.

Orskov, F. and Orskov, I. (1983). From the national institutes of health. Summary of a workshop on the clone concept in the epidemiology, taxonomy, and evolution of the enterobacteriaceae and other bacteria. *J. Infect. Dis.* **148**, 346–357.

Pease, A. C., Solas, D., Sullivan, E. J., Cronin, M. T., Holmes, C. P. and Fodor, S. P. (1994). Light-generated oligonucleotide arrays for rapid DNA sequence analysis. *Proc. Natl Acad. Sci. USA* **91**, 5022–5026.

Perna, N. T., Plunkett, G., 3rd, Burland, V., Mau, B., Glasner, J. D., Rose, D. J., Mayhew, G. F., Evans, P. S., Gregor, J., Kirkpatrick, H. A., Posfai, G., Hackett, J., Klink, S., Boutin, A., Shao, Y., Miller, L., Grotbeck, E. J., Davis, N. W., Lim, A., Dimalanta, E. T., Potamousis, K. D., Apodaca, J., Anantharaman, T. S., Lin, J., Yen, G., Schwartz, D. C., Welch, R. A. and Blattner, F. R. (2001). Genome sequence of enterohaemorrhagic *Escherichia coli* O157:H7. *Nature* **409**, 529–533.

Salama, N., Guillemin, K., McDaniel, T. K., Sherlock, G., Tompkins, L. and Falkow, S. (2000). A whole-genome microarray reveals genetic diversity among *Helicobacter pylori* strains. *Proc. Natl Acad. Sci. USA* **97**, 14668–14673.

Sosnowski, R. G., Tu, E., Butler, W. F., O'Connell, J. P. and Heller, M. J. (1997). Rapid determination of single base mismatch mutations in DNA hybrids by direct electric field control. *Proc. Natl Acad. Sci. USA* **94**, 1119–1123.

Straus, D. and Ausubel, F. M. (1990). Genomic subtraction for cloning DNA corresponding to deletion mutations. *Proc. Natl Acad. Sci. USA* **87**, 1889–1893.

Westin, L., Miller, C., Vollmer, D., Canter, D., Radtkey, R., Nerenberg, M. and O'Connell, J. P. (2001). Antimicrobial resistance and bacterial identification utilizing a microelectronic chip array. *J. Clin. Microbiol.* **39**, 1097–1104.

Winzeler, E. A., Richards, D. R., Conway, A. R., Goldstein, A. L., Kalman, S., McCullough, M. J., McCusker, J. H., Stevens, D. A., Wodicka, L., Lockhart, D. J. and Davis, R. W. (1998). Direct allelic variation scanning of the yeast genome. *Science* **281**, 1194–1197.

Wu, L., Thompson, D. K., Li, G., Hurt, R. A., Tiedje, J. M. and Zhou, J. (2001). Development and evaluation of functional gene arrays for detection of selected genes in the environment. *Appl. Environ. Microbiol.* **67**, 5780–5790.

8 Gene Expression during Host–Pathogen Interactions: Approaches to Bacterial mRNA Extraction and Labelling for Microarray Analysis

Joseph A Mangan, Irene M Monahan and Philip D Butcher

Department of Medical Microbiology, St George's Hospital Medical School, Cranmer Terrace, London SW17 0RE, UK

CONTENTS

Introduction
Bacterial RNA extraction methods: applicability to microarrays
Strategies for sample RNA labelling and detection
Amplification of mRNA
Conclusions
Appendix

◆◆◆◆◆◆ INTRODUCTION

The expression of genes in response to signals associated with environmental stimuli is a prerequisite for the survival of bacterial pathogens within the host and is, therefore, the underlying basis of infectious disease. With the increasing amount of information about gene complement of bacterial pathogens and the development of functional genomics technologies, such as microarrays, it is now possible to monitor bacterial gene expression at a whole genome level both *in vitro* and during host–pathogen interactions. Transcriptome analysis using whole-genome microarrays has the potential to exploit the simultaneous measurement of expression of every bacterial gene under specific conditions. However, the whole-genome biology of bacterial pathogenesis necessitates the investigation of gene expression patterns at multiple time points from complex models of infection (such as cell lines or animal models), as well as *in vitro* conditions designed to mimic *in vivo* environments.

Until very recently, the transcriptional analysis of bacterial genes had been carried out by such methods as reverse transcriptase polymerase chain reaction (RT-PCR) and Northern blotting, which involved examining the expression of individual genes. In theory, microarray analysis replaces these as the method of choice for monitoring global gene expression, although verification of array data will certainly still require such gene-specific techniques. The majority of genomewide studies of intracellular or *in vivo* gene expression thus far analysed has used techniques such as IVET (*in vivo* expression technology) (Mahan *et al.*, 1993), subtractive hybridization (Plum and Clark-Curtiss, 1994; Li *et al.*, 2001) and differential fluorescence induction (DFI) (Valdivia and Falkow, 1997). Although these procedures have led to the identification of putative virulence factors, they are time-consuming, technically challenging, difficult to apply to many genetically intractable organisms, such as the slow-growing pathogenic mycobacteria, and only partially probe the genome. Microarrays will provide substantial advantages over these techniques, in terms of the capacity to identify essentially all differentially expressed genes by direct detection of mRNA in a simple hybridization format.

Paramount to the examination of virulence-associated bacterial gene expression, is the recovery of sufficient organisms from infection models for microarray analysis. Previously, this had proved to be extremely difficult, and much research has, therefore, focused on *in vitro* culture conditions that may mimic aspects of the intracellular environment. This may include bacterial cultures exposed to a variety of reactive oxygen metabolites, or growth in low iron or low oxygen. Selection of appropriate models of infection is important both in terms of numbers of recoverable bacteria and to reflect the true complexities of host–pathogen interactions during natural infection. Therefore, *in vitro* models may have technical advantages related to RNA extraction, but also many biological limitations.

Such considerations impact on the choice of an experimental approach to extracting bacterial RNA for microarray analysis from different models of infection, e.g. axenic culture, cell monocultures infected with bacteria *in vitro* or whole tissues from infected animal models. This chapter presents experiences and methods that address both the biological and technical issues involved in extracting biologically 'meaningful' RNA from the pathogenic mycobacterium, *Mycobacterium tuberculosis*, for the purposes of studying whole-genome expression during host–pathogen interactions using microarray analysis.

◆◆◆◆◆◆ BACTERIAL RNA EXTRACTION METHODS: APPLICABILITY TO MICROARRAYS

Consideration of mRNA stability and transcription-induced artefacts

Crucial to the successful application of microarray technology is the extraction of adequate quantities of intact bacterial RNA. This is directly dependent on the number of bacteria that can be recovered from any particular model. For spotted PCR-product glass slide microarrays, at least 20 µg of total bacterial RNA is normally required per hybridization to give good signal intensities for array analysis. Considering that the distribution in abundance of mRNA species varies from low to high, enough mRNA must be bound to the probe element on the array to give a signal sufficiently above background for the detection of low abundance mRNA species. This signal is variably taken as > 3 standard deviations of background value, or twofold above local background for a particular spot. With recent multiple refinements in array hybridization techniques and more efficient sample labelling methods, the amount of total RNA required may well fall to as little as 1 µg, depending partly on the genome complexity of the organism (i.e. number of genes and hence the absolute amount of any one mRNA species in a population). For *M. tuberculosis*, we use a minimum of 5 µg of total RNA per array, with 8–10 µg being routinely preferable. This gives good signal intensity after indirect incorporation of fluorescent dye-nucleotide analogues (Cy-3 or Cy-5 dCTP) by random primed cDNA synthesis (see later). The yield of total RNA from different bacteria also varies, depending mainly on efficiency of bacterial lysis and subsequent RNA recovery. For *M. tuberculosis*, we extract 10–20 µg of total RNA from 10^9 bacilli and this equates well for a range of different extraction methods for mycobacteria (Butcher *et al.*,1998).

Many prokaryotic RNA extraction methods have been described, reflecting the broad chemical diversity of both the Gram-positive and Gram-negative bacterial cell wall. All methods for mycobacteria, for example, rely on some form of cell disruption followed by various RNA purification stages (Butcher *et al.*, 1998). There are some special considerations to bear in mind when choosing an RNA isolation technique, which are peculiar to bacteria and which may ultimately have a profound effect on the quality of microarray data generated. In this context it may be said that microarray data quality (i.e. biological meaning) is only as good as the structural integrity and biological quality of the RNA used in the experiment. Thus, if RNA is extracted from bacteria that have first been centrifuged to recover them from culture medium (or separated from infected cells by various procedures), then the mRNA population will reflect gene expression induced by the preparative procedure rather than the experimental condition under test. In such cases, although the RNA will be structurally intact and array data will still be generated, the resultant gene expression profile will not be biologically meaningful. The reason for this is that bacteria are capable of rapid transcriptional responses to changes in their environment. The need to prevent transcriptional changes associated with preparative procedures of bacterial RNA is, therefore, obvious but often difficult to implement in the experimental design for complex biological models, and is thus prone to compromise.

Thus, the two critical parameters in bacterial RNA extraction are the extreme instability of most bacterial mRNAs with half-lives of < 2 min (Lewin, 1990), and the potential for rapid transcriptional change. As such, methods of bacterial RNA extraction must involve near instantaneous cell wall permeation under denaturing conditions, with subsequent RNA stabilization. Many prokaryotic RNA extraction protocols have shown little regard for the potential transcriptional changes, and often involved lengthy centrifugations, followed by cell washing and enzymatic digestions (e.g. Bashyam and Tyagi, 1994; Kikuta-Oshima *et al.*, 1995). Bacterial isolation protocols tend to have been based on techniques adapted from eukaryotic tissue RNA isolation protocols (Chomczynski and Sacchi, 1987) and were often developed for either structural studies of the RNAs or for the monitoring of relatively small numbers of genes by Northern or slot blotting. With the advent of more global gene expression screens, such as microarrays, methods that accurately reflect the condition under study are required. Microarray techniques, because they globally interrogate the whole genome, are exquisitely sensitive to any perturbations in the extraction procedure, which, upon array analysis, could easily be interpreted erroneously as changes in gene expression.

Methods for rapid mRNA stabilization

Bacterial RNA isolation techniques show a growing trend toward rapid lysis of bacterial cells and stabilization of the mRNA (Yim and Rubens, 1997; Smoot *et al.*, 2001). This is often achieved through the rapid chilling of the bacterial culture under examination, using either crushed ice or freezing solvents to effect RNA stabilization (Luo and Stevens, 1997; Yim and Rubens, 1997). This is then followed by harsh mechanical lysis using glass or ceramic beads and a reciprocating shaking device or bead-beater, in phenol and/or guanidine containing chaotropic solutions (Mangan *et al.*, 1997; Wei *et al.*, 2001). We have previously argued against the use of chilling on ice and centrifugation at 4°C for microarray-based gene expression analysis on the grounds that both cold shock gene induction and a more generalized global stress response may be seen.

From our own and other studies, we know that bacteria can respond to environmental stimuli within seconds, and, accordingly, we have developed RNA extraction procedures that obviate the need for rapid chilling and lysis by providing an immediate chemical stabilization of the RNA (for detailed method, see Appendix). This method has been developed for mycobacterial species and relies on the addition of a high concentration (final concentration of 4 M) of guanidine isothiocyanate (GTC) solution to a liquid bacterial culture. The GTC immediately penetrates the mycobacterial cell wall and enters into the cytoplasm, where it quenches the intracellular ribonucleases and stabilizes the RNA, but does not cause mycobacterial cell lysis (Monahan *et al.*, 2001). The mycobacteria may then be harvested by centrifugation from the GTC. Pelleted and GTC-stabilized bacteria are resuspended in phenol/GTC-based lysis buffers and the bacilli disintegrated by 100 μm ceramic bead beating to release the nucleic acids. The RNA is then recovered by precipitation.

There are two main advantages in using this particular method of RNA extraction. Firstly, GTC rapidly penetrates the bacterial cells, thereby immediately stabilizing the short half-life mRNA. Secondly, the GTC prevents further transcription from taking place, thus controlling for preparative procedures that could potentially induce artefactual changes in non-protected bacteria. This means, therefore, that the bacterial gene expression profile generated is likely to be a true representation of the experimental environment under examination and not simply a result of processing procedures involved in the extraction and recovery of RNA. We have validated this technique both at the single gene level by Northern blotting and using whole-genome *M. tuberculosis* microarrays. Figure 8.1 shows the results of a slot blot analysis with equal loadings of total RNA extracted from *M. bovis* BCG, probed for the *dnaK* (hsp70) mRNA, to show immediate stabilization of mRNA (Figure 8.1A) and blocking of heat-inducible transcription (Figure 8.1B) with GTC. Heat shock of *M. bovis* BCG induces large changes in *dnaK* mRNA (> 30-fold; see control lanes −/+ heat shock). On addition of GTC to heat-shocked bacteria, *dnaK* mRNA levels remained high for 30 min (Figure 8.1A, duplicate samples) and remained stable for days. The half-life for *dnaK* mRNA as measured by blocking transcription with 10 µg/ml rifampicin after heat shock was < 10 min (not shown). Heat-shock induction of *dnaK* mRNA was abolished after addition of GTC to *M. bovis* BCG within 1 min (Figure 8.1B). The results indicate that mycobacterial mRNA is immediately stabilized and

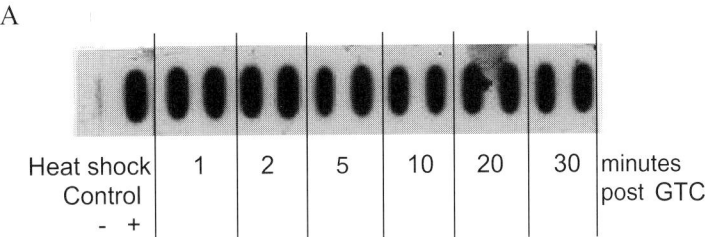

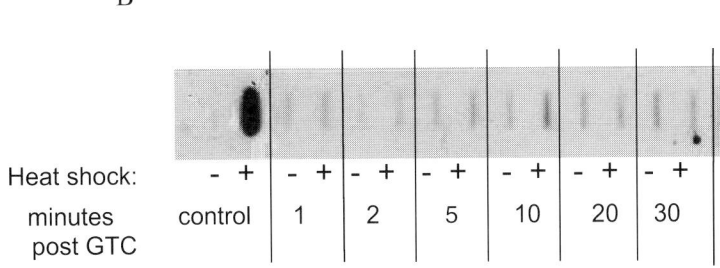

Figure 8.1. Slot blot analysis of *dnaK* mRNA from *M. bovis* BCG. Total RNA was extracted from BCG pre- and post-heat shock at 45°C for 45 min using the guanidine isothiocyanate (GTC) method described in the text. One microgram of total RNA was used per slot and hybridized with a ^{32}P-labelled gene-specific probe. Heat shock induces >30-fold upregulation of *dnaK* mRNA: see −/+ heat shock controls. (A) BCG was first heat-shocked to induce measurable levels of *dnaK* mRNA and then exposed to GTC solution after 1 min and up to 30 min, and slot-blotted in duplicates. Heat-shock levels of *dnaK* mRNA remain stable after GTC addition. (B) BCG −/+ heat shock before (control) and 1–30 min after addition of GTC, showing blocking of heat-shock response within 1 min.

that subsequent transcription is blocked within seconds of addition of the GTC reagent.

A limitation of this method, however, is that it only works on mycobacteria and the more robust Gram-positive bacteria (and maybe other organisms with resistant cell walls including yeast and fungi) that will not lyse under the highly denaturing conditions. A major advantage in terms of studying host–pathogen interactions is that, since mycobacteria do not lyse in GTC, mRNA-stabilized bacteria can be recovered from infected tissues or cells that readily lyse under such conditions (e.g. cultured macrophages or endothelial cells). The method may also be used directly on large volumes of liquid culture media, the chemically stabilized bacilli being recovered by centrifugation whilst maintaining the integrity of the mRNA. However, Gram-negative bacteria lyse in this GTC solution, so although the method will yield good-quality and representational RNA populations, it cannot be used as a means of separating intact and stabilized Gram-negative bacteria from lysed tissue. Attempts to recover such organisms from infected tissues using the GTC method would result in the RNA preparation being a mixture of bacterial and host RNA, with the host RNA in large excess of bacterial RNA. This will create problems of sensitivity when using random-primed cDNA labelling, owing to the limiting amounts and specific activity of the bacterial RNA portion of the total RNA preparations. Approaches to overcome these limitations are explained below. The GTC method can be used for axenic liquid cultures of Gram-negative bacteria without problems.

A commercial solution (RNA*later*™; see Ambion, 2002) has been developed to chemically protect mRNA from degradation in tissues, without tissue lysis, for subsequent mRNA isolation; it can also be used on bacteria. This reagent is not chemically related to GTC, although the manufacturers have not publicly disclosed the constituents. Together, GTC and RNA*later*™ form the basis of a methodological approach to bacterial RNA extraction from infected tissues or cells that may permit *in vivo* gene expression analysis of bacteria during infection. We have successfully used the GTC approach on *M. tuberculosis* recovered from macrophages (Butcher *et al.*, 1998; Li *et al.*, 2001; Monahan *et al.*, 2001). Figure 8.2 shows an ethidium bromide stained non-denaturing agarose gel of total RNA preparations from *M. bovis* BCG grown axenically (lane 3), compared to *M. bovis* BCG recovered from THP-1 monocyte cell lines 24 h after phagocytosis using the GTC selective lysis approach (lane 2) and total RNA from THP-1 cells (lane 1). This result demonstrates that selective lysis and enrichment of GTC-stabilized bacteria, yield almost pure bacterial RNA, as shown by the presence of ribosomal 16S and 23S RNA and 5S RNA. Such purified bacterial RNA preparations recovered from a large excess of host cells can thus readily be used for microarray analysis by random-primed reverse transcription to study intracellular mycobacterial gene expression profiles.

Although many RNA extraction methods yield array-compatible RNA, we have noticed that some detergent-based lysis procedures require extensive postpurification of RNA prior to hybridization, otherwise unacceptably high background signals may be encountered upon scanning.

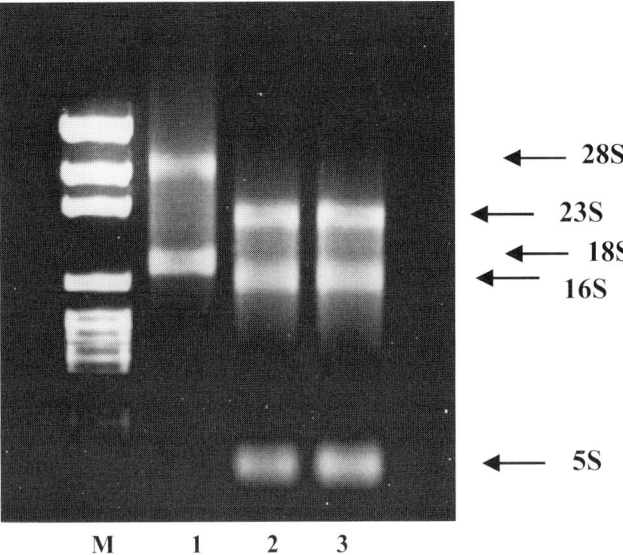

Figure 8.2. Selective lysis and enrichment of *M. bovis* BCG RNA from infected macrophages. Ethidium bromide stained non-denaturing agarose gel of 5 μg of total RNA from: Lane 1, THP-1 human monocyte cells; Lane 2, *M. bovis* BCG recovered from THP-1 monocyte cell lines 24 h after phagocytosis using the GTC selective lysis approach; Lane 3, *M. bovis* BCG grown axenically. Lane M: molecular weight indicators. Eukaryotic ribosomal 18S and 28S RNA and prokaryotic 5S, 16S and 23S RNA are indicated with arrows.

This was noticeable with our previously described detergent method (Mangan *et al.*, 1997), where carryover of some component of the lysis solution affected the arrays. Simple purification of the RNA preparation through a RNA clean-up column (e.g. Qiagen RNeasy), produced RNA compatible with microarrays.

Separation of mRNA from ribosomal RNA

Several novel methods have been reported that appear to overcome one of the apparently intractable hurdles associated with prokaryotic mRNA isolation, namely the absence of a robust procedure for the separation of mRNA from total RNA. Removal of rRNA (>95% of total RNA) from mRNA might be considered advantageous, since it would allow higher specific activity mRNA labelling with greater sensitivity on the array. Bacterial mRNA lacks a poly-A tail and is not, therefore, amenable to purification by oligo-(dT) affinity chromatography, although there are reports that some mRNA species may be oligo-adenylated. Purification on this basis would necessarily involve some selection with resultant changes in representation of the mRNA population.

One method for bacterial mRNA purification relies on the use of biotinylated antisense oligonucleotides targeted against conserved regions of the bacterial 16S and 23S ribosomal operon. These oligonucleotides are mixed with a total RNA preparation and allowed to hybridize. The oligonucleotides bind to their complementary sequences on the ribosomal RNAs and the resultant hybrids are captured with streptavidin-coated paramagnetic particles. Using a magnetic particle separator, the hybrids are captured leaving the enriched mRNA in solution. Ribosomally depleted RNA is then recovered by ethanol precipitation (MICROB*express*, Bacterial mRNA Purification Kit, Ambion™, Inc., Austin, TX, USA). The manufacturers claim that up to 98% of the ribosomal RNA can be removed using this system, and as the conserved ribosomal regions are targeted by the capture oligonucleotides, it is broadly applicable to a wide range of eubacterial species. Such rRNA-depleted mRNA preparations are thus minimally biased and can be utilized in the construction of bacterial cDNA libraries (without the >98% ribosomal DNA clones) and in producing higher specific activity probes for use in microarray hybridizations. Our own experiences, using biotinylated 23S-16S-5S ribosomal operon anti-sense runoff transcripts as hybridization capture probes to remove rRNA and enrich for mRNA from *M. bovis* BCG RNA preparations, reproducibly achieved no greater than 73% of rRNA depletion, measured using ^{32}P-labelled sense rRNA transcript spikes. Although these methods provide enrichment for mRNA, the amount of total RNA required often exceeds that available from biological models of host–pathogen interactions.

◆◆◆◆◆◆ STRATEGIES FOR SAMPLE RNA LABELLING AND DETECTION

Direct incorporation of Cy-dNTPs

The majority of publications using dual-labelled (Cy3 or Cy5 dCTP conjugate) samples are based around the direct incorporation of the relevant fluorescent nucleotide analogue into first-strand cDNA using random hexamers and reverse transcriptase, followed by removal of the unincorporated fluors prior to hybridization (Schena *et al.*, 1995). This has been the mainstay over the last 7 years. More recently, alternatives have been introduced that overcome some of the criticisms associated with cDNA synthesis, namely the bias in incorporation rates of the different dyes due to steric factors, and the high cost of the fluorescent nucleotide analogues.

Aminoallyl-dUTP coupling of RNA

One method to challenge the standard direct incorporation of Cy dye nucleotide analogues into cDNA, was the amino allyl-dUTP (AA-dUTP) method. The fundamental difference lies in the fact that this method incorporates the same modified nucleotide into each RNA population during a standard random hexamer primed reverse transcription reaction. The fact that the nucleotide being incorporated into both populations is identical eliminates the steric bias associated with different Cy-dNTP incorporation. Each RNA sample is individually chemically reacted with a monofunctional *N*-hydroxysuccinimidyl (NHS) ester of either Cy3 or Cy5 (NHS-Cy3, NHS-Cy5); the NHS ester reacts with the free amino groups on the modified nucleotide analogues, forming cDNA containing stable conjugates of both Cy3 and Cy5 (see Brown, 2001; laboratory protocols website). This method also has the added advantage of being considerably cheaper than the direct incorporation procedure due to the cost differential between fluorescent nucleotide analogues and NHS esters. Some evidence also suggests an increased sensitivity in the order of tenfold over direct Cy-dNTP incorporation methods (Schroeder *et al.*, 2002).

Antibody detection of RNA : DNA hybrids on microarrays

Another methodological approach to prevent bias between different cDNAs has been developed and involves a hybridization and detection technology, which utilizes a patented antibody specific for RNA : DNA hybrids coupled with fluorescently labelled secondary antibody reagents (HC ExpressArray™ Kit; Digene Corp., USA; http://www.DiGene.com/lifesciences_1.html). Total RNA or purified mRNA can be hybridized directly to the DNA microarray forming RNA : DNA hybrids. Only hybrids that are perfectly formed are detected, and the signal generated depends on the length of the hybrid formed. No representational biases are introduced, since there is minimal sample processing, no labelling and no enzymatic reactions. This method may be used for glass-slide spotted microarrays comprised of cDNA, PCR products, or oligonucleotides of 70 nucleotides in length or greater.

Genome-directed primers

A technique that circumvents the use of random hexamer primed cDNA synthesis is the use of genome-directed primers (GDP) (Talaat *et al.*, 2000). The rationale behind this method is that it is possible, using a bioinformatic approach, to create a minimum set of short specific oligonucleotides that theoretically will prime within the 3'-end of every gene in a given bacterial genome. The algorithm aligns all the predicted open reading frames (ORFs) in a sequenced genome and chooses a user specified proximity region close to the 3'-end of all the aligned ORFs, and then chooses the minimum number of seven base oligonucleotides that prime within that region in all genes. Because such a set will prime only minimally within the ribosomal genes, the relative specific activity of the cDNA produced during reverse transcription should be increased. This technique is of particular relevance in situations where host cell or tissue contamination of the RNA under study is likely to be problematic, such as *in vitro* cell infection and animal infection models. GDPs designed for a specific bacterial species will prime cDNA synthesis predominately from bacterial mRNAs, even in the presence of excess host RNA. A truly random nonomer would be expected to occur about 10 000 times in the human genome. So GDPs may randomly prime host RNA to some limited extent, but will drive an increased specific activity of bacterial RNA sufficient for hybridization analysis. Host-derived cDNAs are unlikely to hybridize on a bacterial gene-specific array, so signal detection will not be compromised. Such an approach will thus permit bacterial whole-genome analysis in models of infection using total RNA (bacteria plus tissue) where bacteria cannot easily be separated from host tissue without transcriptional changes occurring. This should prove to be a particularly useful approach for Gram-negative bacteria in host–pathogen models.

A second approach to genome-specific selective cDNA synthesis arises from the availability of sets of primer pairs used in the construction of a PCR product-based DNA microarray. To make an array consisting of each ORF in a genome, a pair of PCR primers is synthesized and used to generate an ORF-specific PCR product probe for spotting. If the 3'-primer from each of the ORF pairs were combined to create a 3'-primer pool, this pool would represent an ORF-specific set of primers capable of acting as genomewide primers for cDNA synthesis. This has the advantage that cDNA is driven from the 3'-end of each PCR product probe on the array, thereby limiting priming events to parts of the genome not represented by the probes on the array. In this way, better signal sensitivity and specificity may result. The efficiency of the process will be dependent on the mixed primer concentration and this will need to be experimentally determined for each genome set of 3'-primers. The ORF-specific 3'-primer pool may also, in theory, be used to drive cDNA synthesis selectively from bacterial mRNA targets in mixed RNA preparations from infected cell models of infection where host RNA may predominate. The ORF 3'-primers are 20–24 bases long and will, therefore, randomly prime at very low frequencies within the host genome, offering advantages over the GDP approach above.

◆◆◆◆◆◆ AMPLIFICATION OF mRNA

One of the current major challenges impeding the more widespread use of microarrays for bacterial expression profiling in models of infection is the requirement for large amounts of total RNA, often in the order of tens of micrograms. As such, the use of amplification methods that reduce the amount of starting material is urgently needed. This is particularly true when studying small numbers of bacteria interacting with eukaryotic cells or in infected tissues. All of the existing methods that aim to amplify mRNA to appropriate levels for microarray hybridizations are based on *in vitro* transcription reactions driven by T7 polymerase from a T7 promoter introduced as an oligo(dT)-T7 primer during reverse transcription of cDNA (Erberwine et al., 1992; Erberwine, 1996). Antisense RNA runoff transcripts (aRNA) represent a >1000-fold amplification of mRNA in a single round of transcription and reactions typically use 1 µg of total RNA and variably produce 10 µg of aRNA. The method produces representational amplification of small amounts of RNA for array analysis by maintaining a linear proportionality between template and aRNA product. This is unlike RT-PCR, which produces exponential amplification with well-described plateau effects. Further rounds of aRNA amplification are possible using aRNA as template for cDNA synthesis and a second round of transcript amplification. It is not known to what extent further rounds of amplification may alter the mRNA representation, but it has the advantage of reducing the starting amount of total RNA to <100 ng. Amplified aRNA may be labelled directly by reverse transcription with Cy-labelled dNTPs or amino allyl dUTP for microarray hybridization. Alternatively, biotin-rNTPs or amino allyl UTP may be incorporated during the amplification reaction.

Such methods have little utility in prokaryotic systems in which the mRNAs are only partially oligoadenylated and, as such, not amenable to oligo-dT based capture or oligo(dT)-T7 priming for cDNA synthesis (Van Gelder *et al.*, 1990). No amplification methods have as yet been described for use in bacterial systems. However, several theoretical approaches may be possible, combining selective genome priming with T7 transcript amplification:

1. ORF-specific 3'-primer pool with T7 tags. The major disadvantage of this approach would be the large expense of synthesizing a set of whole genome ORF specific 3'-primers, each with a 24-base T7 tag.
2. Genome-directed primers (Talaat *et al.*, 2000) with T7 tags. For *M. tuberculosis* this would involve approximately 40 7'mer oligonucleotides each with a T7 tag. This would be much more economical whilst maintaining bacterial genome specificity.
3. 3'-Degenerate oligonucleotides (N_{6-8}) biased towards %GC and codon usage for a particular genome, with a 5'-T7 tag. This will not retain the genome selectivity as would (1) or (2) above, but represents a random priming skewed towards bacterial mRNA.
4. Combinations of bacterial RNA selection and enrichment methods, together with mRNA separation from total RNA, followed by direct aRNA techniques using either random or genome-directed priming.

These approaches, although so far untested, are based on standard molecular biology technologies, and will no doubt require substantial validation and experimental refinement before they may be used as robust approaches to bacterial transcriptome analysis on microarrays.

◆◆◆◆◆◆ CONCLUSIONS

Much of the interest in bacterial whole-genome microarrays lies in the identification of gene expression profiles associated with a particular environmental condition or stage of infection. In this way, correlation between expression and function may be derived for the many genes of bacterial genomes with no known function. Although the potential for microarrays is great and will contribute to a non-reductionist view of the functioning of a whole genome, the function of individual genes will still require traditional reductionist molecular biology to elucidate its contribution to pathogenesis and disease. Furthermore, the transcriptome is only a partial output of the genome and it is important not to lose sight of proteomics as a powerful and complementary tool to microarrays in bacterial functional genomics. Proteomics allows for the analysis of differential gene expression at the translational (proteins) and post-translational (protein modification) level. As with any global approach to gene expression profiling, whole genome data can be readily generated. However, the quality and biological validity of that data is only as good as the input to the experiment, which in the case of microarrays is mRNA. If you put rubbish in you get rubbish out! The challenges of bacterial mRNA extraction in the context of studying host–pathogen interactions are often technically difficult. To overcome them requires: an understanding of the biology of bacterial mRNA; great care with and consideration of the methods used in the extraction of RNA; the use of appropriate pathogenesis models; development of representational amplification methods for small starting amounts of RNA; and robust statistical analysis of the microarray data so generated since 'One array does not a summer make'!

◆◆◆◆◆◆ APPENDIX

The protocol described below is a robust and reproducible method to extract total RNA from mycobacteria and, in theory at least, is applicable to other organisms with resistant cell walls, such as some Gram-positive bacteria as well as fungi and yeast. We have only tested this method for mycobacteria. We have also used this method extensively to recover mycobacteria selectively from infected macrophages (Butcher *et al.*, 1998; Li *et al.*, 2001; Monahan *et al.*, 2001) and animal tissues (Hu *et al.*, 2001), whilst stabilizing the labile mRNA and preventing transcriptional changes during harvesting. For these purposes the infected cells/tissues are lysed directly in 4 M GTC solution, the host DNA viscosity is reduced by vortexing and passing through a narrow gauge needle, prior to spinning out the unlysed but now GTC-stabilized mycobacteria. Recovered mycobacteria are then treated as from step 2 below.

Protocol for RNA extraction from *in vitro* grown broth cultures of *M. tuberculosis*

1. Add 4 vols of 5 M GTC lysis solution (5 M guanidinium isothiocyanate [Fluka; ultrapure grade], 0.5% sodium *N*-lauryl sarcosine, 25 mM trisodium citrate, pH 7.0, 0.1 M 2-mercaptoethanol, 0.5% Tween 80) to the mycobacteria in liquid culture, i.e. 400 ml of 5 M GTC to 100 ml of bacteria in broth culture, and mix rapidly by swirling. This produces a final concentration of 4 M GTC.
2. Transfer to 30 ml plastic V-bottom Universals (Bibby Sterilin Ltd, UK) and centrifuge at 5000*g* (~3000 rpm in bench-top centrifuge) for 15 min to pellet the mycobacteria. Tip away or decant off the supernatant carefully without disturbing the pellet and resuspend the mycobacterial pellet from one universal (which is often very small

and sometimes difficult to see) in 1 ml of 4 M GTC solution. Use this 1 ml GTC to resuspend the pellets from other universals. Thus, all the pellets are now resuspended in 1 ml of 4 M GTC solution. Avoid the temptation to use larger centrifuge tubes in an attempt to minimize the number of tubes to spin. In our experience, other tubes do not result in such a tight pellet of bacteria and these are easily lost when decanting the supernatant.

3. Pellet the pooled GTC-treated bacteria in a microcentrifuge at ~13 000 rpm for 20 s.
4. Rapidly remove the GTC and resuspend pellet in 1.2 ml of TRIzol (Life Technologies) and add to a 2 ml screw-cap skirted microcentrifuge tube containing 0.5 ml of 100 μm ceramic beads.
5. Using a reciprocal shaker (Hybaid Ribolyser), shake the sample at maximum speed setting (6.5) for 45 s.
6. The tubes are removed from the Ribolyser and left at room temperature for 5 min.
7. 200 μl of chloroform is added to the tube and it is vortexed for 30 s. The tube is then spun at 13 000 rpm for 10 min in a microcentrifuge.
8. Upon centrifugation the sample separates into two phases, the upper phase is removed to a fresh tube and an equal volume of chloroform is added to it. It is again vortexed and spun in a microfuge for 2 min.
9. The upper aqueous phase is precipitated with 500 μl of isopropanol at −80°C for at least an hour or for long-term storage, if required.
10. The RNA is pelleted at 13 000 rpm in a microfuge for 15 min after which it is washed with 70% ethanol and repelleted.
11. The residual ethanol is removed with a pipette and the pellet briefly air-dried.
12. Each RNA pellet is resuspended in 100 μl of RNAase free water.
13. The resuspended RNA is treated with Amplification Grade RNAase free DNAase I (Life Technologies) as per the manufacturers' instructions. The RNA preparations are then cleaned up using the Qiagen RNeasy Mini Kit (Cat. No. 74104).
14. Concentration and integrity is assessed by non-denaturing agarose gel electrophoresis.

References

Ambion (2002). Ambion Technical Note 9(1): Isolating bacterial RNA for Array Analysis. http://www.ambion.com/techlib/tn/91/9110.html.

Bashyam, M. D. and Tyagi, A. K. (1994). An efficient and high yielding method for isolation of RNA from mycobacteria. *BioTechniques* **17**, 834–836.

Brown, P. (2001). http://cmgm.stanford.edu/pbrown/protocols.

Butcher, P. D., Mangan J. A. and Monahan, I. M. (1998). Intracellular gene expression: analysis of RNA from mycobacteria in macrophages using RT-PCR. In *Methods in Molecular Biology*, Vol. 101: *Mycobacteria Protocols* (Parish, T. and Stoker, N. G., eds), pp. 285–306. Totowa, NJ: Humana Press Inc.

Butcher, P. D., Sole, K. M. and Mangan, J. A. (1998). RNA extraction. In *Molecular Mycobacteriology: Techniques and Clinical Applications* (Ollar, R. A. and Connell, N. D., eds), pp. 325–350. New York: Marcel Dekker Inc.

Chomczynski, P. and Sacchi, N. (1987). Single-step method of RNA isolation by acid guanidine thiocyanate-phenol-chloroform extraction. *Analyt. Biochem.* **162**, 156–159.

Eberwine, J. (1996). Amplification of mRNA populations using aRNA generated from immobilized oligo(dT)-T7 primed cDNA. *Biotechniques* **20**, 584–591.

Eberwine, J., Yeh, H., Miyashiro, K., Cao, Y., Nair, S., Finnell, R., Zettel, M. and Coleman, P. (1992). Analysis of gene expression in single live neurons. *Proc. Natl Acad. Sci. USA* **89**, 3010–3014.

Hu, Y., Mangan, J. A., Dhillon, J., Sole, K. M., Mitchison, D. A., Butcher, P. D. and Coates, A.R. (2000). Detection of mRNA transcripts and active transcription in persistent *Mycobacterium tuberculosis* induced by exposure to rifampin or pyrazinamide. *J. Bacteriol.* **182**, 6358–6365.

Kikuta-Oshima, L. C., Quinn, F. D., Butler, W. R., Shinnick, T. M. and King, C. H. (1995). Isolation of RNA from *Mycobacterium tuberculosis* using a Nitrogen decompression chamber. *BioTechniques* **18**, 987–990.

Lewin, B. (1990). The messenger RNA template. In *Genes IV* (Lewin, B., ed.), pp. 171–184. Oxford: Oxford University Press.

Li, M-S., Monahan, I. M., Waddell, S. J., Mangan, J. A., Martin, S. L., Everett, M. J. and Butcher, P. D. (2001). cDNA–RNA subtractive hybridisation reveals increased expression of mycocerosic acid synthase in intracellular *Mycobacterium bovis* BCG. *Microbiology* **147**, 2293–2305.

Luo, X. Z. and Stevens, S. E., Jr (1997). Isolation of full length-length RNA from a thermophilic cyanobacterium. *BioTechniques* **5**, 904–906.

Mahan, M. J., Slauch, J. M. and Mekalanos, J. J. (1993). Selection of bacterial virulence genes that are specifically induced in host tissues. *Science* **259**, 686–688.

Mangan, J. A., Sole, K. M., Mitchison D. A. and Butcher, P. D. (1997). An effective method of RNA extraction from bacteria refractory to disruption, including mycobacteria. *Nucleic Acids Res.* **25**, 675–676.

Monahan, I. M., Mangan, J. A. and Butcher, P. D. (2001). Extraction of RNA from intracellular *M. tuberculosis*; methods, considerations and applications. In *Methods in Molecular Medicine* Vol. 54, Mycobacterium tuberculosis *Protocols* (Parish, T. and Stoker, N. G., eds), pp. 31–42. Totowa, NJ: Humana Press, Inc.

Plum, G. and Clark-Curtiss, J. E. (1994). Induction of *Mycobacterium avium* gene expression following phagocytosis by human macrophages. *Infect. Immun.* **62**, 476–483.

Schena, M., Shalon, D., Davis, R. W. and Brown, P. O. (1995). Quantitive monitoring of gene expression patterns with a complementary DNA microarray. *Science* **270**, 467–470.

Schroeder, B. G., Peterson, L. M. and Fleischmann, R. D. (2002). Improved quantitation and reproducibility in *Mycobacterium tuberculosis* DNA microarrays. *J. Microbiol. Biotechnol.* **4**, 123–126.

Smoot, L. M., Smoot, J. C., Graham, M. R., Somerville, G. A., Sturdevant, D. E., Migliaccio, C. A., Sylva, G. L. and Musser, J. M. (2001). Global differential gene expression in response to growth temperature alteration in group A Streptococcus. *Proc. Natl Acad. Sci. USA* **98**, 10416–10421.

Talaat, A. M., Hunter, P. and Johnston, S. A. (2000). Genome-directed primers for selective labeling of bacterial transcripts for DNA microarray analysis. *Nature Biotechnol.* **18**, 679–682.

Valdivia, R. H. and Falkow, S. (1997). Fluorescent-based isolation of bacterial genes expressed within host cells. *Science* **277**, 2007–2011.

Van Gelder, R. N., von Zastrow, M. E., Yool, A., Dement, W. C., Barchas, J. D. and Eberwine, J. H. (1990). Amplified RNA synthesized from limited quantities of heterogeneous cDNA. *Proc. Natl Acad. Sci. USA* **87**, 1663–1667.

Wei, Y., Lee, J-M., Richmond, C., Blattner, F. D., Rafalski, J. A. and LaRossa, R. A. (2001). High-density microarray mediated gene expression profiling of *Escherichia coli*. *J. Bacteriol.* **183**, 545–556.

Yim, H. H. and Rubens, C. E. (1997). Use of a dental amalgamator to extract RNA from the Gram-positive bacterium *Streptococcus agalactiae*. *BioTechniques* **23**, 229–231.

9 High Throughput *In Vivo* Screens: Signature-tagged Mutagenesis

Nicholas West[1], Philippe Sansonetti[2] and Christoph M Tang[1]

[1]*Centre for Molecular Microbiology and Infection, The Flowers Building, Imperial College of Science, Technology and Medicine, Armstrong Road, London SW7 2AZ, UK*
[2]*Unité de Pathogénie Microbienne Moléculaire, Institut Pasteur, 28 rue du Dr Roux F-75724, Paris, Cédex 15, France*

◆◆

CONTENTS

Introduction
Application of STM to *Shigella flexneri*
Selecting an animal model for shigellosis
Bacterial strains, vectors and culture conditions
Construction of a mutant library
Preparation of inoculum
Inoculation of the rabbit ligated loop
Extraction of genomic DNA
PCR labelling of tags
Identification of attenuated mutants
Identification of transposon insertion sites

◆◆◆◆◆◆ INTRODUCTION

The advent of microbial genomics has posed a new set of challenges for research in prokaryotic biology. One of the most important problems is how the overwhelming mass of information coming from whole-genome sequencing projects can be assimilated and effectively exploited to understand the function of gene products (Strauss and Falkow, 1997; Tang *et al.*, 1997; McDevitt and Rosenberg, 2001), including their role in the progression of infectious disease. And, ultimately, how this knowledge can be used to develop a novel range of diagnostic, preventative and therapeutic measures to protect susceptible individuals from invasive pathogens (Tang and Holden, 1999; Rosamond and Allsop, 2000).

Although the availability of whole-genome sequences undoubtedly represents a major advance, the data are a mere starting point for further work, a point emphasized by the considerable proportion of each genome encoding proteins of unknown function. Even for a bacterium as intensively studied as *Escherichia coli*, over a third of the genome comprises genes of unknown function (Blattner *et al.*, 1997; Perna *et al.*, 2001). This is a striking feature of all genomes sequenced to date (Strauss and Falkow, 1997; Tang *et al.*, 1997) and demonstrates the deficiency of current knowledge of microbial biology. Therefore, there remains a need to develop rapid, high-throughput methods, sometimes referred to as 'functional genomics', for taking full advantage of the opportunity afforded by the sequence data.

Signature-tagged mutagenesis (STM) was not developed with genomes in mind (Hensel and Holden, 1996). Rather, STM was devised to identify bacterial genes required for disease progression, and has been successful as it circumvents the constraint of having to test mutants individually in animal models of pathogenesis (Hensel and Holden, 1996; Unsworth and Holden, 2000). Instead, mutants are tagged with a unique identifying DNA sequence, allowing analysis of multiple mutants in parallel, opening the way for high-throughput mutational analysis *in vivo*. Thus, STM is as applicable to the problems of 'functional genomics' as the search for pathogenicity determinants. Furthermore, STM has distinct advantages over other methods; especially those based on transcript analysis, as it involves the generation of knockout strains, allowing a direct attribution of genotype to phenotype.

In this chapter, we outline approaches for using tagging mutagenesis and some of the technical refinements that have been developed. We illustrate some of the issues by discussing our recent work with *Shigella flexneri*, the causative agent of bacillary dysentery.

STM was originally used to identify genes involved in *Salmonella typhimurium* pathogenicity (Hensel *et al.*, 1995a). The methodology is based around the construction of a set of mini-Tn5 transposons, each of which carry a unique 'signature tag', that can be reliably identified by DNA hybridization. Signature tags are 40 bp long, and flanked by 20 bp arms common to all tags, which allow the amplification and labelling of the tag via the polymerase chain reaction (PCR) (Figure 9.1). The mutant library was constructed via conjugation between a pool of *E. coli* donors carrying tagged transposons and *S. typhimurium*, which resulted in tagged *S. typhimurium* mutants. PCR-based radiolabelling and hybridization can detect the tags, and thus the corresponding mutants.

The fundamental methodology of STM, following production of the mutant library, is separated into two phases, an 'input' and an 'output' phase (Hensel *et al.*, 1995b). The two phases of the experiment are essentially identical except that, prior to the output phase, a round of selection is incorporated (Figure 9.2). The presence of mutants is detected at each stage by extracting genomic DNA from the pool of mutants, and amplifying the tags by PCR with primers based on regions flanking the tag. The flanking regions are removed from the resultant product by endonuclease digestion and the labelled tag is hybridized to a dot blot comprised of individual tags from each mutant. Comparison of the output blot and the

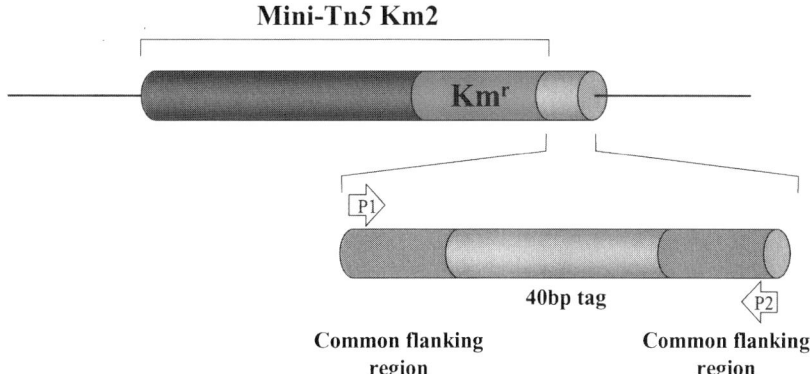

Figure 9.1. Structure of tagged transposon utilized for the mutagenesis. The 40 bp variable tags are flanked by two common regions allowing amplification of the tag by PCR using primers P1 and P2. Reproduced in colour between pages 178 and 179.

input blot can reveal attenuated mutants. If a mutant is absent from the output (no hybridization seen to the corresponding tag) but is present in the input (hybridization detected), then the mutant is presumed to be attenuated.

In work with *S. typhimurium*, a library of 1500 mutants was screened for attenuated strains in a murine model of systemic infection, and led to the identification of the second pathogenicity island (SPI-2) in *S. typhimurium*, which is required for systemic but not mucosal infection (Hensel *et al.*, 1997). The advantage of screening a library in which the identity of mutants is initially unknown is that it can be performed rapidly to generate biologically relevant information. However, this approach is not comprehensive. A library of over 20 000 mutants is needed to have a 98% chance that it contains a mutant in each open reading frame (ORF) of a microbe with 2000 genes. Furthermore, the approach does not address essential genes, which are of particular interest as drug and vaccine targets. Alternatively, systematic tagged mutagenesis of individual ORFs can be undertaken, resulting in an ordered library of strains in which the identity of the disrupted gene is known for all mutants (see Chapter 10). Furthermore, failure to obtain an insertional mutant in a given ORF would provide presumptive evidence for its essentiality (Akerley *et al.*, 1998; Sassetti *et al.*, 2001). However, as construction of

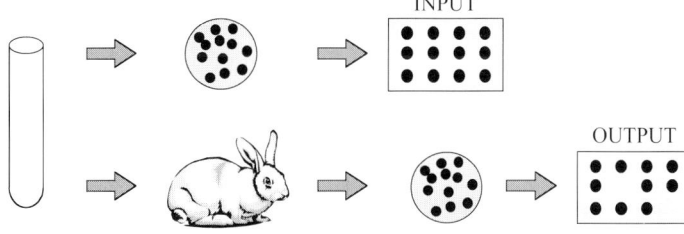

Figure 9.2 Identification of attenuated *Shigella* mutants by comparison of mutants inoculated into the rabbit ileal loop (input) with those that survive passage through the animal (output).

ordered libraries is labour intensive, genomewide tagged knockouts have only been constructed in the budding yeast, *Saccharomyces cerevisiae* (Ross-Macdonald *et al.*, 1999; Winzeler *et al.*, 1999).

STM has been performed on a range of bacterial and fungal microbes (Mei *et al.*, 1997; Polissi *et al.*, 1998; Cormack *et al.*, 1999; Edelstein *et al.*, 1999; Martindale *et al.*, 2000), and a number of different mutagenesis protocols employed. Bacterial transposons are widely used tools for constructing libraries of insertional mutants. In general, the transposon is delivered by conjugation into the host cell where it excises from a donor plasmid, and integrates into the target chromosome. Mini-Tn5 was used in the original STM screen in *S. typhimurium*, and while Tn5 is an effective mutagen for many Gram-negative bacteria, other transposons must be used for Gram-positive organisms. For example, a derivative of Tn916 was used for mutagenesis of *Staphylococcus aureus* (Mei *et al.*, 1997). The recent development of transposons for *Mycobacterium tuberculosis* allowed STM to be carried out by two groups (Camacho *et al.*, 1999; Cox *et al.*, 1999). Both groups identified a gene cluster necessary for the biosynthesis of a cell surface lipid, mycolic acid. This molecule is specifically required for the survival of the bacterium in the lung and not other tissues (Cox *et al.*, 1999).

Other techniques must be devised for bacteria in which *in vivo* transposition is ineffective. For *Streptococcus pneumoniae*, insertion–duplication was used. In this method, short (200–400 bp) fragments of pneumococcal DNA were ligated into vectors in *E. coli*. The resultant plasmids were transferred into *S. pneumoniae* and integrated into the chromosome by targeting genomic sequences homologous to those in the plasmid inserts (Polissi *et al.*, 1998). *Neisseria meningitidis* is naturally competent for the uptake of linear DNA. This allowed *in vitro* modification of genomic DNA by Tn10 transposition prior to introduction into the bacterium (Sun *et al.*, 2000). This avoids the problems of multiple restriction–modification systems (Stein *et al.*, 1995) that affect the efficiency of insertion–duplication.

Fungal pathogens present particular problems. For example, some medically important fungi, such as *Candida albicans*, are diploid and, therefore, construction of null mutants cannot be performed in a single step (Brown *et al.*, 1998). Instead, to further understand *Candida* pathogenicity, a haploid organism, *Candida glabrata*, was studied (Cormack *et al.*, 1999). Results from this microbe, which is an infrequent cause of human disease, identified an adhesin also present in *C. albicans*. However, a significant number of mutants were isolated that were disrupted for synthesis of this adhesin, indicating the non-random nature of the mutagenesis (Cormack *et al.*, 1999). In a further study, with another approach completely, of *Aspergillus fumigatus*, restriction enzyme mediated insertion was used to generate libraries of mutants (Brown *et al.*, 2000).

◆◆◆◆◆◆ APPLICATION OF STM TO *SHIGELLA FLEXNERI*

Shigellae are Gram-negative, non-sporulating, facultative anaerobic bacteria that belong to the family Enterobacteriaceae. *Shigellae* cause bacillary dysentery, an invasive infection of the human colon that, in its classical form, presents as a bloody, mucopurulent diarrhoea accompanied by fever and intestinal cramps. These symptoms reflect the capacity of the causative microorganisms to disrupt, invade, colonize, and cause inflammatory destruction of the intestinal epithelial barrier (Sansonetti, 2001a).

The molecular and cellular pathogenesis of *Shigella* infection has been extensively studied in cell assays *in vitro* and has been recently reviewed (Nhieu and Sansonetti, 1999; Sansonetti, 2001b). In *S. flexneri*, a 214 kb virulence plasmid (Sansonetti *et al.*, 1982; Buchrieser *et al.*, 2000) contains a 30 kb pathogenicity island (PAI) that is required for entry into epithelial cells. This PAI comprises the *mxi–spa* operons encoding a type III secretion system that is able to inject into the epithelial cell membrane and cytoplasm effector proteins encoded by the *ipa* operon also present on the PAI, as well as proteins encoded elsewhere on the virulence plasmid. The result is that the bacteria gain access to the intracellular compartment where they lyse their phagocytic membranes and escape into the cytoplasm of the cell (Sansonetti *et al.*, 1986). Direct contact with cytosolic components then allows *Shigella* to express efficient actin-based intracellular motility that eventually promotes cell-to-cell spread of the microorganism (Bernardini *et al.*, 1989).

These *in vitro* data have led to an understanding of the steps leading to invasion and inflammatory destruction of the intestinal barrier. However, these data need to be put into the perspective of the *in vivo* situation. For instance, the factors that the bacterium needs to survive within the lumen of the gastrointestinal tract are not well understood.

◆◆◆◆◆◆ SELECTING AN ANIMAL MODEL FOR SHIGELLOSIS

The pathogenic scheme of shigellosis is a complex one for which there is no optimal model that reproduces the entire series of events in a relevant background, possibly with the exception of the intragastric infection of macaque monkeys that leads to acute rectocolitis with clinical symptoms (Fontaine *et al.*, 1988). However, owing to its cost and the ethical issues involved, monkeys cannot be considered as a suitable model for routine studies. All parameters considered, the rabbit ligated loop model of *Shigella* infection has turned out to be the most suitable model to study shigellosis *in vivo* (Sansonetti *et al.*, 1983, 1995; Perdomo *et al.*, 1994). In this model, it has been possible to confirm that *Shigella* preferentially crosses the intestinal barrier at the level of the follicle associated epithelium (FAE) and subsequently induces apoptosis of the infected macrophages (Zychlinsky *et al.*, 1996).

The rabbit model does have limitations as it requires general anaesthesia and abdominal surgery, thus making it more than just a routine test. Also, cell and tissue-specific probes, particularly antibody markers for immune cells, are limited, compared to the wealth of reagents available for mice. Last but not least, shigellosis does not naturally affect the rabbit, therefore, the species-specific dimension of the disease is lost.

The advantages, however, largely outweigh the limitations. Among them are: (1) the flexibility to vary the inoculum size and the duration of infection; (2) the sensitivity offered by a tissue that can discriminate between the plasmid and chromosomally encoded virulence factors of *Shigella* (Sansonetti *et al.*, 1983); (3) the ability to perform multiple ileal loops per animal, thereby analysing several mutants in a similar background; (4) the ease and accuracy of the read-outs reflecting the degree of pathogenicity of the respective strains; and (5) the possibility to perform intestinal loops containing a Peyer's patch and to compare the behaviour of the bacteria in the FAE, compared to the villous epithelium (Sansonetti *et al.*, 1996). Thus, for screening the STM bank of *Shigella* mutants, we considered that the rabbit ligated intestinal loop assay was the most suitable model.

◆◆◆◆◆◆ BACTERIAL STRAINS, VECTORS AND CULTURE CONDITIONS

Shigella flexneri strain M90T, which is resistant to nalidixic acid (Nal^r), was utilized as the parental–recipient strain for the transposon mutagenesis. The donor strain utilized was *Escherichia coli* S17-1 λ*pir* that harboured one of the 96 possible suicide vectors based on pUT-mini Tn5 Km2 (de Lorenzo *et al.*, 1990) (Figure 9.1). For general culture of *E. coli* strains Luria Bertani (LB) media was used. *Shigella* strains were cultured with trypticase soy (TCS) broth or TCS agar plates supplemented with 50 μg/ml kanamycin (Km), 40 μg/ml nalidixic acid (Nal), and 0.01% congo red (Sigma), when necessary.

◆◆◆◆◆◆ CONSTRUCTION OF A MUTANT LIBRARY

The mutant library was constructed by standard conjugation of *S. flexneri* M90T Nal^r (recipient) and *E. coli* S17-1 λ *pir* (donor) containing the tagged mini-Tn5 construct (Figure 9.1). Selected tagged mini-Tn5 transposons containing independent signature tags that do not cross-hybridize are available (see Chapter 10). These are used to perform individual matings with the recipient strain.

The protocol for the construction of the mutant library is as follows.

1. Prepare overnight cultures of the recipient *S. flexneri* M90T and the donor *E. coli*.
2. Combine by spotting 10 μl samples of each strain onto a non-selective LB plate and allow to dry.
3. After 2 h at 37°C remove the resultant growth and resuspend into 500 μl of physiological saline (0.7%).
4. 100 μl of this cell suspension when spread on to selective media and incubated overnight should yield 50–100 colony-forming units (CFU), which can now be picked into 96-well plates for storage. Colonies

should be organized in the following manner. Mutants produced with transposon A1 should be stored in the same location of another plate containing media suitable for freezing. Mutants produced with Tn A2 should be suspended in position A2 of storage plate and so on.
5. Continue this procedure until enough storage plates have been filled to give the required number of mutants, with 96 individual mutants in each multiwell plate.

◆◆◆◆◆◆ PREPARATION OF INOCULUM

The following inoculum was chosen on the basis of preliminary experiments in which the total dose, the number of different mutants and time to harvesting were optimized.

1. Using a 96-pin replicator, seed a 96-well plate containing TCS broth and the appropriate antibiotic from the mutant library storage plates.
2. Grow at 37°C to mid-exponential phase, determined by plate spectroscopy (OD A_{600} approximately 0.4), and pool mutants.
3. Dilute culture in physiological saline to yield 2×10^5 CFU/ml knowing that an OD of 1.0 at A_{600} contains approximately 4×10^8 CFU/ml.
4. Keep inoculum on ice until required.

◆◆◆◆◆◆ INOCULATION OF THE RABBIT LIGATED LOOP

1. Following anaesthesia of the rabbit, make midline abdominal incision and locate the appendix.
2. Working toward the stomach, create isolated sections of the intestine, measuring approximately 5 cm in length separated by a 'spacer', also of 5 cm. Sections (loops) to be inoculated and spacer loops are to be identified by alternating the colour of thread used to create the loops.
3. Continue until the required number of loops has been reached (up to a maximum of 12 per rabbit).
4. Inoculate 500 µl of cell suspension prepared earlier in order to deliver 1×10^5 CFU into test loops leaving spacers empty.
5. Retain an aliquot of inoculum for determination of 'input' mutants.
6. Suture peritoneum and skin and allow rabbit to recover for 16 h.
7. Following 16 h incubation, euthanize rabbit and recover contents of test loops, recording volume of fluid accumulation if necessary.

◆◆◆◆◆◆ EXTRACTION OF GENOMIC DNA

The following quick method can be performed in order to extract the genomic DNA from the mutants of the input inoculum and those recovered from the loops, i.e. output. The resultant DNA is of sufficient quality for use as a template for PCR.

1. Dilute aliquots of the input and output samples from each loop in order to achieve a semiconfluent growth when cultured on TCS agar.
2. Harvest bacteria from entire plate and suspend in 200 µl of dH$_2$O.
3. To 50 µl of the cell suspension, add 50 µl of 0.25 M KOH and boil for 5 min.
4. Add 50 µl of 0.25 M HCl and mix.
5. Add 50 µl of 0.5 M Tris (pH 7.6) and mix.
6. The addition of 675 µl of dH$_2$O at this stage will cause a white precipitate to form. Avoiding the precipitate, transfer an aliquot of the supernatant to clean tube.
7. This genomic DNA solution can now be used as PCR template.

◆◆◆◆◆◆ PCR LABELLING OF TAGS

The labelling of the tags is a process involving two rounds of PCR. The first is to amplify the tags from the genomic DNA extracted above. This is done for the input pool and for the output pool. This PCR should be set up as shown below. Following this amplification, the products can be labelled. The traditional label used was ^{32}P-dCTP; however, we have found fluorescein labelling to be particularly efficient and have, therefore, replaced the radiolabel. The parameters for the second round PCR are also outlined below.

The PCR parameters for the amplification of the input and output tags are listed below.

First-round PCR

1. Prepare a 50 µl reaction comprised of 5 µl of genomic DNA template extracted as above, 50 µM each of dATP, dCTP, dGTP and dTTP, 2 mM MgCl$_2$, 100 ng each of primers P1 (5'-TACCTACAACCTCAAGCT-3') and P2 (5'-TACCCATTCTAACCAAGC-3'), the appropriate volume of PCR buffer, and 1 U of *Taq* DNA polymerase (Life Technologies Ltd).
2. Following an initial denaturation period of 5 min at 94°C, 30 cycles of amplification can be carried out as follows, 94°C for 30 s, 50°C for 45 s and 72°C for 10 s.
3. The product of the PCR should now be purified (Qiaspin, Qiagen, Hilden, Germany).

Second-round PCR

The second-round PCR is used to label the product of the first round only. Virtually no amplification occurs in this reaction due to the steric hindrance of the fluorescein molecule.

1. Together with 5 µl of the purified first round PCR product combine 50 µM each of dATP, dGTP, dTTP and 15 µM fluorescein-12-dCTP (NEN, Hounslow, UK), 2 mM $MgCl_2$, 100 ng each of primers P1 and P2, the appropriate volume of PCR buffer and 0.4 U of *Taq* DNA polymerase. Bring the volume of the reaction to 20 µl with dH_2O.
2. Perform thermal cycling as outlined for the first round PCR with the exception that only 20 cycles are necessary.
3. Following the labelling reaction, purify the tags from their invariable flanking regions by restriction endonuclease digestion.
4. Add the following directly to the PCR tube: 20 µl of restriction enzyme buffer, 50 U *Hin*dIII, and 155 µl of dH_2O.
5. Incubate at 37°C for 2 h.

Subsequent purification of the tags is not necessary due to the low cytosine concentration of the flanking regions. This results in negligible interference of the label.

◆◆◆◆◆◆ IDENTIFICATION OF ATTENUATED MUTANTS

Attenuated mutants are identified by the comparison of plasmid blots individually hybridized with the labelled PCR products of the input and output DNA samples. The method described below is based on hybridization in aqueous solutions; however, other standard hybridization protocols could be used. All hybridization and washing steps are performed at 65°C in a roller bottle hybridization oven.

Preparation of plasmid dot blots

1. Cut the required number of positively charged nylon membranes (Hybond-N+, Amersham) to fit a 96-well vacuum manifold applicator.
2. Place the dH_2O pre-wet membrane between the two halves of the applicator and fasten securely.
3. Fill the wells of the applicator with 10 µg of plasmid DNA, denatured with a solution of 0.5 M NaOH, 1.5 M NaCl. When all wells are full, apply the vacuum until all the solution has passed through.
4. With the vacuum on, rinse the wells twice with 1.5 M NaCl, 0.5 M Tris–HCl, pH 7.6.
5. Remove the membrane, rinse with 2 × saline sodium citrate (SSC) and ultraviolet cross-link the DNA to the membrane for 30 s.

Hybridization of dot blots

1. Prehybridize membrane in 20–30 ml hybridization buffer [1 mM EDTA, 0.5 M $NaHPO_4$, 7% sodium dodecyl sulphate (SDS)] containing 0.5% non-fat milk as blocking agent, for 1 h.

2. Denature probe by the addition of 12 µl of 2 M NaOH.
3. Replace prehybridization solution with 20 ml of fresh hybridization buffer and add probe. Hybridize probe overnight.
4. Remove probe solution (fluorescein-labelled probes can be retained and reused if necessary) and add 100 ml wash buffer 1 (1 mM EDTA, 40 mM $NaHPO_4$, 5% SDS). Incubate for 30 min.
5. Replace wash buffer 1 and again incubate for 30 min.
6. Remove wash buffer 1 and add 100 ml wash buffer 2 (1 mM EDTA, 40 mM $NaHPO_4$, 1% SDS). Repeat washing steps as for wash buffer 1.

Detection of the fluorescein-labelled probe is most easily achieved with a commercially available alkaline phosphatase conjugated antifluorescein antibody and an alkaline phosphatase detection kit (Amersham Pharmacia Biotech).

Detection of non-radioactive probes is achieved using the following method.

1. Detection of hybridized probe begins with a 1 h incubation of the membrane with an antifluorescein–alkaline phosphatase conjugated antibody diluted to the recommended concentration in phosphate-buffered saline (PBS).
2. Following three 10-min washes with PBS continue detection with an alkaline phosphatase chemiluminescent detection kit according to manufacturer's instructions.
3. Expose membrane to chemiluminescent or X-ray film. Modern chemiluminescent substrates for alkaline phosphatase such as CDP-Star (Amersham Pharmacia Biotech) may require exposure times of just a few seconds after the luminescence has been allowed to develop.
4. A direct comparison of blots produced from the input phase and those from the output phase will reveal any attenuated mutants for survival within the model tested. Mutants that fail to give rise to a hybridization signal on the output blot where one is evident on the input blot of the same pool are considered attenuated.

◆◆◆◆◆◆ IDENTIFICATION OF TRANSPOSON INSERTION SITES

The optimum approach for recovering transposon insertion sites depends on the mutagenesis method and vector used. For instance, in work with *N. meningitidis*, the transposon contained a low copy number origin of replication, facilitating the isolation of flanking sequences (Sun et al., 2000). However, mini-Tn5 does not harbour an origin and, therefore, other methods are required. We have found that arbitrary PCR is an efficient and rapid way of establishing the disrupted gene in a mutant of interest. The method (see below) routinely generates 200–400 bp of flanking DNA, which is sufficient to perform database searches, especially if the complete genome sequence of the organism is available.

1. Obtain genomic DNA from the mutant of interest and the parental organism.
2. Prepare a first-round PCR using the concentration of reagents described above. Include the transposon specific primer P9 (5'-CGCAGGGCTTTATTGATTC-3') and either ARB4 (5'-GGCCACGCGTCGACTAGTACNNNNNNNNNNNTGACG-3') or ARB6 (5'-GGCCACGCGTCGACTAGTACNNNNNNNNNNNACGCC-3'). Use approximately 100 ng of target genomic DNA.
3. Cycling conditions: six cycles of 94°C for 30 s, 30°C for 45 s, 72°C for 1 min; followed by 30 cycles of 94°C for 30 s, 50°C for 45 s, 72°C for 1 min.
4. Take 1 µl of the initial PCR and use as target in a second PCR with the primers P7 (5'-GCACTTGTGTATAAGAGTCAG-3') and ARB2 (5'-GGCCACGCGTCGACTAGTAC-3').
5. Cycling conditions: 30 cycles of 94°C for 30 s, 50°C for 45 s, 72°C for 1 min.
6. Analyse the amplification products by gel electrophoresis using a 3% metaphor gel (FMC, Rockland, Germany).
7. Purify any bands amplified using DNA from the mutant but not the wild-type by the qiaquick method (Qiagen, Hilden, Germany).
8. Ligate the purified products into pCR2.1 TOPO (Invitrogen). Isolate recombinants by standard methods, and determine the nucleotide sequence of the inserts using oligonucleotides annealing to sequences flanking the multiple cloning site.

An alternative method for identifying sequence data that flank transposon insertion sites using inverse PCR is described in Chapter 10.

References

Akerley, B. J., Rubin, E. J., Camilli, A., Lampe, D. J., Robertson, H. M. and Mekalanos, J. J. (1998). Systematic identification of essential genes by *in vitro* mariner mutagenesis. *Proc. Natl Acad. Sci. USA* **95**, 8927–8932.

Bernardini, M. L., Mounier, J., d'Hauteville, H., Coquis-Rondon, M. and Sansonetti, P. J. (1989). Identification of icsA, a plasmid locus of *Shigella flexneri* that governs bacterial intra- and intercellular spread through interaction with F-actin. *Proc. Natl Acad. Sci. USA* **86**, 3867–3871.

Blattner, F. R., Plunkett, G., 3rd, Bloch, C. A., Perna, N. T., Burland, V., Riley, M., Collado-Vides, J., Glasner, J. D., Rode, C. K., Mayhew, G. F., Gregor, J., Davis, N. W., Kirkpatrick, H. A., Goeden, M. A., Rose, D. J., Mau, B. and Shao, Y. (1997). The complete genome sequence of *Escherichia coli* K-12 [see comments]. *Science* **277**, 1453–1474.

Brown, A. J., Cormack, B. P., Gow, N. A., Kvaal, C., Soll, D. R. and Srikantha, T. (1998). Advances in molecular genetics of *Candida albicans* and *Candida glabrata*. *Med. Mycol.* **36**(Suppl. 1), 230–237.

Brown, J. S., Aufauvre-Brown, A., Brown, J., Jennings, J. M., Arst, H., Jr and Holden, D. W. (2000). Signature-tagged and directed mutagenesis identify PABA synthetase as essential for *Aspergillus fumigatus* pathogenicity. *Mol. Microbiol.* **36**, 1371–1380.

Buchrieser, C., Glaser, P., Rusniok, C., Nedjari, H., D'Hauteville, H., Kunst, F., Sansonetti, P. and Parsot, C. (2000). The virulence plasmid pWR100 and the repertoire of proteins secreted by the type III secretion apparatus of *Shigella flexneri*. *Mol. Microbiol.* **38**, 760–771.

Camacho, L. R., Ensergueix, D., Perez, E., Gicquel, B. and Guilhot, C. (1999). Identification of a virulence gene cluster of *Mycobacterium tuberculosis* by signature-tagged transposon mutagenesis. *Mol. Microbiol.* **34**, 257–267.

Cormack, B. P., Ghori, N. and Falkow, S. (1999). An adhesin of the yeast pathogen *Candida glabrata* mediating adherence to human epithelial cells [see comments]. *Science* **285**, 578–582.

Cox, J. S., Chen, B., McNeil, M. and Jacobs, W. R., Jr (1999). Complex lipid determines tissue-specific replication of *Mycobacterium tuberculosis* in mice. *Nature* **402**, 79–83.

de Lorenzo, V., Herrero, M., Jakubzik, U. and Timmis, K. N. (1990). Mini-Tn5 transposon derivatives for insertion mutagenesis, promoter probing, and chromosomal insertion of cloned DNA in gram-negative eubacteria. *J. Bacteriol.* **172**, 6568–6572.

Edelstein, P. H., Edelstein, M. A., Higa, F. and Falkow, S. (1999). Discovery of virulence genes of *Legionella pneumophila* by using signature tagged mutagenesis in a guinea pig pneumonia model. *Proc. Natl Acad. Sci. USA* **96**, 8190–8195.

Fontaine, A., Arondel, J. and Sansonetti, P. J. (1988). Role of Shiga toxin in the pathogenesis of bacillary dysentery, studied by using a Tox-mutant of *Shigella dysenteriae* 1. *Infect. Immun.* **56**, 3099–3109.

Hensel, M. and Holden, D. W. (1996). Molecular genetic approaches for the study of virulence in both pathogenic bacteria and fungi. *Microbiology* **142**, 1049–1058.

Hensel, M., Shea, J. E., Gleeson, C., Jones, M. D., Dalton, E. and Holden, D. W. (1995). Simultaneous identification of bacterial virulence genes by negative selection. *Science* **269**, 400–403.

Hensel, M., Shea, J. E., Baumler, A. J., Gleeson, C., Blattner, F. and Holden, D. W. (1997). Analysis of the boundaries of *Salmonella* pathogenicity island 2 and the corresponding chromosomal region of *Escherichia coli* K-12. *J. Bacteriol.* **179**, 1105–1111.

Martindale, J., Stroud, D., Moxon, E. R. and Tang, C. M. (2000). Genetic analysis of *Escherichia coli* K1 gastrointestinal colonization [In Process Citation]. *Mol. Microbiol.* **37**, 1293–1305.

McDevitt, D. and Rosenberg, M. (2001). Exploiting genomics to discover new antibiotics. *Trends Microbiol.* **9**, 611–617.

Mei, J. M., Nourbakhsh, F., Ford, C. W. and Holden, D. W. (1997). Identification of *Staphylococcus aureus* virulence genes in a murine model of bacteraemia using signature-tagged mutagenesis. *Mol. Microbiol.* **26**, 399–407.

Nhieu, G. T. and Sansonetti, P. J. (1999). Mechanism of *Shigella* entry into epithelial cells. *Curr. Opin. Microbiol.* **2**, 51–55.

Perdomo, O. J., Cavaillon, J. M., Huerre, M., Ohayon, H., Gounon, P. and Sansonetti, P. J. (1994). Acute inflammation causes epithelial invasion and mucosal destruction in experimental shigellosis. *J. Exp. Med.* **180**, 1307–1319.

Perna, N. T., Plunkett, G., 3rd, Burland, V., Mau, B., Glasner, J. D., Rose, D. J., Mayhew, G. F., Evans, P. S., Gregor, J., Kirkpatrick, H. A., Posfai, G., Hackett, J., Klink, S., Boutin, A., Shao, Y., Miller, L., Grotbeck, E. J., Davis, N. W., Lim, A., Dimalanta, E. T., Potamousis, K. D., Apodaca, J., Anantharaman, T. S., Lin, J., Yen, G., Schwartz, D. C., Welch, R. A. and Blattner, F. R. (2001). Genome sequence of enterohaemorrhagic *Escherichia coli* O157:H7. *Nature* **409**, 529–533.

Polissi, A., Pontiggia, A., Feger, G., Altieri, M., Mottl, H., Ferrari, L. and Simon, D. (1998). Large-scale identification of virulence genes from *Streptococcus pneumoniae*. *Infect. Immun.* **66**, 5620–5629.

Rosamond, J. and Allsop, A. (2000). Harnessing the power of the genome in the search for new antibiotics. *Science* **287**, 1973–1976.

Ross-Macdonald, P., Coelho, P. S., Roemer, T., Agarwal, S., Kumar, A., Jansen, R., Cheung, K. H., Sheehan, A., Symoniatis, D., Umansky, L., Heidtman, M., Nelson, F. K., Iwasaki, H., Hager, K., Gerstein, M., Miller, P., Roeder, G. S. and Snyder, M. (1999). Large-scale analysis of the yeast genome by transposon tagging and gene disruption. *Nature* **402**, 413–418.

Sansonetti, P. J. (2001a). Microbes and microbial toxins: paradigms for microbial–mucosal interactions III. Shigellosis: from symptoms to molecular pathogenesis. *Am. J. Physiol. Gastrointest. Liver Physiol.* **280**, G319–323.

Sansonetti, P. J. (2001b). Rupture, invasion and inflammatory destruction of the intestinal barrier by *Shigella*, making sense of prokaryote–eukaryote cross-talks. *FEMS Microbiol. Rev.* **25**, 3–14.

Sansonetti, P. J., Kopecko, D. J. and Formal, S. B. (1982). Involvement of a plasmid in the invasive ability of *Shigella flexneri*. *Infect. Immun.* **35**, 852–860.

Sansonetti, P. J., Hale, T. L., Dammin, G. J., Kapfer, C., Collins, H. H., Jr and Formal, S. B. (1983). Alterations in the pathogenicity of *Escherichia coli* K-12 after transfer of plasmid and chromosomal genes from *Shigella flexneri*. *Infect. Immun.* **39**, 1392–1402.

Sansonetti, P. J., Ryter, A., Clerc, P., Maurelli, A. T. and Mounier, J. (1986). Multiplication of *Shigella flexneri* within HeLa cells: lysis of the phagocytic vacuole and plasmid-mediated contact hemolysis. *Infect. Immun.* **51**, 461–469.

Sansonetti, P. J., Arondel, J., Cavaillon, J. M. and Huerre, M. (1995). Role of interleukin-1 in the pathogenesis of experimental shigellosis. *J. Clin. Invest.* **96**, 884–892.

Sansonetti, P. J., Arondel, J., Cantey, J. R., Prevost, M. C. and Huerre, M. (1996). Infection of rabbit Peyer's patches by *Shigella flexneri*: effect of adhesive or invasive bacterial phenotypes on follicle-associated epithelium. *Infect. Immun.* **64**, 2752–2764.

Sassetti, C. M., Boyd, D. H. and Rubin, E. J. (2001). Comprehensive identification of conditionally essential genes in mycobacteria. *Proc. Natl Acad. Sci. USA* **98**, 12712–12717.

Stein, D. C., Gunn, J. S., Radlinska, M. and Piekarowicz, A. (1995). Restriction and modification systems of *Neisseria gonorrhoeae*. *Gene* **157**, 19–22.

Strauss, E. J. and Falkow, S. (1997). Microbial pathogenesis: genomics and beyond. *Science* **276**, 707–712.

Sun, Y. H., Bakshi, S., Chalmers, R. and Tang, C. M. (2000). Functional genomics of *Neisseria meningitidis* pathogenesis. *Nature Med.* **6**, 1269–1273.

Tang, C. and Holden, D. (1999). Pathogen virulence genes – implications for vaccines and drug therapy. *Br. Med. Bull.* **55**, 387–400.

Tang, C. M., Hood, D. W. and Moxon, E. R. (1997). *Haemophilus* influence: the impact of whole genome sequencing on microbiology. *Trends Genet.* **13**, 399–404.

Unsworth, K. E. and Holden, D. W. (2000). Identification and analysis of bacterial virulence genes *in vivo*. *Phil. Trans. R. Soc. Lond. B Biol. Sci.* **355**, 613–622.

Winzeler, E. A., Shoemaker, D. D., Astromoff, A., Liang, H., Anderson, K., Andre, B., Bangham, R., Benito, R., Boeke, J. D., Bussey, H., Chu, A. M., Connelly, C., Davis, K., Dietrich, F., Dow, S. W., El Bakkoury, M., Foury, F., Friend, S. H., Gentalen, E., Giaever, G., Hegemann, J. H., Jones, T., Laub, M., Liao, H., Davis, R. W. *et al.* (1999). Functional characterization of the *S. cerevisiae* genome by gene deletion and parallel analysis. *Science* **285**, 901–906.

Zychlinsky, A., Thirumalai, K., Arondel, J., Cantey, J. R., Aliprantis, A. O. and Sansonetti, P. J. (1996). *In vivo* apoptosis in *Shigella flexneri* infections. *Infect. Immun.* **64**, 5357–5365.

10 Further Strategies for Signature-tagged Mutagenesis and the Application of Oligonucleotide Microarrays for the Quantification of DNA-tagged Strains

Andrey V Karlyshev[1], Nick Dorrell[1], Elizabeth Winzeler[2] and Brendan W Wren[1]

[1]Department of Infectious and Tropical Diseases, London School of Hygiene and Tropical Medicine, London WC1E 7HT, UK
[2]Novartis Institute for Functional Genomics, 3115 Merryfield Row, Suite 200, San Diego, CA 92121-1125, USA

CONTENTS

Introduction
Microarray-based STM of *Y. pseudotuberculosis*
Signature-tagged allele replacement in *Helicobacter pylori*

◆◆◆◆◆◆ INTRODUCTION

The previous chapter described the principles of signature-tagged mutagenesis (STM) and how this can be applied to study to identify potential virulence determinants in the causative agent of bacillary dysentery, *Shigella flexneri*. The original STM technique has certain technical limitations, including high variation in signal intensities after hybridization owing to the variability of the tag sequences, necessitating further rounds of screening. Additionally, mutants are frequently found to contain unamplifiable tags or no tags at all (Zhao *et al.*, 1999). Moreover, the need to remove the invariable arms of the labelled probes prior to hybridization often necessitates the use of radioactively labelled probes. Plasmids carrying transposons with preselected tags and dot blotting with purified plasmid DNA have helped to overcome some of these problems (Mei et al., 1997) and the technique has been further improved by using membranes containing tags amplified by the polymerase chain reaction (PCR) (Polissi et al., 1998). However, hybridization on membranes can be insensitive and labour intensive. We reasoned that detection of DNA tagged strains could be improved by using: (1) DNA sequence tags designed to have similar hybridization properties; (2) two tags for each mutant; (3) sequences corresponding to both strands of the tags arrayed on an Affymetrix 'bar coding' microarray containing immobilized oligonucleotide tag sequences.

This chapter will describe how we have modified the basic STM approach to identify potential virulence determinants in *Yersinia pseudotuberculosis* and to monitor the survival rate of *Helicobacter pylori* mutants under different environmental conditions. In principle, this technology could be more widely applied to measure the relative abundance of tagged microbial strains in any complex environment.

◆◆◆◆◆◆ MICROARRAY-BASED STM OF Y. PSEUDOTUBERCULOSIS

Construction of an optimally designed double-tagged mini-Tn5 library

Initial experiments using the random tagging technique (Hensel et al., 1995) resulted in variable hybridization signals in the input pools, causing difficulties in data interpretation. Sequencing the tags from different clones revealed that weak signals were due to large deletions in the variable regions, possibly a result of errors in primer synthesis, whereas exceptionally strong signals were due to the presence of above average GC/AT ratio in the variable regions. To overcome these difficulties, we introduced preselected tag sequences into a transposon. Tags were chosen from the *Saccharomyces cerevisiae* mutagenesis project (Shoemaker et al., 1996). The 20 base sequences lack secondary structure, have similar melting temperatures and a minimum of five base difference between each tag. From this initial set of 6000 tag sequences, 192 (two per transposon) that had demonstrated strong hybridization potential were chosen. Sequences of all tags and oligonucleotides used are presented in Table 10.1 and on our website at http://www.lshtm.ac.uk/itd/units/pmbbu/karlyshev/tags.htm respectively. The strategy for constructing double-tagged transposons is presented in Figure 10.1. Two sets of 96 long primers containing tag sequences were used in two sequential PCR reactions. A third PCR was used to fix the ends of the PCR products. The cloning step involved restoration of the kanamycin resistance cassette into the cloning vector, allowing positive selection of the recombinant plasmids.

Three *Escherichia coli* libraries with double-tagged transposons were constructed. Clones from each library were combined and the tags for each pool were PCR amplified using biotinylated primers. Subsequent

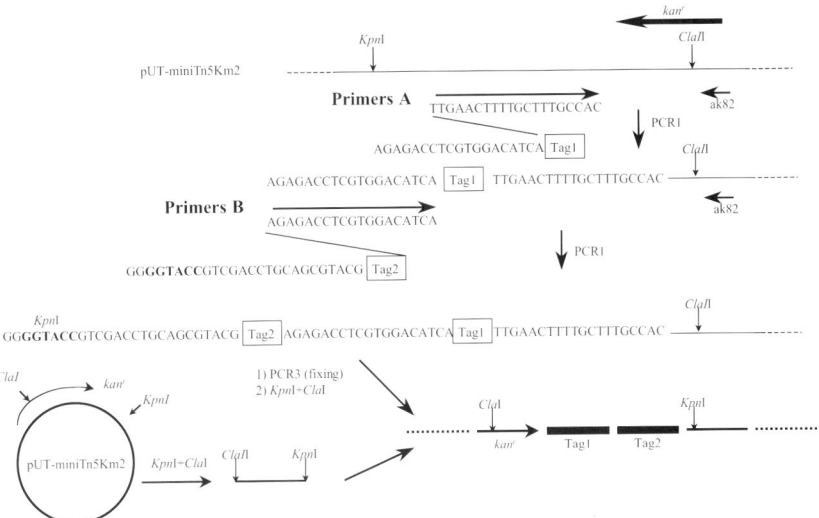

Figure 10.1. Construction of double-tagged mini-Tn5 transposons. The 192 different tags were incorporated into the mini-Tn5 vector by means of two sequential sets of PCR reactions using two sets of long tag-containing primers.

hybridization to the microarray revealed some variation in signal intensities for the same tags in different pools (Figure 10.2). Sequence analysis identified single base deletions in the tags derived from the clones with reduced signals (data not shown). Such microdeletions may have resulted from errors during chemical synthesis of the long PCR primers used for the library construction. The master library of 96 donor *Escherichia coli*

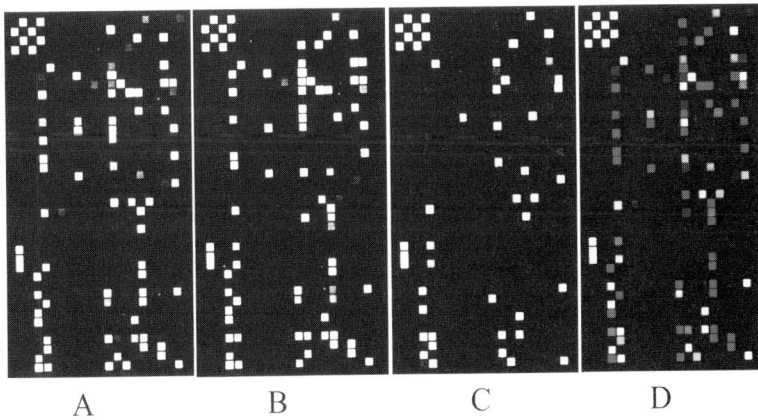

Figure 10.2. Comparative analysis of scanned microarrays using imaging software. (A, B and C) Fragments of hybridization images using probes derived from pools of three different libraries of tagged transposons. (D) Images from (A), (B) and (C) were converted into red, blue and green colour images, respectively, and overlaid using Adobe PhotoShop 5 software. White spots indicate similar relative abundance of respective tags in all pools. Such image overlaying technique was also used for the detection of attenuated mutants (see text). Reproduced in colour between pages 178 and 179.

Table 10.1. Sequences of tag1 and tag2 present on 96 tagged transposons labelled with double tags (D-Tag)

D-Tag	tag1	Sequence	tag2	Sequence
1	1	TGCTAGGTTGTCTGTGCCCA	97	CGGCGTCATGGGCATCATAT
2	2	TTCGAGATTGTTGATGCCGC	98	CATTCATGGCATCGGTGCGA
3	3	TTCGCGCTTGTTGGTCAGCA	99	GATACGATGTGGCGTCTCCT
4	4	TAGCCAAGTGTTGATGATGC	100	AATAGCCTGAAGAGCAGTCG
5	5	TTGTGCATTGTTGGACGCCC	101	GAGAATGCTGTATAGTGTCC
6	6	GTGCTGTTGAATCATGTCGC	102	TCGCGTCATTTATCGAATCG
7	7	TATATGCAGTGGTGTTCGCC	103	GTCGTATATTCTCAGCATGG
8	8	TGCTTAGTTGTGAGTCGCCA	104	ATCTGTGAACATGAATCGGC
9	9	TGCTACATTGTGTCGCTGGC	105	TGCTGTGATTTATACCAGGG
10	10	GTGCCTATTGATGATCTGGC	106	GTCGCACGTTCTGCTGGTAT
11	11	TCGGCATCTGTTCTGATCGC	107	ATGAGATGCAAGTGAGCCCT
12	12	GCATCAGATCGTATGTGGCA	108	TCGCATCTTGTCATATAGGC
13	13	TATATGTGTTCCTGGAGCGC	109	CTGTATCATCTGTAGAGTGC
14	14	TGGACTGTTGTGATATGCCA	110	CACGGAGCTGGCAGACTAAT
15	15	TCTATTGCTGTTCGGCGGCA	111	GATAGGCTTATATGCGTGCT
16	16	CCAGTCAGTGCGGTCATCAT	112	GATCTCTGTGAAGTTAGTGC
17	17	TTCGCGCATTTGGCACTATG	113	GCCAGATTCGAGAGGTCTAT
18	18	GAAAGCTGCGACTGTGATTT	114	GCCTGCTGTGGCTGTATATC
19	19	CTGGTGTGAAGTGATCTCAT	115	TCGGATCGTTTACATCATGG
20	20	GTGCTGGCTTCTCTGCATGA	116	GTCGCCTATTCTCATGTCAG
21	21	GCCATGTATGCTGATATGTC	117	GATAACGATGGGTGCAATCT
22	22	GACCTATATGTCGTGAGTGT	118	TGCTGGTGTTGTATGCCCAC
23	23	TTCGGTCGTTTGACATGAGC	119	GCCTGTAATGGTGGATCTCA
24	24	CACGGCACTGGCACATGAAT	120	TGGATTCATTTGCCGCATGG
25	25	GCCGATGGTGCGTCTACTAT	121	TTGGCTCATTTGCATGGCGC
26	26	GACGAGTCTGGCTGATCTCT	122	TTGCGATATTTGGGCCATGC
27	27	GCCATGCGTGCTGTCAGTAT	123	GCCATGTGTGTGAGCATATA
28	28	CTTGAGGCGCGTGTCCATAT	124	CTGTCAACTGTACTGTCATC
29	29	GATATTAAGTCATGGTGCGC	125	GATCTGTCTGACGCTGTATG
30	30	GGCATCACTGGTTACGTCTG	126	GTGCCGTATGTGTTAATGCA
31	31	GACGCGGGTGCTCATCATAT	127	CGGATCTATGGCATCTACTG
32	32	GACTCATTCAGAAGGTGTGC	128	TCGCCGTTGGTCTGTATGCA
33	33	ATGATGAGCCGTGATGACCC	129	TCGTGCAATTTAGAGGTGTC
34	34	TGCTCGCTTGTGATCGACTG	130	TCCGCATATCTTGATTGACC
35	35	GTCGGCGATTCTCTAGTCTG	131	CTGTCCCGTGTGCTGAGATA
36	36	AGAGCAGATGCACGAGACTC	132	CGGCTGGATGGCATACTATA
37	37	CACAGAACTGCGGGTGACAT	133	CCAATAGGTGCTCACGTCAT
38	38	TCGTGAGTTGTCCTGCTGCA	134	TAGCACTGTTGGAGAGTATC
39	39	TGCATTTGTTGGAGTGCCAC	135	TAGCAGTCTTGGCATACATG
40	40	GATACGAGTTCAGCAGATGC	136	CTGTGACTGACTGTAGCTCT
41	41	ATGAGCAATGATGGCAGATC	137	GTGTGATTGAGCTTACTGCA
42	42	GATAGATGTGAGTGCATACC	138	AACGCACATCGCAAGCAGTG
43	43	TGGATTATTGTGAGTGCCCA	139	GATATGCGTTACGTGAGTCT
44	44	CTGTGCCCTGCTCTGATGAT	140	TTCATGCTTGTGTGGAGCCA
45	45	GAAAGGTTATGCCACATGCT	141	TCCCAGAATTGTCAGCGATT
46	46	CATGAATCTGACTGCGCTCT	142	CCAATAGTGATGAGTCGCCT
47	47	TCGGCATATTTAGTGTCAGG	143	GATCTCGATTATGCTCAAGG
48	48	GCTGATGTGTCGAGTCCTAT	144	CTGTCAGATTAGTGAGCATG

D-Tag	tag1	Sequence	tag2	Sequence
49	49	GCCTGCGCTGGTCTGAATAT	145	CTGGCTCTGGATCTAGTCTC
50	50	CTGGTGGCTGTCCTACATCA	146	AGCGCACATGACATCCAGTG
51	51	ATCATATAGGATGCGGCGAT	147	TTGGCAGATTTGGTATGCAC
52	52	CTGTGAGCTGCGGTATCTAT	148	CCATATCATGTAAGGGCGTG
53	53	GCATCATGTTCACGGTGGTT	149	CACGGATCTGCCGCTAGAAT
54	54	CTGTGATCCGCTGTACTGAG	150	ATCCAGCATCTCGGGTACTG
55	55	CTGTAGCTGCTGTCCGGTAT	151	CACACATGGGCAGGCGTATA
56	56	GATCACTGTGGTCCCTGTCT	152	CATGAGAGTGCGGCTAACTA
57	57	ATGAATCCTCGGCTCTGGTG	153	GATCTGACTTAGTAGTGGCT
58	58	GCCTGATTCCCGCTGATTAT	154	TCGGCAGCTTTAGGACATGC
59	59	CGGCTGGGTAGCATCATCTA	155	AGCAGAGATGGACAGACCTC
60	60	AATACTATGAGCAAGCCGTG	156	TATAGCAATTCGGTATGCGG
61	61	TTGTCTCATTTGACGGGAGC	157	TGCTAACGTGTGCGTGTCTC
62	62	AGAGCGAATCGGTCAGCCAT	158	GGCATAAGTTCGCATAGATG
63	63	CTGTATCTGGTATGGTGCAA	159	GATCTGGCTTCAACTGTATG
64	64	TATCTGAATTGCGTGCTAGG	160	CTGTGCTCAGATCGGATTAT
65	65	CTGTCGTCTAGTCTCTGAGG	161	TTGGTTCATTTGAGAGCAGC
66	66	CTTGGGACGAGTGCTCACAT	162	AGCACTCTGGCAGCAGTCAT
67	67	GTGCGAGCTTCTGATTACTG	163	GCCGATATTGAATCAGATGG
68	68	GACTAGGTGCATGTCTCACT	164	AATAGACATCAGTAGCTCCG
69	69	GCTGATGGTGGTCAATATCT	165	CACACGGCTGGAGAGCATAT
70	70	TGGATGGATTTCACTGTAGC	166	TGGATGCTTGTGATTCAGCC
71	71	CTGTATATCGAGGTGAGTCT	167	CGCATACGTGTGGACTGATA
72	72	CACTCACTGGCACGGTATAG	168	CGCATGGCTGGGCATACATA
73	73	CCACATGATCGCGTCTGATG	169	ATCTGGCACTGACTCTCAAT
74	74	TATCTGCGTTGCTGACGTGC	170	AGCGGGCATCGTAGCATACT
75	75	CTGTAGCATGGTGACGCATC	171	TGCCAGAGTTTCAGCTTCTC
76	76	GATAGAGGTTCACGCACTGC	172	GCATCCCGTGTGGTCAGATA
77	77	CGCTGGGCTAGGATCACTAT	173	TAGTGACATTGGATGCAAGC
78	78	TGCTGACTTGTGATGCTGCC	174	TGCGTTGGTTGTCGCTCACA
79	79	CACTTCCGGCATCTATGTTG	175	GTGCGTCCTTCTGCAATCTG
80	80	TCCATGTCTTGGGTGATGCC	176	GGCAGCTTGAGACGCATTCT
81	81	GGAATCTGTGTAGTATGCCT	177	GATAGGCATTCAGCTCACTT
82	82	TAGTGGCTTGGCGTCAATTC	178	GATATGGTGGGATACATCTC
83	83	GACGCCTATGGTATCTGTTG	179	GTGCCTGTTGACATATAGTG
84	84	CCAATGCAGGTGAGTGAAAT	180	ATCTGGATCAACGTCGCGCT
85	85	CCACGATGTAGAGGATGCTC	181	AAGAGAGCTGAATCACGTCT
86	86	GGAATGTCTGCCGTGCCATA	182	ATCATGTCAAGATCAGTGGG
87	87	TGCTGTGTTGGTGTCAAACC	183	GAAACACTTATCAGATCGCG
88	88	CTGGTGACTGGTAGATCATC	184	GCTGCATATTCGTAACCATG
89	89	GACGTATTCCCGCCATGATG	185	TGCTTCATTGTGCAGGATGC
90	90	GTCGATGCTTCTCATGCTGG	186	ATCTCCACTGCATCAGGTGA
91	91	TATCCCGTTGCGTGGCATTG	187	TCGCAATGTTGTTCACCTGG
92	92	CTGTAGTCTCATGGTCGAGT	188	TGCGTTGCTTGGTGTCCACA
93	93	TCTATGAATTGGTATCGCGG	189	TATATTAAGTGCTGTGGCGC
94	94	TTCGCAGATTTGCCCTCTAG	190	CAGACGTATCTCTTGTTGCA
95	95	CGTCTCGCTGGTCACTAATG	191	CTTATGGCGCTGTCGGCTAT
96	96	GATCTGATTACGTCTCTCGC	192	CTGTCATCTCTGTCAATGGA

Strategies for Signature-tagged Mutagenesis

strains containing double-tagged transposons was assembled via selection of the clones producing the strongest signal. This resulted in more uniform signal intensities among the tags.

Improved conjugal DNA transfer method for transposon mutagenesis

During conjugation experiments, problems are often encountered by frequent lysogenization of the recipient strain by the spontaneous release of λ*pir* prophage in the donor strain. Similar problems during conjugation have been described previously (Schweizer, 1994). The frequency of ampicillin-resistant (Amp^r) exconjugants carrying markers of the delivery plasmid can be highly variable depending on the recipient strain. For example, as many as 52% of exconjugants in *Vibrio cholerae* were found to be ampicillin-resistant (Chiang and Mekalanos, 1998). For different *E. coli* donor strains, the frequency was shown to vary from 10% to 90% (de Lorenzo *et al.*, 1990). We attempted to use sodium citrate to inhibit possible phage adsorption during mating; however, this completely inhibited conjugation. The addition of sodium citrate to the selective media at a later stage increased the relative proportion of the Amp^s colonies, but resulted in a significant decrease in the conjugation efficiency. These results imply that divalent cations are possibly required for both phage infection and conjugation. To avoid complications caused by potential prophage induction, we employed an *E. coli* 19851 donor strain with a stable *pir*$^+$ gene (Metcalf *et al.*, 1996). Use of the donor library based on this strain resulted in a sufficient number of exconjugants. Almost all of the exconjugants in this case were Amp^s, which implies that prophage induction was the reason for the observed high frequency of Amp^r colonies obtained in previous experiments. Southern hybridization and PCR experiments with the Amp^r clones confirmed random insertion of the delivery plasmid, possibly due to illegitimate recombination (data not shown).

Amp^s Kan^r Nal^r exconjugants were assembled on microtitre plates. Using CRMOX selection plates (Riley and Toma, 1989) and agarose gel analysis of the extracted DNA, no loss of the virulence plasmid in the randomly selected exconjugants was observed. Total DNA extracted from arbitrarily selected exconjugants was used in Southern blot experiments and the random integration of the transposon into the *Yersinia* chromosome was confirmed (Figure 10.3).

Testing of tagged *Yersinia* mutant pools in murine infection model

Determination of the appropriate infectious dose and selection of the complexity of the input pool is critical to obtain meaningful STM data (Holden and Hensel, 1998). The optimal infection dose and pool complexity depends on the route of infection, the animal model and the microorganism used (Chiang and Mekalanos, 1998). To determine the optimum infection dose and the number of mutants in each dose, we varied the

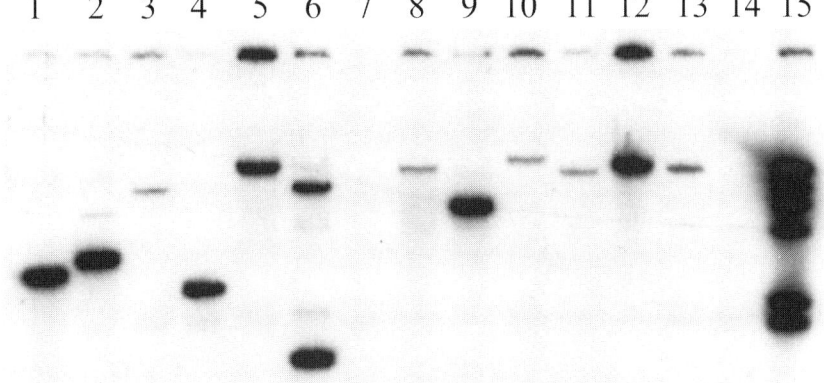

Figure 10.3. Southern blotting of DNA extracted from random mutants of *Y. pseudotuberculosis*. Lanes 1–6 and 8–13 contain different mutants numbered 1–6, respectively; lanes 7 and 14 contains wild-type DNA (recipient); in lanes 1–7 the DNA is digested with *Bam*HI and in lanes 8–14 with *Pvu*II; lane 15 contains DIG-labelled control (DNA of bacteriophage λ digested with *Hind*III).

pool complexity between 30 and 60 clones and tested two infection doses of 10^5 and 5×10^5 colony-forming units (cfu). According to published data, the LD_{50} for *Y. pseudotuberculosis* via the intravenous route in mice is between 20 (Une and Brubaker, 1984) and 100 cfu (Simonet *et al.*, 1984). Our figure in a similar test was 51.5 cfu. We, therefore, chose an infectious dose significantly (20–100 times) higher than the LD_{50} in order to select mutants with severely impaired virulence properties. In *Yersinia enterocolitica* STM experiments, the optimum infectious dose leading to reproducible results, via the intraperitoneal infection route, was 100 times greater than the LD_{50} (Darwin and Miller, 1999).

Quantitative analysis of tagged mutants

Genomic DNA was isolated from aliquots of the bacterial culture used for infection (input pools) and from bacteria recovered from spleens (output pools) (Karlyshev *et al.*, 2001). Mutants with reduced survival *in vivo* were visually identified by comparing the scanned images from arrays that had been hybridized with tags amplified from the input pools with images obtained from two independent output pools (see legend to Figure 10.2). The hybridization intensities of different oligonucleotide sequences on the array could be quantified. Quantitative analysis is important not only for comparison of different mutants, but also for statistical analysis of the results. The relative signal intensity for the same mutant may vary between different animals, the tags and may depend on infection dose and other factors. However, when these variables are taken into account, it is possible to obtain reliable data on the mutants in a complex mixture. An example of such quantitative analysis is presented in Figure 10.4.

The hybridization patterns were found to be reproducible, regardless of the infection dose (10^5 or 5×10^5 cfu) or the number of colonies in the

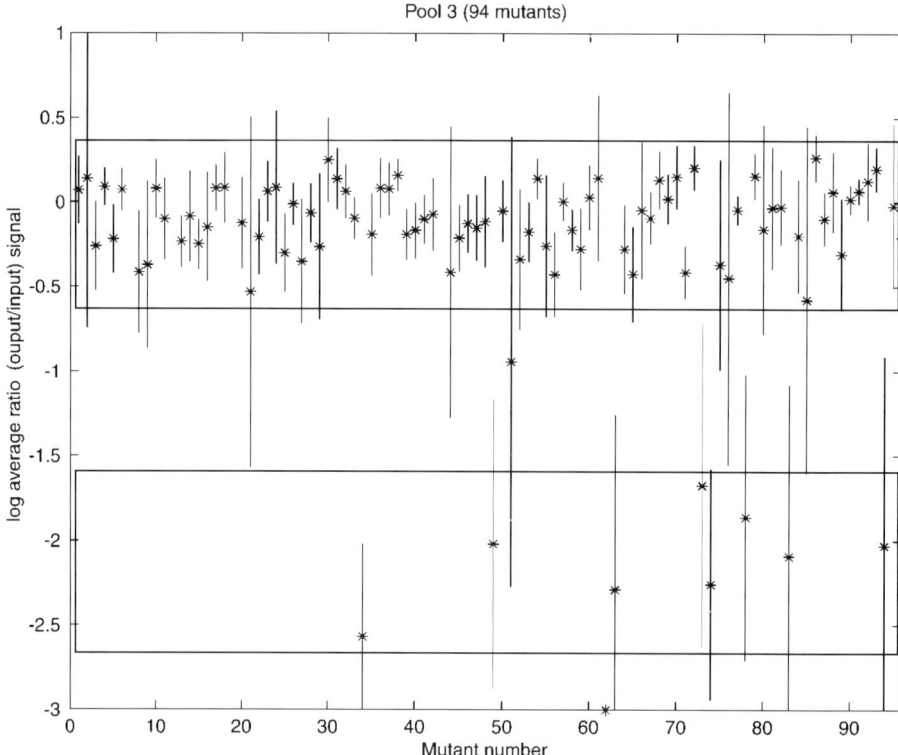

Figure 10.4. Example of quantitative analysis of hybridization results. Signals corresponding to two tags in three experiments for each of the 94 mutants are plotted. Signals for the majority of the mutants (upper box) remain unchanged, whereas several mutants clearly reveal reduced signals (lower box) and are attenuated.

output pool (10^4 or 5×10^4). However, less variability was found in smaller (lower complexity) input pools (data not shown). For larger (or higher complexity) input pools, animal to animal variability was greater than tag to tag variability, suggesting that there are biological limits on pool size. Mutants that showed reduced signals in the output pool for both tags in duplicate mice were selected for further analysis. Authenticity of the mutants was verified via direct sequencing of both tags after PCR amplification (data not shown).

Sequencing of transposon insertion sites and data analysis

Thirty-one out of 603 exconjugants exhibited a statistically valid reduction of signal intensity and were selected for further analysis using the single primer PCR sequencing procedure for identifying DNA sequences flanking a transposon insertion site (Karlyshev et al., 2000). Prediction of gene function in *Y. pseudotuberculosis* was assisted by availability of the complete sequence of the closely related genome of *Y. pestis* (Parkhill et al., 2001). We found that, in those cases where a gene counterpart in *Y. pestis* was available, the nucleotide sequence identity was higher than 90%.

Although only a partial gene sequence can be derived when analysing transposon insertion sites, the corresponding full gene sequence from *Y. pestis* could be used for function prediction.

Almost a third of the sequenced mutants had transposon insertions in genes related to polysaccharide biosynthesis (mainly a lipopolysaccharide core or O-antigen biosynthesis). All these genes, except a gene encoding O-antigen transporter, are present in *Y. pestis*. Putative interrupted virulence genes include those encoding a phospholipase A, a sensory transducer histidine kinase VirA, a putative adhesin, a Pro-dipeptidase, a negative regulator of sigma 24 transcription factor and a transcription regulator flanked by a *vspC* gene essential for secretion of virulence factors. Two mutations were in genes located in the operons involved in ammonia and nitrate transport and encode ABC transporters. As stated above, one of the genes found encoded outer membrane phospholipase PldA. The gene encoding this enzyme has also been identified in other pathogenic bacteria (Brok *et al.*, 1998), although the exact biological significance of this virulence factor remains unknown. Our work extended the list of bacteria producing this enzyme to *Yersinia* species and confirmed its importance for virulence. *In vivo* competition studies revealed significant attenuation of the *Y. pseudotuberculosis pldA* mutant. The competitive index (CI = 0.01655) obtained in the mixed infection experiment confirmed that the strain was severely attenuated. The median lethal dose (MLD) of the mutant (1.04×10^4 cfu per animal) was approximately 200-fold higher than that of the wild-type strain (51.5 cfu per animal).

Some of the transposon-mutated genes were found to have similarities to the *Y. pestis* genome but had no homologues in a non-redundant database. Other genes revealed a similarity to hypothetical *E. coli* genes, but did not have counterparts in *Y. pestis*. Genes encoding hypothetical proteins found in both *Y. pestis* and other bacteria were also identified.

The identification of 31 attenuating strains from 603 tagged transposon mutants is higher than that reported in the majority of STM studies. The higher hit rate achieved in this study may result from the modifications to the standard STM procedure. These include the use of optimized double-tagged minitransposons and quantitative hybridization analysis using Affymetrix oligonucleotide microarray technology.

◆◆◆◆◆◆ SIGNATURE-TAGGED ALLELE REPLACEMENT IN *HELICOBACTER PYLORI*

Transposon mutagenesis is not possible for many microbial species. In an effort to circumvent the requirement for transposons to introduce DNA tags into bacterial strains we developed signature-tagged allele replacement (STAR), where the DNA tags are incorporated into the cloned mutated gene prior to allelic replacement (Figure 10.5). This directed mutagenesis approach has several advantages over STM and is suitable as systematic mutagenesis strategy for the comprehensive functional characterization of a microbial genome. A similar approach has been attempted

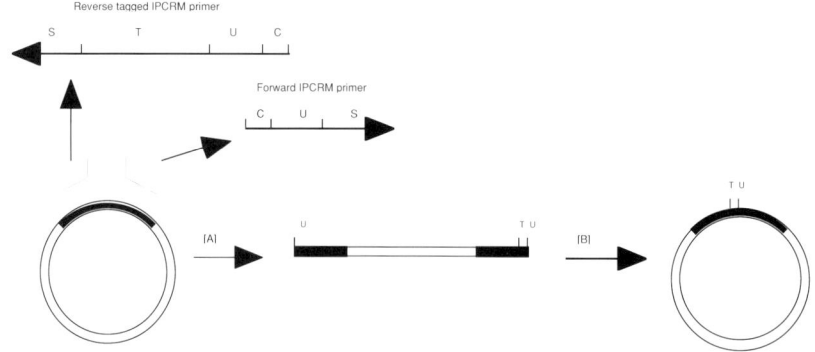

Figure 10.5. Diagrammatic representation of tagged inverse PCR mutagenesis (IPCRM) including primer design. Solid boxes represent cloned DNA. [A] represents the inverse PCR reaction and [B] represents digestion of the inverse PCR product with the unique restriction enzyme and self-ligation. U represents the unique restriction site, C a random 3-bp nucleotide clamp, T the unique 56-bp tag sequence including a unique 20-bp tag (see pages 170–1) and S the gene sequence.

for *S. cerevisiae* (Winzeler *et al.*, 1999). To develop an unbiased systematic tagged mutagenesis strategy to determine gene function in bacteria, we piloted this procedure for the gastric pathogen *Helicobacter pylori*. *H. pylori* was chosen owing to the availability of the whole-genome sequences (Tomb *et al.*, 1997), the ease of construction of defined deletion mutants by allelic replacement (Allan *et al.*, 1999) and because of its importance as a human pathogen (Cover and Blaser, 1996; Blaser, 1997).

Mutagenesis and tagging strategy

Defined isogenic *H. pylori* SS1 mutants were constructed using inverse PCR mutagenesis (IPCRM) (Dorrell *et al.*, 1999). *H. pylori* genes were amplified by PCR with primers designed from the genome sequence (Tomb *et al.*, 1997) using *H. pylori* 26695 chromosomal DNA as a template. The PCR products were cloned into pUC18 and sequenced. Defined deletions, a unique restriction site and an individual DNA tag sequence were introduced into the cloned gene fragments by a modified IPCRM procedure (see Figure 10.5). The forward IPCRM primer consists of a 3-bp clamp to assist restriction endonuclease binding, the 6-bp unique restriction site and a 15-bp sequence from the cloned gene, downstream of the desired deletion. The reverse IPCRM primer consists of a 3-bp clamp, the 6-bp unique restriction site, a 56-bp sequence including a unique 20-bp tag (Shoemaker *et al.*, 1996) and a 15-bp sequence from the cloned gene, upstream of the desired deletion. The unique 20-bp tag sequences are flanked by two invariant sequences of 18-bp, which allow amplification of any tag using the same pair of fluorescein- or biotin-labelled oligonucleotide primers (Shoemaker *et al.*, 1996). A kanamycin resistance cassette (Trieu-Cuot *et al.*, 1985) was then cloned into the unique restriction site and these constructs introduced into the wild-type *H. pylori* SS1 strain by

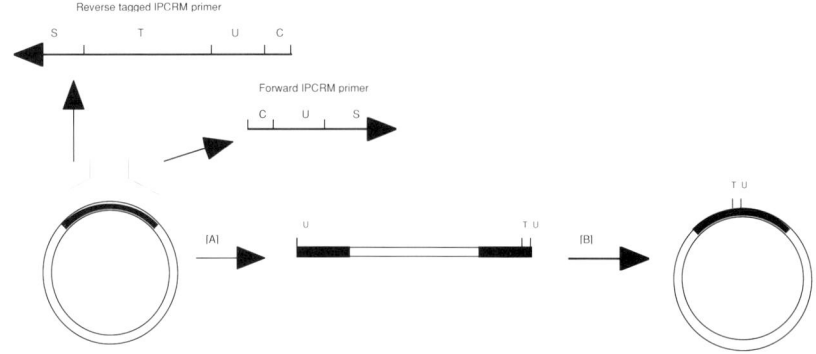

electroporation to produce defined isogenic mutants, as described previously (Dorrell *et al.*, 1999; Foynes *et al.*, 1999).

Tagged defined isogenic mutants were constructed using STAR methodology for 11 genes potentially important for survival in low pH, low concentration of iron ions and pathogenesis (see Table 10.2) identified from the *H. pylori* 26695 genome sequence (Tomb *et al.*, 1997). Homologous recombination between the wild-type gene and the STAR constructs results in the insertion of the kanamycin cassette and deletion–disruption of the wild-type gene. Defined mutations were confirmed by PCR and Southern blotting as described previously (Dorrell *et al.*, 1999). Incorporation of the tag sequences was confirmed by sequencing (data not shown).

Tag detection

The tags were detected by PCR amplification from genomic DNA isolated from the tagged defined isogenic *H. pylori* mutants using one biotin-labelled primer and one unlabelled primer complementary to the common regions flanking each tag sequence (Winzeler *et al.*, 1999). The amplicons were then hybridized to high-density oligonucleotide arrays (Affymetrix, Santa Clara, CA, USA) containing the tag complements. Hybridization at specific locations on the array was detected by an Affymetrix GeneChip™ Scanner as described previously (Winzeler *et al.*, 1999). The fluorescent intensities for different positions on the array corresponding to the tags were determined using the 75th percentile method using the Affymetrix GeneChip™ software package. The relative abundance of each mutant was calculated by first normalizing the hybridization intensity for each scan above background and then computing the fold change for each feature relative to time zero. Relative

Table 10.2. Targeted *H. pylori* mutants by STAR

Gene	HP number[a]	Function/predicted function	Reference
fliQ	1419	Flagellar	Foynes *et al.* (1999), Porwollik *et al.* (1999)
cheAY2	0392	Chemotaxis histidine kinase	Foynes *et al.* (2000)
feoB	0687	Fe^{2+} and Fe^{3+}	Velayudhan *et al.* (2000a)
fliS	0753	Flagellar biosynthesis	Allan *et al.* (2000)
fliF	0351	Basal body M ring protein	Allan *et al.* (2000)
cheY1	1067	Chemotaxis response regulator	Foynes *et al.* (2000)
cagP	0536	*Cag* pathogenicity island	Censini *et al.* (1996)
pldA	0499	Phospholipase	Dorrell *et al.* (1999)
fliI	1420	Flagellar	Jenks *et al.* (1997), Porwollik *et al.* (1999)
cagF	0543	*Cag* pathogenicity island	Censini *et al.* (1996)
ppiB	1441	Peptidyl-proplyl cis-trans isomerase B	This study

[a] According to Tomb *et al.* (1997).

abundance was plotted against time to show individual growth curves for each mutant within the pool.

Growth studies under *in vitro* conditions

The growth of the 11 different mutants in broth was investigated separately and in parallel using the tags, as proof of principle of the ability to monitor different levels of individual mutants in a pooled culture. The relative abundance of each mutant was calculated by comparing the recorded intensity for each tag at each time point to the initial intensity, then plotted against time to show individual growth curves for each mutant within the pool (see Figure 10.6A). The growth of the pooled culture and a control culture (wild-type strain SS1) was also monitored by

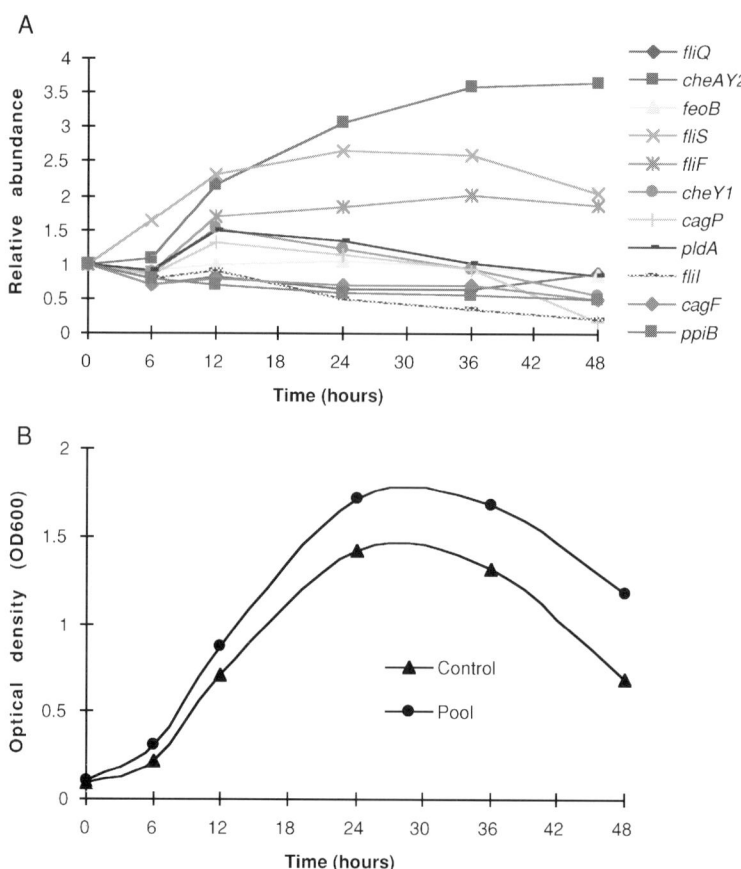

Figure 10.6. Growth of a pooled culture containing 11 different tagged *H. pylori* mutants under batch culture conditions. (A) Growth of individual tagged *H. pylori* mutants within the pooled culture. (B) Growth of pooled and control (wild-type SS1) cultures recorded by OD_{600} readings. The relative abundance was calculated by first normalizing the hybridization intensity for each scan above background, and then computing the fold change for each feature relative to time zero. Reproduced in colour between pages 178 and 179.

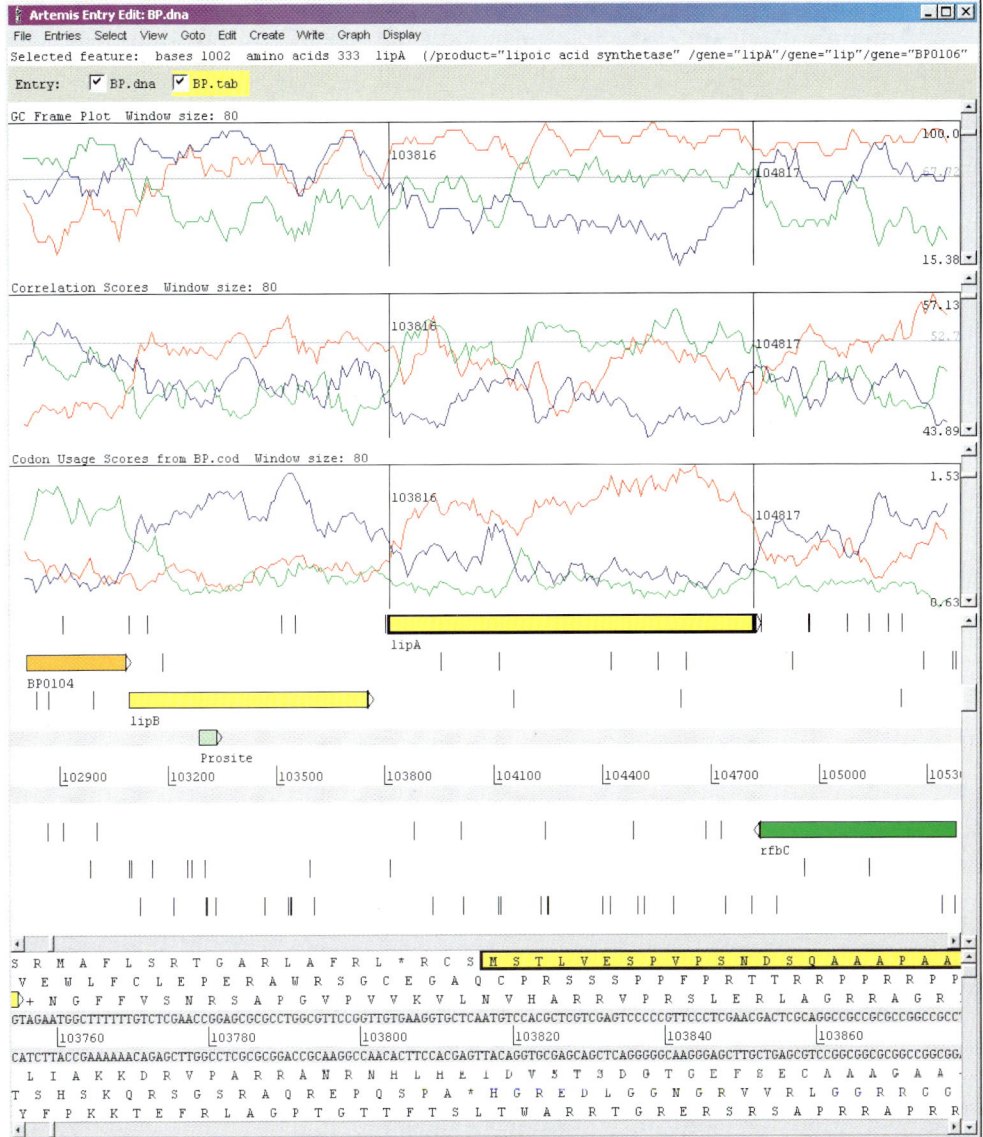

Figure 1.1. Illustration of coding parameters in Artemis. A short section of the genome of *Bordetella pertussis* is shown. In the bottom two panels is the six-frame translation of the sequence at two zoom levels, with stop codons represented as vertical bars in the zoomed-out upper panel, and the predicted genes marked as open boxes. The top graph (GC frameplot) shows a position-specific G + C content plot (FramePlot; Bibb *et al.*, 1984). The coloured lines represent the G + C content of the first, second and third positions of the triplet code, relative to the start of the sequence. For example, red represents the third position for a CDS in frame 1, while green represents the third position for a CDS in frame 2, and blue the third position for a CDS in frame 3. In G + C-rich organisms, the third position of the codon tends to be more G + C-rich than the first and second. The second graph (correlation scores) indicates the results of a calculation of the correlation between the amino-acid usage of the coding frame within the specified window, and the general amino-acid usage calculated from genes in EMBL. The green line indicates the score for frame 1, blue for frame 2 and red for frame 3. Values close to the line at 52.7 indicate a strong likelihood of coding. This calculation is independent of codon usage and G + C content. The third plot indicates the codon usage scores compared with a codon usage table generated from *B. pertussis* genes in the public databases (http://www.kazusa.or.jp/codon/; Nakamura *et al.*, 2000). In this case the red line indicates a good codon usage score in frame 1, green in frame 2 and blue in frame 3.

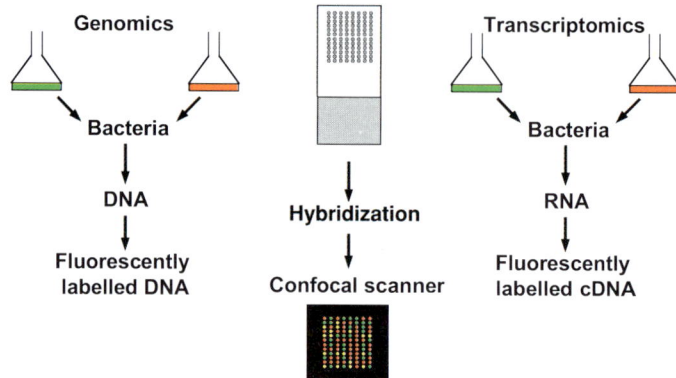

Figure 5.1. Schematic of experimental approach for the use of a bacterial whole-genome microarray to study comparative genomics and gene expression profiles. The two DNA or RNA preparations to be compared on the microarray are extracted from the bacteria and each of the samples are labelled separately with one or other of the two fluorescent dyes, commonly Cy3 and Cy5. The labelled samples are mixed and co-hybridized under a coverslip on the array. When scanned using a dual laser scanner the relative intensities of the two fluors hybridized to each reporter indicate the relative abundance of the particular DNA or mRNA species in the two samples.

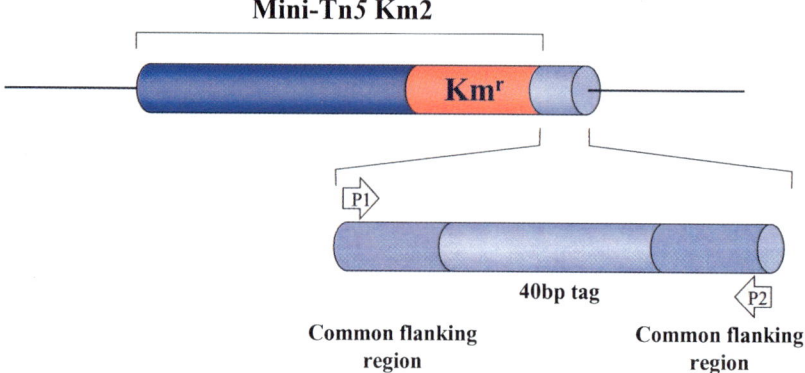

Figure 9.1. Structure of tagged transposon utilized for the mutagenesis. The 40 bp variable tags are flanked by two common regions allowing amplification of the tag by PCR using primers P1 and P2.

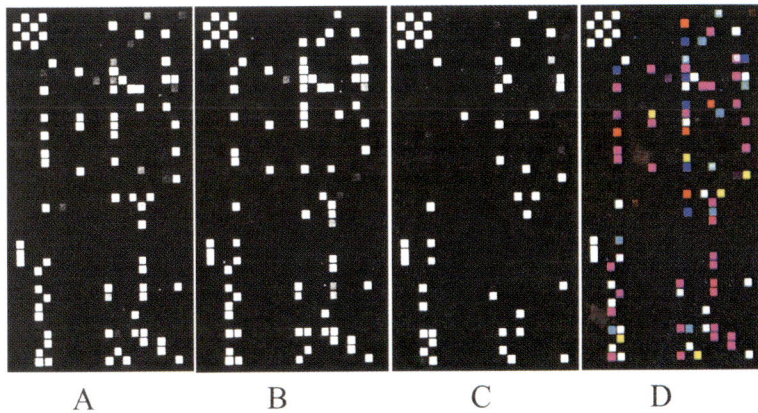

Figure 10.2. Comparative analysis of scanned microarrays using imaging software. (A, B and C) Fragments of hybridization images using probes derived from pools of three different libraries of tagged transposons. (D) Images from (A), (B) and (C) were converted into red, blue and green colour images, respectively, and overlaid using Adobe PhotoShop 5 software. White spots indicate similar relative abundance of respective tags in all pools. Such image overlaying technique was also used for the detection of attenuated mutants (see text).

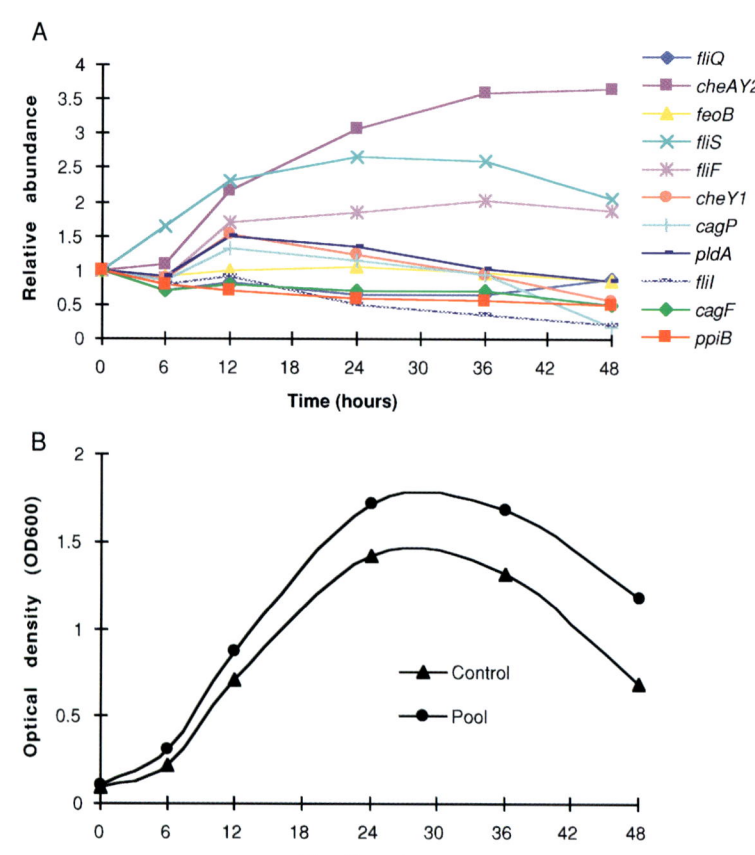

Figure 10.6. Growth of a pooled culture containing 11 different tagged *H. pylori* mutants under batch culture conditions. (A) Growth of individual tagged *H. pylori* mutants within the pooled culture. (B) Growth of pooled and control (wild-type SS1) cultures recorded by OD_{600} readings. The relative abundance was calculated by first normalizing the hybridization intensity for each scan above background, and then computing the fold change for each feature relative to time zero.

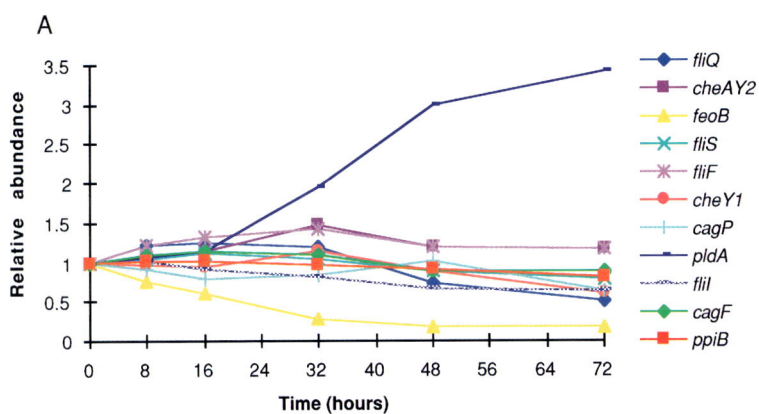
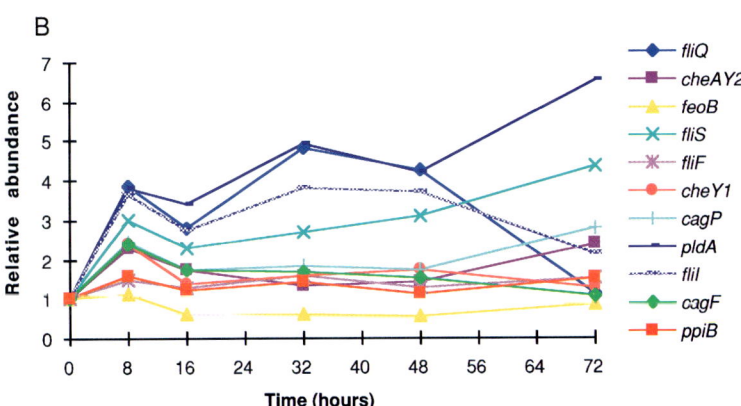

Figure 10.7. Growth of a pooled culture containing 11 different tagged *H. pylori* mutants under *in vitro* stress conditions. (A) Growth of individual tagged *H. pylori* mutants within the pooled culture under low pH growth conditions and (B) under iron-limiting growth conditions.

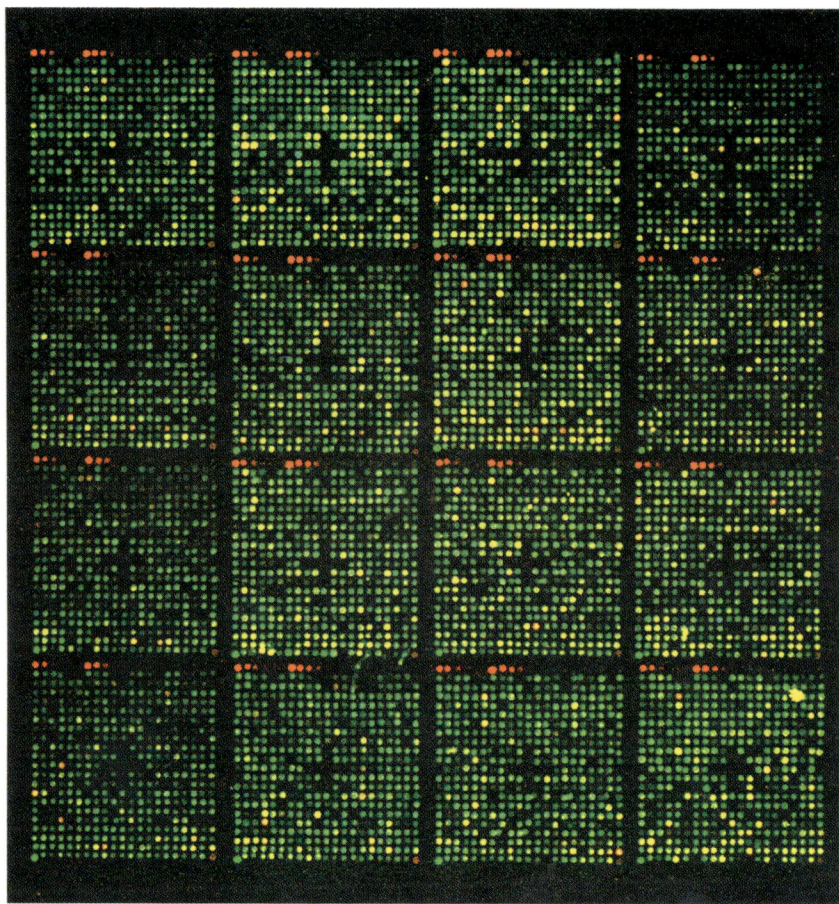

Figure 17.3. Version 1 of the 'CoelicolorArray' representing ca. 6000 ORFs. Cy5-labelled cDNA (red) generated from RNA isolated from the prototrophic *S. coelicolor* MT1110 grown on cellophane-coated Oxoid Nutrient Agar for 25 h was co-hybridized with Cy3-labelled *S. coelicolor* M145 genomic DNA (green). In each subarray the first row contains a rRNA dilution series and negative control (*Pseudomonas* ORF) spots; a number of carry-over negative controls are also included and empty spots form a cross in the middle of each subarray.

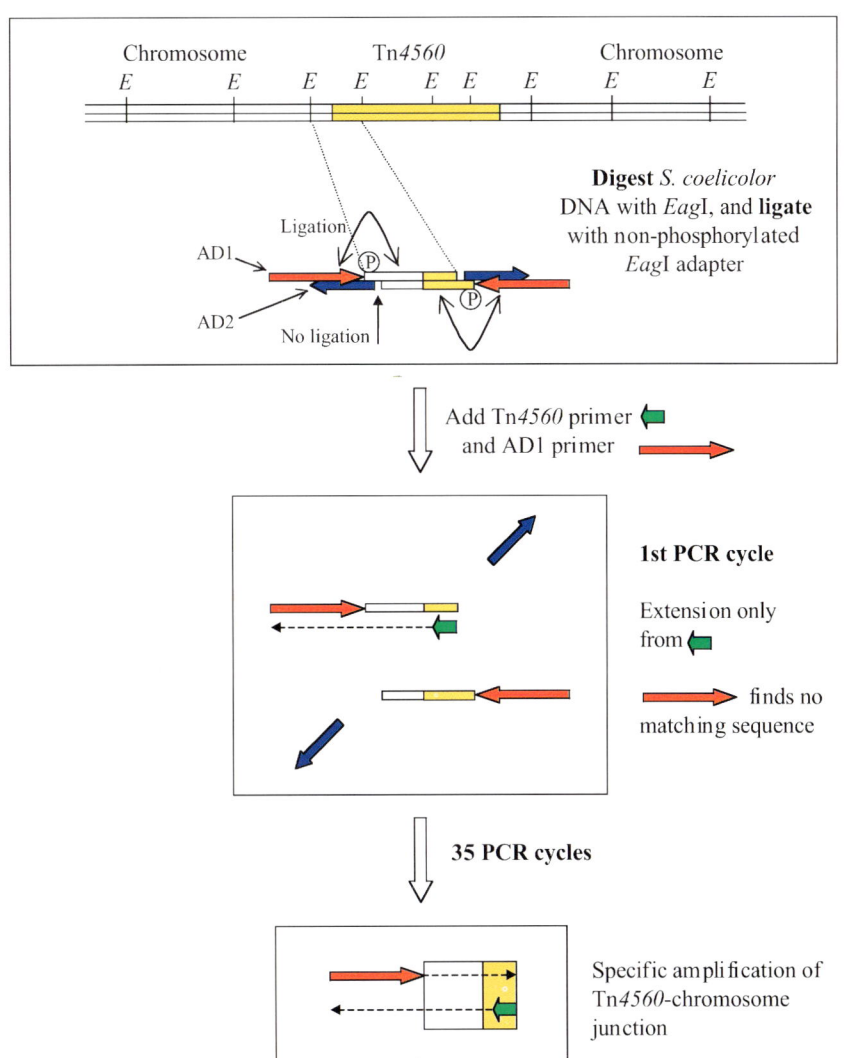

Figure 17.6. Use of ligation-mediated PCR to detect a Tn4560 insertion in a specific gene (applicable to 10^5-complexity libraries).

recording OD_{600} readings at each time point. The two growth curves showed little difference (see Figure 10.6B), suggesting that the growth of a pooled culture of tagged *H. pylori* mutants as a whole is very similar to the growth of a single strain. High variation in the growth rate of different mutants within the pooled culture was observed. The growth rate of the pooled culture as a whole mimics that of a single strain, whether it be the wild-type or a single mutant, as individually each mutant had a growth rate in broth showing no significant difference to that of the wild-type strain SS1 (data not shown). This indicates that in a pooled culture, certain mutants appear to be 'fitter' than other mutants and are able to grow at a more rapid rate.

Growth studies under *in vitro* stress conditions

Following the successful monitoring of individual mutants in a pooled culture, two experiments were performed under the physiologically relevant *in vitro* stresses of low pH and iron-limiting conditions over a 72-hour period. The data from the low pH growth study (see Figure 10.7A) show nine of the mutants remaining at a fairly consistent level within the pool. However, the abundance of the *pldA* mutant increases dramatically with time relative to the rest of the pool, whilst the *feoB* mutant decreases to a point where the tag is undetectable at 72 h. The data from the low iron growth study (see Figure 10.7B) are more complex. The abundance of the *pldA*, *fliI* and *cagF* mutants increases relative to the rest of the pool for 48 h, but then the *fliI* and *cagF* mutants decline while the *pldA* mutant continues to increase. The abundance of the *fliS* mutant also increases with time, but to a lesser extent than the *pldA* mutant. The remaining mutants stay at a constant level within the pool, apart from the *feoB* mutant, the abundance of which decreases.

No explanation for the faster relative growth rate of the *pldA* mutant under either low pH or iron-limiting conditions is immediately apparent. The *pldA* mutant shows no differences in growth rate in broth when compared individually to the wild-type strain SS1 or any of the other mutants, and the *pldA* mutant did not dominate the pooled broth culture (see Figure 10.6A). The growth of the *feoB* mutant in broth in comparison to the rest of the pooled mutants is not severely affected in contrast to the situation under *in vitro* stress conditions (see Figures 10.6 and 10.7). FeoB has been shown to be important in both Fe^{2+} and Fe^{3+} transport (Velayudhan *et al.*, 2000b). Thus a decrease in the relative abundance of the *feoB* mutant within the pool under both *in vitro* stress conditions is not surprising.

Discussion on STAR technology

The results presented here show the feasibility of incorporating unique DNA tags into a systematic mutagenesis strategy for the functional analysis of a bacterium and using these tags to provide rapid quantitative fitness data under various growth conditions. Indeed, the ease of monitoring tagged mutants during the time course of experiments, particularly

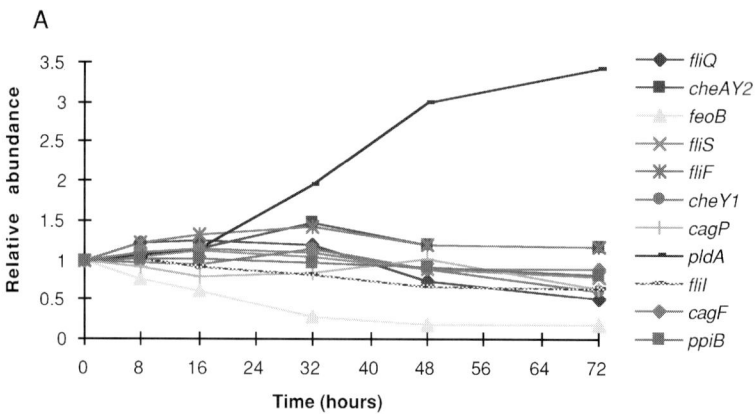

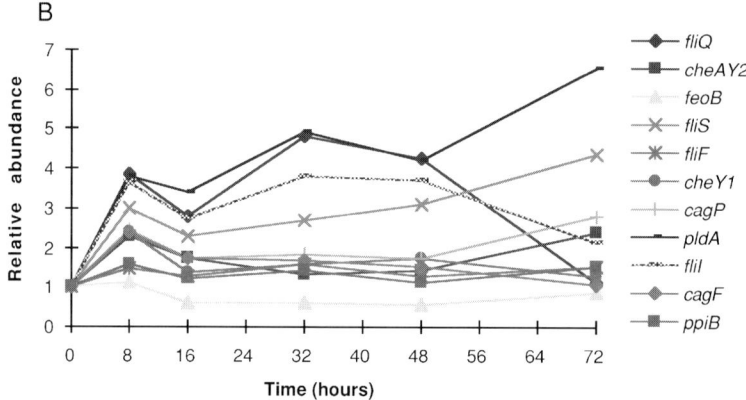

Figure 10.7. Growth of a pooled culture containing 11 different tagged *H. pylori* mutants under *in vitro* stress conditions. (A) Growth of individual tagged *H. pylori* mutants within the pooled culture under low pH growth conditions and (B) under iron-limiting growth conditions. Reproduced in colour between pages 178 and 179.

those using more complex environments or animal models, is a major advantage of this technique. A reduction in the number of animals required to perform *in vivo* screens would be another major benefit of using pools of mutants. The integrated tag also serves as a unique permanent identifier for each mutant. STAR avoids the use of transposons, which generally integrate randomly into the genome, thus potentially allowing a systematic unbiased genetic analysis of the complete genome. Furthermore, for several bacteria, such as *H. pylori* and *Campylobacter jejuni*, there have been problems with non-random integration of transposons into the genome, which have only recently started to be addressed (Golden *et al.*, 2000; Hendrixson *et al.*, 2001). As the gene target is predetermined, STAR removes the need to sequence the DNA flanking a transposon to identify the mutated gene. The same generic tag sequences have multiple applications in parallel analysis. The same set of tags can be incorporated into any organism and can also be used for polymorphism detection

(Winzeler *et al.*, 1999). Because of the tag's broad utility, tag detection methods should become widely available. An additional benefit of a whole-genome STAR project is in the systematic approach where non-viable mutants can be readily identified. Such mutations can thus be considered to be lethal, thus the gene products of these essential genes are potential drug targets.

IPCRM has been used to construct defined bacterial mutants in bacteria as diverse as *Brucella suis*, *C. jejuni* and *Yersinia enterocolitica* (Li *et al.*, 1996; Dorrell *et al.*, 1998; Bras *et al.*, 1999). The application of the STAR protocol, combined with the availability of the complete genome sequences of many bacterial pathogens, will prove to be an extremely useful technique for the functional analysis of bacterial pathogens. Most bacterial genome sequencing projects use random libraries of ≈2 kb genome fragments cloned in pUC18, such as with the *C. jejuni* 11168 genome project (Parkhill *et al.*, 2001). After assembly of the sequence, the resulting ordered plasmid library provides an ideal template for a whole-genome STAR approach, as a clone for each open reading frame can be selected from the ordered collection (Dorrell and Wren, 1998). In this light, the construction of a whole-genome STAR library should prove to be a cost-effective and long-term investment for the scientific community.

Acknowledgements

We acknowledge financial support for our research from the BBSRC and the Wellcome Trust. We are grateful to Dr R. W. Titball and Dr P. C. F. Oyston (Defence Evaluation Research Agency, Porton Down, UK) for performing experiments *in vivo*.

References

Allan, E., Foynes, S., Dorrell, N. and Wren, B. W. (1999). Genetic characterization of the gastric pathogen *Helicobacter pylori*. In *Methods in Microbiology: Genetic Methods for Diverse Prokaryotes*, Vol. 29 (Smith, M. C. M. and Sockett, R. E., eds), pp. 329–346, London: Academic Press.

Allan, E., Dorrell, N., Foynes, S., Anyim, M. and Wren, B. W. (2000). Mutational analysis of genes encoding the early flagellar components of *Helicobacter pylori*: evidence for transcriptional regulation of flagellin A biosynthesis. *J. Bacteriol.* **182**, 5274–5277.

Blaser, M. J. (1997). Ecology of *Helicobacter pylori* in the human stomach. *J. Clin. Invest.* **100**, 759–762.

Bras, A. M., Chatterjee, S., Wren, B. W., Newell, D. G. and Ketley, J. M. (1999). A novel *Campylobacter jejuni* two-component regulatory system important for temperature-dependent growth and colonization. *J. Bacteriol.* **181**, 3298–3302.

Brok, R. G., Boots, A. P., Dekker, N., Verheij, H. M. and Tommassen, J. (1998). Sequence comparison of outer membrane phospholipases A: implications for structure and for the catalytic mechanism. *Res. Microbiol.* **149**, 703–710.

Censini, S., Lange, C., Xiang, Z., Crabtree, J. E., Ghiara, P., Borodovsky, M., Rappuoli, R. and Covacci, A. (1996). cag, a pathogenicity island of *Helicobacter pylori*, encodes type I-specific and disease-associated virulence factors. *Proc. Natl Acad. Sci. USA* **93**, 14648–14653.

Chiang, S. L. and Mekalanos, J. J. (1998). Use of signature-tagged transposon mutagenesis to identify *Vibrio cholerae* genes critical for colonization. *Mol. Microbiol.* **27**, 797–805.

Cover, T. L. and Blaser, M. J. (1996). *Helicobacter pylori* infection, a paradigm for chronic mucosal inflammation – pathogenesis and implications for eradication and prevention. *Adv. Intern. Med.* **41**, 85–117.

Darwin, A. J. and Miller, V. L. (1999). Identification of *Yersinia enterocolitica* genes affecting survival in an animal host using signature-tagged transposon mutagenesis. *Mol. Microbiol.* **32**, 51–62.

de Lorenzo, V., Herrero, M., Jakubzik, U. and Timmis, K. N. (1990). Mini-Tn5 transposon derivatives for insertion mutagenesis, promoter probing, and chromosomal insertion of cloned DNA in gram-negative eubacteria. *J. Bacteriol.* **172**, 6568–6572.

Dorrell, N. and Wren, B. W. (1998). From genes to genome biology: a new era in *Helicobacter pylori* research. *Gut* **42**, 451–453.

Dorrell, N., Spencer, S., Foulonge, V., Guigue-Talet, P., O'Callaghan, D. and Wren, B. W. (1998). Identification, cloning and initial characterisation of FeuPQ in *Brucella suis*: a new sub-family of two-component regulatory systems. *FEMS Microbiol. Lett.* **162**, 143–150.

Dorrell, N., Martino, M. C., Stabler, R. A., Ward, S. J., Zhang, Z. W., McColm, A. A., Farthing, M. J. and Wren, B. W. (1999). Characterization of *Helicobacter pylori* PldA, a phospholipase with a role in colonization of the gastric mucosa. *Gastroenterology* **117**, 1098–1104.

Foynes, S., Dorrell, N., Ward, S. J., Zhang, Z. W., McColm, A. A., Farthing, M. J. G. and Wren, B. W. (1999). Functional analysis of the roles of FliQ and FlhB in flagellar expression in *Helicobacter pylori*. *FEMS Microbiol. Lett.* **174**, 33–39.

Foynes, S., Dorrell, N., Ward, S. J., Stabler, R. A., McColm, A. A., Rycroft, A. N. and Wren, B. W. (2000). *Helicobacter pylori* possesses two CheY response regulators and a histidine kinase sensor, CheA, which are essential for chemotaxis and colonization of the gastric mucosa. *Infect. Immun.* **68**, 2016–2023.

Golden, N. J., Camilli, A. and Acheson, D. W. (2000). Random transposon mutagenesis of *Campylobacter jejuni*. *Infect. Immun.* **68**, 5450–5453.

Hendrixson, D. R., Akerley, B. J. and DiRita, V. J. (2001). Transposon mutagenesis of *Campylobacter jejuni* identifies a bipartite energy taxis system required for motility. *Mol. Microbiol.* **40**, 214–224.

Hensel, M., Shea, J. E., Gleeson, C., Jones, M. D., Dalton, E. and Holden, D. W. (1995). Simultaneous identification of bacterial virulence genes by negative selection. *Science* **269**, 400–403.

Holden, D. W. and Hensel, M. (1998). Signature tagged mutagenesis. *Meth. Microbiol.* **27**, 359–369.

Jenks, P. J., Foynes, S., Ward, S. J., Constantinidou, C., Penn, C. W. and Wren, B. W. (1997). A flagellar-specific ATPase (FliI) is necessary for flagellar export in *Helicobacter pylori*. *FEMS Microbiol. Lett.* **152**, 205–211.

Karlyshev, A. V., Pallen, M. J. and Wren, B. W. (2000). Single-primer PCR procedure for rapid identification of transposon insertion sites. *BioTechniques* **28**, 1078–1082.

Karlyshev, A. V., Oyston, P. C., Williams, K., Clark, G. C., Titball, R. W., Winzeler, E. A. and Wren, B. W. (2001). Application of high-density array-based signature-tagged mutagenesis to discover novel yersinia virulence-associated genes. *Infect. Immun.* **69**, 7810–7819.

Li, S. R., Dorrell, N., Everest, P. H., Dougan, G. and Wren, B. W. (1996). Construction and characterization of a *Yersinia enterocolitica* O:8 high-temperature requirement (*htrA*) isogenic mutant. *Infect. Immun.* **64**, 2088–2094.

Mei, J. M., Nourbakhsh, F., Ford, C. W. and Holden, D. W. (1997). Identification of *Staphylococcus aureus* virulence genes in a murine model of bacteraemia using signature-tagged mutagenesis. *Mol. Microbiol.* **26**, 399–407.

Metcalf, W. W., Jiang, J., Daniels, L. L., Kim, S-K., Haldimann, A. and Wanner, B. L. (1996). Conditionally replicative and conjugative plasmids carrying *lacZa* for cloning, mutagenesis, and allele replacement in bacteria. *Plasmid* **35**, 1–13.

Parkhill, J., Wren, B. W., Thomson, N. R., Titball, R. W., Holden, M. T., Prentice, M. B., Sebaihia, M., James, K. D., Churcher, C., Mungall, K. L., Baker, S., Basham, D., Bentley, S. D., Brooks, K., Cerdeno-Tarraga, A. M., Chillingworth, T., Cronin, A., Davies, R. M., Davis, P., Dougan, G., Feltwell, T., Hamlin, N., Holroyd, S., Jagels, K., Karlyshev, A. V., Leather, S., Moule, S., Oyston, P. C., Quail, M., Rutherford, K., Simmonds, M., Skelton, J., Stevens, K., Whitehead, S. and Barrell, B. G. (2001). Genome sequence of *Yersinia pestis*, the causative agent of plague. *Nature* **413**, 523–527.

Polissi, A., Pontiggia, A., Feger, G., Altieri, M., Mottl, H., Ferrari, L. and Simon, D. (1998). Large-scale identification of virulence genes from *Streptococcus pneumoniae*. *Infect. Immun.* **66**, 5620–5629.

Porwollik, S., Noonan, B. and O'Toole, P. W. (1999). Molecular characterization of a flagellar export locus of *Helicobacter pylori*. *Infect. Immun.* **67**, 2060–2070.

Riley, G. and Toma, S. (1989). Detection of pathogenic *Yersinia enterocolitica* by using congo-red magnesium oxalate agar medium. *J. Clin. Microbiol.* **27**, 213–214.

Schweizer, H. P. (1994). A method for construction of bacterial hosts for *lac*-based cloning and expression vectors: alpha-complementation and regulated expression. *Biotechniques* **17**, 452–454, 456.

Shoemaker, D. D., Lashkari, D. A., Morris, D., Mittmann, M. and Davis, R. W. (1996). Quantitative phenotypic analysis of yeast deletion mutants using a highly parallel molecular bar-coding strategy. *Nature Genet.* **14**, 450–456.

Simonet, M., Mazigh, D. and Berche, P. (1984). Growth of *Yersinia pseudotuberculosis* in mouse spleen despite loss of a virulence plasmid of mol wt 47×10. *J. Med. Microbiol.* **18**, 371–375.

Tomb, J. F., White, O., Kerlavage, A. R., Clayton, R. A., Sutton, G. G., Fleischmann, R. D., Ketchum, K. A., Klenk, H. P., Gill, S., Dougherty, B. A., Nelson, K., Quackenbush, J., Zhou, L. X., Kirkness, E. F., Peterson, S., Loftus, B., Richardson, D., Dodson, R., Khalak, H. G., Glodek, A., McKenney, K., Fitzegerald, L. M., Lee, N., Adams, M. D., Hickey, E. K., Berg, D. E., Gocayne, J. D., Utterback, T. R., Peterson, J. D., Kelley, J. M., Cotton, M. D., Weidman, J. M., Fujii, C., Bowman, C., Watthey, L., Wallin, E., Hayes, W. S., Borodovsky, M., Karp, P. D., Smith, H. O., Fraser, C. M. and Venter, J. C. (1997). The complete genome sequence of the gastric pathogen *Helicobacter pylori*. *Nature* **388**, 539–547.

Trieu-Cuot, P., Gerbaud, G., Lambert, T. and Courvalin, P. (1985). *In vivo* transfer of genetic information between gram-positive and gram-negative bacteria. *EMBO J.* **4**, 3583–3587.

Une, T. and Brubaker, R. R. (1984). *In vivo* comparison of avirulent Vwa- and Pgm- or Pstr phenotypes of *Yersiniae*. *Infect. Immun.* **43**, 895–900.

Velayudhan, J., Hughes, N. J., McColm, A. A., Bagshaw, J., Clayton, C. L., Andrews, S. C. and Kelly, D. J. (2000a). Iron acquisition and virulence in *Helicobacter pylori*: a major role for FeoB, a high-affinity ferrous iron transporter. *Mol. Microbiol.* **37**, 274–286.

Winzeler, E. A., Shoemaker, D. D., Astromoff, A., Liang, H., Anderson, K., Andre, B., Bangham, R., Benido, R., Boeke, J. D., Bussey, H., Chu, A. M., Connelly, C., Davis, K., Dietrich, F., Dow, S. W., El Bakkoury, M., Foury, F., Friend, S. H., Gentalen, E., Giaever, G., Hegemann, J. H., Jones, T., Laub, M., Liao, H.,

Liebundguth, N., Lockhart, D. J., Lucau-Danila, A., Lussier, M., M'Rabet, N., Menard, P., Mittmann, M., Pai, C., Rebischung, C., Revuelta, J., Riles, L., Roberts, C., Ross-MacDonald, P., Scherens, B., Snyder, M., Storms, R. K., Veronneau, S., Voet, M., Volckaert, G., Ward, T. R., Whelen, S., Wysocki, R., Yen, G. S., Yu, K., Zimmermann, K., Philippsen, P., Johnston, M. and Davis, R. W. (1999). Functional characterization of the *S. cerevisiae* genome by gene deletion and parallel analysis. *Science* **285**, 901–906.

Zhao, H., Li, X., Johnson, D. E. and Mobley, H. L. T. (1999). Identification of protease and *rpoN*-associated genes of uropathogenic *Proteus mirabilis* by negative selection on a mouse model of ascending urinary tract infection. *Microbiology* **145**, 185–195.

Part IV
Proteome Analysis

11 Advances in Bacterial Proteome Analysis

Stuart J Cordwell
Australian Proteome Analysis Facility, Sydney, Australia 2109

◆◆◆

CONTENTS

Introduction – the proteome approach
The 2DGE/MS approach
Applications
Complementarity and 'new' proteome technologies
'Functional' proteomics
Conclusions

◆◆◆◆◆◆ INTRODUCTION – THE PROTEOME APPROACH

The microbial genome era has heralded a renewed sense of perspective for functional and molecular biology. Since 25–50% of the open reading frames (ORFs) sequenced in individual bacterial sequencing projects are of unknown function, the challenge that remains is to utilize novel technologies to decipher the individual biology of each microorganism. Such hopes rely on methods that can provide an initially 'global' view of gene expression to pinpoint those genes that are expressed, and the biological significance of their expression under a variety of experimental situations. The technologies encompassed under the term 'proteomics' are a means of performing this task at the protein, rather than mRNA (see Chapter 8), level. The relationship of proteomics to traditional biochemistry is a reflection of the relationship between genomics and molecular biology. It simply refers to a scale that could not be dreamt of 10 years ago. Traditional proteomics experiments generally involve protein solubilization, separation, visualization, comparison and, finally, identification (Figure 11.1). In all five areas significant improvements have been introduced in the last 5–10 years including highly reproducible two-dimensional gel electrophoresis (2DGE), automation for high-throughput

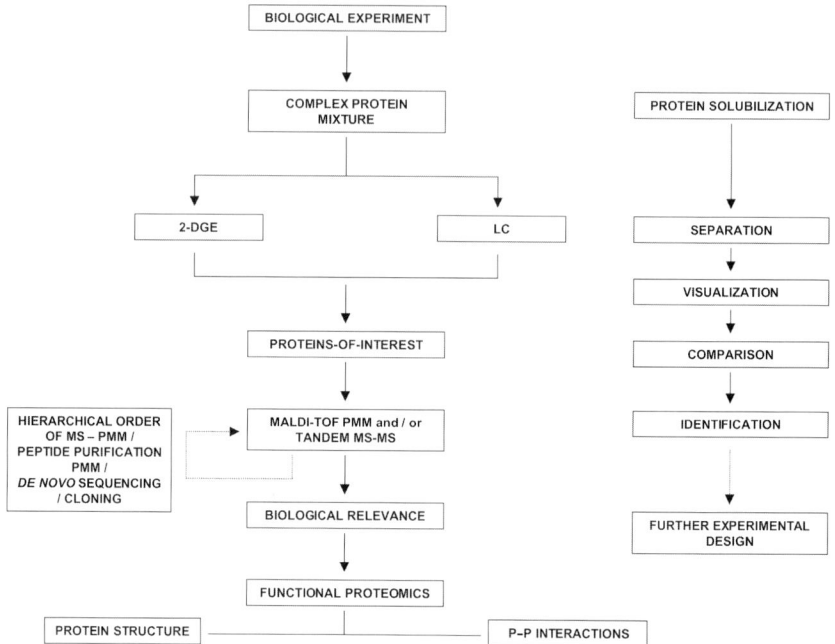

Figure 11.1. The proteome approach. Each experiment includes a 'global' phase and a 'functional' phase. The global approach aims to determine a set of 'proteins-of-interest' from amongst a complex mixture. This is performed experimentally in five steps: protein solubilization, separation (2-DGE or LC), visualization, comparison (spots or peaks of significance) and identification (via MS). The 'proteins-of-interest' are then characterized functionally to determine their interactions with other proteins and ligands.

chemical delivery to excised 2D gel 'spots', and the sensitivity and utility of mass spectrometry (MS) for protein characterization. Such developments make proteomics an attractive tool for expression analyses of microbial genomes.

Microbes are particularly suitable to proteomic analyses owing to their relatively small genomes, and the ease with which large culture volumes and hence protein can be obtained. The proteome approach allows three significant questions to be highlighted: (1) is a gene expressed; (2) when and under what conditions (genetic or environmental) is that gene expressed; and (3) is the gene-product post-translationally modified and under what conditions does that modification occur? Very basic, yet highly informative, experiments can be performed to provide answers to these questions. For example, our work on several bacteria has revealed that approximately 15–20% of the most abundant expressed bacterial proteins are of unknown function. This now equates to nearly 500 expressed 'hypotheticals' worthy of further consideration. Furthermore, the analysis of bacterial proteomes also raises interesting new perspectives with regard to the functions of previously characterized proteins. From this data point, further questions need to be asked to begin the process of deciphering the functions of 'hypothetical' proteins, or the additional functions of previously characterized proteins. Differential display proteomics allows the monitoring of global protein expression influenced by a genetic (genes and proteins belonging to a 'regulon') or an environmental (genes and proteins belonging to a 'stimulon') challenge. Each such experiment provides a new set of proteins whose expression is in some way modified (induced or repressed) by the particular stimulus. Further functional experimentation is thus directed by the results achieved using the 'global' approach. These include so-called 'functional proteomics' experiments that include the analysis of protein–protein interactions and complexes, ligand and co-factor binding assays, high-throughput enzyme assay, and the definition of tertiary structure (Pandey and Mann, 2000).

This chapter describes in detail methodology for producing high-quality 2D gels from microbial species and the identification of proteins from these gels by mass spectrometry. Finally, the latest approaches in non-2D gel-based proteomics will be discussed as the basis for a complementary strategy aimed at deciphering the 'total' microbial proteome.

◆◆◆◆◆◆ THE 2DGE/MS APPROACH

Standard proteome analysis begins with 2DGE and MS to determine 'proteins of interest'. 2DGE is still the most powerful tool available for separating complex mixtures of bacterial proteins. While this remains true, several classes of protein are excluded from 2DGE. These include very hydrophobic proteins, basic proteins, high and low molecular mass proteins and low abundance proteins that cannot be visualized owing to the restricted dynamic range afforded by complex cell lysates (reviewed in Cordwell et al., 2001). Without doubt the most significant group of proteins poorly isolated using 2DGE are hydrophobic proteins. In *Helicobacter pylori*, approximately 14% of the genome encodes proteins that have theoretical Kyte–Doolittle values that render them inaccessible to solubilization for 2DGE, while codon adaptation index values predict that only 5% of the ORFs encode products that are outside the dynamic range accessible to 2DGE for this organism (unpublished data). Surprisingly, our studies have shown that even sodium dodecyl sulphate (SDS) does not considerably improve the solubilization of such hydrophobic proteins, even in conjunction with sequential extraction protocols.

Sample preparation

The single most important step in any proteomics procedure is the preparation of the protein mixture. This step is also the most variable between different organisms and to some extent must be arrived at by a means of 'trial-and-error' in each particular laboratory. However, good advice is not to spend large periods of time or reagents perfecting methodology that has been proven elsewhere. This is especially the case for microbial research where several standard protocols exist. The principle for sample preparation is simple – determine a method that solubilizes a maximum number of proteins without determinants that interfere with isoelectric focusing (e.g. ionic detergents such as SDS) that is performed in the first dimension separation. Sample buffers for 2DGE contain several constituents and their roles are described in Table 11.1.

Besides the chemical disruption of bacterial cells, physical disruption will also aid in maximizing the capability of a solubilizing solution. Such methods include French press, tip-probe sonication or 'breaking' cells with small-diameter glass or zirconium beads ('bead-beating'). In each case, the optimal time for physical disruption must be derived on a case-by-case basis, for example, large cell volumes of *Escherichia coli* can be disrupted in a tip-probe sonicator in less than 2 min, while the same relative amount of *Mycobacterium tuberculosis* may take ten times longer. In such cases, care should be taken to keep the sample on ice to avoid urea carbamylation that leads to artifactual 'isoforms' on 2D gels. Finally, the whole cell lysate should be centrifuged to remove any remaining insoluble material.

Separation

The method of choice for isoelectric focusing applications is now immobilized pH gradient (IPG) technology owing to the high reproducibility, resolving power, higher protein load and ease-of-use of these commercially available strips. The chemistry of IPG technology has been described in detail elsewhere (Görg *et al.*, 1988, 2000; O'Connor *et al.*, 1998) and will not be reviewed here. Since IPGs can be cast with a high degree of reproducibility, several pH ranges are now commercially available. These cover wide (pH 3–10), mid (pH 3–6, 4–7, 5–8) and narrow (e.g. pH 4–5) pH ranges, as well as gradients designed specifically for the separation of basic proteins (pH 7–10, 6–11). IPGs are packaged as thin acrylamide strips bonded to plastic for ease of handling. These strips must be rehydrated in sample buffer prior to isoelectric focusing in one of two methods. The 'passive' rehydration method uses the sample itself, diluted in sample buffer to a volume suitable for rehydrating the IPG strip (approximately 500 µl for a 17–18 cm IPG strip). The alternative method is to rehydrate the strip in sample buffer alone and then introduce the protein lysate via an applicator 'cup' placed at one end of the IPG strip. In either method, depending on the pH range of the IPG strip, approximately 0.25–1 mg of total protein should be loaded. As a general rule, the wider the pH gradient, the lower the amount of protein that should be applied to the strip to obtain quality separation. For basic IPGs, sample should always be applied in a sample cup to the anodic end of the strip.

Isoelectric focusing is usually performed in a stepwise or 'ramped' manner, beginning with low (100–300 V) and stepping up to higher (3500–5000 V) voltages over a period of 10–16 h. The low initial voltages ensure that the proteins do not precipitate while removing interfering salts from the IPG strip. As the voltage is ramped higher, proteins migrate through the strip according to their charge until equilibrium (the p*I* of the protein) is reached. For optimal separation, the total kilovolt hour value is

Table 11.1 Components of 2DGE-compatible sample solubilization buffers.

Component	Chemicals	Amounts	Function
Denaturants/ detergents	CHAPS SB3-10 Triton X-100 ASB-14 Octyl-glucoside C8Ø	4% 1–2% 2–4% 1–2% 2–4% 1–2%	1. Aid in fully denaturing proteins 2. Improve solubilization 3. Use cocktails of detergents with a total concentration of 4% or less Note: only non-ionic or zwitterionic detergents can be used in isoelectric focusing
Denaturants/ chaotropes	Urea Thiourea	3–9 M 1–2 M	1. Unfold proteins and aids solubilization 2. Destroy tertiary structure 3. Expose ionizable groups
Reducing agent	2-Mercaptoethanol Dithiothreitol (DTT) Tributyl-phosphine (TBP)	1–2% 20–100 mM 2–5 mM	1. Reduce disulphide bonds Note: DTT is ionic and migrates in IEF, hence higher concentrations are needed
Carrier Ampholytes	Pharmalytes™ BioLytes™	0.5–2.0%	1. Minimize charge–charge interactions 2. Act as salt buffer for ionic strength
Tris	Tris	40–50 mM	1. Aid in solubilizing proteins 2. Inhibit protease activity
Nuclease	Endonuclease	150–200 U	1. Cleave large nucleic acids that interfere with gel patterns
Protease inhibitors	Cocktails PMSF	— 5–10 mM	1. Inhibit protease activity Note: not always necessary where urea and Tris are present; peptide protease inhibitors may be visible on 2D gels
Dye	Bromophenol Blue	Trace	1. Follow progress of IEF

usually between 30 and 80. After isoelectric focusing, proteins in the IPG strip must be reduced, alkylated and detergent exchanged so that they are coated in SDS prior to SDS–polyacrylamide gel electrophoresis (PAGE). This step is often also called 'equilibration'. A typical equilibration buffer contains a reducing agent [5 mM tributyl phosphine (TBP) or 20 mM dithiothreitol (DTT)] plus SDS (1–2%) in an initial step and then a second buffer containing the alkylating agent (e.g. 4–5% iodoacetamide) and a running dye (e.g. bromophenol blue). A simpler version is to perform a single buffer step containing the reducing agent as well as free acrylamide monomer (2.5% v:v final concentration) as the alkylating agent. Following incubation in this buffer, the IPG strip is placed on to an SDS-PAGE slab gel and embedded with 0.5% agarose (prepared in Tris/glycine running buffer). The gel is electrophoresed until the bromophenol blue dye runs off the end of the gel.

Visualization

Following 2DGE the proteins in the gel are fixed with methanol (10–40%) and acetic acid (5–10%) and then stained using a variety of methods. The simplest of these is to use highly sensitive fluorescent dyes such as Sypro Ruby (Lopez et al., 2000). Silver staining is labour intensive and, owing to the developing process, poorly reproducible. Silver stains, unlike fluorescent stains, are not linear in their binding capability and thus do not always provide an accurate reflection of the relative amounts of each protein separated by 2DGE. Non-colloidal Coomassie has very low sensitivity and thus will only stain very abundant proteins within a whole cell lysate. Colloidal Coomassie stains are tenfold more sensitive. Most stains are compatible with mass spectrometry, provided glutaraldehyde or other fixatives are not used in the pre-staining process. As a general rule, we utilize Sypro Ruby stain prior to image analysis and then 'double-stain' the same gel in colloidal Coomassie Blue (Cordwell, 2003) enabling protein spots to be easily excised from the gel without the need for specialized equipment. These dyes appear to have a high affinity for each other (unpublished data) and hence staining trays for each stain should be kept separate at all times. Furthermore, this protocol, combined with the superior sensitivity of mass spectrometry, removes the need to run the traditional parallel 'analytical' (low protein load for optimal separation and pattern reproducibility for image analysis) and 'preparative' (high protein load for protein characterization) 2D gels.

Once the 2DGE images have been acquired via high-resolution densitometry or fluorescence scanning, experimental data sets need to be visually compared to pinpoint protein spots of significance related to the biological question under examination. While several groups prefer to use commercially available image analysis software packages to perform this task, it is a better option to visually scan magnified gel regions by eye to determine whether differences are real, overall gel quality and reproducibility, and then use software to determine the statistical parameters associated with each up- or down-regulated protein (Figure 11.2). The

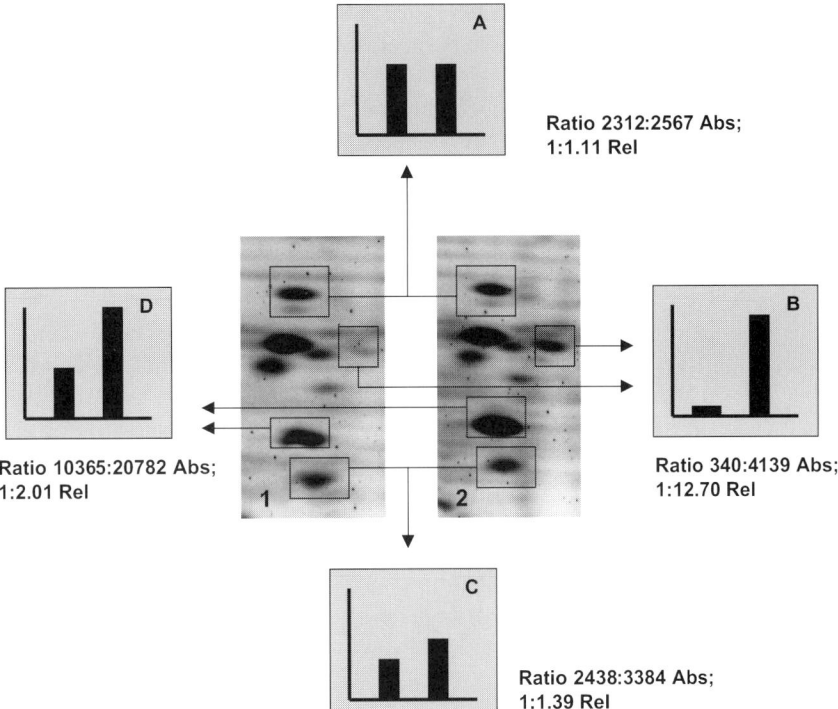

Figure 11.2. Pattern recognition and image analysis in proteomics. Four corresponding spots are compared between control (1) and test (2) experiments. 'Abs' refers to the absolute intensity of the spot compared to all spots visible on each 2D gel. For example, there is approximately four times more spot 'D' than spot 'C' in the control 2D gel. 'Rel' refers to the relative intensity of that particular spot between control and test experiments. Spot 'A' is identical between the control and test, statistically up-regulated by a relative value of 1.11 in the test 2D gel, but neither visible to the eye nor statistically significant over a series of three gel replicates. Spot 'B' is induced in the test 2D gel by a factor of 12.70. Spot 'C' is induced by a relative value of 1.39, below the significance threshold of most 2DGE experiments (induction or repression of 1.5 times) but visible to the eye. Spot 'D' is induced twofold in the test 2D gel.

human eye is unparalleled in its ability to perform pattern recognition over small areas. After the images have been compared visually, software such as PD-Quest (Bio-Rad), Melanie 3 (Swiss Institute for Bioinformatics) or Z3 (Compugen) can be used to determine statistically, and over a number of gel replicates, the exact amount of relative up- or down-regulation as dictated by spot volume and intensity (Figure 11.2).

Identification

Mass spectrometry has become the method of choice for identification of gel-separated proteins. It is now standard to apply a hierarchical order for protein characterization that depends on mass spectrum quality and the amount of sequence information present in genome databases. Gel-purified spots are excised and the protein is washed (acetonitrile/50 mM ammonium bicarbonate, 60 : 40 v : v) and cleaved with a protease (generally trypsin) to generate peptide fragments suitable for an initial 'global' analysis using matrix-assisted laser desorption ionization time-of-flight mass spectrometry (MALDI-TOF MS). In the first pass, a small aliquot of eluted peptide mixture is placed on the MALDI target plate with matrix (generally α-cyano-hydroxycinnamic acid; 4-HCCA). This 'dried droplet' (Karas and Hillenkamp, 1988) method is readily automated and thus several hundred spots can be processed in a single day. MALDI plates are then processed, the data automatically calibrated (using either internal trypsin autolysis peaks or external calibrants such as Substance P or Gramicidin S that have molecular masses between 1000 and 2000 Da) and the resulting peptide masses used to perform automated database searches (Figure 11.3). This 'peptide mass mapping' (PMM) or 'fingerprinting' technique was first described by several independent groups in 1993–1994 (Henzel *et al.*, 1993; James *et al.*, 1993; Mann *et al.*, 1993; Pappin *et al.*, 1993; Yates *et al.*, 1993). PMM depends heavily on the presence of the corresponding protein sequence within a sequence database, as even a single amino acid substitution across species boundaries will, in most cases, dramatically alter the molecular mass of a given peptide. Following the initial screen, two questions need to be addressed: firstly, 'What is the quality of the PMM spectrum?' and, then, 'Is there confidence in the determined database match?'. If the quality of the spectrum is poor, it is likely that a low level of analyte is present in the protein spot or that salts, such as SDS, are interfering with the signal. In either case, the remaining peptide mixture can be concentrated and desalted using narrow-diameter pipette tips packed with C_{18} chromatography resin (Jensen *et al.*, 1998; Gobom *et al.*, 1999). Such tips are also available commercially (e.g. ZipTips™, Millipore, Bedford, MA, USA). The tips are activated in low volumes (10–20 µl) of 100% acetonitrile and then acidified with 5% formic acid. The remaining peptide mixture is then slowly passed through the column and the bound peptides washed with a further aliquot of 5% formic acid. Finally, peptides are eluted from the column in 0.5–1.0 µl matrix (4-HCCA in 70% acetonitrile; 10 mg/ml) solution, directly on to the MALDI target plate. PMM is then performed again to determine possible matches. Specialized software is not always necessary for searching with MS-derived data as several public databases can be interrogated, as has been reviewed elsewhere (O'Connor *et al.*, 1998).

When high-quality spectra have been obtained but no match could be achieved following database search, it is most likely that the corresponding protein is not present in the database. In this scenario, the remaining peptide mixture is concentrated and desalted as above and then subjected to electrospray-ionization tandem MS (ESI-MS/MS; Wilm *et al.*, 1996). The C_{18}-bound peptides are extracted in low volumes of solvent. Traditionally ESI-MS is performed using a triple quadrupole analyser. A precursor ion scan in MS mode uses the final quadrupole to scan the *m/z* (mass : charge) area for total peptides in the solution. In MS/MS mode, peptide ions with a characteristic *m/z* can be selected in the first quadrupole and collided

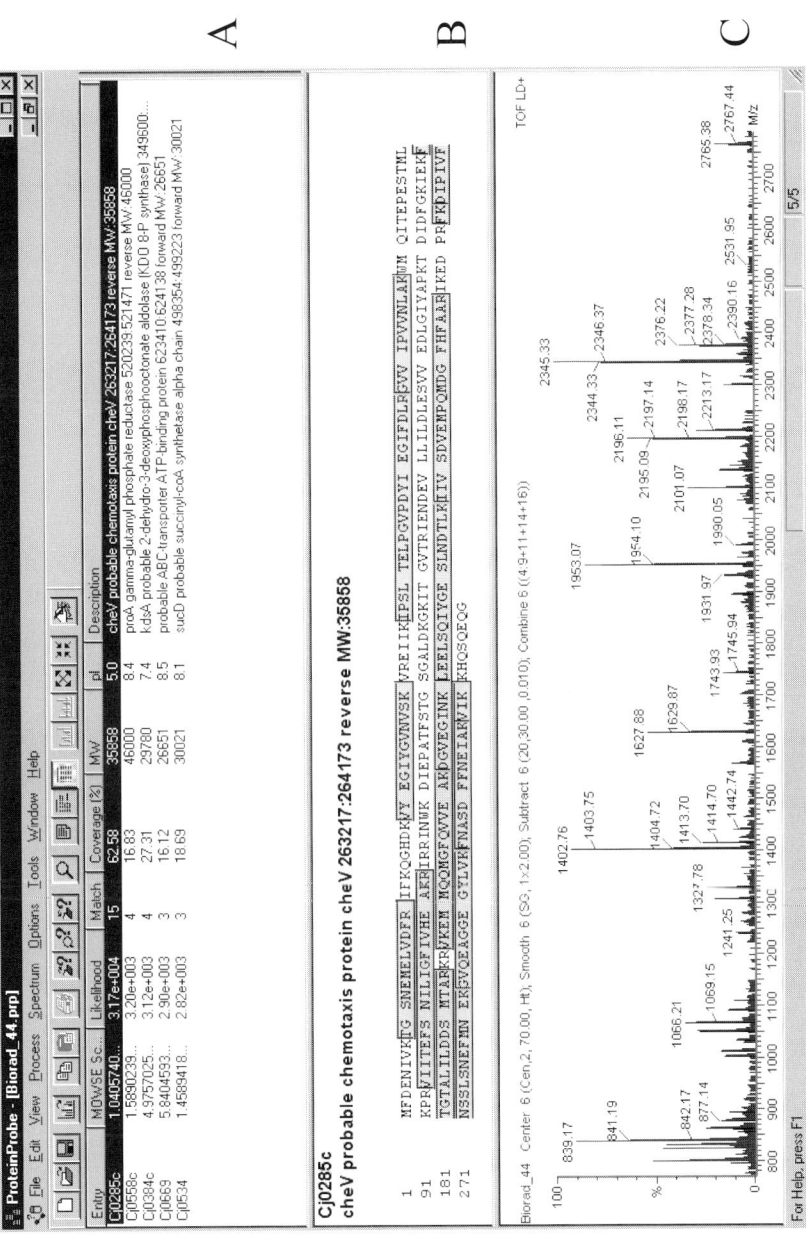

Figure 11.3. Automated database search using PMM (MassLynx software, Micromass, Manchester, UK). Identification of *Campylobacter jejuni* chemotaxis protein CheV following tryptic digestion and MALDI-TOF MS. (A) Potential matches listed by the number of matching peptide masses (15 peptides) and the total percentage sequence covered by these matching peptides (62.58%). (B) Matching peptides highlighted on the corresponding protein sequence. (C) PMM spectrum showing matching peptide masses.

with an inert gas (typically argon) in the second. The resulting fragment ions are recorded in the third quadrupole. Peptides preferentially fragment at amide bonds when subjected to this collision energy, thus the fragment ion spectrum contains a series of specific ion signals that differ in mass by one amino acid residue. Series of complementary N- and C-terminal sequence 'ladder' ion signals ('b' and 'y' ions, respectively) can be observed. Doubly and triply charged ions are preferentially selected for MS/MS, since they fragment more readily than singly charged species. Post-translational modifications, such as phosphorylation, or adducts caused by oxidation, sodium or methylation, can also be detected as these bonds are similarly unstable and will fragment when subjected to collision energy. Each sequenced peptide can then be used to perform a BLAST search (Altschul *et al.*, 1990) to determine whether homologous proteins from other species are present in the database. If matching proteins cannot be detected with a high degree of confidence, the *de novo* peptide sequence can form the basis for degenerate primers aimed at cloning and sequencing the gene of interest.

Prefractionation

The need to visualize a greater proportion of a 'functional' (i.e. expressed) proteome has led many researchers to explore methodology to enrich for particular classes of proteins prior to their separation by SDS-PAGE or 2-DGE. These fall into three broad classes: (1) a pI or molecular mass 'window'; (2) cellular location; and (3) post-translational modification or ligand-binding affinity (Figure 11.4).

Prefractionating devices (often those based on isoelectric focusing) can be used as a pre-2DGE separation to examine proteins within a particular pI or mass range, and are thus particularly useful as a front-end prior to the application of protein sample to narrow-range ('zoom') single pH unit IPG 2D gels. As an example, 1 mg of total protein lysate loaded on to a pH 4–5 IPG gel will actually result in the separation of only approximately 20–30% of that sample, the remainder focusing beyond the pH range of the strip. This is a waste of resources, however, pre-fractionation of 10 mg of protein into several pH ranges prior to isoelectric focusing allows 1 mg of protein within the pH range 4–5 to be separated on that particular gel. On all strips, especially those in the basic pH range, this reduces the possibility of high sample loads precipitating at the ends of narrow-range IPGs. By maximizing the amount of protein that can be added to single pH unit 2D gels, the number of resolved spots will also increase, as 2D gel resolution is directly proportional to the resolving area of the gel and the width of the pH separation. Hence, more protein can be loaded and more spots will be resolved when using 17–18 cm pH 4–5 gels than in the same range on 17–18 cm pH 4–7 IPG 2D gels (Cordwell *et al.*, 2000; Figure 11.5). Proteins outside the scope of current 2DGE techniques (e.g., basic proteins with pI > 10.0; Cordwell *et al.*, 2001) may also be separated into a discrete fraction and then analysed by SDS-PAGE or coupled to liquid chromatography (LC)–MS. Prefractionating devices have been

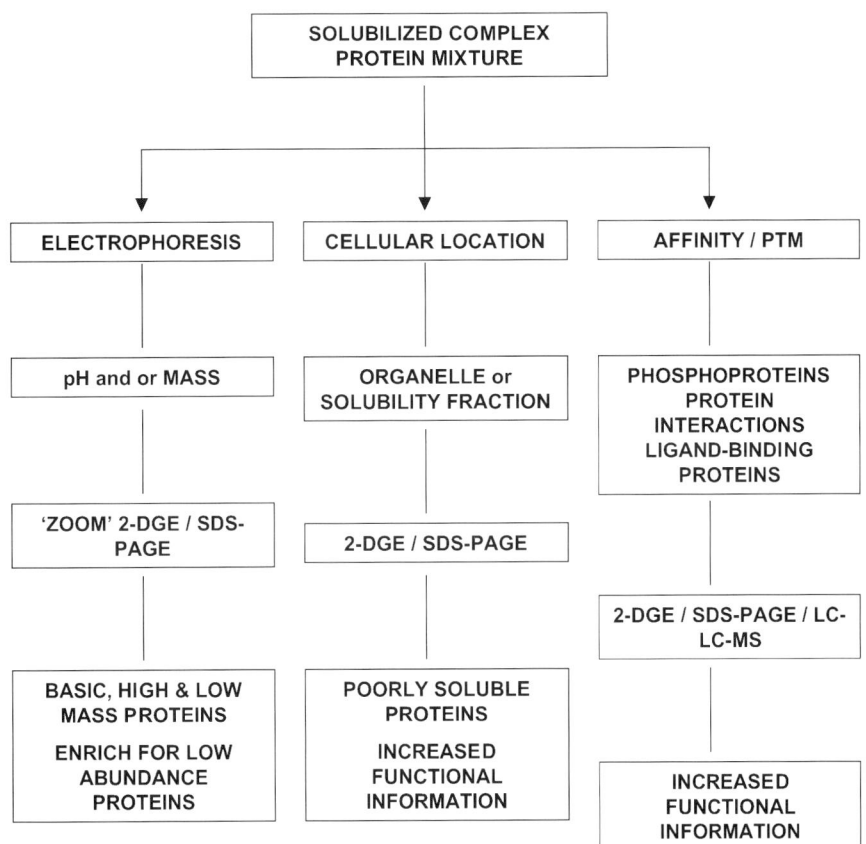

Figure 11.4. Prefractionation approaches in proteomics. Proteins can be enriched using physical (p*I*, mass), biological (cellular location) or functional (post-translational modification, ligand-binding) properties.

well described elsewhere and include commercial instruments such as Rotofor (Bio-Rad), Gradiflow (Gradipore; Rylatt *et al.*, 1999), and systems currently under development (Righetti *et al.*, 1997; Herbert and Righetti, 2000; Zuo and Speicher, 2000).

The second method for prefractionation aims to use a 'subproteomics' approach based on discrete cellular compartments or organelles, rather than taking a whole cell approach. This achieves two outcomes: firstly, less complex 2DGE patterns can be resolved; and, secondly, the identification of novel or hypothetical proteins can be correlated with a distinct subcellular location. For bacterial proteome analysis, great success has been achieved through the analysis of Gram-negative outer membrane protein (OMP) fractions, which are readily enriched by precipitating membranes in ice-cold sodium carbonate (Molloy *et al.*, 2000). Once the membrane fraction has been washed of cellular contaminants, a cocktail of detergents, including amidosulphobetaine-14 (ASB-14; Rabilloud *et al.*, 1999), is used to solubilize membrane-associated proteins. This methodology has been used to map OMPs from *E. coli* (Molloy *et al.*, 2000) and

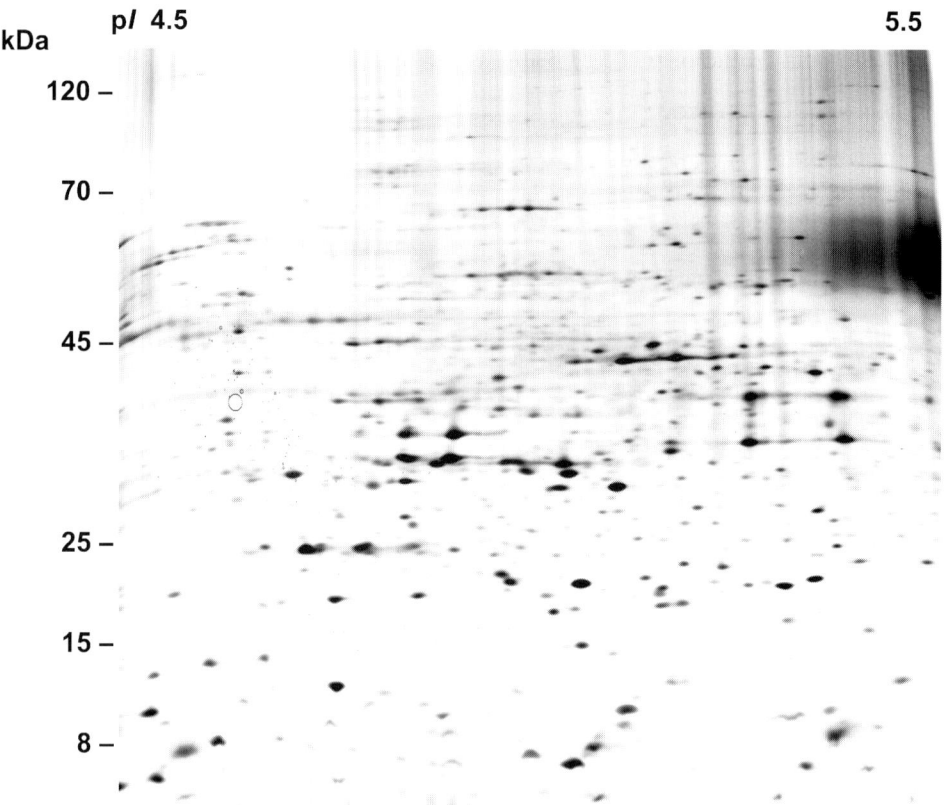

Figure 11.5. Prefractionated whole cell lysate of *Staphylococcus aureus* separated on a narrow-range pH 4.5–5.5 2D gel. Approximately 300 more spots were resolved on this gel than were resolved on the entire pH 4–7 range (Cordwell *et al.*, 2000).

Pseudomonas aeruginosa (Nouwens *et al.*, 2000). Importantly, the method appears to be quite specific with only a single cytoplasmic protein (the ubiquitous cytosolic chaperone GroEL) identified amongst 189 unique *P. aeruginosa* OMPs. However, the method does not appear additionally to solubilize highly hydrophobic proteins as only three of the identified OMPs have a positive GRAVY or Kyte–Doolittle value.

A second bacterial fraction that can provide much information is the culture supernatant (CSN), which is significantly enriched in secreted proteins, including proteases, toxins and other pathogenicity factors. Cultures are generally grown to mid-stationary phase for optimal recovery of secreted proteins, and then the CSN is filtered to remove whole cells and remaining proteins precipitated with trichloroacetic acid (TCA)/methanol (Hirose *et al.*, 2000). Recently, we have applied this methodology to the examination of *P. aeruginosa* CSN proteins (unpublished data), while others have examined extracellular fractions from *Staphylococcus aureus* (Ziebandt *et al.*, 2001), *Bacillus subtilus* (Hirose *et al.*, 2000) and *Mycobacterium tuberculosis* (Jungblut *et al.*, 1999). However, unlike purified Gram-negative outer membrane fractions, bacterial CSNs contain both secreted and cytosolic proteins. While it cannot be ruled out that some of these cytosolic proteins have dual function that is necessitated by secretion, a more likely explanation is that cell turnover and autolysis, releasing cytosolic components into the extracellular milieu, is occurring (Tullius *et al.*, 2001). Furthermore, the bacterial CSN is also highly active with proteases, leading to the appearance of several protein spot mass and p*I* variants visible on 2D gels. While this method allows the identification of proteins that would not be detected via examination of whole cell lysates, it should also be noted that bacteria with complex growth requirements, such as the presence of serum, are poorly amenable to this approach, owing to the overabundance of serum proteins, especially albumin.

Sequential solubilization may also aid in the visualization of a greater complement of a bacterial proteome (Molloy *et al.*, 1998; Cordwell *et al.*, 2000). This technique aims to provide solubility, rather than cellular, fractions over a series of extraction steps with rounds of 2DGE. In the first step, highly soluble proteins are extracted with 40 mM tris buffer in conjunction with methanol precipitation. The insoluble pellet is then extensively washed and resolubilized with standard 2DGE buffers including CHAPS, sulphobetaines, urea and reducing agents. The remaining insoluble pellet is then washed and solubilized in a final round of SDS sample buffer and separated using SDS-PAGE. While this protocol has been shown effectively to fractionate some classes of membrane proteins from readily soluble cellular constituents, many highly hydrophobic proteins remain insoluble in even the harshest SDS-containing buffers. We currently believe that up to 15% of the theoretical ORFs encoded by a bacterial genome may encode proteins that are too hydrophobic to be solubilized and resolved using available 2DGE technologies.

The final prefractionation technique is the use of functional parameters or post-translational modifications to examine sets of proteins with those particular properties. For example, 'phosphoproteomics' can be performed via simple radiolabelling experiments in conjunction with 2DGE. However, the protein spots are then often difficult to identify owing to the low levels of sample present. Conversely, several commercial antiphospho antibodies are also available (antiphosphoserine, threonine and tyrosine) and these can be used in conjunction with 2DGE to define the global phosphoproteome (Kaufmann *et al.*, 2001). Once again, the specificity of antibody reactions requires low starting amounts of total protein to maintain resolution, while the identification of the phosphorylated proteins is also difficult following Western blotting. Furthermore, the reliability of these antibodies has often been questioned. More recently, several unique methods have been introduced to enrich for phosphorylated proteins using affinity tags to replace phosphate and thus allowing selective binding of these proteins from amongst whole cell lysates to chromatography columns (Oda *et al.*, 2001). Once phosphoproteins have been enriched using prefractionation, they can be separated by LC, SDS-PAGE or 2DGE and analysed by mass spectrometry to determine their phosphorylation sites. Several reviews have dealt with this topic (Larsen *et al.*, 2001; McLachlin and Chait, 2001).

Affinity chromatography is rapidly becoming a technique of choice for prefractionating proteins prior to their separation and identification. This technique is also particularly amenable to characterizing protein–protein interactions and complexes (reviewed later). As for phosphoproteins, several other properties can be used to enrich for particular classes of proteins. These include ATP (ATP-binding proteins) and several cations (e.g. calcium and magnesium). More importantly, for pathogenic bacteria, mammalian target proteins, such as heparin, fibronectin, fibrinogen and laminin, can also be used to determine which proteins interact with the host during infection. Furthermore, the targets of drugs can also be determined by directly analysing proteins that interact with a particular antibiotic or chemical agent (Boaretti and Canepari, 2000; Sinha Roy et al., 2001). Thus, this method of prefractionation provides a high degree of functional information about proteins of interest and can stimulate much further experimentation.

◆◆◆◆◆◆ APPLICATIONS

Reference mapping

The 2DGE/MS approach is particularly amenable to generating large amounts of protein characterization information from an organism under a standard set of culture conditions. Combined with prefractionation techniques described above, it is now possible, at least theoretically, to decipher up to 75% of a genome using 2DGE technologies. In practice, however, this is rarely the case, as culture conditions themselves create an artificial environment that allows the organism to utilize minimal physiological effort. However, some insight can be gained into which biochemical pathways are required and, more importantly, on the number and genetic location of expressed hypothetical or unique proteins. This approach provides a positive response to the first question of genome sequencing, that is, does a particular gene sequence lead to an expressed protein. In combination with subcellular fractions, further information about functionally unknown proteins can be derived. Where bacteria can be grown in defined, minimal media, it is possible to dictate which pathways are switched on or off, for example, by simply changing sole carbon sources or depleting the media of trace nutrients such as iron or phosphate. The reference mapping approach allows large amounts of data to be acquired rapidly utilizing the information contained within genome sequences and the throughput of MALDI-TOF MS. As an example, we recently analysed 97 proteins from whole cell lysates of the gut pathogen *Campylobacter jejuni* in less than 2 days (Figure 11.6). From this simple work, at least 16 hypothetical proteins were shown to be expressed for the first time. Furthermore, the relative amounts of each protein can also provide useful information. For example, the major outer membrane protein (MOMP or PorA) and the fibronectin-binding adhesin CadF are amongst the most abundant soluble proteins in *C. jejuni*, while enzymes of the glycolytic pathway (amongst the most abundant proteins in *E. coli*, *P. aeruginosa* and *S. aureus*, for example)

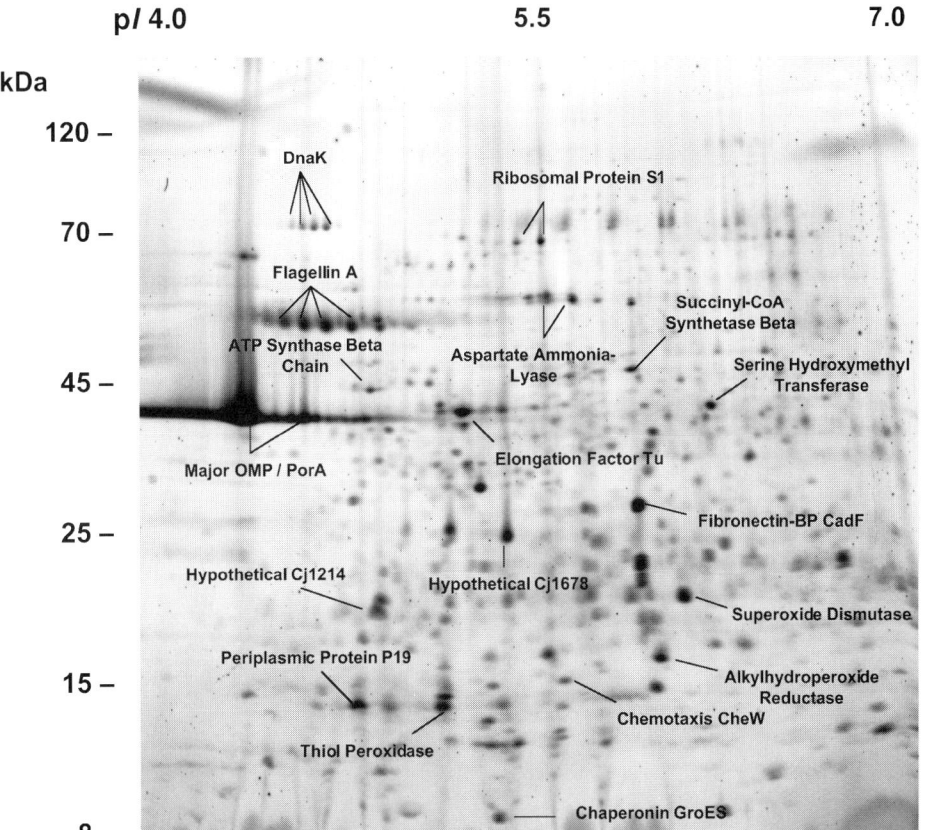

Figure 11.6 Reference 2DGE 'map' of soluble proteins from *Campylobacter jejuni* showing landmark proteins and their location. Proteins were identified by PMM.

are absent or at lower expression levels. This most likely reflects the complexity of the growth medium and the use of alternate pathways to generate ATP. 2DGE/MS maps can also be used as a starting point for comparative studies (differential display proteomics), where changes in the genetic or environmental structure can be analysed by tracing the levels of protein expression across a series of 2D gels. Other studies aim to compare protein expression across phenotypically diverse bacterial strains or species (Betts *et al.*, 2000; Jungblut *et al.*, 2000; Mattow *et al.*, 2001).

Differential display proteomics

The major utility of 2DGE/MS proteomics is to uncover proteins of interest that respond to a biological, genetic or environmental stimulus. Basically, any experimental model can be applied to proteome analysis. Such systems include the global analysis of: (1) regulatory gene knockout (in combination with phenotypic and functional studies); (2) changes in the environment (these include nutrient limitation, heat and cold stress, growth temperature, the presence of inhibitory chemicals, oxygen availability, etc.); (3) bacterial strains; (4) the presence of antibiotics (especially in comparisons where resistant and sensitive strain phenotypes are known); and (5) response to host factors. At present, the proteome of *E. coli* is best understood via such experimentation, mainly due to the pioneering work of VanBogelen and Neidhardt (reviewed in VanBogelen *et al.*, 1997). Significant progress has also been made in *Bacillus subtilis*, especially on proteins of the general stress regulon (reviewed in Hecker and Volker, 2001; Hecker and Engelmann, 2000). The microbial response to host factors (and indeed the host response itself) is more difficult to apply to this technology as the complexity of the analysis is increased when pathogens are internalized by the host. However, some success has been achieved in a symbiotic system (*Rhizobium* and subterranean clover during root nodulation; Morris and Djordjevic, 2001), and the success of such experiments suggests that host–pathogen interactions may be studied by three-way analyses involving the host, the pathogen and the changes occurring in each during infection.

◆◆◆◆◆◆ COMPLEMENTARITY AND 'NEW' PROTEOME TECHNOLOGIES

While the 2DGE/MS approach has been, and will continue to be, successful in identifying many proteins from bacterial species, large subsets of proteins still retain physical characteristics that make them non-amenable to such analyses. These include proteins that are hydrophobic, basic, and/or high molecular mass, as well as those under the dynamic range of expression and thus of low enough abundance to remain undetected using stains that visualize spots on 2D gels. Therefore, many researchers have aimed to replace 2DGE with alternative separation technologies in the hope of 'covering' a greater proportion of the total proteome. Such technologies have included isotope-coded affinity tags (ICAT; Gygi *et al.*, 1999) combined with MS, and multidimensional LC-MS (MuD LC-MS; Washburn *et al.*, 2001). The ICAT approach is quantitative and amenable to comparative analysis as 'light' and 'heavy' reagents (approximately 4 Da apart when measured by MS) can be used to label control and test experiments prior to mixing the sample and generating the complex peptide mixture. Cysteine-containing peptides (which have the isotope tag attached) are then selectively bound to an affinity support. This reduces the complexity of the peptide pool approximately 10–20-fold. The levels of each peptide from both the control and test experiment are then detected by MS (each peptide 'pair' separated by 4 Da) and those showing significant quantitative differences selected for MS/MS to identify the corresponding protein. While this technique is robust and obviously a significant throughput improvement on 2DGE/MS approaches, several issues still remain. Proteins must be solubilized prior to their enzymatic digestion and thus many of the problems occurring with 2DGE and poorly soluble proteins are also apparent. Furthermore, proteins that contain no cysteine residues are completely excluded. However, we believe that the combination of ICAT and/or MuD LC-MS with 2DGE/MS approaches are necessary to optimize the number of proteins amenable to identification. For example, ICAT technology is capable of discerning many basic, high and low mass proteins not accessible to 2DGE, while 2DGE can separate to purity many of those proteins that do not contain cysteine. Furthermore, chromatography-based approaches are not limited by the dynamic range of protein expression and hence very low abundance proteins can be monitored using such technologies.

◆◆◆◆◆◆ 'FUNCTIONAL' PROTEOMICS

Once proteins of interest have been identified using a global approach, several methods have become standard for beginning to elucidate their biological function. An area that has received much interest is determining protein–protein interactions. These can be examined on a hypothesis basis via the yeast 2-hybrid system, or on a single protein or complex basis using affinity chromatography. The yeast 2-hybrid system is discussed in Chapter 12; however, it remains highly predictive and can only suggest potentially interacting proteins rather than test for them *in vivo*. Alternatively, affinity chromatography approaches are lower throughput, but provide accurate interaction data on single proteins or protein complexes (Figure 11.7; Neubauer *et al.*, 1997; Houry *et al.*, 1999; Link *et al.*, 1999; Pandey and Mann, 2000). Essentially, the protein of interest is bound to an affinity matrix via a removable tag and proteins from cell

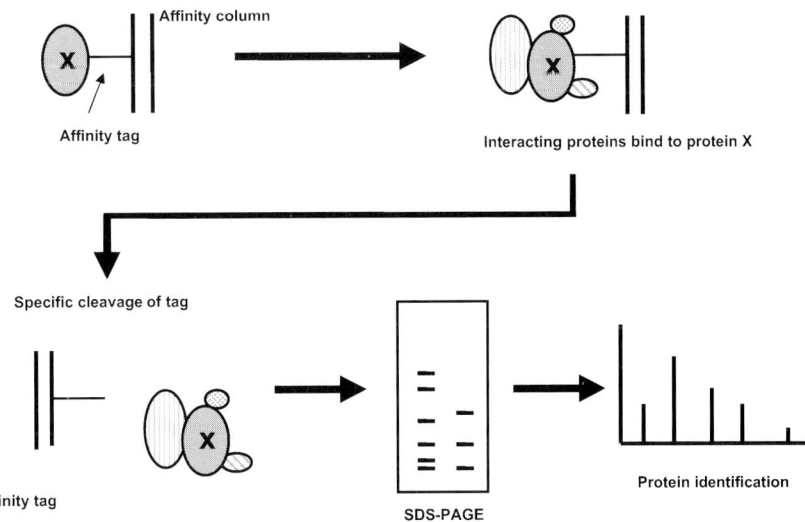

Figure 11.7. Schematic showing the affinity approach for characterizing protein–protein interactions.

lysates are washed through. Interacting proteins bind the protein of interest while non-interacting proteins are removed by extensive washing. Alternatively, nearest-neighbour protein analyses can be performed *in vivo* using cross-linkers prior to selecting the protein of interest by immunoaffinity chromatography (Layh-Schmitt *et al.*, 2000). The tag binding the protein of interest to the affinity support is then cleaved, releasing the newly formed protein complexes. These proteins are then separated by SDS-PAGE or 2DGE and the individual components identified by MS. This methodology has been used most extensively to find interacting partners of the ubiquitous chaperone GroEL (Houry *et al.*, 1999).

◆◆◆◆◆◆ CONCLUSIONS

The technologies encompassed under the term 'proteomics' are widening to reflect our ability to perform firstly global analyses (including 2DGE/MS and quantitative MS) to discover proteins of interest from amongst a complex bacterial protein set, and then to characterize these proteins functionally within a biological context. Outside the scope of this review, high-throughput enzyme assays and cellular localization via fluorescent tags are also methodologies that are receiving much attention for elucidating biological function. Combined with the methods described above including genetic disruption, ligand-binding and characterization of post-translational modifications, we are now beginning to understand the roles of previously functionally unknown proteins discovered during genome sequencing projects.

Acknowledgements

This work was facilitated by access to the Australian Proteome Analysis Facility (APAF) established under the Australian Government Major National Research Facility (MNRF) program. The author wishes to thank Bio-Rad Laboratories and Micromass Ltd for financial and instrumentation support.

References

Altschul, S. F., Gish, W., Miller, W., Myers, E. W. and Lipman, D. J. (1990). Basic local alignment search tool. *J. Mol. Biol.* **215**, 403–410.

Betts, J. C., Dodson, P., Quan, S., Lewis, A. P., Thomas, P. J., Duncan, K. and McAdam, R. A. (2000). Comparison of the proteome of *Mycobacterium tuberculosis* strain H37Rv and clinical isolate CDC1551. *Microbiology* **146**, 3205–3216.

Boaretti, M. and Canepari, P. (2000). Purification of daptomycin-binding proteins (DBPs) from the membrane of *Enterococcus hirae*. *New Microbiol*. **23**, 305–317.

Cordwell, S. J. (2003). Acquisition and archiving of information for bacterial proteomics – from sample preparation to databasing. *Meth. Enzymol.* (in press).

Cordwell, S. J., Nouwens, A. S., Verrills, N. M., Basseal, D. J. and Walsh, B. J. (2000). Subproteomics based upon protein cellular location and relative solubilities in conjunction with composite two-dimensional electrophoresis gels. *Electrophoresis* **21**, 1094–1103.

Cordwell, S. J., Nouwens, A. S. and Walsh, B. J. (2001). Comparative proteomics of bacterial pathogens. *Proteomics* **1**, 461–472.

Gobom, J., Nordhoff, E., Mirgorodskaya, E., Ekman, R. and Roepstorff, P. (1999). Sample purification and preparation technique based on nano-scale reversed-phase columns for the sensitive analysis of complex peptide mixtures by matrix-assisted laser desorption/ionization mass spectrometry. *J. Mass Spectrom.* **34**, 105–116.

Görg, A., Postel, W. and Günther, S. (1988). Two-dimensional electrophoresis. *Electrophoresis* **9**, 531–546.

Görg, A., Obermaier, C., Boguth, G., Harder, A., Scheibe, B., Wildgruber, R. and Weiss, W. (2000). The current state of two-dimensional electrophoresis with immobilized pH gradients. *Electrophoresis* **21**, 1037–1053.

Gygi, S. P., Rist, B., Gerber, S. A., Turecek, F., Gelb, M. H. and Aebersold, R. (1999). Quantitative analysis of complex protein mixtures using isotope-coded affinity tags. *Nature Biotechnol.* **17**, 994–999.

Hecker, M. and Engelmann, S. (2000). Proteomics, DNA arrays and the analysis of still unknown regulons and unknown proteins of *Bacillus subtilis* and pathogenic Gram-positive bacteria. *Int. J. Med. Microbiol.* **290**, 123–134.

Hecker, M. and Völker, U. (2001). General stress response of *Bacillus subtilis* and other bacteria. *Adv. Microb. Physiol.* **44**, 35–91.

Henzel, W. J., Billeci, T. M., Stults, J. T., Wong, S. C., Grimley, C. and Watanabe, C. (1993). Identifying proteins from two-dimensional gels by molecular mass searching of peptide fragments in protein sequence databases. *Proc. Natl Acad. Sci. USA* **90**, 5011–5015.

Herbert, B. and Righetti, P. G. (2000). A turning point in proteome analysis: sample prefractionation via multicompartment electrolyzers with isoelectric membranes. *Electrophoresis* **21**, 3639–3648.

Hirose, I., Sano, K., Shioda, I., Kumano, M., Nakamura, K. and Yamane, K. (2000). Proteome analysis of *Bacillus subtilis* extracellular proteins: a two-dimensional protein electrophoretic study. *Microbiology* **146**, 65–75.

Houry, W. A., Frishman, D., Eckerskorn, C., Lottspeich, F. and Hartl, F. U. (1999). Identification of *in vivo* substrates of the chaperonin GroEL. *Nature* **402**, 147–154.

James, P., Quadroni, M., Carafoli, E. and Gonnet, G. (1993). Protein identification by mass profile fingerprinting. *Biochem. Biophys. Res. Commun.* **195**, 58–64.

Jensen, O. N., Larsen, M. R. and Roepstorff, P. (1998). Mass spectrometric identification and microcharacterization of proteins from electrophoretic gels: strategies and applications. *Proteins* (Suppl. 2), 74–89.

Jungblut, P. R., Schaible, U. E., Mollenkopf, H. J., Zimny-Arndt, U., Raupach, B., Mattow, J., Halada, P., Lamer, S., Hagens, K. and Kaufmann, S. H. (1999). Comparative proteome analysis of *Mycobacterium tuberculosis* and *Mycobacterium bovis* BCG strains: towards functional genomics of microbial pathogens. *Mol. Microbiol.* **33**, 1103–1117.

Jungblut, P. R., Bumann, D., Haas, G., Zimny-Arndt, U., Holland, P., Lamer, S., Siejak, F., Aebischer, A. and Meyer, T. F. (2000). Comparative proteome analysis of *Helicobacter pylori*. *Mol. Microbiol.* **36**, 710–725.

Karas, M. and Hillenkamp, F. (1988). Laser desorption ionization of proteins with molecular masses exceeding 10,000 daltons. *Analyt. Chem.* **60**, 2299–2301.

Kaufmann, H., Bailey, J. E. and Fussenegger, M. (2001). Use of antibodies for detection of phosphorylated proteins separated by two-dimensional gel electrophoresis. *Proteomics* **1**, 194–199.

Larsen, M. R., Sorensen, G. L., Fey, S. J., Larsen, P. M. and Roepstorff, P. (2001). Phospho-proteomics: evaluation of the use of enzymatic de-phosphorylation and differential peptide mass mapping for site specific phosphorylation assignment in proteins separated by gel electrophoresis. *Proteomics* **1**, 223–238.

Layh-Schmitt, G., Podtelejnikov, A. and Mann, M. (2000). Proteins complexed to the P1 adhesin of *Mycoplasma pneumoniae*. *Microbiology* **146**, 741–747.

Link, A. J., Eng, J., Schieltz, D., Carmack, E., Mize, G., Morris, G., Garvik, B. and Yates, J. R., III (1999). Direct analysis of protein complexes by mass spectrometry. *Nature Biotechnol.* **17**, 676–682.

Lopez, M. F., Berggren, K., Chernokalskaya, E., Lazarev, A., Robinson, M. and Patton, W. F. (2000). A comparison of silver stain and SYPRO Ruby Protein Gel Stain with respect to protein detection in two-dimensional gels and identification by peptide mass profiling. *Electrophoresis* **21**, 3673–3683.

Mann, M., Højrup, P. and Roepstorff, P. (1993). Use of mass spectrometric molecular weight information to identify proteins in sequence databases. *Biol. Mass. Spectrom.* **22**, 338–345.

Mattow, J., Jungblut, P. R., Schaible, U. E., Mollenkopf, H. J., Lamer, S., Zimny-Arndt, U., Hagens, K., Muller, E. C. and Kaufmann, S. H. (2001). Identification of proteins from *Mycobacterium tuberculosis* missing in attenuated *Mycobacterium bovis* BCG strains. *Electrophoresis* **22**, 2936–2946.

McLachlin, D. T. and Chait, B. T. (2001). Analysis of phosphorylated proteins and peptides by mass spectrometry. *Curr. Opin. Chem. Biol.* **5**, 591–602.

Molloy, M. P., Herbert, B. R., Walsh, B. J., Tyler, M. I., Traini, M., Sanchez, J.-C., Hochstrasser, D. F., Williams, K. L. and Gooley, A. A. (1998). Extraction of membrane proteins by differential solubilization for separation using two-dimensional gel electrophoresis. *Electrophoresis* **19**, 837–844.

Molloy, M. P., Herbert, B. R., Slade, M. B., Rabilloud, T., Nouwens, A. S., Williams, K. L. and Gooley, A. A. (2000). Proteomic analysis of the *Escherichia coli* outer membrane. *Eur. J. Biochem.* **267**, 2871–2881.

Morris, A. C. and Djordjevic, M. A. (2001). Proteome analysis of culitvar-specific interactions between *Rhizobium leguminosarum* biovar *trifolii* and subterranean clover cultivar Woogenellup. *Electrophoresis* **22**, 586–598.

Neubauer, G., Gottschalk, A., Fabrizio, P., Seraphin, B., Luhrmann, R. and Mann, M. (1997). Identification of the proteins of the yeast U1 small nuclear ribonucleoprotein complex by mass spectrometry. *Proc. Natl Acad. Sci. USA* **94**, 385–390.

Nouwens, A. S., Cordwell, S. J., Larsen, M. R., Molloy, M. P., Gillings, M., Willcox, M. D. P. and Walsh, B. J. (2000). Complementing genomics with proteomics: the membrane subproteome of *Pseudomonas aeruginosa* PAO1. *Electrophoresis* **21**, 3797–3809.

O'Connor, C. D., Farris, M., Hunt, L. G. and Wright, J. N. (1998). The proteome approach. *Meth. Microbiol.* **27**, 191–204.

Oda, Y., Nagasu, T. and Chait, B. C. (2001). Enrichment analysis of phosphorylated proteins as a tool for probing the phosphoproteome. *Nature Biotechnol.* **19**, 379–382.

Pandey, A. and Mann, M. (2000). Proteomics to study genes and genomes. *Nature* **405**, 837–846.

Pappin, D. J. C., Højrup, P. and Bleasby, A. J. (1993). Rapid identification of proteins by peptide-mass fingerprinting. *Curr. Biol.* **3**, 327–332.

Rabilloud, T., Blisnick, T., Heller, M., Luche, S., Aebersold, R., Lunardi, J. and Braun-Breton, C. (1999). Analysis of membrane proteins by two-dimensional electrophoresis: comparison of the proteins extracted from normal or *Plasmodium falciparum*-infected erythrocytes. *Electrophoresis* **20**, 3603–3610.

Righetti, P. G., Bossi, A., Wenisch, E. and Orsini, G. (1997). Protein purification in multicompartment electrolyzers with isoelectric membranes. *J. Chromatogr. B Biomed. Sci. Appl.* **699**, 105–115.

Rylatt, D. B., Napoli, M., Ogle, D., Gilbert, A., Lim, S. and Nair, C. H. (1999). Electrophoretic transfer of proteins across polyacrylamide membranes. *J. Chromatogr. A* **865**, 145–153.

Sinha Roy, R., Yang, P., Kodali, S., Xiong, Y., Kim, R. M., Griffin, P.R., Onishi, H. R., Kohler, J., Silver, L. L. and Chapman, K. (2001). Direct interaction of a vancomycin derivative with bacterial enzymes involved in cell wall biosynthesis. *Chem. Biol.* **8**, 1095–1106.

Tullius, M. V., Harth, G. and Horwitz, M. A. (2001). High extracellular levels of *Mycobacterium tuberculosis* glutamine synthetase and superoxide dismutase in actively growing cultures are due to high expression and extracellular stability rather than to a protein-specific export mechanism. *Infect. Immun.* **69**, 6348–6363.

VanBogelen, R. A., Abshire, K. Z., Moldover, B., Olson, E. R. and Neidhardt, F. C. (1997). *Escherichia coli* proteome analysis using the gene–protein database. *Electrophoresis* **18**, 1243–1251.

Washburn, M. P., Wolters, D. and Yates, J. R., III (2001). Large-scale analysis of the yeast proteome by multidimensional protein identification technology. *Nature Biotechnol.* **19**, 242–247.

Wilm, M., Shevchenko, A., Houthaeve, T., Breit, S., Schweiger, L., Fotis, T. and Mann, M. (1996). Femtomole sequencing of proteins from polyacrylamide gels by nano-electrospray mass spectrometry. *Nature* **379**, 466–469.

Yates, J. R., III, Speicher, S., Griffin, P. R. and Hunkapiller, T. (1993). Peptide mass maps: a highly informative approach to protein identification. *Analyt. Biochem.* **214**, 397–408.

Ziebandt, A. K., Weber, H., Rudolph, J., Schmid, R., Hoper, D., Engelmann, S. and Hecker, M. (2001). Extracellular proteins of *Staphylococcus aureus* and the role of SarA and sigma B. *Proteomics* **1**, 480–493.

Zuo, X. and Speicher, D. W. (2000). A method for global analysis complex proteomes using sample prefractionation by solution isoelectrofocusing prior to two-dimensional electrophoresis. *Analyt. Biochem.* **284**, 266–278.

12 Discovery of Protein–Protein Interaction Using Two-hybrid Systems

Amit Patel, Kenneth H Mellits and Ian F Connerton
Division of Food Sciences, School of Biosciences, University of Nottingham, Sutton Bonington Campus, Loughborough LE12 5RD, UK

◆◆◆

CONTENTS

Introduction
History and evolution
Methodology
Variations on the two-hybrid theme

◆◆◆◆◆◆ INTRODUCTION

The basic yeast two-hybrid system as developed in the bakers yeast *Saccharomyces cerevisiae* detects the interaction between two proteins via the transcriptional activation of one or more reporter genes (Fields and Song, 1989). The two-hybrid system is based on the notion that eukaryotic transcriptional activators contain two essential domains in the form of a DNA binding domain (BD) that binds to specific promoter DNA sequences and an activation domain (AD) that recruits the transcription machinery and that, although these domains can be physically separated, they will retain their ability to activate transcription if brought into close proximity (Figure 12.1). In the yeast two-hybrid system, these two domains are configured as two distinct polypeptides that are translationally fused to alternative polypeptides, X and Y. The concept of the assay is that transcription will occur only if X and Y interact to bring together the BD and AD. In a wider context, the yeast two-hybrid system has been used to most effect as a tool to search for unknown partners (prey) for a known protein (bait), where the bait is fused to the BD and prey to the AD. The original technique and those that evolved from it have since been used to identify the protein–protein interactions in many organisms.

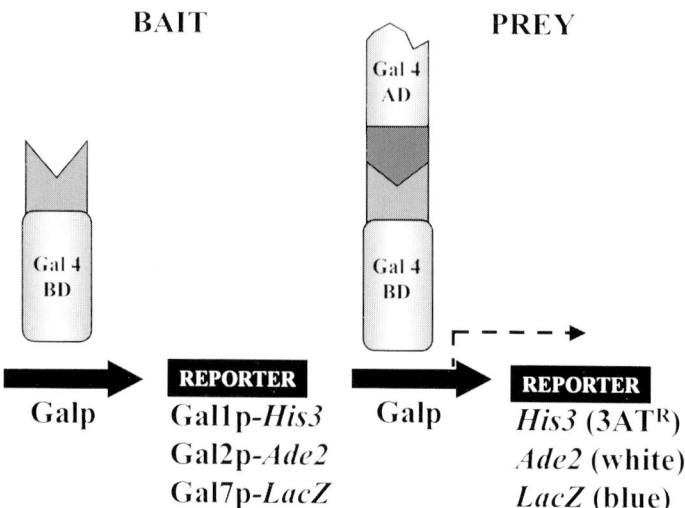

Figure 12.1. Schematic representation of the yeast two-hybrid system. In this scheme three alternative Gal4-activable promoters (*Gal1*, *Gal2* and *Gal7*) are configured to drive the transcription of three separate reporter gene activities. The reporter genes are activated (as denoted by the dashed line) when an interaction occurs between the bait protein and its prey to bring together the DNA-binding (BD) and activation (AD) domains to which they are tethered. The reporter genes *His3* (conferring histidine growth independence and resistance to 3-amino triazole), *Ade2* (conferring adenine growth independence and a visible white colony phenotype amongst red *ade2* mutants) and *LacZ* (conferring a growth-independent visible blue colony phenotype in the presence of X-gal, the chromogenic substrate of the product enzyme β-galactosidase) are scored independently to assess the validity of the putative interaction.

◆◆◆◆◆◆ HISTORY AND EVOLUTION

Gene cloning by functional complementation was demonstrated more than 20 years ago in yeast (Nasmyth and Reed, 1980), subsequently cDNAs from other species including humans were also identified through the use of relevant yeast mutations (Lee and Nurse, 1987; Elledge and Spotswood, 1991). The genetic complementation of functionally characterized mutants of yeast led to the successful cloning of many genes from alternative species where biochemical methods had failed. However, a clear limitation to the use of yeast as a test tube for genetic experiments was the availability of physiologically relevant yeast mutants. To overcome this, functional assays were designed in order to identify and characterize genes in yeast without the need for any particular mutant. A prime example is the development of the yeast two-hybrid system, which can detect any protein–protein interaction independent of the function of the corresponding proteins.

The importance of protein–protein interactions within biological systems has been established over many years in virtually all fields of molecular biology. In recent years, the pivotal role of protein–protein interactions have been highlighted in signal transduction pathways, enzymatic cascades and cellular macromolecular structures (Lane and Crawford, 1979; Linzer and Levine, 1979; Xu et al., 1996). Prior to the widespread use of yeast two-hybrid, the identification of protein–protein interactions and their characterization relied heavily upon biochemical methods which were both labour and time intensive. Critical to the use of these traditional biochemical methods is the assumption that the interacting components can function in isolation or that the specific requirements can be supplemented as key ingredients. The introduction of the yeast two-hybrid and its derivatives has overcome some of these problems.

The main advantages of the yeast two-hybrid over other assays used in gene identification are its simplicity and broad applicability. The *in vivo* nature of the assay provides the user with the ability to select and then dissect a wide selection of protein–protein interactions from a single experimental route before recourse to more time-consuming biochemical methods to confirm and refine the molecular details of the interactions.

The basis of the yeast two-hybrid system emerged from the analysis of transcription factors, notably yeast Gal4p (Keegan et al., 1986). Transcription factors such as Gal4p increase transcriptional rate by binding to specific nucleotide sequences termed upstream activating sequences (UAS) in the promoters of target genes and thereby recruit activated RNA polymerase II. Within the Gal4p it had been shown that the DNA binding and transcriptional activating functions were located in physically separable domains, and that the DNA-binding and activation domains could be independently reassorted by Brent and Ptashne (1985) demonstrated that a hybrid protein, bacterial LexA BD fused to Gal4p AD, could activate transcription in yeast carrying a bacterial reporter gene containing the LexA operator within its promoter.

The reporter gene of choice in the early yeast two-hybrid systems was the bacterial *lacZ* gene encoding β-galactosidase, which cleaves substrates such as X-gal to produce a blue pigment and provide a means of visualization. Growth selection markers using auxotrophic reporters such as *LEU2*, *HIS3* and *ADE2*, have subsequently been introduced and allow growth on medium lacking leucine, histidine and adenine, respectively (cf Figure 12.1). These provide powerful growth selection methods for the detection of protein–protein interactions (Durfee et al., 1993; Gyuris et al., 1993; James et al., 1996). The yeast two-hybrid system relies upon the interaction of proteins X and Y (bait and prey), to activate transcription, which as an indirect assay is prone to false-positive and false-negative results. This inherent flaw has stimulated the development of several variations on the yeast two-hybrid system, which have been designed to reduce the frequency of false positives through increased specificity. One such development is the use of centromeric vectors to maintain the haploid copy number of the two-hybrid genes within the cell and control the level of two-hybrid protein expression (Chevray and Nathans, 1992). Another significant step was the introduction of multiple reporter genes with very different promoter configurations, so as to reduce the number of promoter context-dependent false positives. Clearly the specificity of the assay is increased by scoring all the phenotypes conferred by independent reporter genes in the same cell (James et al., 1996).

The first use of the yeast two-hybrid as a basis to screen for interacting partners was performed with the yeast protein Sir4p (Chien et al., 1991). Here the Sir4p was the bait and an AD-Y library was generated using partially restricted yeast genomic DNA. A screen of over 200 000 independent transformants produced two Sir4p-interacting AD-Y fusion clones, both of which were identified as Sir4p itself and demonstrated the potential significance of homodimerization to the function of this protein. Mammalian protein–protein interactions were also identified with the yeast two-hybrid system using baits of the bZIP (basic domain leucine zipper motif) region of c-jun and the c-fos serum response element (SRE), which was found to interact with SRF accessory protein-1 (SAP-1; Chevray and Nathans, 1992; Dalton and Treisman, 1992). These screens and many others highlighted that the *lac* Z-based colony colour assay was not an efficient selection procedure, since these experiments required pools of over 10^6 transformants.

To address the problem of screening vast numbers of transformants containing mammalian cDNAs, growth-based positive selection strategies were implemented. The specificity and dependence on the interacting partners could be demonstrated by initially selecting a positive clone and then testing it again through the activation of an independent second reporter gene for which growth had not been selected for in the first instance (Durfee et al., 1993; Gyuris et al., 1993; Vojtek et al., 1993). Such multiple reporter systems established the tools required to identify potential protein–protein interactions from biochemical pathways originating from diverse biological systems.

Despite successes, there were still limitations to these systems, notably the high number of false positives and negatives returned in such screens. False positives were defined as clones lacking in biological relevance and false negatives as an ability to recover participants of expected protein–protein interaction. These false results arise due to possible mutational events as the yeast two-hybrid relies on transcriptional activation of reporter genes, any such events could lead to misinterpretation of BD-X or AD-Y interaction.

The yeast two-hybrid screen can inadvertently select examples of fusion proteins showing activation, independent of X–Y contact. An example of this is BD-lamin (Bartel et al., 1993). The cloning of RNA-coding, non-coding, out of frame and antisense DNA sequences can result in fusions of hybrid proteins containing polypeptides that are not naturally expressed in the organism of interest (Fromont-Racine et al., 1997). There is also scope for out of frame fusions to generate bona fide proteins through frame-shifting events.

The reason for an inability to detect two-hybrid interactions arising due to false negatives is still unclear in many cases. A lack of transformed cells though may be due to a particular fusion being toxic and thus affecting its viability, alternatively the protein stability and/or folding of BD-X or AD-Y fusions could be responsible.

Table 12.1. Commonly used yeast two-hybrid systems

System	DNA-binding domain	Activation domain	Reporter selection	Origin
Matchmaker 3	GAL4	GAL4	HIS3, ADE3, LacZ	Clontech
Two hybrid	GAL4	GAL4	HIS3, LacZ	S. Fields, S. Elledge
Interaction trap	LexA	B42	LEU2, LacZ	R. Brent
Modified two hybrid	LexA	VP16	HIS3, LacZ	S. Hollenberg
ProQuest	GAL4	GAL4	HIS3, URA3, LacZ	Gibco BRL
Dual bait	LexA cI	B42	LEU2, LacZ, LYS2, GusA	E. Golemis
Two bait	LexA TetR	B42	LEU2, URA3	R. Brent
Grow and glow GFP; one and two Hybrid	LexA	B42	LEU2, GFP	MoBiTec
Hybrid hunter	LexA	B42	LacZ HIS3	Invitrogen

AD, activation domain; *ADE3*, gene encoding adenine; B42, acidic activator; BD, binding domain; cI, bacteriophage repressor protein; GAL4, transcription factor; GFP, green fluorescent protein; GusA, β-glucuronidase; *HIS3*, gene encoding histidine; *LacZ*, gene encoding β-galactosidase; *LEU2*, gene encoding leucine; LexA, bacterial protein; *LYS2*, gene encoding lycine; TetR, gene conferring tetracycline resistance; *URA3*, gene encoding uracil; VP16, viral protein 16.

◆◆◆◆◆◆ METHODOLOGY

When screening for protein–protein interactions, the initial decision to take is the choice of the most appropriate vector system. Some of the most commonly used gene associations constituting two-hybrid systems are detailed in Table 12.1. More specific information regarding the vectors that bear BD and AD fusion sequence can be found in Table 12.2. The first commercially available vectors were based in the yeast GAL4 protein and since have been used the most extensively. Another widely used system utilizes the BD of bacterial LexA protein and a cannibalized AD originating from the Herpes simplex virus (HSV) virion protein 16 (VP16; Thompson and McKnight, 1992; O'Hare, 1993) or an 88-residue acidic sequence selected for its intermediate transactivation (B42AD; Gyuris *et al.*, 1993). LexA protein is an *E. coli* protein that acts as an SOS gene repressor by binding LexA operator sequences that are an integral part of the promoter (Ebina *et al.*, 1983). In the yeast two-hybrid system, the LexA does not act as a repressor but its operators are integrated upstream of the minimal promoter in order to recruit the yeast transcription machinery when brought into contact with an activator domain.

Table 12.2. Commonly used two-hybrid plasmid vectors

Vector	Selection	Functional domain	Promoter	Reference
GAL4				
pMA424	HIS3	GAL4 BD	Original vector, 12 kb	Chien et al. (1991)
pGBT9	TRP1	GAL4 BD	ADH1 (truncated)	Clontech
pAS1	TRP1	GAL4 BD + HA	ADH1 (full length), CYH2	Durfee et al. (1993)
pAS2	TRP1	GAL4 BD + HA	ADH1 (full length) CYH2	Clontech
pAS2-1	TRP1	GAL4 BD	ADH1 (full length) CYH2	Clontech
pGAD2F	LEU2	GAL4 AD	Original vector, 13 kb	Chien et al. (1991)
pGAD424	LEU2	GAL4 AD	ADH1 (truncated)	Clontech
pGAD10	LEU2	GAL4 AD	ADH1 (truncated)	Clontech
pGAD-GL	LEU2	GAL4 AD	ADH1 (truncated)	Clontech
pGAD-GH	LEU2	GAL4 AD	ADH1 (full length)	Clontech
pGAD1318	LEU2	GAL4 AD	ADH1 (full length)	Clontech
PGTB9	LEU2	GAL4 AD	ADH1	Clontech
pGBKT7	LEU2	GAL4 DB	ADH1 (truncated)	Clontech
pGADT7	LEU2	GAL4 AD	ADH1 (full length)	Clontech
pSE1107	LEU2	GAL4 AD		Durfee et al. (1993)
pACT1	LEU2	GAL4 AD	ADH1 (truncated)	Durfee et al. (1993)
pACT2	LEU2	GAL4 AD + HA	ADH1 (truncated), medium expression	Clontech
LexA				
PBTM116	TRP1	LexA	ADH1 (truncated)	Vojtek et al. (1993)
pLexA	HIS3	LexA	ADH1 (full length) = pEG202	Clontech
pB42 AD	TRP1	B42 + SV40 NLS + HA	GAL1 (full length), inducible promoter = pJG4-5	Clontech
pHybLex/Zeo	Zeocin	LexA	ADH1 (truncated)	Biowire/Invitrogen
pYESTrp	TRP1	V5 epitope + SV40 NLS + B42	GAL1 (full length), inducible promoter	Biowire/Invitrogen
pGilda	HIS3	LexA	GAL1 (full length), inducible promoter centromeric vector	Clontech
pEG202	HIS3	LexA	ADH	Golemis et al. (1996)
pCGLex/p2GLex	Zeocin	LexA	GAL1	Huang and Schreiber (1997)

AD, activation domain; ADH1, Alcohol dehydrogenase 1 promoter; B42, acidic activator; BD, Binding domain; CYH2, gene encoding cycloheximide sensitivity; GAL1, galactose 1 promoters; GAL4, transcription factor; HA, haemagglutinin epitope tag; HIS3, gene encoding histidine; LEU2, gene encoding leucine; LexA, bacterial protein; NLS, nuclear localization sequence; SV40, viral large T antigen; TRP1, gene encoding tryptophan; V5, epitope tag.

Promoter

The full-length ADH1 (alcohol dehydrogenase 1) promoter is 1500-bp long and produces relatively high levels of expression for the sequences under its control. It is found within the pAS2(-1) and pLexA plasmids among others, which are used to clone the reading frame to generate the target fusion protein. The promoter is also found within pGAD-GH, a plasmid that can be used to clone DNA libraries to create potential AD fusion genes. ADH1 is a strong promoter but its expression can be repressed up to tenfold on non-fermentable carbon sources (Denis et al., 1983). Vectors such as pGBT9 (BD) and pGAD424 (AD) utilize a 410-bp truncated ADH1 promoter, which leads to lower levels of fusion protein expression. This can be useful, if the target is toxic at higher levels or has the potential to interact with the normal metabolic function of the host. Such interactions can interfere with the growth selection criteria employed for the two-hybrid assay, whether they be to enhance growth or cause inhibition. The influence of the expression levels of the interacting partners has been investigated by Legrain et al. (1994), who concluded that, although some interactions were unaffected, they showed a marked reduction in their sensitivity when the constituent partners were expressed at lower levels.

Two-hybrid vectors

Two-hybrid host strains of yeast contain a range of auxotrophic mutations necessary to effect genetic selection of the plasmids required to introduce the bait and prey fusion genes. The host yeast strains are transformed to prototrophy with respect to plasmid markers but this may be performed together (co-transformation) or sequentially, depending on the relative stability of the fusion protein and the host strain to be transformed. Alternatively, the yeast host may come in opposite mating types, in which case, the bait and prey can be independently transformed into each mating type and then brought together to enable the protein interaction through the formation of diploids upon mating the haploid strains.

The choice of vector to be used will depend upon the configuration of the two-hybrid screen and the host employed. It is also as well to consider at this early stage what subsequent experiments may be performed in yeast to characterize the two-hybrid interacting partners. For example, if a panel of related baits are to be screened to test the validity of the two-hybrid interaction, then the available auxotrophic markers in the host should be matched with the vector choice to perform co-transformation or interaction mating experiments. More mundane considerations will evolve around the available restriction site to secure the correct fusion product. However, the vector series produced by Roder et al. (1996), in which restriction sites are available in all three reading frames, have alleviated some of these constraints. In any event, the construction of the baits should be confirmed by nucleotide sequencing.

A nuclear localization sequence needs to be associated with the fusion proteins in yeast, and to affect the transcriptional assay GAL4 needs to bind specifically to the UAS, this is conferred by a zinc-cluster motif found within the N-terminal 64 residues (Johnston et al., 1986; Keegan et al., 1986; Marmorstein et al., 1992). The binding of GAL4 to DNA occurs as a dimer and is partially mediated through an α-helical region within residues 65–94 and form a coiled-coil interaction (Carey et al., 1989). Within this binding region GAL4 carries an integral nuclear localization signal (NLS) (Silver et al., 1984, 1986, 1988).

The LexA DNA-binding protein is of bacterial origin and thus has no NLS. Instead, the NLS of the SV40 large T antigen is fused in-frame to ensure nuclear localization. Nuclear localization of LexA fusions may be assayed using repression or blocking assays. The basis of these assays is the observation that LexA and non-activating LexA fusions are able to repress yeast reporter gene transcription, if the genes have LexA operators located between the TATA binding protein site and the target UAS (Brent and Ptashne, 1984). However, most problems concerning bait construction can easily be avoided by cloning known interacting partners in opposite vectors of the two-hybrid system, to create positive controls to determine whether specific reporter gene activation occurs as a faithful response.

To be able to ensure the expression of the fusion proteins in yeast required to form the two-hybrid interaction, many vectors include an in-frame epitope tag, which can be detected by specific antibodies. For example, the pAS vectors encode a haemagglutanin (HA) tag, which places the epitope peptide YPYDVPDYA in frame with the GAL4BD to enable the protein to be visualized with commercial anti-HA antibodies. The HA tag is also to be found in pACT2, a successor of the Clontech vector pGAD424.

Autoactivation

The use of bait-only or prey-only controls may highlight the problem of autoactivation. This occurs when the target fusion protein provides a cryptic activation domain or binding domain partner to trigger protein transcription of the reporter gene independent of any protein–protein interaction. This situation requires serious consideration when using the two-hybrid system. In practice, the two-hybrid system is often configured that the fusion protein library is linked to the activation domain because many proteins have the ability to provide the activation function when coupled to a DNA-binding domain. As a consequence this configuration should be avoided, as it would lead to an unacceptable level of false-positive clones. Therefore, at the onset, the main concern is that the bait fused to the DNA-binding domain does not contain a serendipitous activation sequence. In some cases these problems can be overcome, for example, with the use of the *HIS3* gene as an activatable reporter with increasing concentrations of the inhibitor 3-aminotriazole (3-AT). The effect of incorporating 3-AT into the plates is to suppress the background growth. With the LexA system, however, the use of a less sensitive reporter host or the use of plasmids expressing *LacZ* that contain a variable DNA-binding region provide a genetic control of signal sensitivity. A more extreme but direct method to prevent autoactivation is to delete the offending region of the protein that activates transcription, whilst retaining other functional properties to validate the screen. However, this approach is far from ideal without a prior knowledge of the domain structure of the target protein. Some success has also been reported by using an N-terminal fusion to the BD instead of the normal C-terminal fusion, which failed to prevent autoactivation (Beranger *et al.*, 1997).

After the library screen, it is equally important to check the candidate clones do not encode DNA-binding domains, which may trigger transcription independent of the bait. These may be identified by single plasmid transformations into the host but are often noted in translations of the nucleotide sequences obtained from positive clones (see the later section on the analysis of positive clones).

Library choice

Commonly used plasmids for library construction are shown in Table 12.2. It is often good practice to begin with a library prepared from a tissue in which the target protein is known to be biologically relevant. The relevant strength of activation domains is also worth considering, as it will provide an indication of their ability to initiate transcription. VP16 and the AD of GAL4 are strong activators and thus make the system more sensitive. This can be useful in the detection of weak interactions but will obviously result in higher background transcription levels. Alternatively B42AD is a random protein fragment that demonstrates intermediate transactivation; fusion to this sequence will lower the background transcription at the expense of some sensitivity but may be of more use when setting up a screen.

In the case of the library plasmids in use, major differences exist between the promoters in the LexA and GAL4 systems. Generally, the GAL4 system employs weak constitutive promoters, whereas in the LexA system the library fusions are cloned behind strong, inducible promoters. The advantage of using inducible gene expression is that there is a reduced chance of the AD fusion to exert a toxic effect upon the host in the absence of the inducer. Using this strategy will help maintain the representation of the library in the yeast host by preventing the elimination of toxic products from the pool of potential interacting proteins. Plasmids carrying toxic products are also prone to genetic rearrangement or mutation to eliminate the expression of the activation domain hybrid. These products will be disadvantaged in any library screen, as the loss of the activation domain will mean the reporter gene cannot be activated and, if a growth selective marker is employed, these clones will not survive.

Selecting a yeast strain

Reporter gene selection and sensitivity

Upstream activating and TATA regions form the basis of yeast promoters. A TATA box-containing region precedes all yeast genes and many are associated with *cis*-acting transcription elements and sequences. These can alter the transcription levels by binding to other *trans*-acting proteins. Genetic regulation within yeast involves the *cis*-regulatory elements closely associated with the TATA box. The most common type of *cis*-acting transcription elements found within yeast are the UASs. Such sequences promote transcription through their recognition by specific transcriptional activators. UASs often work independently of the orientation but are distance sensitive within a few hundred basepairs of their proximity to a TATA box. The genes used for galactose metabolism are under the control of the pathway specific regulatory proteins GAL4 and GAL80, in addition to general glucose repression mediated through the MIG1 repressor (Guthrie and Fink, 1991; Frolova *et al.*, 1999). The presence of galactose allows GAL4 to bind to the GAL-responsive elements found within the UAS of at least 20 known galactose-responsive genes. The absence of galactose causes GAL80 to bind to GLA4. This subsequent interaction has the effect of blocking transcription. The transcription of the galactose genes is also repressed in the presence of glucose (Carlson, 1999). The 17 nucleotide consensus sequence that constitutes UAS_{GAL} will function in an additive manner, so multiple sites will lead to higher transcription levels until titration of the relevant transcription factors (Giniger and Ptashne, 1988). The manipulation of the number of consensus DNA-binding sites available in the promoter sequence of the reporter gene is, therefore, a mechanism to alter the detection threshold of the reporter activity. The yeast strains employed with the GAL4 system often carry deletions of the GAL4 and GAL80 genes in order to prevent any endogenous interference in the assay. This can reduce the growth rate of the yeast; however, using the bacterial LexA system maintains normal growth rate. There are two ways to incorporate the reporter genes into the host: by integration at permissive chromosomal locations, or by maintaining the reporter genes on an autonomously replicating plasmid. Although it may seem inconvenient to carry an additional plasmid and utilize an additional auxotrophic marker, using this configuration it is possible to increase the high copy number of reporter gene, and therefore, offer the opportunity to boost weak signals. The higher expression levels will allow the relative β-galactosidase activities to be assessed directly on indicator plates containing the chromogenic substrate X-GAL.

The GAL4-based Matchmaker™ system of Clontech utilizes the GAL1–GAL-10 intergenic region containing four GAL4-binding sites or an artificial construction containing three UAS_{GAL} sites to drive expression of the reporter genes. Hence the construction of the promoter will entirely determine the sensitivity of the reporter genes. It is possible to set these promoters with different expression levels so as to refine the selection of positive clones by testing increasingly stringent promoters and thereby reduce the number of false-positive clones. In the case of the LexA-based system, the DNA-binding domain is provided by the bacterial LexA protein.

Reporter plasmid constructs for this system place 1–8 copies of the LexA operator within the context of a minimal GAL1 promoter to drive LacZ expression (Estojack *et al.*, 1995). This minimal reporter will operate independently of the physiological status of the host yeast, as the UAS_{GAL} and MIG1 binding sequences have been removed so as not to regulate galactose or glucose (West et al., 1984). Alternatively, there are several so-called 'hybrid two hybrid' systems that combine both the GAL4 and LexA systems, such as the L40 reporter strain, which allows the use of the same reporter genes present in the GAL4 system, but these are under the control of LexA operators. This opens up the use of a number of commercially available libraries designed for the GAL4 system to alternative use with the LexA system.

Library transformation

Introducing the plasmid DNA into yeast can be achieved through three well-defined methods: the spheroplast, electroporation or lithium acetate method (LiAc) (Gietz and Schiestl, 1991; Guthrie and Fink,

Protocol I

1. Inoculate the yeast strain transformed with the target BD fusion protein into 25 ml of the appropriate synthetic complete (SC)-omission medium and incubate at 30°C overnight with shaking at 200 rpm.
2. Determine the cell titre and calculate the volume of cells that yields 2.5×10^8 cells (usually about 10 ml).
3. Centrifuge this at 3000g for 5 min.
4. Resuspend in 50 ml of prewarmed (30°C) yeast peptone dextrose (YPD).
5. Incubate at 30°C with shaking at 200 rpm for 3–4 h until the cell titre reaches 2×10^7 cells/ml.
6. Harvest the cells by centrifugation at 3000g for 5 min.
7. Resuspend with half the volume of sterile distilled water followed by centrifugation as before.
8. Transformations proceed following the addition of (the volumes for a single transformation are scaled up 10×):
   2.4 ml 50% PEG
   360 µl 1.0 M LiAc
   250 µl salmon sperm DNA (2 mg/ml)
   50 µl of library plasmid DNA
   450 µl sterile double distilled water.
9. Vortex well for at least 1 min to ensure DNA distribution.
10. Incubate at 30°C for 30 min.
11. Heat shock at 42°C for 30 min with mixing by inversion for 15 s after every 5 min.
12. Centrifuge as above to collect the cells and gently resuspend in an appropriate volume of sterile distilled water and plate on to SC-omission medium. Incubate the plates for 3–5 days at 30°C or until colony formation. HIS3 transformants can be plated directly on to SC-omission medium lacking trp, leu or his.

Recipes for SC and YPO media are detailed in Guthrie and Fink (1991).

1991; Gietz et al., 1992, 1995). Of the three, the LiAc method (see Protocol 1) is the most suited for the yeast two-hybrid applications as it provides a quick method for introducing foreign DNA into yeast. The LiAc method involves preparing competent yeast cells and resuspending them in a LiAc solution that contains the plasmid DNA and an excess of carrier DNA. Polyethylene glycol (PEG) containing the appropriate amount of LiAc is then added to the mix and incubated at 30°C, followed by heat shock to introduce the DNA to the cells prior to plating on the appropriate medium to select the correct plasmid-containing transformants. It is often a good idea to use frozen stocks of competent cells that are known to give reliable transformation efficiencies. It is also advisable to use yeast cells harvested from log phase growth. The growth of yeast can be maintained by measuring optical density at 600 nm or by microscopic inspection of the fraction of cells actively budding. The volume of cell culture to be used needs careful consideration with respect to the final cell density in the transformation protocol. Too high a cell density will result in a significant overgrowth of yeast on the transformation plates making selection of the transformants impossible. Conversely, too low a cell density will result in a poor transformation efficiency. Excess plasmid DNA will increase the likelihood that multiple library plasmids will enter the same host cell, which may be deleterious to the host and problematic later with respect to the recovery and identification of the positive interacting clone. The carrier DNA must be of high molecular weight and entirely denatured. This can be achieved by repeated boiling and chilling of the DNA prior to use. The competence of DNA uptake at the cell wall is induced by treatment with both PEG and lithium ions. To check that all the plasmids present are capable of transformation and expression, then plasmids carrying partners that are known to interact may be transformed with the same competent cells to generate a positive control with each experiment.

The plasmids carrying the bait and prey may be either transformed singly or simultaneously into the yeast host. Simultaneous transformations with the bait and library plasmids that possess different selection markers is often undertaken, although the efficiency of co-transformation may be lower. The co-transformation strategy, however, avoids repeated rounds of growth following the first transformation that could result in counterselection of the bait fusion protein.

If doubts exist as to the efficiency of recovery of clones expressing interacting partners, then the library transformation may be spiked with a clone containing a fusion protein known to interact with the bait. However, care is required with these experiments as it is tempting to set the sensitivity threshold of the reporter activity too high, based on the recovery of clones containing the known partner.

Selection procedures for the yeast two-hybrid interactions

In two-hybrid screens the first round of selection is often dependent on a growth phenotype associated with the reporter gene. Commonly used

Table 12.3. Commonly used yeast strains

Strain	Reporter gene	Number of operators	Comments	Reference
H7Fc	LacZ HIS3	GAL4: 3 × UAS$_{G17mer}$ GAL4, GAL1: 4 × UAS$_{G17mer}$	Tighter regulation of HIS3 reporter than Y190	Feilotter et al. (1994)
CG-1945	LacZ HIS3	GAL4: 3 × UAS$_{G17mer}$ GAL4, GAL1: 4 × UAS$_{G17mer}$	Tighter regulation of HIS3 reporter than Y190	Clontech
YRG2	LacZ HIS3	GAL4: 3 × UAS$_{G17mer}$ GAL4, GAL1: 4 × UAS$_{G17mer}$		Stratagene
SFY526	LacZ	GAL4, GAL1: 4 × UAS$_{G17mer}$		Q.Biogene
Y190	LacZ HIS3	GAL4, GAL1: 4 × UAS$_{G17mer}$ GAL4, GAL1: 4 × UAS$_{G17mer}$	High expression level but significant constitutive leaky expression	Harper et al. (1993)
Y187	LacZ	GAL4, GAL1: 4 × UAS$_{G17mer}$		Clontech
EGY48	HIS3 URA3	6 × LexA op	Basic strain for cDNA cDNA library screening	Estojack et al. (1995)
EGY191	HIS3 URA3	2 × LexA op	More stringent than EGY48. Gives lower background with transcriptionally active baits	Estojack et al. (1995)
L40	HIS3 LacZ	4 × LexA op 8 × LexA op	Prototrophy selection	Triolo and Sternglanz (1996)
PJ69-4A	HIS3 ADE2 LacZ		High sensitivity, low background. Three different reporter genes driven by different promoters	James et al. (1996)

ADE2, gene encoding adenine; GAL4, transcription factor; *HIS3*, gene encoding histidine; *LacZ*, gene encoding β-galactosidase; LexA, bacterial protein; LexA op, LexA operator; UAS, upstream activating sequence; *URA3*, gene encoding uracil.

yeast and their respective reporter genes are listed in Table 12.3. The *HIS3* gene has proven useful in this context either by plating on supplemented medium less histidine or through the application of the counterselective drug 3-AT. Because the positive interacting clones affect a growth phenotype, the size of the resulting colonies may reflect the strength of the interaction but smaller ones should not be considered any less biologically significant. The *GAL4*-based selection system of James *et al.* (1996) employs a second phenotypic marker in *ADE2*. This will confer a growth phenotype (adenine independence) or provide a non-growth selective readout of the putative interaction through the visible appearance of colonies: *ade2* mutants accumulate the red pigment derived from the polymerization of the intermediate phosphoribosylamino-imidazole (AIR), whereas the colonies expressing the reporter *ADE2* will be white. Growth-selective screens are often followed by monitoring a growth-independent reporter. The reporter of choice is often the E. coli *lacZ* gene, which produces the enzyme β-galactosidase. β-Galactosidase activity can be monitored *in situ* with prospective yeast colonies or quantified in cellular extracts. As with the advice on growth selection, the weakly expressing β-galactosidase-positive colonies should not be ignored.

Primary transformants will take between 3 and 6 days to appear on the transformation plates. These should be replica plated to assess their true phenotypes with respect to the reporter gene. In the case of the HIS3$^+$ prospective two-hybrid positive colonies, these may be titrated against increasing concentrations of 3-AT, which may provide an insight into the stability and strength of the interacting partners. However, be wary at this stage, as false positives that induce gratuitous transcription can sometimes appear strong. The use of multiple reporters using independent promoters will alleviate this problem. Note that not all the colonies showing a positive growth phenotype will be positive for the alternative reporters. For example, the *HIS3* reporter can give rise to leaky transcription in several host yeast strains. This can be countered using 3-AT, which is a known inhibitor of the *HIS3* gene product, thereby inhibiting the basal expression of the gene. The amount of 3-AT used is important as too much will eliminate weak interactions and too little will result in a huge excess of false positives. The optimal concentration of 3-AT should be determined by titration in pilot transformation experiments. Typically this would involve transforming two or more strong, intermediate and weak interacting proteins with the target co-transformed against a background of plasmid library DNA. The primary transformants may then be patched and replica plated on to a dilution series of 3-AT (0.3–30 mM). The 3-AT concentration should be sufficient to suppress background but still allow the growth of weakly interacting partners.

To assay for β-galactosidase, it is necessary to permeate the yeast colonies. This is accomplished by freeze–thaw in liquid nitrogen. The yeast colonies are replica transferred on to a membrane filter and allowed to grow while the filter is resting on an appropriate selective agar plate, then subjected to freeze–thaw before application of the chromogenic β-galactosidase substrate X-gal. Fresh-growing yeast will provide the best β-galactosidase activities, where strong interactions will appear after 2 h, intermediate around 6 h and weak interactions may need to be incubated overnight. Both positive and negative controls should be incorporated on to the same filter.

Library plasmid rescue

DNA from yeast indicating potential two-hybrid interaction needs to be extracted and the plasmids rescued by transformation into bacteria, where they can be propagated to enable subsequent manipulation and DNA sequence analysis. The positive yeast strain will contain two plasmids, one encoding the bait and a second the prey to be characterized in the form of a cDNA or genomic DNA library insert. Rescuing plasmid DNAs from yeast can be troublesome. In particular picking the plasmid containing the library insert can be time consuming. However, conveniently the yeast *LEU2* gene will complement the *LeuB* deficiency present in *E. coli* strains such as HB101. Thus yeast plasmids carrying the library insert on a LEU2 plasmid can be selected directly upon transformation into HB101. Under these conditions the HB101 strain is quite slow growing and transformants can take up to 3 days to grow on minimal supplemented medium. Owing to the difficulties of preparing large quantities of DNA from this host, the plasmids once rescued can be transformed into an alternative host to facilitate genetic manipulation. Some two-hybrid plasmids carry the *CYH2* gene of yeast, which confers cycloheximide sensitivity on the host strain. This addition provides a method to counter-select the plasmid-carrying *CYH2* gene by adding cycloheximide to the medium. The resulting strains will carry only the plasmid derived from the library. Alternative systems have been developed in which the plasmid recovery and separation has been made simpler by providing the plasmids harbouring the bait and library insert with different bacterial origins of replication, which enables them to be preferentially rescued in permissive *E. coli* hosts upon transformation.

Analysis of the positive clones

To be certain of the specificity of the two-hybrid interaction, it is necessary to retransform the positive clones on their own and with a series of control plasmids. The controls should include transformation with empty vectors and null baits that could have no physiological relevance. Depending upon the representation within the library screened, a specific interacting partner can be selected many times. To prevent time-consuming replication, it is best to sequence the candidate clones first to eliminate the redundancy in the clones to be tested. In recent years, DNA sequencing has become cost effective and efficient, and so probably represents the quickest route forward to reduce the number of clones requiring confirmation. Based on the DNA sequence analysis, out of frame and inverted clones may be eliminated from the search. Other genes that can be excluded from additional study are those that encode RNA or retroposons as well as known false positives. A useful database of commonly associated false positives can be found at http://www.fccc.edu/research/labs/golemis.

Genomics–bioinformatics

There is currently widespread use of the yeast two-hybrid system, due mainly to its relative simplicity and low cost as a means to detect direct physical interaction between novel predicted proteins and those previously characterized. The source of novel protein interactions being used this way comes from the vast quantification of sequence data for genes and predicted proteins that has been achieved as a result of genome sequencing projects. The necessity for these assays is evident considering that the majority of the sequenced genomic DNA lacks any significant homology to functionally characterized proteins or protein domains (Legrain et al., 2001). It is clear that the yeast two-hybrid system has become an important tool in the postgenomic era to perform proteomic studies of protein–protein interactions (Graves and Haystead, 2002).

Interaction mating

One of the first advantages of the yeast two-hybrid in genomic screening was that of 'interaction mating' (Finlay and Brent, 1994). Panels of prey or bait are mated in defined grids to allow possible protein–protein interactions to be rapidly assessed from a large pool. The use of panels is well suited to the high throughput analysis required for genomic studies as they can be arranged in microtitre plates. In addition, formatting the cDNA libraries into pooled subsets allows the data to be handled electronically and reduces the process of tracing the identity of clones derived from highly complex arrays such as normalized EST (Hua et al., 1998).

Buckholz et al. (1999) have recently described an automated method for screening protein–protein interactions in yeast two-hybrid libraries. The method involves using interaction mating of pooled cDNA library subsets in a liquid format. Using 96-well plates, the bait is mated with arrayed subsets of cDNA prey library. The interactions themselves are selected for by prototrophy and the same array is used in detecting reporter activation. In addition, the formatting of the cDNA library will allow the functional subtraction of promiscuous, positive class interactors together with the use of a liquid array, and the electronically automated handling of screens will greatly reduce time and allow for its widespread use in genomic screens.

A high throughput yeast two-hybrid system has recently been described for the study of protein–protein interactions within plants (Fang et al., 2002). The screen incorporates yeast gap-repair cloning and multiple positive and negative selection to reduce the incidence of false-positive negative clones and is automated to increase throughput. Although described for the study of rice genes, the method can actually be used to map protein–protein interaction networks and signal transduction pathways in any system. The high-throughput method was developed for *E. coli* transformation, using 96-well plates (deep well). Here the purified plasmid DNA is used to transform DH5α competent cells in the plates. LB medium is then inoculated from this for the isolation of plasmid DNA that serves as the sequencing template for identifying positive two-hybrid clones. Incorporated automation includes the use of Q-Bot (Genetix, Hampshire, UK) for both picking colonies from the selection to the multiwell plates and also for spotting cultures on to selection plates. Biomek (Beckman, California, USA) is used for liquid transfer between 96-well plates and BioRobot (Qiagen) is employed for plasmid DNA isolation from both yeast and *E. coli*.

Protein–protein interaction maps

The use of protein–protein interaction maps (termed the matrix approach) uses a collection of full-length proteins that are predefined open reading frames (ORFs) as both the prey and bait for the assay. The ORFs are cloned into two hybrid vectors (prey or bait specific) via the polymerase chain reaction (PCR). The fusion proteins are then expressed in yeast cells of opposite mating type and all combinations brought together as diploids by mating in a large matrix. Combinations of yeast and prey are then assessed individually or following a pooling of cells that express different bait and prey proteins (Legrain et al., 2001). The major drawback here is that this strategy only tests predefined proteins. Successful uses of this approach began with *Drosophila* protein–protein interaction within the cell cycle (Finlay and Brent, 1994). More recently, large-scale approaches have been used for the vaccinia virus and yeast proteomes (McCraith et al., 2000; Uetz et al., 2000; Ito et al., 2001). Reproducibility is a problem with this yeast two-hybrid approach. In the case of the vaccinia study, it was countered by assaying the interacting partners in quadruplicate and screening interactions with at least three independent positive clones (McCraith et al., 2000).

The yeast two-hybrid is also now widely used as a method of screening libraries for protein interaction with other known bait proteins. When screens are repeated with proteins within the same biochemical context, specific functional protein-interaction maps result. These in turn are used to further identify other proteins of that pathway, as was the case with the T7 phage proteome (Bartel et al., 1996).

There are several examples of groups working to produce comprehensive maps of protein interactions via two-hybrid genomic analysis, at the level of the whole genome and also for specific cellular processes. Bartel *et al.* (1996) produced a protein–protein interaction map of the ~55 known *E. coli* bactereriophage T7 proteins, using a combination of interaction mating and library screening of defined and randomly generated bait and prey fusion proteins. A great deal of data has been collected on protein–protein interactions of T7 using this approach. However, of concern was the omission of the protein–protein interactions of the T7 proteome that are already known to exist.

Screening randomly generated protein fragments, allows the determination of interacting domains. However, the orientation, location of the fragment or reading frame means that only a small fraction of cDNA fragments will actually encode genuine protein–protein interacting domains, and thus large libraries are required (Siomi et al., 1998). Alternatively Framont-Racine *et al.* (1997) identified a specific set of protein–protein interactions involved in spliceosome function. The target baits were selected from the *S. cerevisiae* proteome thought to be involved in pre-mRNA splicing. Then, for each of the baits, a library screen was performed. This method was useful in predicting networks of interactions of known and novel putative splicing factors.

Large-scale proteomic studies have also been used with genomes from viruses and bacterial species, such as hepatitis C virus (Flagolet et al., 2000) and that of *Helicobacter pylori* (Rain *et al.*, 2001) (see Chapter 16). In this study, complexes were identified that had first either been demonstrated or thought to exist in other organisms.

The experimentally derived interaction maps have been exploited *in silico* by proposing algorithms that aim to assign function to uncharacterized gene products. The principle is based on 'guilt by association', where function is assigned to a protein by transposing existing annotations from its interacting partners. Thus it relies heavily upon the completeness of the interaction map. Such analysis has been performed for the proteome of *S. cerevisiae* (Fallenberg *et al.*, 2000; Schinkowski *et al.*, 2000). Wojcik and Schächter (2001) in fact recently described a computational approach aimed at predicting the protein interaction map of a given organism by 'inference' from collective interaction maps found in a large database that also includes interaction domain information. The technique has been referred to as the interacting domain profile pairs (IDPP). It is based upon a combination of interaction and sequence data, and uses a combination of homology searches and clustering. The researchers have applied this approach in the prediction of an interaction map of *E. coli* from the published map of *H. pylori*.

Problems with the library approach include the preparation of highly complex prey fragment libraries and the high cost of interaction screens. Specialized technological platforms are required, such as robots, specialized computer software and algorithms, to allow the library to be covered completely and to identify the prey fragments using high-throughput sequencing (Legrain *et al.*, 2001).

Independent confirmation

Once the positive interacting clones have been selected from yeast it is essential that the interaction be confirmed through independent experimentation. Confirmation of the interaction should be tested in the most relevant biological system available. The most commonly used methods for confirming protein–protein interactions rely upon *in vitro* assays such as co-immunoprecipitation and 'pull down' assays involving epitope tagged versions of the proteins.

◆◆◆◆◆◆ VARIATIONS ON THE TWO-HYBRID THEME

Since its advent, the yeast two-hybrid system has spawned a multitude of related strategies to detect not only protein–protein interactions, but also protein–nucleic acid interactions and other protein–ligand interactions.

Post-translational modifications in the yeast two-hybrid system

Post-translational modification of proteins may be essential to ensure their correct function. This is particularly true of proteins found in higher eukaryotic organisms. Application of the yeast two-hybrid system under these circumstances may only yield limited data sets. For example, if the post-translational modifications are required to facilitate some of the protein–protein interactions and the relevant modifying enzymes are absent within *S. cerevisiae*. This will be especially true when the baits are components of a signal transduction pathway that rely upon their cognate binding to sites created by post-translational modifications. Chervitz *et al.* (1998) compared the proteome of *S. cerevisiae* and *C. elegans* to find that yeast possessed notably fewer regulatory and signal transduction domains. In particular, it was shown that phosphotyrosine-dependent signalling pathways were under-represented in yeast due to a paucity of tyrosine kinase, phosphotyrosinephosphatase and phosphotyrosine binding domains. This problem was overcome by co-expressing the relevant tyrosine kinase with the bait and prey (Osborne *et al.*, 1995, 1996). However, it is not clearly predictable how the additional modifying enzyme will affect a wide range of cellular functions. Expression of the *src* family kinase, for example, is toxic in yeast, owing to the kinases broad substrate recognition (Kornbluth *et al.*, 1987).

Three-hybrid system

The three-hybrid system evolved to accommodate the fact that many cellular proteins actually function as part of a large multimeric complex, where the cohesion is maintained by the cumulative effect of a number of relatively low-affinity interactions. The three-hybrid system is a direct extension of the two-hybrid system that relies upon a third protein component. The third protein may act directly to create or stabilize a physical bridge between the fusion proteins carrying the binding domain and activation domains. Alternatively, the third protein may act to induce a conformational change in either the binding domain or activation domain fusion proteins such that they are capable of association, which will, in turn, trigger transcription of the reporter genes. The three-hybrid system can also be used in reverse, that is, to find a protein or ligand that will disrupt a resident two-hybrid interaction (Vidal, 1996a, 1996b; Vidal and Endoh, 1999). This may be achieved either by blocking the interacting domains or by post-translational modification of one of the components. The pBridge system of Clontech takes advantage of the Met25 promoter, which is repressed upon the addition of methionine, such that the expression of the third protein can be controlled from the same plasmid carrying the binding domain fusion plasmid. The three-hybrid approach has been used successfully to demonstrate an increase in ternary interactions by the co-expression of the third member 'auxiliary bait' (Zhang and Lauter 1996; Tirode *et al.*, 1997).

The yeast three-hybrid system has been used as a method to identify enzymatic substrates within the context of a ternary complex (Van Criekinge *et al.*, 1998) and, more recently, the technique is being applied to that of quarternary protein complexes (Pause *et al.*, 1999).

Dual-bait systems

Dual-bait systems allow for the screening of proteins that bind one bait and not the other by the use of two independent baits that are fused to different BDs to activate parallel but separate reporter genes upon interaction with prey. Thus, one can also screen for mutations that remove or increase binding of prey to one or two closely related baits (Toby and Golemis, 2001). Other possible applications include using co-expressed proteins or pharmaceutical agents as competitors (Fashena et al., 2000).

Swapped systems

The swapped system was developed to overcome one of the limitations of the yeast two-hybrid system, which allowed only the study of proteins that are transcriptionally silent in RNA polymerase II-based assays. Here, the transcriptionally active protein of interest is fused with the AD as opposed to the BD, then is used to screen a library of BD-fused proteins (Du et al., 1996). However, the incidence of non-specific DNA-binding activities in these libraries can often result in the selection of a large number of false-positive clones.

Modification of the transcription response

In this system, the reporter genes have been coupled to an RNA polymerase III-based transcription unit in place of the normal RNA polymerase II transcription unit activated in the yeast two-hybrid system. This assay provides the benefit of being able to study proteins that would be transcriptionally active in an RNA polymerase II-dependent assay. The RNA polymerase III-dependent assay provides a sensitive detection system for protein–protein interactions and is characterized by the low frequency at which false-positive clones are recovered (Marsolier et al., 1997; Marsolier and Sentenac, 1999).

SOS recruitment system

The SOS recruitment system (SRS) evolved to study and examine protein–protein interactions that cannot be studied using the standard yeast two-hybrid screen, such as proteins that are transcriptionally active or those that associate with membranes and, therefore, do not localize to the nucleus (Aronheim, 1997; Aronheim et al., 1997). The system makes use of the Ras signal transduction network for positive selection. The SRS approach results in a high false-positive background, but this has been reduced by the advent of the Ras recruitment system (RRS) which is based on the absolute requirement of plasma membrane localization for Ras function (Broder et al., 1998).

Ubiquitin-based split protein sensor (USPS)

This USPS system, like the SRS, is another yeast two-hybrid derivative, which is not reliant upon third-party reporters. Here the two-hybrid partner interaction induces a positive phenotype via the generated catalytic activity (Toby and Golemis, 2001). The system is based on the reconstitution of ubiquitin enzymatic activity upon interaction of bait and prey (Johnsson and Varshavsky, 1994; Dunnwald et al., 1999).

Detecting protein–nucleic acid interactions

The one-hybrid system was developed from the yeast two-hybrid to study protein–DNA interaction (Wilson et al., 1991; Wang and Reed, 1993; Li and Herskowitz, 1993). In this system, proteins are screened from an AD-fused library for transcriptional activation, via a regulatory motif of interest, previously inserted upstream of the reporter genes in yeast. Here the BD fusion protein is not required and the DNA GAL4 or LexA binding sites are replaced by the specific DNA sequence of interest. The AD library is then screened to identify proteins that bind to the regulatory motif and activate transcription. Successfully identified proteins include the yeast origin of replication complex Orc6 protein (Li and Herskowitz, 1993), and also the mammalian olfactory neuronal transcription factor Olf1 (Chong et al., 1995).

The 'one and a half' hybrid system allows for the identification of proteins that bind DNA motifs only in the presence of a co-expressed auxilliary protein, as it will cause a stable ternary complex. The first example of this was with the identification of SAP-1, which binds the c-fos serum response element when complexed with serum response factor (SRF) (Dalton and Treisman, 1992).

A three-component system has been developed to identify RNA-binding proteins in yeast (Putz et al., 1996; SenGupta et al., 1996, 1999; Wang et al., 1996; Park et al., 1999). In these systems, the DNA-binding domain is coupled to a characterized RNA-binding protein that is integrated into a chromosomal location, the cognate RNA-binding structure for which is provided *in trans* by an RNA polymerase III transcription unit located on a plasmid. The bait RNA is fused to the characterized RNA motif and used to screen an activation domain library for binding partners to the target RNA. However, the system requires the use of careful controls, first to ensure the RNA structure of the bait is not impaired when fused to the characterized RNA motif and then to ensure the elimination of false-positive clones, since non-specific RNA binding is a common feature found in activation domain libraries.

Protein–ligand detection systems

A system to detect the binding of small ligands has been developed based on the chemical conjugation of target drugs to a known ligand of the glucocorticoid receptor, dexamethasome. The synthetic chemical conjugate provides a bridge between the prey protein and the hormone-binding domain of the glucocorticoid receptor, which is translationally fused to a DNA-binding domain. The screen against the activation domain, therefore, requires that the synthetic conjugate can permeate actively growing yeast to affect the reporter gene transcription (Licitra and Liu, 1996).

Bacterial two-hybrid systems

Two *E. coli*-based two-hybrid systems have been reported. These alternatively use the transcriptional activation/repression of reporter genes (Kornacker *et al.*, 1998) or the functional reconstitution of an enzyme (Karimova *et al.*, 1998). *E. coli* systems benefit from the high transformation rates achievable to allow the full coverage of complex genomic and cDNA libraries (Legrain and Selig, 2000). The circumstances under which the two-hybrid systems are applied will dictate the most appropriate choice of *E. coli* system. Similarly, the physiological background to the interaction sought will dictate the choice between *E. coli* and the model eukaryote *S. cerevisiae*.

Yeast two-hybrid systems at the bacteria–eukaryote interface

Pathogenic mechanisms of disease caused by Gram-negative bacteria employ a type III secretion system to deliver effector proteins into animal and plant host cells (Lahaye and Bonas, 2001; Plano *et al.*, 2001; Statskawicz *et al.*, 2001). The yeast two-hybrid system is proving to be a powerful tool in the identification of protein–protein interactions involved in the delivery system and between the microbial effectors and their eukaryotic targets.

Salmonella species utilize a type III secretion system to deliver a panel of effector proteins, of which the SipA protein is necessary to promote efficient entry into host cells. SipA binds actin to facilitate the typical appearance of membrane ruffling associated with *Salmonella* internalization. Zhou *et al.* (1999) have utilized the yeast two-hybrid system to identify T-plastin as one of the interacting partners of SipA, and together they modulate the actin-bundling activity at the site of entry. Later Scherer *et al.* (2000) identified SspC as an essential component of the translocation apparatus that interacts with eukaryotic intermediate filament protein. Yet more recently, Carlson *et al.* (2002) employed a yeast two-hybrid screen, using the *Salmonella* secreted protein SipC as the bait to demonstrate an interaction between cytokeratin-18 and SipC as a critical process in *Salmonella* invasion. Cytokeratin proteins are submembranous, cytoplasmic intermediate filament phosphoproteins that form an anastomotic network that provides integrity and support to the cytoskeleton. These studies highlight direct links between the invading bacteria and the host cytoskeleton, a process which is becoming a recurrent theme within the context of pathogen–eukaryotic infections.

Host infection by enterovirulent *E. coli* proceed following initial attachment to gut enterocytes, via a type III secretion and delivery of bacterial effector proteins into the host cell to produce a characteristic pedestal upon which the bacterium is held in intimate attachment. This process requires a gross cytoskeletal reorganization of the host cell to produce the attaching and effacing lesion. Intimate attachment occurs through the binding of the bacterial outer membrane protein intimin, to its cognate translocated receptor tir (translocated intimin receptor). The tir protein is inserted in the host cell membrane and is one of a number of effector molecules translocated from the bacterium. The yeast two-hybrid system has been used to detect and locate the protein regions involved in the intimin–tir interaction (de Grado *et al.*, 1999; Hartland *et al.*, 1999). These and complementary biochemical studies provided the information necessary before employing the tools of structural biology (high-resolution nuclear magnetic resonance and protein crystallography) to establish that tir binds intimin as a dimer at the interface of the bacterial and host membranes (Batchelor *et al.*, 2000; Luo *et al.*, 2000).

In the case of *Shigella flexneri*, the yeast two-hybrid system has been used by Page *et al.* (2001, 2002) to identify interaction partners with domains of its virulence proteins. Prey libraries were created of the DNA fragments spanning the entry region of the virulence plasmid. These were then screened against the secreted proteins and baits. Interactions were detected between the effectors IpaB and IpaC with their chaperone IpgC, and the association of the secreted proteins IpA, IpgB1 and Ospc3 with the chaperone Spa15.

Pathogenic *Yersinia* species contain a plasmid, which encodes a contact-dependent type III secretion system that acts to deliver *Yersinia* outer membrane proteins or 'Yops' into host cells. The Yops subvert host cell function and confound the development of a cell-mediated immune response. One of these effector proteins, known as YopJ in *Y. pseudotuberculosis* and YopP in *Y. enterocolitica*, has been shown to interfere with a number of host signalling pathways. These include inhibition of extracellular signal-regulated kinase (ERK), c-Jun amino-terminal kinase (JNK), the p38 mitogen-activated protein kinase (MAPK) and nuclear factor kappa-B (NF-κB) signalling pathways. YopJ has orthologues in other animal and plant bacterial pathogens, including AvrA of *Salmonella typhimurium* and AvrRxv of *Xanthomonas campestris* pv. *vesicatori*, which may have some functions in common. Orth *et al.* (1999) were able to demonstrate using YopJ as bait in a yeast two-hybrid screen against a HeLa cell cDNA library to identify the MAPK kinase superfamily (MKKs) as binding partners, which would explain the diverse effects of the YopJ protein on signalling pathways.

Pseuodomonas syringae pv tomato is the causative agent of bacterial speck disease of tomatoes. However, the plant host has a resistance mechanism consistent with the gene-for-gene hypothesis, where the avirulence signal is the delivery of the bacterial avrPto protein, which prompts the plant resistance response mediated by the host Pto protein (Scofield *et al.*, 1996). The resistance response requires action of several signal transduction pathways to bring about the expression of defence genes, rapid generation of an oxidative burst and rapid localized cell death, a process termed the hypersensitive response. The Pto protein has been identified as a serine/threonine protein kinase that has direct interaction with the avrPto protein (Tang *et al.*, 1996). The yeast-two hybrid system has proved instrumental in the identification of the Pto-interacting proteins (Pti). These include Pti1, a serine/threonine protein kinase that is phosphorylated by Pto (Zhou et al., 1995) and proteins Pti4–6, which are putative transcription factors that may be involved in the activation of defence-related genes (Zhou et al., 1997; Gu et al., 2000). More recently, a modified yeast two-hybrid screen has been devised, in which host proteins were screened that interact with Pto only in the presence of the bacterial effector avrPto (Bogdanove and Martin, 2000). This screen identified five new protein classes: Adi1 – tomato catalase 1; Adi2 – a serine/threonine protein kinase homologue of Pti1; Adi3 – a protein similar to serine/threonine protein kinase; Adi4 – a protein of unknown function; and Adi5 – a full-length cDNA of Pti2, which was previously identified as a truncated proteosome α-subunit that interacts with Pto independent of avrPto. Further, an avrPto bait identified four host proteins that interact with the effector protein. These include a stress-induced protein that is known to respond to wound-induced and virus-induced stress, two putative Ras-related small GTP-binding proteins and protein similar to N-myristyltransferase (Bogdanove and Martin, 2000).

The yeast two-hybrid system has also been employed to screen for host proteins that interact with the major virulence factor vacuolating cytotoxin (VacA) protein of the gastric pathogen H. pylori (Hennig et al., 2001). The VacA-containing baits were screened against a human gastric mucosa cDNA library. Positive clones retrieved and confirmatory pull-down assays confirmed interaction with the RACK1 protein, a receptor for activated C kinase and a homologue of the heterotrimeric G-protein β-subunit. Further experiments within this study showed the interacting fragment to be located within the C-terminal portion of p33 VacA subunit and six residues from the p55 subunit. The overall significance of these interactions are yet to be determined but, as experienced with many biological systems, the elucidation of the interacting proteins using two-hybrid technology will provide a crucial starting point for further analysis into the interaction and subsequent cellular effects.

References

Aronheim, A. (1997). Improved efficiency sos recruitment system: expression of the mammalian GAP reduces isolation of Ras GTPase false positives. *Nucleic Acids Res.* **25**, 3373–3374.

Aronheim, A., Zandi, E., Hennemann, H., Elledge, S. J. and Karin, M. (1997). Isolation of an AP-1 repressor by a novel method for detecting protein–protein interactions. *Mol. Cell. Biol.* **17**, 3094–3102.

Bartel, P., Chien, C. T., Sternglanz, R. and Fields, S. (1993). Elimination of false positives that arise in using the two-hybrid system. *BioTechniques* **14**, 920–924.

Bartel, P. L., Roecklein, J. A., SenGupta, D. and Fields, S. (1996). A protein linkage map of *Escherichia coli* bacteriophage T7. *Nature Genet.* **12**, 72–77.

Batchelor, M., Prasannan, S., Daniell, S., Reece, S., Connerton, I., Bloomberg, G., Dougan, G., Frankel, G. and Matthews, S. (2000). Structural basis for recognition of the translocated intimin receptor (Tir) by intimin from enteropathogenic *Escherichia coli*. *EMBO J.* **19**, 2453–2464.

Beranger, F., Aresta S., de Gunzburg J. and Camonis J. (1997). Getting more from the two-hybrid system: N-terminal fusions to LexA are efficient and sensitive baits for two-hybrid studies. *Nucleic Acids Res.* **25**, 2035–2036.

Bogdanove, A. J. and Martin, G. B. (2000). AvrPto-dependent Pto-interacting proteins and AvrPto-interacting proteins in tomato. *Proc. Natl Acad. Sci. USA* **97**, 8836–8840.

Brent R. and Ptashne, M. (1985). A eukaryotic transcriptional activator bearing the DNA specificity of a prokaryotic repressor. *Cell* **43**, 729–736.

Broder, Y. C., Katz, S. and Aronheim, A. (1998). The *ras* recruitment system, a novel approach to the study of protein–protein interactions. *Curr. Biol.* **8**, 1121–1124.

Buckholz, R. G., Simmons, C. A., Stuart, J. M. and Weiner, M. P. (1999). Automation of yeast two-hybrid screening. *J. Mol. Microbiol. Biotechnol.* **1**, 135–140.

Carey, M., Kakidani, H., Leatherwood, J., Mostashari, F. and Ptashne, M. (1989). An amino-terminal fragment of GAL4 binds DNA as a dimer. *J. Mol. Biol.* **209**, 423–432.

Carlson, M. (1999) Glucose repression in yeast. *Curr. Opin. Microbiol.* **2**, 202–207.

Carlson, S. A., Omary, M. B. and Jones, B. D. (2002). Identification of cytokeratins as accessory mediators of *Salmonella* entry into eukaryotic cells. *Life Sci.* **70**, 1415–1426.

Chervitz, S. A., Aravind, L., Sherlock, G., Ball, C. A., Koonin, E. V., Dwight, S. S., Harris, M. A., Dolinski, K., Mohr, S., Smith, T., Weng, S., Cherry, J. M. and Botstein, D. (1998). Comparison of the complete protein sets of worm and yeast: orthology and divergence. *Science* **282**, 2022–2028.

Chevray, P. M. and Nathans, D. (1992). Protein interaction cloning in yeast: identification of mammalian proteins that react with the leucine zipper of Jun. *Proc. Natl Acad. Sci. USA* **89**, 5789–5793.

Chien, C. T., Bartel, P. L., Sternglanz, R. and Fields, S. (1991). The two-hybrid system: a method to identify and clone genes for proteins that interact with a protein of interest. *Proc. Natl Acad. Sci. USA* **88**, 9578–9582.

Chong, J. A., Tapia-Ramirez, J., Kim, S., Toledo-Aral, J. J., Zheng, Y., Boutros, M. C., Altshuller, Y. M., Frohman, M. A., Kraner, S. D. and Mandel, G. (1995). REST: a mammalian silencer protein that restricts sodium channel gene expression to neurons. *Cell* **80**, 949–957.

Dalton, S. and Treisman, R. (1992). Characterization of SAP-1, a protein recruited by serum response factor to the c-fos serum response element. *Cell* **68**, 597–612.

de Grado, M., Abe, A., Gauthier, A., Steele-Mortimer, O., DeVinney, R. and Finlay, B. B. (1999). Identification of the intimin-binding domain of Tir of enteropathogenic *Escherichia coli*. *Cell Microbiol.* **1**, 7–17.

Denis, C. L., Ferguson, J. and Young, E. T. (1983). mRNA levels for the fermentative alcohol dehydrogenase of *Saccharomyces cerevisiae* decrease upon growth on a non-fermentable carbon source. *J. Biol. Chem.* **258**, 1165–1171.

Du, W., Vidal, M., Xie, J. E. and Dyson, N. (1996). RBF, a novel RB-related gene that regulates E2F activity and interacts with cyclin E in Drosophila. *Genes Dev.* **10**, 1206–1218.

Dunnwald, M., Varshavsky, A. and Johnsson, N. (1999). Detection of transient in vivo interactions between substrate and transporter during protein translocation into the endoplasmic reticulum. *Mol. Biol. Cell.* **10**, 29–44.

Durfee, T., Becherer, K., Chen, P. L., Yeh, S. H., Yang, Y., Kilburn, A. E., Lee, W. H. and Elledge, S. J. (1993). The retinoblastoma protein associates with the protein phosphatase type 1 catalytic subunit the major. *Genes Dev.* **7**, 555–569.

Ebina, Y., Takahara, Y., Kishi, F. and Nakazawa, A. (1983). LexA protein is a repressor of colicin E1 gene. *J. Biol. Chem.* **258**, 13258–13261.

Elledge, S. J. and Spottswood, M. R. (1991). MR nucleotide, Protein A new human p34 protein kinase, CDK2, identified by complementation of a cdc28 mutation in *Saccharomyces cerevisiae*, is a homolog of Xenopus Eg1. *EMBO J.* **10**, 2653–2659.

Estojak, J., Brent, R. and Golemis, E. A. (1995). Correlation of two-hybrid affinity data with in vitro measurements. *Mol. Cell. Biol.* **15**, 5820–5829.

Fallenberg, M., Albermann, K., Zollner, A., Mewes, H. W. and Hani, J. (2000). Integrative analysis of protein interaction data. *Intelligent Systems Mol. Biol. Proc.* **8**, 152–161.

Fang, Y., Macool, D. J., Xue, Z., Heppard, E. P., Hainey, C. F., Tingey, S. V. and Miao, G-H. (2002). Development of a high-throughput yeast two-hybrid screening system to study protein–protein interactions in plants. *Mol. Gen. Genomics* **267**, 142–153.

Fashena, S. J., Serebriiskii, I. and Golemis, E. A. (2000). The continued evolution of two-hybrid screening approaches in yeast: how to outwit different preys with different baits. *Gene* **250**, 1–14.

Feilotter, H. E., Hannon, G. J., Ruddell, C. J. and Beach, D. (1994). Construction of an improved host strain for two hybrid screening. *Nucleic Acids Res.* **22**, 1502–1503.

Fields, S. and Song, O. (1989). A novel genetic system to detect protein–protein interactions. *Nature* **340**, 245–246.

Finlay, R. and Brent, R. (1994). Interaction mating reveals binary and ternary connections between Drosophila cell cycle regulators. *Proc. Natl Acad. Sci. USA* **91**, 12980–12984.

Flajolet, M., Rotondo, G., Daviet, L., Bergametti, F., Inchauspe, G., Tiollais, P., Transy, C. and Legrain, P. (2000). A genomic approach of the hepatitis C virus generates a protein interaction map. *Gene* **242**, 369–379.

Frolova, E., Johnston, M. and Majors, J. (1999). Binding of the glucose-dependent Mig1p repressor to the GAL1 and GAL4 promoters in vivo: regulation by glucose and chromatin structure. *Nucleic Acids Res.* **27**, 1350–1358.

Fromont-Racine, M., Rain, J. C. and Legrain, P. (1997). Toward a functional analysis of the yeast genome through exhaustive two-hybrid screens. *Nature Genet.* **16**, 277–282.

Gietz, R. D. and Schiestl, R. H. (1991). Applications of high efficiency lithium acetate transformation of intact yeast cells using single-stranded nucleic acids as carrier. *Yeast* **7**, 253–263.

Gietz, R. D., St Jean, A., Woods, R. A. and Schiestl, R. H. (1992). Improved method for high efficiency transformation of intact yeast cells. *Nucleic Acids Res.* **20**, 1425.

Gietz, R. D., Schiestl, R. H., Willems, A. R. and Woods, R. A. (1995). Studies on the transformation of intact yeast cells by the LiAc/SS-DNA/PEG procedure. *Yeast* **11**, 355–360.

Giniger, E. and Ptashne, M. (1988). Cooperative DNA binding of the yeast transcriptional activator GAL4. *Proc. Natl Acad. Sci. USA* **85**, 382–386.

Golemis, E. A., Gyuris, J. and Brent, R. (1996). In *Current Protocol in Molecular Biology*, (Ausubel, F. M., Brent, R., Kingston, R., Moore, D., Seidman, J., Smith, J. A. and Struhl, K. eds) Wiley: New York, p. 20.1.

Graves, P. R. and Haystead, T. A. (2002). Molecular biologist's guide to proteomics. *Microbiol. Mol. Biol. Rev.* **66**, 39–63.

Gu, Y. Q., Yang, C., Thara, V. K., Zhou, J. and Martin, G. B. (2000). Pti4 is induced by ethylene and salicylic acid, and its product is phosphorylated by the Pto kinase. *Plant Cell* **12**, 771–786.

Guthrie, C. and Fink, G. R. (1991). Guide to yeast genetics and molecular biology. *Meth. Enzymol.* **194**, 1–932.

Gyuris, J., Golemis, E. A., Chertkov, H. and Brent, R. (1993). Cdi1 a human G1 and S phase protein phosphatase that associates with Cdk2. *Cell* **75**, 791–803.

Harper, J. W., Adami, G. R., Wei, N., Keyomarsi, K. and Elledge, S. J. (1993). The p21 Cdk-interacting protein Cip1 is a potent inhibitor of G1 cyclindependent kinases. *Cell* **75**, 805–816.

Hartland, E. L., Batchelor, M., Delahay, R. M., Hale, C., Matthews, S., Dougan, G., Knutton, S., Connerton, I. and Frankel, G. (1999). Binding of intimin from enteropathogenic *Escherichia coli* to Tir and to host cells. *Mol. Microbiol.* **32**, 151–158.

Hennig, E. E., Butruk, E. and Ostrowski, J. (2001). RACK1 protein interacts with *Helicobacter pylori* VacA cytotoxin: the yeast two-hybrid approach. *Biochem. Biophys. Res. Commun.* **289**, 103–110.

Huang, J. and Schreiber, S. L. (1997). A yeast genetic system for selecting small molecule inhibitors of protein–protein interactions in nanodroplets. *Proc. Natl Acad. Sci. USA* **94**, 13396–13401.

Hua, S. B., Luo, Y., Qiu, M., Chan, E., Zhou, H. and Zhu, L. (1998). Construction of a modular yeast two-hybrid cDNA library from human EST clones for the human genome protein linkage map. *Gene* **215**, 143–152.

Ito, T., Chiba, T., Ozawa, R., Yoshida, M., Hattori, M. and Sakaki, Y. (2001). A comprehensive two-hybrid analysis to explore the yeast protein interactome. *Proc. Natl Acad. Sci. USA* **98**, 4569–4574.

James, P., Halladay, J. and Craig, E. A. (1996). Genomic libraries and a host strain designed for highly efficient two-hybrid selection in yeast. *Genetics* **144**, 1425–1436.

Johnsson, N. and Varshavsky, A. (1994). Split ubiquitin as a sensor of protein interactions in vivo. *Proc. Natl Acad. Sci. USA* **91**, 10340–10344.

Johnston, S. A., Zavortink, M. J., Debouck, C. and Hopper J. E. (1986). Functional domains of the yeast regulatory protein GAL4. *Proc. Natl Acad. Sci. USA* **83**, 6553–6557.

Karimova, G., Pidoux, J., Ullman, A. and Ladant, D. (1998). A bacterial two-hybrid system based on a reconstituted signal transduction pathway. *Proc. Natl Acad. Sci. USA* **95**, 5752–5756.

Keegan, L., Gill, G. and Ptashne, M. (1986). Separation of DNA binding from the transcription-activating function of a eukaryotic regulatory protein. *Science* **231**, 699–704.

Kornacker, M. G., Remsburg, B. and Menzel, R. (1998). Gene activation by the AraC protein can be inhibited by DNA looping between AraC and a LexA repressor that interacts with AraC: possible applications as a two-hybrid system *Mol. Microbiol.* **30**, 615–624.

Kornbluth, S., Jove, R. and Hanafusa, H. (1987). Characterization of avian and viral p60src proteins expressed in yeast. *Proc. Natl Acad. Sci. USA* **84**, 4455–4459.

Lahaye, T. and Bonas, U. (2001). Molecular secrets of bacterial type III effector proteins. *Trends Plant Sci.* **6**, 479–485.

Lane, D. P. and Crawford, L. V. (1979). T antigen is bound to a host protein in SV40-transformed cells. *Nature* **278**, 261–263.

Lee, M. G. and Nurse, P. (1987). OMIM, protein complementation used to clone a human homologue of the fission yeast cell cycle control gene cdc2. *Nature* **327**, 31–35.

Legrain, P. and Selig, L. (2000). Genome-wide protein interaction maps using two-hybrid systems. *FEBS Lett.* **480**, 32–36.

Legrain, P., Dokhelar, M. C. and Transy, C. (1994). Detection of protein–protein interactions using different vectors in the two-hybrid system. *Nucleic Acids Res.* **22**, 3241–3242.

Legrain, P., Wojcik, J. and Gauthier, J. M. (2001). Protein–protein interaction maps: a lead towards cellular functions. *Trends Genet.* **17**, 346–352.

Li, J. J. and Herskowitz, I. (1993). Isolation of ORC6, a component of the yeast origin recognition complex by a one-hybrid system. *Science* **262**, 1870–1874.

Licitra, E. J. and Liu, J. O. (1996). A three-hybrid system for detecting small ligand–protein receptor interactions. *Proc. Natl Acad. Sci. USA* **93**, 12817–12821.

Linzer, D. I. and Levine, A. J. (1979). Characterization of a 54K dalton cellular SV40 tumor antigen present in SV40-transformed cells and uninfected embryonal carcinoma cells. *Cell* **17**, 43–52.

Luo, Y., Frey, E. A., Pfuetzner, R. A., Creagh, A. L., Knoechel, D. G., Haynes, C. A, Finlay, B. B. and Strynadka, N. C. (2000). Crystal structure of enteropathogenic *Escherichia coli* intimin-receptor complex. *Nature* **405**, 1073–1077.

Marmorstein, R., Carey, M., Ptashne, M. and Harrison, S. C. (1992). DNA recognition by GAL4: structure of a protein–DNA complex. *Nature* **356**, 408–414.

Marsolier, M. C. and Sentenac, A. (1999). RNA polymerase III-based two-hybrid system. *Meth. Enzymol.* **303**, 411–422.

Marsolier, M. C., Prioleau, M. N. and Sentenac, A. (1997). A RNA polymerase III-based two-hybrid system to study RNA polymerase II transcriptional regulators. *J. Mol. Biol.* **268**, 243–249.

McCraith, S., Holtzman, T., Moss, B. and Fields, S. (2000). Genome-wide analysis of vaccinia virus protein–protein interactions. *Proc. Natl Acad. Sci. USA* **97**, 4879–4884.

Nasmyth, K. A. and Reed, S. I. (1980). Isolation of genes by complementation in yeast: molecular cloning of a cell-cycle gene. *Proc. Natl Acad. Sci. USA* **77**, 2119–2123.

Naya, F. J., Stellrecht, C. M. M. and Tsai, M-J. (1995). Tissue-specific regulation of the insulin gene by a novel basic helix–loop–helix transcription factor. *Genes Dev.* **9**, 1009–1019.

O'Hare, P. (1993). The virion transactivator of herpes simplex virus. *Semin. Virol.* **4**, 145–155.

Orth, K., Palmer, L. E., Bao, Z. Q., Stewart, S., Rudoph, A. E., Bliska, J. B. and Dixon, J. E. (1999). Inhibition of the mitogen-activated protein kinase superfamily by a *Yersinia* effector. *Science* **285**, 1920–1923.

Osborne, M., Dalton, S. and Kochan, J. P. (1995). The yeast tribrid system: genetic detection of trans-phosphorylated ITAM–SH2 interactions. *BioTechnol.* **13**, 1474–1478.

Osborne, M. A., Zenner, G., Lubinus, M., Zhang, X., Songyang, Z., Cantley, L. C., Majerus, P., Burn, P. and Kochan, J. P. (1996). The inositol 5' phosphatase SHIP binds to immunoreceptor signaling motifs and responds to high affinity IgE receptor aggregation. *J. Biol. Chem.* **271**, 29271–29278.

Page, A. L., Fromont-Racine, M., Sansonetti, P., Legrain, P. and Parsot, C. (2001). Characterization of the interaction partners of secreted proteins and chaperones of *Shigella flexneri*. *Mol. Microbiol.* **42**, 1133–1145.

Page, A. L., Sansonetti, P. and Parsot, C. (2002). Spa15 of *Shigella flexneri*, a third type of chaperone in the type III secretion pathway. *Mol. Microbiol.* **43**, 1533–1542.

Park, Y. W, Willusz, J. and Katze, M. G. (1999). Regulation of eukaryotic protein synthesis: selective influenza viral mRNA translation is mediated by the cellular RNA-binding protein GRSF-1. *Proc. Natl Acad. Sci. USA* **96**, 6694–6699.

Pause, A., Peterson, B., Schaffar, G., Stearman, R. and Klausner, R. D. (1999). Studying interactions of four proteins in the yeast two-hybrid system: structural resemblance of the pVHL/elongin BC/hCUL-2 complex with the ubiquitin ligase complex SKP1/cullin/F-box protein. *Proc. Natl Acad. Sci. USA* **96**, 9533–9538.

Plano, G. V., Day, J. B. and Ferracci, F. (2001). Type III export: new uses for an old pathway. *Mol. Microbiol.* **40**, 284–293.

Putz, U., Skehel, P. and Kuhl, D. (1996). A tri-hybrid system for the analysis and detection of RNA–protein interactions. *Nucleic Acids Res.* **24**, 4838–4840.

Rain, J. C., Selig, L., De Reuse, H., Battaglia, V., Reverdy, C., Simon, S., Lenzen, G., Petel, F., Wojcik, J., Schächter, V., Chemama, Y., Labigne, A. and Legrain, P. (2001). The protein–protein interaction map of *Helicobacter pylori*. *Nature* **409**, 211–215.

Roder, K. H., Wolf, S. S. and Schweizer, M. (1996). Refinement of vectors for use in the yeast two-hybrid system. *Analyt. Biochem.* **241**, 260–262.

Scherer, C. A., Cooper, E. and Miller, S. I. (2000). The *Salmonella* type III secretion translocon protein SspC is inserted into the epithelial cell plasma membrane upon infection. *Mol. Microbiol.* **37**, 1133–1145.

Schinkowski, B., Uetz, P. and Fields, S. (2000). A network of protein–protein interactions in yeast. *Nature Biotechnol.* **18**, 1257–1261.

Scofield, S. R., Tobias, C. M., Rathjen, J. P., Chang, J. H., Lavelle, D. T., Michelmore, R. W. and Staskawicz, B. J. (1996). Molecular basis of gene-for-gene specificity in bacterial speck disease of tomato. *Science* **274**, 2063–2065.

SenGupta, D. J., Zhang, B., Kraemer, B., Pochart, P., Fields, S. and Wickens, M. (1996). A three-hybrid system to detect RNA–protein interactions in vivo. *Proc. Natl Acad. Sci. USA* **93**, 8496–8501.

SenGupta, D. J., Wickens, M. and Fields, S. (1999). Identification of RNAs that bind to a specific protein using the yeast three-hybrid system. *RNA* **5**, 596–601.

Silver, P. A., Keegan, L. P. and Ptashne, M. (1984). Amino terminus of the yeast GAL4 gene product is sufficient for nuclear localization. *Proc. Natl Acad. Sci. USA* **81**, 5951–5955.

Silver, P. A., Brent, R. and Ptashne, M. (1986). DNA binding is not sufficient for nuclear localization of regulatory proteins in *Saccharomyces cerevisiae*. *Mol. Cell. Biol.* **6**, 4763–4766.

Silver, P. A., Chiang, A. and Sadler, I. (1988). Mutations that alter both localization and production of a yeast nuclear protein. *Genes Dev.* **2**, 707–717.

Siomi, M. C., Fromont, M., Rain, J. C., Wan, L., Wang, F., Legrain, P. and Dreyfuss, G. (1998). Functional conservation of the transportin nuclear import pathway in divergent organisms. *Mol. Cell. Biol.* **18**, 4141–4148.

Staskawicz, B. J., Mudgett, M. B., Dangl, J. L. and Galan, J. E. (2001). Common and contrasting themes of plant and animal diseases. *Science* **292**, 2285–2289.

Tang, X., Frederick, R. D., Zhou, J., Halterman, D. A., Jia, Y. and Martin, G. B. (1996). Initiation of plant disease resistance by physical interaction of AvrPto and Pto kinase. *Science* **274**, 2060–2063.

Thompson, C. and McKnight, S. (1992). Anatomy of an enhancer. *Trends Genet.* **8**, 232–236.

Tirode, F., Malaguti, C., Romero, F., Attar, R., Camonis, J. and Egly, J. M. (1997). A conditionally expressed third partner stabilizes or prevents the formation of a transcriptional activator in a three-hybrid system. *J. Biol. Chem.* **272**, 22995–22999.

Toby, G. G., and Golemis, E. A. (2001). Using the yeast interaction trap and other two-hybrid-based approaches to study protein–protein interactions. *Methods* **24**, 201–217.

Triolo, T. and Sternglanz, R. (1996). Role of interactions between the origin recognition complex and SIR1 in transcriptional silencing. *Nature* **381**, 251–253.

Uetz, P., Giot, L., Cagney, G., Mansfield, T. A., Judson, R. S., Knight, J. R., Lockshon, D., Narayan, V., Srinivasan, M., Pochart, P. *et al.* (2000). A comprehensive analysis of protein–protein interactions in *Saccharomyces cerevisiae*. *Nature* **403**, 623–627.

Van Criekinge, W., van Gurp, M., Decoster, E., Schotte, P., Van de Craen, M., Fiers, W., Vandenabeele, P. and Beyaert, R. (1998). Use of the yeast three-hybrid system as a tool to study caspases. *Analyt. Biochem.* **263**, 62–66.

Vidal, M. and Endoh, H. (1999). Prospects for drug screening using the reverse two-hybrid system. *Trends Biotechnol.* **17**, 374–381.

Vidal, M., Brachmann, R. K., Fattaey, A., Harlow, E. and Boeke, J. D. (1996a). Reverse two-hybrid and one-hybrid systems to detect dissociation of protein–protein and DNA–protein interactions. *Proc. Natl Acad. Sci. USA* **93**, 10315–10320.

Vidal, M., Braun, P., Chen, E., Boeke, J. D. and Harlow, E. (1996b). Genetic characterization of a mammalian protein–protein interaction domain by using a yeast reverse two-hybrid system. *Proc. Natl Acad. Sci. USA* **93**, 10321–10326.

Vojtek, A. B., Hollenberg, S. M. and Cooper, J. A. (1993). Mammalian Ras interacts directly with the serine/threonine kinase Raf. *Cell* **74**, 205–214.

Wang, M. M. and Reed, R. R. (1993). Molecular cloning of the olfactory-neuronal transcription factor Olf-1 by genetic selection in yeast. *Nature* **364**, 121–126.

Wang, Z. F., Whitfield, M. L., Ingledue, T. C., 3rd, Dominski, Z. and Marzluff, W. F. (1996). The protein that binds the 3' end of histone mRNA: a novel RNA-binding protein required for histone pre-mRNA processing. *Genes Dev.* **10**, 3028–3040.

West, R. W., Jr, Yoccum, R. R. and Ptashne, M. (1984). Saccharomyces cerevisiae GAL1–GAL10 divergent promotor region: location and function of the upstream activating sequence UASG. *Mol. Cell. Biol.* **4**, 2467–2478.

Wilson, T. E., Fahrner, T. J., Johnston, M. and Milbrandt, J. (1991). Identification of the DNA binding site for NGFI-B by genetic selection in yeast. *Science* **252**, 1296–1300.

Wojcik, J. and Schächter, V. (2001). Protein–protein interaction map inference using interacting domain profile pairs. *Bioinformatics* **17**, 296–305.

Xu, G., Jansen, G., Thomas, D. Y., Hollenberg, C. P. and Ramezani Rad, M. (1996). Ste50p sustains mating pheromone-induced signal transduction in the yeast *Saccharomyces cerevisiae*. *Mol. Microbiol.* **20**, 773–783.

Zhang, J. and Lautar, S. (1996). A yeast three-hybrid method to clone ternary protein complex components. *Analyt. Biochem.* **242**, 68–72.

Zhou, D., Mooseker, M. S. and Galan, J. E. (1999). An invasion-associated *Salmonella* protein modulates the actin-bundling activity of plastin. *Proc. Natl Acad. Sci. USA* **96**, 10176–10181.

Zhou, J., Loh, Y. T., Bressan, R. A. and Martin, G. B. (1995). The tomato gene Pti1 encodes a serine/threonine kinase that is phosphorylated by Pto and is involved in the hypersensitive response. *Cell* **83**, 925–935.

Zhou J., Tang X. and Martin G. B. (1997). The Pto kinase conferring resistance to tomato bacterial speck disease interacts with proteins that bind a cis-element of pathogenesis-related genes. *EMBO J.* **16**, 3207–3218.

Part V
Applications of Microbial Genomics

Part V
Applications of Microbial Genomics

13 Cloning the Metagenome: Culture-independent Access to the Diversity and Functions of the Uncultivated Microbial World

Jo Handelsman, Mark Liles, David Mann, Christian Riesenfeld and Robert M Goodman

Department of Plant Pathology, University of Wisconsin, Madison, WI 53706, USA

CONTENTS

Introduction
Overview of metagenomics
Isolation of environmental DNA
Vectors
Host cells
Storage and clone management
Characterizing metagenomic libraries
Concluding remarks

◆◆◆◆◆◆ INTRODUCTION

For more than a century, the gold standard in microbiology has been to achieve pure cultures of microorganisms for laboratory study. The central role of cultivation in microbiology derives from the origins of the field, which were rooted in the study of disease etiology and the uses of microorganisms for processes related to making foods and beverages. The ecological study of microorganisms has advanced notably where a single microbial taxon performs an ecological role of major effect.

With the revelation over the past two decades of a vast diversity of previously unsuspected, and largely uncultivated, microflora inhabiting diverse natural environments throughout the biosphere, microbiology has undergone a revolution. The application of molecular biology methods first revealed convincing evidence for this new view of microbial diversity in the biosphere. Now molecular biology methods are providing powerful new approaches to address what now looms as one of the major challenges in microbiology, which is to understand the functions of microorganisms that have not been cultivated.

The use of molecular, culture-independent methods for assessing microbial diversity has shown that most microorganisms in the environment are not readily cultivated (Stahl et al., 1985; Giovannoni et al., 1990; Ward et al., 1990; Pace, 1996; Suzuki et al., 1997; Hugenholtz et al., 1998), and that the diversity of the uncultivated majority (Whitman et al., 1998) is vast (Stahl, 1993; Pace, 1996; Head et al., 1998; Beja et al., 2002b). For example, the readily cultivatable prokaryotes from soil represent 1% or less of the total (Griffiths et al., 1996). Estimates indicate that soil contains more than 4000 species/g (Torsvik et al., 1990, 1996), many of which fall into novel divisions of Bacteria and Archaea (Bintrim et al., 1997; Borneman et al., 1996). Similarly, sea water harbors diverse prokaryotes that defy cultivation attempts (Ward et al., 1990; DeLong et al., 1999). Analysis of both soil and sea water has revealed new groups of Archaea and an unanticipated abundance of Archaea, which may numerically dominate the open oceans (DeLong, 1992; DeLong et al., 1994, 1999; McInerney et al., 1995; Bintrim et al., 1997; Simon et al., 2000). The emerging image of the microbial world in natural environments is quite different from the one constructed by cultivation alone.

The challenge that long stumped microbiologists was to learn about the biology associated with the 16S rRNA sequences that indicate the high degree of diversity of uncultivated microorganisms. The solution lies, in part, in the application of genomics to assemblages of microorganisms using methods that circumvent the need for cultivation.

◆◆◆◆◆◆ OVERVIEW OF METAGENOMICS

Metagenomics is the genomic analysis of the collective genomes of an assemblage of organisms, or the 'metagenome'. The approach has also been termed 'environmental genomics' (Stein et al., 1996; DeLong, unpublished) and 'community genomics' (Banfield, unpublished). In this chapter, we use the term 'metagenomics' to apply to all studies that fall within the definition, whether or not the original study used this term.

Metagenomic libraries containing DNA extracted directly from an environmental sample provide genomic sequences, and phylogenetic and functional information. Libraries can be screened for functions, and genomic sequence surrounding genes required for those functions can provide insight into the organism from which the function was derived. Phylogenetic markers such as 16S rRNA, *recA*, *radA* and genes encoding DNA polymerases or other conserved genes provide indications of the origin of the DNA (Schleper et al., 1997; Eisen, 1998; Vergin et al., 1998; Sandler et al., 1999; Rondon et al., 2000; Beja et al., 2000b, 2002a). Linking phylogeny, sequence and functional analysis provides a multifaceted approach to dissecting and reassembling the uncultivated microbial community.

Metagenomic libraries have been constructed from assemblages of microorganisms in seawater (Stein *et al.*, 1996; Cottrell *et al.*, 1999; Beja *et al.*, 2000a, b;), soil (Henne *et al.*, 1999, 2000; Rondon *et al.*, 2000; Entcheva *et al.*, 2001), and associated with a marine sponge (Schleper *et al.*, 1998). Metagenomic analysis has been applied to diverse problems in microbiology and has yielded insight into the physiology of uncultivated organisms as well as access to the potentially useful enzymes and secondary metabolites they produce (Table 13.1).

Table 13.1. Applications of metagenomic analysis

Characterization of phylogenetic diversity
Characterization of new genome organizations
Elucidation of new biochemical pathways for primary metabolism or energy transduction
Identification of reservoirs of resistance to environmental pollutants
Discovery of enzymes
Discovery of secondary metabolites and other biologically active small molecules
Discovery of polymers
Elucidation of trophic structure of microbial assemblages

◆◆◆◆◆◆ ISOLATION OF ENVIRONMENTAL DNA

The physical characteristics of a metagenomic library, such as insert size, number of clones and storage method, are dictated by the kind of natural products to be identified from the library. The method used to isolate metagenomic DNA from a natural environment must reflect the desired physical characteristics as well as the factors that will affect recovery of DNA from the targeted microorganism(s). For example, to identify lipases (or other enzymes) present within a microbial community, a library of DNA inserts less than 10 kb in size would be sufficient, but the number of clones must be sufficiently large to compensate for the relatively low frequency of clones that are active on indicator media (Henne *et al.*, 2000). For such small insert metagenomic libraries, harsh extraction methods (e.g. bead-beating) may be used to achieve genomic DNA from diverse microorganisms in soil or other natural environments (Krsek and Wellington, 1999).

The recovery of high molecular weight metagenomic DNA from natural environments can be a significant challenge, especially from soil or sediments. Metagenomic libraries containing large inserts from marine picoplankton have been prepared by lysing bacteria embedded in agarose (Stein *et al.*, 1996; Beja *et al.* 2000b), which is the same method that is commonly used to construct BAC libraries from eukaryotes or prokaryotes in pure culture (Wang and Warren, 1996; Rondon *et al.*, 1999b). The use of agarose plugs to protect the DNA from shearing during cell lysis and restriction digestion facilitates construction of libraries containing large inserts, although we have found that cell lysis within agarose plugs containing extracted soil microbes is inefficient, at least from certain soils (unpublished data). The choice of method for DNA extraction will balance maintaining the DNA in large pieces with the diversity of the community to be represented in the library and the resistance of the targeted population to lysis (Krsek and Wellington, 1999; Stach *et al.*, 2001). We have found that multiple rounds of freezing and thawing combined with hot phenol extraction is effective for recovery of high molecular weight genomic DNA from diverse soil microorganisms (Zhou *et al.*, 1996; Rondon *et al.*, 2000). Efficiency of restriction digestion is enhanced by further purification of DNA by separation of high molecular weight DNA on, and recovery from, a preparative agarose gel. Typically, the largest genomic DNA recovered in this manner is less than 400 kb, and after restriction digestion the majority of restricted genomic DNA is sized less than 50 kb, but size selection can produce inserts as large as 80–155 kb (Beja *et al.*, 2000b; Rondon *et al.*, 2000). We are investigating cloning methods that do not require restriction digestion and methods for DNA recovery from cells physically separated from soil (Lindahl, 1996) to produce metagenomic libraries with inserts greater than 100 kb.

◆◆◆◆◆◆ VECTORS

Large-insert vectors

The design and choice of vectors for construction of metagenomic libraries is driven largely by the desired insert size and the relative importance of heterologous gene expression. For some analyses, isolation of the DNA is sufficient and gene expression is not required. Libraries constructed with DNA isolated from sea water (Stein *et al.*, 1996; Vergin *et al.*, 1998; Beja *et al.*, 2000b, 2002a) and soil (Rondon *et al.*, 2000) have provided a source of genomic DNA linked to 16S rRNA genes. The rRNA genes were identified by polymerase chain reaction (PCR) analysis, and the flanking DNA was analyzed by sequencing, thereby obviating the need for gene expression until after the gene of interest was identified. For these applications, large inserts are an advantage because larger fragments enhance the probability of the same clone containing a phylogenetic marker and a gene encoding an activity of interest (Stein *et al.*, 1996) and provide insights into genome organization. Large inserts are also required to capture large genes or operons encoding biosynthetic pathways for secondary metabolite or for genomic analysis that requires large segments of genomes (Handelsman *et al.*, 1998; Rondon *et al.*, 1999a; August *et al.*, 2000). These applications are facilitated by bacterial artificial chromosomes (BAC vectors), which were developed for use in eukaryotic genomics and can stably maintain inserts as large as 350 kb (Shizuya *et al.*, 1992). If medium-sized fragments are needed, then fosmids (Stein *et al.*, 1996) and cosmids (Brady and Clardy, 2000; Brady *et al.*, 2001) are appropriate. An advantage of using these vectors is that the efficiency of cloning in fosmids and cosmids is typically 100–1000-fold greater than BACs, but they accept fragments up to 40 kb. BACs and cosmids are maintained in single copy in *E. coli*, which is an advantage when cloning genes that may be detrimental to the host cell when maintained in high copy. Genes for antibiotic synthesis have been cloned successfully from soil in the BAC vector, pBeloBAC11, and the cosmid, superCos1 (Brady and Clardy, 2000; Brady *et al.*, 2001; MacNeil *et al.*, 2001; Gillespie *et al.*, 2002).

While low copy number is an advantage for preventing cell death due to genes that are toxic in high copy and for stability of large inserts, it is a disadvantage for vector purification, insert sequencing and detection of products from poorly expressed genes. To combine the advantages of high and low copy plasmids, we developed a new vector, superBAC1 (Figure 13.1). This plasmid is maintained at low copy number in strain JW366, a genetically modified version of *E. coli* strain DH10B (Wild *et al.*, 1998, 1999) and can be induced with arabinose to 50–100 copies per cell. superBAC carries the pBeloBAC11 backbone with its *oriS* origin of replication as well as *oriV* from the IncP plasmid RK2. *oriV* requires the transcription factor TrfA, and the *trfA* gene is under control of an arabinose-inducible promoter in strain JW366. Therefore, replication of superBAC1 in the absence of arabinose is mediated by the *oriS* origin, which maintains the plasmid at one copy per cell. When arabinose is added to the medium, *trfA* is expressed and replication is mediated by *oriV*. As expression from the arabinose-inducible promoter can be modulated by titrating glucose and arabinose concentrations, the copy number can be modulated between 1 and 100 copies per cell. The low copy plasmid facilitates construction and maintenance of large insert libraries, while the ability to attain high copy has several advantages for sequence and expression analysis of the resulting libraries. The difficulties associated with the purification of large, low copy BACs for sequencing, subcloning and other manipulations is alleviated by the induction of high copy. In addition, activities undetected on a low copy BAC may be identified more readily when maintained at high copy. Increased levels of antimicrobial agent synthesis might be obtained if the genes encoding the pathway were maintained in high cell copy number, thereby facilitating detection of activity in a functional assay that might be below the level of detection if the genes were maintained at single copy. However, the clones can be grown in the absence of arabinose prior to screening, thereby reducing the likelihood of cell death due to high level production of a toxic product. If a copy of the clone bank is always maintained without arabinose, then self-poisoning by the copy of a clone that produces high levels of an antimicrobial agent under arabinose induction does not prevent further study of the clone.

Expression vectors

Although there is ample evidence that diverse genes from diverse organisms are expressed in *E. coli* (Ding and Yelton, 1993; Ferreya *et al.*,

1993; Black et al., 1995; Chávez et al., 1995), expression of some traits in metagenomic libraries requires transcription or translation machinery or chemical substrates that are lacking in E. coli. While the genetic code is nearly universal, transcription factors are specialized for organism and gene type, and translation is affected by organism-specific codon usage, which is co-ordinated with the availability of the cognate tRNAs in the cell. Moreover, biosynthesis of some secondary metabolites is dependent on products of primary metabolism that are not universal.

Expression vectors can be used to overcome the barrier of recognition of foreign promoters by E. coli's transcription machinery (Newman and Fuqua, 1999). To avoid the detrimental effects of constitutive expression of toxic products, many expression vectors contain inducible promoters whose expression can be regulated by amendments to the medium. Expression vectors typically accept small fragments of DNA and the promoter in the vector will drive only genes adjacent to it and in the correct orientation. Cloning small fragments of DNA from environmental sources is appropriate when the goal is isolation of single genes or very small operons. Many expression vectors are available for cloning small fragments (<10 kb) that are maintained at low, medium or high copy number. For example, libraries constructed with soil DNA cloned into pBluescript SK(+) contain clones expressing lipase, esterase, and 4-hydroxybutyrate dehydrogenase activities (Henne et al., 1999, 2000).

Shuttle vectors

Although expression vectors effectively address limitations on transcription, many other factors are required for expression of traits in a foreign host. Translation and post-translational processing must be faithfully conducted (Joseph et al., 2001; Karlin et al., 2001). To perform new biosynthetic functions, the substrates must be available in the cell. If the gene product must be secreted to be detected, then secretion signals must be recognizable to E. coli. Lack of co-ordination of any of these processes will result in no observable activity.

For certain pathways, expression of the authentic product is unlikely. For example, synthesis of polyketides in E. coli with genes derived from actinomycetes is unlikely because of the differences in promoters and codon usage and the lack of the appropriate building blocks (Cane et al., 1998; Tang et al., 2000; Pfeifer et al., 2001). To provide a different cellular environment and, therefore, facilitate a different range of gene expression, an alternative approach is to deliver the cloned genes to a different host species. This requires a vector that can be mobilized from E. coli to another species.

To take advantage of potential host-specific abilities for heterologous expression of metagenomic DNA constructs, we have further modified superBAC1 to make it amenable to transfer and maintenance of BAC libraries in alternative host species. The inclusion of the origin of transfer, *oriT*, from RK2 on superBAC1 affords conjugal mating of the plasmid into alternative host species, including *Bacillus, Enterococcus, Staphylococcus, Streptococcus* and *Listeria* strains when the *tra* gene is supplied *in trans* (Guiney and Yakobsen, 1983; Yakobson and Guiney, 1984; Trieu-Cuot et *al.*, 1987). Conjugal transfer circumvents transformation, which occurs at low efficiency in many Gram-positive species. To maintain the transferred BACs in an alternative host, the inclusion of a replication origin that is functional in the chosen host species is required. Our initial shuttle vector is designed to transfer to *Bacillus* species, using *rep60* from *Bacillus* plasmid pTA1060 to mediate replication (Uozumi *et al.*, 1980) and a chloramphenicol-resistance gene from *Bacillus* plasmid pC194 for selection (Leonhardt and Alonso, 1988; Steinmetz and Richter, 1994).

Wang *et al.* constructed metagenomic libaries in an *E. coli*–*Streptomyces* shuttle vector and discovered an antibiotic biosynthetic pathway that was likely derived from a relative of *Streptomyces* (Wang *et al.*, 2000). Other replication origins that would expand the host range for BAC libraries include the broad host range origin host range pAMβ1, (3–5 copies per cell) (Trieu-Cuot *et al.*, 1987; Poyart and Trieu-Cuot, 1997) and the *repA* gene from pSH71 for streptococci (deVos, 1987; deVos *et al.*, 1997).

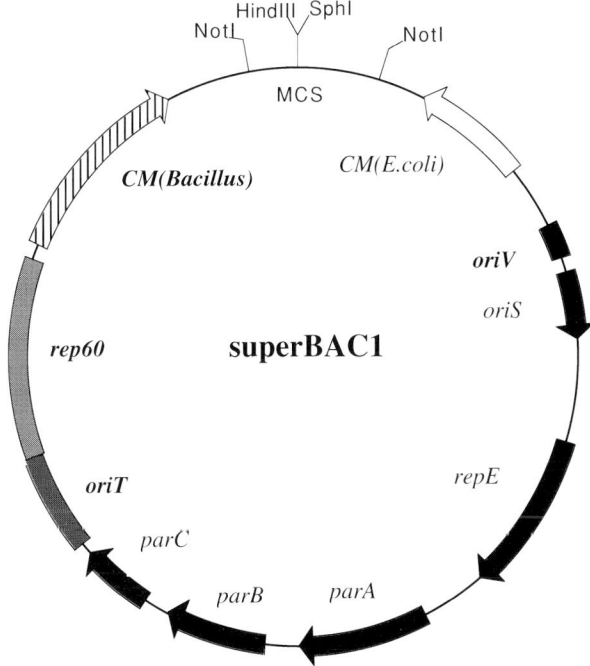

Figure 13.1. superBAC1: a shuttle vector with inducible copy number in *E. coli* strain JW366.

◆◆◆◆◆◆ HOST CELLS

The choice of the host cell in which to construct the library is guided in part by the requirements for transformation and stability of the vector. Choice of the host cell is also determined by the phenotypes of interest in the library. For functional screening or selections, it is essential that the host cell itself does not express the characteristic of interest.

◆◆◆◆◆◆ STORAGE AND CLONE MANAGEMENT

The intrinsic complexity of a metagenome, whether derived from a few species or from thousands, requires that metagenomic libraries contain many clones. The resulting storage and management challenges can be daunting if addressed with the standard techniques typically used for maintenance of genomic libraries of organisms in pure culture. Libraries are ideally maintained with each clone in pure culture in a separate well of a 96-well. The cells can be preserved with cryoprotectants, such as glycerol or dimethyl sulfoxide and stored at −80°C. This format is amenable to high throughput screening with manual or robotic replicators that are designed for the 96-well format. However, for study of complex microbial assemblages in typical laboratories that lack facilities to handle more than 10 000–20 000 clones, this format is impractical. For example, the soil is thought to contain at least 4000 different genomes of the size and complexity of that of *E. coli*. A metagenomic library containing clones with inserts of an average of 80 kb would require 250 000 clones to provide singlefold coverage of the metagenome. To obtain adequate representation of all sequences in the metagenome would require 1–10 million clones. Therefore, for many applications, maintaining libraries in pools is the only feasible approach for laboratories lacking extensive robotic capabilities.

◆◆◆◆◆◆ CHARACTERIZING METAGENOMIC LIBRARIES

Phylogenetic analysis

Once a library has been constructed, it is important to determine which microbial genomes are represented within the library. Phylogenetic information, in the form of 16S rRNA or other highly conserved gene sequences, reveal potential biases in the preparation or cloning of genomic DNA, and may suggest alternative hosts for functional screening. Although a small proportion of the clones in a library is likely to carry a rRNA operon, each one that does provides the opportunity to link phylogenetic and functional information, and through genomic sequence annotation and analysis permits a partial assessment of the metabolic capabilities for uncultured microorganisms (Stein *et al.*, 1996; Vergin *et al.*, 1998; Beja *et al.*, 2000a, b, 2002; Rondon *et al.*, 2000).

Identification of rRNA gene-containing BAC clones has been accomplished by several methods. Arrays of BAC clones have been probed with Archaea-specific probes (Stein *et al.*, 1996), and pools of BAC clones have been screened by PCR-based techniques using universal primer sets (Beja *et al*, 2000b; Rondon *et al.*, 2000). A principal challenge in the PCR amplification of rRNA genes from BAC library template is the contamination of host chromosomal DNA in each BAC preparation. The presence of non-*E. coli* rRNA genes within a pool of BAC clones may be demonstrated by rRNA intergenic spacer analysis (Beja *et al.*, 2000b), denaturing gradient gel electrophoresis or restriction fragment polymorphism analysis (Rondon *et al.*, 2000). Furthermore, the amplification of host rRNA genes may be prevented by exonuclease digestion of chromosomal DNA (often only partially effective) and by incorporating host template-specific, terminally modified oligonucleotides ('terminator' oligos) within a PCR, allowing preferential amplification of non-host rRNA genes (Rondon *et al.*, 2000). When screening for rRNA genes from a pool of BAC clones representing an entire library, a combination of exonuclease treatment and incorporation of terminator oligos provides sufficient reduction of interference by *E. coli* rRNA genes. When screening large numbers of BAC clones for unique rRNA genes, division- or taxa-specific primer sets provide selectivity for particular phylogenetic groups, eliminating the problem of interference of the host cell's rRNA genes. The use of superBAC1, described above, which has inducible copy permits a greater ratio of BAC vector to host chromosomal DNA within BAC preparations.

Alternative phylogenetic markers can be used to assess the diversity of microorganisms that contributed DNA to a metagenomic library. Genes that have revealed phylogenetic relationships and unexpected diversity include genes involved in DNA repair (*recA* and *radA*), replication and photosynthesis (Schleper *et al.*, 1997; Sandler *et al.*, 1999; Beja *et al.*, 2002b). To identify evolutionarily conserved genes, a strategy must be developed so that the *E. coli* copies of the genes do not complicate the screen. In some cases, it is possible to use a functional complementation, taking advantage of *E. coli* mutations. Currently, we are screening metagenomic libraries for the presence of *recA* genes, which can be identified on the basis of sequence conservation (Eisen, 1998; Sandler *et al.*, 1999) as well as functional complementation of the *E. coli recA* gene, selecting BAC clones by their ability to confer resistance to ultraviolet radiation or chemical mutagenesis. Using a combination of markers provides a larger arsenal of tools that assess microbial diversity, potentially identifying BAC clones that might not be identified by functional or sequence-based analysis that are of particular interest owing to the identity of the organism from which there were derived.

Sequence analysis

The tremendous amount of genomic sequence information contained within metagenomic libraries can provide a window into the 'black box' of natural environments. The steep increase in sequencing capacity in recent years has reduced the sequencing of microbial genomes to a daily routine, and holds the key to unlocking the mysteries of the natural microbial world. At present, the sheer scale of metagenomic libraries necessary to represent a microbial assemblage in soil prevents the sequencing of entire BAC libraries, although for less complex natural environments this is a viable option. For the moment, generation of sequence data from soil metagenomic BAC clones relies upon a variety of screening methods to identify BAC clones that have a phylogenetic and/or functional linkage, as well as random BAC end-sequencing to acquire a crude census of the BAC clones within a metagenomic library. Sequence-based analysis has led to surprising discoveries, including the identification of a bacterial rhodopsin gene in a bacterial species, which contrasted with all work on cultured organisms that indicated that these genes were confined to Archaea (Beja et al., 2000a). Analysis of clones from uncultivated Archaea has indicated a functional diversity that is not necessarily reflected in the 16S rRNA genes (Beja et al., 2002a).

Functional analysis

Most of the functional screening of metagenomic libraries reported to date has involved traditional low-technology screens. Enzymes, such as lipases, amylases and esculin hydrolases, have been identified with colorimetric or other visually discernible indicators in the medium on which the clones are plated (Henne et al., 1999, 2000; Brady and Clardy, 2000; Rondon et al., 2000; Brady et al., 2001; MacNeil et al., 2001). Pigmented antibiotics have been identified by visual inspection of arrayed libraries followed by antibiosis assays of selected clones (Brady et al., 2001; MacNeil et al., 2001; Gillespie et al., 2002). Antibiotics that lack visible color have been identified by visualizing zones of inhibition around clones plated on sensitive indicator organisms (Brady and Clardy, 2000). Clones that either lyse or alter the appearance of red blood cells have been identified by replicating libraries on to lawns of sheep red blood cells (Gillespie et al., 2002).

The barriers of heterologous expression and perhaps other factors have resulted in a low frequency of metagenomic clones expressing any given characteristic for which libraries have been screened. For example, in one study of 700 000 clones, 65 were found to produce antimicrobial activity (Brady and Clardy, 2000); in a library of 3648 clones, one expressed DNase, one antibacterial, two lipase and eight amylase activity (Rondon et al., 2000); and in another study, a library of 286 000 clones yielded three lipolytic clones (Henne et al., 2000). Therefore, the ability to screen many clones is essential to discovery in metagenomic libraries. Use of selectable phenotypes enhances discovery because only clones expressing the desired characteristic will grow in culture. Selections can be designed to complement a mutation for a normal function in E. coli, such as was done to identify new antiporters (Majernik et al., 2001), or to identify a new function, such as antibiotic resistance (Riesenfeld et al., unpublished).

❖❖❖❖❖❖ CONCLUDING REMARKS

Microbiology has seen three major revolutions, each coupled to development of a new powerful technique for 'seeing' microorganisms. The invention of the microscope, followed by improvements in optics and manufacturing, first revealed the existence of a microbial world. The microscope remains a microbiologist's indispensable tool. In vitro cultivation opened up a vast suite of powerful experimental approaches for microbiology. The diversity of microorganisms (based on features such as colony morphology, pigmentation and growth rates) that were responsive to cultivation from many environments was an immediate result of the cultivation technologies that today we call plating. Especially in etiology, isolation in pure culture of microorganisms that were candidate pathogens and their reintroduction as inoculum into healthy hosts resulting in disease (followed by reisolation – the four canonical steps of Kochs' Postulates), was a major influence on microbiology that continues to dominate microbiology 150 years after its introduction. The third revolution stemmed from the insight of Carl Woese and his colleagues that the sequences of ribosomal RNA genes revealed evolutionary relationships among organisms, and the application of Woese's insight by Norm Pace and his colleagues to microbial ecology, confirming previous hints that environmental samples contained previously unknown microbial diversity. In each of these revolutions, the technology innovation provided direct access to a previously inaccessible view of microorganisms. It is a statement that can rarely be made about technological innovations in science to note that none of these innovative technologies replaced a previous less powerful technology, but instead each added to the toolbox of the microbiologist.

Each of the three innovations also led to further developments that built upon the power of the technology. In the case of the application of molecular biology tools to evolutionary and ecological microbiology, pioneered by Woese and Pace, the further developments have included powerful new tools for phylogenetic inference based on ribosomal RNA gene sequences, and *in situ* methods (marrying molecular biology to microscopy) for detection and ecological analysis of microorganisms in the environments, and at the scales, in which they function.

We view metagenomics as a further example of a powerful new tool that derives from the application of molecular methods to microbial ecology. Metagenomics presents an opportunity to derive new insights about the microbial world by uniting the power of genomics with the study of the vast phylogenetic and functional diversity of microorganisms in natural environments.

Acknowledgements

We thank Laurie Luther for assistance in preparing the manuscript. This work was supported by the National Science Foundation, the National Institutes of Health, the McKnight Foundation, the Packard Foundation, and the University of Wisconsin–Madison College of Agricultural and Life Sciences. DAM was supported by the Damon Runyon Cancer Research Foundation Fellowship, DRG1594.

References

August, P. R., Grossman, T. H., Minor, C., Draper, M. P., MacNeil, I. A., Pemberton, J. M., Call, K. M., Holt, D. and Osburne, M. S. (2000). Sequence analysis and functional characterization of the violacein biosynthetic pathway from *Chromobacterium violaceum*. *J. Mol. Microbiol. Biotechnol.* **2**, 513–519.

Beja, O., Aravind, L., Koonin, E. V., Suzuki, M., Hadd, A., Nguyen, L. P., Jovanovich, S. B., Gates, C. M., Feldman, R. A., Spudich, J. L., Spudich, E. N. and DeLong, E. F. (2000a). Bacterial rhodopsin: evidence for a new type of phototrophy in the sea. *Science* **289**, 1902–1906.

Beja, O., Suzuki, M. T., Koonin, E. V., Aravind, L., Hadd, A., Nguyen, L. P., Villacorta, R., Amjadi, M., Garrigues, C., Jovanovich, S. B., Feldman, R. A. and DeLong, E. F. (2000b). Construction and analysis of bacterial artificial chromosome libraries from a marine microbial assemblage. *Environ. Microbiol.* **2**, 516–529.

Beja, O., Suzuki, M. T., Heidelberg, J. F., Nelson, W. C., Preston, C. M., Hamada, T., Eisen, J. A., Fraser, C. M. and DeLong, E. F. (2002b). Unsuspected diversity among marine aerobic anoxygenic phototrophs. *Nature* **415**, 630–633.

Beja, O., Koonin, E. V., Aravind, L., Taylor, L. T., Seitz, H., Stein, J. L., Bensen, D. C., Feldman, R. A., Swanson, R. V. and DeLong, E. F. (2002a). Comparative genomic analysis of archaeal genotypic variants in a single population and in two different oceanic provinces. *Appl. Environ. Microbiol.* **68**, 335–345.

Bintrim, S. B., Donohue, T. J., Handelsman, J., Roberts, G. P. and Goodman, R. M. (1997). Molecular phylogeny of Archaea from soil. *Proc. Natl Acad. Sci. USA* **94**, 277–282.

Black, C., Fyfe, J. A. M. and Davies, J. K. (1995). A promoter associated with the neisserial repeat can be used to transcribe the *uvrB* gene from *Neisseria gonorrhoeae*. *J. Bacteriol.* **177**, 1952–1958.

Borneman, J., Skroch, P. W., O'Sullivan, K. M., Palus, J. A., Rumjanek, N. G., Jansen, J. L., Nienhuis, J. and Triplett, E. W. (1996). Molecular microbial diversity of an agricultural soil in Wisconsin. *Appl. Environ. Microbiol.* **62**, 1935–1943.

Brady, S. F. and Clardy, J. (2000). Long-chain N-acyl amino acid antibiotics isolated from heterologously expressed environmental DNA. *J. Am. Chem. Soc.* **122**, 12903–12904.

Brady, S. F., Chao, C. J., Handelsman, J. and Clardy, J. (2001). Cloning and heterologous expression of a natural product biosynthetic gene cluster from eDNA. *Organic Lett.* **3**, 1981–1984.

Cane, D. E., Walsh, C. T. and Khosla, C. (1998). Biochemistry – harnessing the biosynthetic code: combinations, permutations, and mutations. *Science* **282**, 63–68.

Chávez, S., Reyes, J., Cahuvat, C. F., Florencio, F. J. and Candau, P. (1995). The NADP–glutamate dehydrogenase of the cyanobacterium *Synechocystis* 6803: cloning, transcriptional analysis and disruption of the *gdhA* gene. *Plant Mol. Biol.* **28**, 173–188.

Cottrell, M. T., Moore, J. A. and Kirchman, D. L. (1999). Chitinases from uncultured marine microorganisms. *Appl. Environ. Microbiol.* **65**, 2553–2557.

DeLong, E. F. (1992). Archaea in coastal marine environments. *Proc. Natl. Acad. Sci. USA* **89**, 5685–5689.

DeLong, E. F., Wu, K. Y., Prezelin, B. B. and Jovine, R. V. M. (1994). High abundance of Archaea in Antarctic marine picoplankton. *Nature (London)* **371**, 695–697.

DeLong, E. F., Taylor, L. T., Marsh, T. L. and Preston, C. M. (1999). Visualization and enumeration of marine planktonic archaea and bacteria using polyribonucleotide probes and fluorescence in situ hybridization. *Appl. Environ. Microbiol.* **65**, 5554–5563.

deVos, W. M. (1987). Gene cloning and expression in lactic streptococci. *FEMS Microbiol. Rev.* **46**, 281–295.

deVos, W. M., Kleerebezem, M. and Kuipers, O. (1997). Expression systems for industrial Gram-positive bacteria with low guanine and cytosine content. *Curr. Opin. Biotechnol.* **8**, 547–553.

Ding, M. and Yelton, D. B. (1993). Cloning and analysis of the *leuB* gene of *Leptospira interrogans* serovar *pomona*. *J. Gen. Microbiol.* **139**, 1093–1103.

Eisen, J. A. (1998). Phylogenomics: improving functional predictions for uncharacterized genes by evolutionary analysis. *Genome Res.* **8**, 163–167.

Entcheva, P., Liebl, W., Johann, A., Hartsch, T. and Streit, W. R. (2001). Direct cloning from enrichment cultures, a reliable strategy for isolation of complete operons and genes from microbial consortia. *Appl. Environ. Microbiol.* **67**, 89–99.

Ferreyra, R. G., Soncini, F. C. and Viale, A. M. (1993). Cloning, characterization, and functional expression in *Escherichia coli* of chaperonin (*groESL*) genes from the prototrophic sulfur bacterium *Chromatium vinosum*. *J. Bacteriol.* **175**, 1514–1523.

Fiandt, M. (2000). Construction of an environmental genomic DNA library from soil using the EpiFOS fosmid library production kit. *Epicentre Forum* **7**, 6.

Gillespie, D. E., Brady, S. F., Bettermann, A. D., Cianciotto, N. P., Liles, M. R., Rondon, M. R., Clardy, J., Goodman, R. M. and Handelsman, J. (2002). Isolation of antibiotics turbomycin A and B from a metagenomic library of soil microbial DNA. *Appl. Environ. Microbiol.* (in press).

Giovannoni, S. J., Britschgi, T. B., Meyer, C. L. and Field, K. G. (1990). Genetic diversity in Sargasso Sea bacterioplankton. *Nature* **345**, 60–63.

Griffiths, B. S., Ritz, K. and Glover, L. A. (1996). Broad-scale approaches to the determination of soil microbial community structure: application of the community DNA hybridization technique. *Microbial Ecol.* **31**, 269–280.

Guiney, D. G. and Yakobsen, E. (1983). Location and nucleotide sequence of the transfer origin of the broad range host plasmid RK2. *Proc. Natl Acad. Sci. USA* **80**, 3595–3598.

Handelsman, J., Rondon, M. R., Brady, S., Clardy, J. and Goodman, R. M. (1998). Molecular biology provides access to the chemistry of unknown soil microbes: a new frontier for natural products. *Chem. Biol.* **5**, R245–R249.

Head, I. M., Saunders, J. R. and Pickup, R. W. (1998). Microbial evolution, diversity and ecology: a decade of ribosomal RNA analysis of uncultivated microorganisms. *Microb. Ecol.* **35**, 1–21.

Henne, A., Daniel, R., Schmitz, R. A. and Gottschalk, G. (1999). Construction of environmental DNA libraries in *Escherichia coli* and screening for the presence of genes conferring utilization of 4-hydroxybutyrate. *Appl. Environ. Microbiol.* **65**, 3901–3907.

Henne, A., Schmitz, R. A., Bomeke, M., Gottschalk, G. and Daniel, R. (2000). Screening of environmental DNA libraries for the presence of genes conferring lipolytic activity on *Escherichia coli*. *Appl. Environ. Microbiol.* **66**, 3113–3116.

Hugenholtz, P., Goebel, B. M. and Pace, N. R. (1998). Impact of culture-independent studies on the emerging phylogenetic view of bacterial diversity. *J. Bacteriol.* **180**, 4765–4774.

Joseph, P., Fantino, J. R., Herbaud, M. L. and Denizot, F. (2001). Rapid orientated cloning in a shuttle vector allowing modulated gene expression in *Bacillus subtilis*. *FEMS Microbiol. Lett.* **205**, 91–97.

Karlin, S., Mrazek, J., Campbell, A. and Kaiser, D. (2001). Characterizations of highly expressed genes of four fast-growing bacteria. *J. Bacteriol.* **183**, 5025–5040.

Krsek, M. and Wellington, E. M. H. (1999). Comparison of different methods for the isolation and purification of total community DNA from soil. *J. Microbiol. Meth.* **39**, 1–16.

Leonhardt, H. and Alonso, J. C. (1988). S1 mapping of the pC194 encoded chloramphenicol gene in *Bacillus subtilis*. *Nucleic Acids Res.* **16**, 1618.

Lindahl, V. (1996). Improved soil dispersion procedures for total bacterial counts, extraction of indigenous bacteria and cell survival. *J. Microbiol. Meth.* **25**, 279–286.

MacNeil, I. A., Minor, T. C., August, P. R., Grossman, T. H., Loiacono, K. A., Lynch, B. A., Phillips, T., Narula, S., Sundaramoorthi, R., Tyler, A., Aldredge, T., Long, H., Gilman, M. and Osburne, M. S. (2001). Expression and isolation of antimicrobial small molecules from soil DNA libraries. *J. Mol. Microbiol. Biotechnol.* **3**, 301–308.

Majernik, A., Gottschalk, G. and Daniel, R. (2001). Screening of environmental DNA libraries for the presence of genes conferring $Na^+(Li^+)/H^+$ antiporter activity on *Escherichia coli*: characterization of the recovered genes and the corresponding gene products. *J. Bacteriol.* **183**, 6645–6653.

McInerney, M. O., Wilkinson, M., Patching, J. W., Embley, T. M. and Powell, R. (1995). Recovery and phylogenetic analysis of novel archaeal rRNA sequences from a deep-sea deposit feeder. *Appl. Environ. Microbiol.* **61**, 1646–1648.

Newman, J. R. and Fuqua, C. (1999). Broad-host-range expression vectors that carry the L-arabinose-inducible *Escherichia coli* araBAD promoter and the araC regulator. *Gene* **227**, 197–203.

Pace, N. R. (1996). New perspective on the natural microbial world: molecular microbial ecology. *ASM News* **62**, 463–470.

Pfeifer, B. A., Admiraal, S. J., Gramajo, H., Cane, D. E. and Khosla, C. (2001). Biosynthesis of complex polyketides in a metabolically engineered strain of *E. coli*. *Science* **291**, 1790–1792.

Poyart, C. and Trieu-Cuot, P. (1997). A broad-host-range mobilizable shuttle vector for the construction of transcriptional fusions to beta-galactosidase in Gram-positive bacteria. *FEMS Microbiol. Lett.* **156**, 193–198.

Rondon, M. R., Goodman, R. M. and Handelsman, J. (1999a). The Earth's bounty: assessing and accessing soil microbial diversity. *Trends Biotechnol.* **17**, 403–409.

Rondon, M. R., Raffel, S. J., Goodman, R. M. and Handelsman, J. (1999b). Toward functional genomics in bacteria: analysis of gene expression in *Escherichia coli* from a bacterial artificial chromosome library of *Bacillus cereus*. *Proc. Natl Acad. Sci. USA* **96**, 6451–6455.

Rondon, M. R., August, P. R., Bettermann, A. D., Brady, S. F., Grossman, T. H., Liles, M. R., Loiacono, K. A., Lynch, B. A., MacNeil, I. A., Osburne, M. S., Clardy, J., Handelsman, J. and Goodman, R. M. (2000). Cloning the soil metagenome: a strategy for accessing the genetic and functional diversity of uncultured microorganisms. *Appl. Environ. Microbiol.* **66**, 2541–2547.

Sandler, S. J., Hugenholtz, P., Schleper, C., DeLong, E. F., Pace, N. R. and Clark, A. J. (1999). Diversity of *radA* genes from cultured and uncultured *Archaea*: comparative analysis of putative RadA proteins and their use as a phylogenetic marker. *J. Bacteriol.* **181**, 907–915.

Schleper, C., Swanson, R. V., Mathur, E. J. and DeLong, E. F. (1997). Characterization of a DNA polymerase from the uncultivated psychrophilic archaeon *Cenarchaeum symbiosum*. *J. Bacteriol.* **179**, 7803–7811.

Schleper, C., DeLong, E. F., Preston, C. M., Feldman, R. A., Wu, K. Y. and Swanson, R. V. (1998). Genomic analysis reveals chromosomal variation in natural populations of the uncultured psychrophilic archaeon *Cenarchaeum symbiosum*. *J. Bacteriol.* **180**, 5003–5009.

Shizuya, H., Birren, B., Kim, U.-J., Mancino, V., Slepak, T., Tachiri, Y. and Simon, M. (1992). Cloning and stable maintenance of 300-kilobase-pair fragments of human DNA in *Escherichia coli* using an F-factor-based vector. *Proc. Natl Acad. Sci. USA* **89**, 8794–8797.

Simon, H. M., Dodsworth, J. and Goodman, R. M. (2000). Crenarchaeota colonize terrestrial plant roots. *Environ. Microbiol.* **2**, 495–505.

Stach, J. E. M., Bathe, S., Clapp, J. P. and Burns, R.G. (2001). PCR-SSCP comparison of 16S rDNA sequence diversity in soil DNA obtained using different isolation and purification methods. *FEMS Microbiol. Ecol.* **36**, 139–151.

Stahl, D. A. (1993). The natural history of microorganisms. *ASM News* **59**, 609–613.

Stahl, D. A., Lane, D. J., Olsen, G. J. and Pace, N. R. (1985). Characterization of a Yellowstone hot spring microbial community by 5S rRNA sequences. *Appl. Environ. Microbiol.* **49**, 1379–1384.

Stein, J. L., Marsh, T. L., Wu, K. Y., Shizuya, H. and DeLong, E. F. (1996). Characterization of uncultivated prokaryotes: isolation and analysis of a 40-kilobase-pair genome fragment from a planktonic marine archaeon. *J. Bacteriol.* **178**, 591–599.

Steinmetz, M. and Richter, R. (1994). Plasmids designed to alter the antibiotic resistance expressed by insertion mutations in *Bacillus subtilis*, through in vivo recombination. *Gene* **142**, 79–83.

Suzuki, M. T., Rappe, M. S., Haimberger, Z. W., Winfield, H., Adair, N., Strobel, J. and Giovannoni, S. J. (1997). Bacterial diversity among small-subunit rRNA gene clones and cellular isolates from the same seawater sample. *Appl. Environ. Microbiol.* **63**, 983–989.

Tang, L., Shah, S., Chung, L., Carney, J., Katz, L., Khosla, C. and Julien, B. (2000). Cloning and heterologous expression of the epothilone gene cluster. *Science* **287**, 640–642.

Torsvik, V., Goksøyr, J. and Daae, F. L. (1990). High diversity in DNA of soil bacteria. *Appl. Environ. Microbiol.* **56**, 782–787.

Torsvik, V., Sorheim, R. and Goksoyr, J. (1996). Total bacterial diversity in soil and sediment communities – a review. *J. Indust. Microbiol. Biotech.* **17**, 170–178.

Trieu-Cuot, P., Carlier, C., Martin, P. and Courvalin, P. (1987). Plasmid transfer by conjugation from *Escherichia coli* to Gram-positive bacteria. *FEMS Microbiol. Lett.* **48**, 289–294.

Uozumi, T., Ozaki, A., Beppu, T. and Arima, K. (1980). New cryptic plasmid of *Bacillus subtilis* and restriction analysis of other plasmids found by general screening. *J. Bacteriol.* **142**, 315–318.

Vergin, K. L., Urbach, E., Stein, J. L., DeLong, E. F., Lanoil, B. D. and Giovannoni, S. J. (1998). Screening of a fosmid library of marine environmental genomic DNA fragments reveals four clones related to members of the order *Planctomycetales*. *Appl. Environ. Microbiol.* **64**, 3075–3078.

Wang, G.L. and Warren, R. (1996). Construction of an Arabidopsis BAC library and isolation of clones hybridizing with disease-resistance, gene-like sequences. *Plant Mol. Biol. Reporter* **14**, 107.

Wang, G.-Y.-S., Graziani, E., Waters, B., Pan, W., Li, X., McDermott, J., Meurer, G., Saxena, G., Andersen, J. and Davies, J. (2000). Novel natural products from soil DNA libraries in a Streptomycete host. *Organic Lett.* **2**, 2401–2404.

Ward, D. M., Weller, R. and Bateson, M. M. (1990). 16S rRNA sequences reveal numerous uncultured microorganisms in a natural community. *Nature* **345**, 63–65.

Whitman, W. B., Coleman, D. C. and Wiebe, W. J. (1998). Prokaryotes: the unseen majority. *Proc. Natl Acad. Sci. USA* **95**, 6578–6583.

Wild, J., Sektas, M., Hradecna, Z. and Szybalski, W. (1998). Targeting and retrofitting pre-existing libraries of transposon insertions with FRT and oriV elements for in-vivo generation of large quantities of any genomic fragment. *Gene* **223**, 55–66.

Wild, J., Hradecna, Z. and Szybalski, W. (1999). On command amplification of large genomic fragments cloned in pBeloBAC/oriV vectors. In: *The 1999 Molecular Genetics of Bacteria and Phages Meeting*, Madison, WI, p. 44.

Yakobson, E. A. and Guiney, D. G. (1984). Conjugal transfer of bacterial chromosomes mediated by the RK2 plasmid transfer origin cloned into transposon Tn5. *J. Bacteriol.* **160**, 451–453.

Zhou, J., Bruns, M. A. and Tiedje, J. M. (1996). DNA recovery from soils of diverse composition. *Appl. Environ. Microbiol.* **62**, 316–322.

14 Reverse Vaccinology: from Genome to Vaccine

John L. Telford*, Mariagrazia Pizza, Guido Grandi and Rino Rappuoli
IRIS, Chiron S.p.A., Via Fiorentina 1, 53100 Siena, Italy

CONTENTS

Introduction
Reverse vaccinology
The meningitis B problem
The complete genome sequence of N. meningitidis serogroup B
Candidate antigen prediction
Candidate screening
Antigenic variation in the protective antigens
Are the conserved antigens cross-protective?
Other protective antigens
Outlook for development of a vaccine
The future of reverse vaccinology
Conclusions

◆◆◆◆◆◆ INTRODUCTION

For nearly 150 years after the first successful vaccinations performed by Edward Jenner, approaches to vaccine discovery barely changed in nature. Even as late as 1980 all vaccines in use were based on immunization with inactivated or attenuated pathogens, partially purified toxins and polysaccharides. Since then, however, two major revolutions in vaccine design have occurred. The first was the use of modern recombinant DNA technology to produce subunit vaccines based on specific antigens. The first of these new vaccines was the highly pure surface antigen of the hepatitis B virus. The second was subunit vaccines against *Bordetella pertussis* containing three highly pure proteins. This latter vaccine also pioneered the use of structure–function studies to produce genetically altered pertussis toxin that lacked toxicity but maintained an unaltered

*To whom correspondence should be addressed. Email: John_Telford@chiron.it

antigenic conformation (Pizza *et al.*, 1989). These novel approaches led to the development of a paradigm for vaccine research, which has persisted for the last two decades. In this approach, the microorganism is studied from the point of view of pathogenicity and immunology in order to identify factors involved in virulence, which may be suitable vaccine candidates. This approach has been used for the development of most new vaccines currently available or in clinical development.

The second revolution is more recent and follows the development in biology brought about by the possibility to obtain complete genome sequences of microbial pathogens. Here we will describe the first vaccine project to take this approach that has led to the definition of a completely new paradigm of more general use in developing the next generation of vaccines for which the traditional approaches have failed.

◆◆◆◆◆◆ REVERSE VACCINOLOGY

Developments in automatic DNA sequencing and the genomic revolution have led to the possibility to approach the design of vaccines from a completely different perspective. No longer is the identification of a candidate antigen, its cloning and characterization a lengthy project of several years duration. With the DNA sequence of the complete genome of a microorganism, every single protein synthesized by the bacterium can be identified. Surface exposed proteins or proteins secreted by the bacterium can be predicted on the basis of our knowledge of microbial biology, and the genes coding for these proteins can be cloned and expressed using polymerase chain reaction (PCR) technology. Hence, the entire complement of surface proteins of a pathogen can be produced in recombinant form and tested as vaccine candidates in appropriate models. We will illustrate this approach with the first example of application of this strategy: the design of a novel vaccine against serogroup B, *Neisseria meningitidis* (Pizza *et al.*, 2000).

◆◆◆◆◆◆ THE MENINGITIS B PROBLEM

Bacterial meningitis is a devastating disease leading frequently to death or serious disability. A major cause of bacterial meningitis is the Gram-negative bacterium, *Neisseria meningitidis*. The incidence of meningococcal disease ranges from 1–3/100 000 in industrialized countries to 10–25/100 000 in developing countries. *N. meningitidis* is an encapsulated, aerobic Gram-negative diplococcus. It is an exclusively human pathogen and its natural habitat is the mucosal membrane, primarily the nasopharynx (for a review, see van Deuren *et al.*, 2000). Approximately 30% of the human population are healthy carriers (Scholten *et al.*, 1993). In the last two decades, vaccines based on the polysaccharide capsule have been developed, which are effective in adults. More recently, capsule polysaccharides conjugated to protein carriers have been developed, which are also effective in children, leading to a drastic decrease in the incidence of disease (Zollinger, 1997).

There are five major capsular serotypes of *N. meningitidis* (A, B, C, Y and W135). Polysaccharide vaccines against four of these (A, C, Y and W135) have been developed. In addition, a conjugate vaccine against serogroup C has been developed and similar conjugates against A, Y and W serogroups are in development. Serogroup B, however, which accounts for 50% of invasive disease in Europe and the USA, has been a particular problem as its polysaccharide capsule is identical to a polysaccharide widely distributed in human tissue. This means that it is almost completely non-immunogenic because it is recognized as a self-antigen. Furthermore, attempts to break tolerance to induce immunity to this polysaccharide are likely to lead to autoimmunity (Hayrinen *et al.*, 1995).

Three decades of research have failed to produce a vaccine against serotype B, *N. meningitidis*. Most approaches have aimed either at producing type B polysaccharide modified to be immunogenic or have targeted surface proteins of the bacteria. To date, these approaches have failed to produce a vaccine capable of inducing protection against circulating strains, mainly due to the variability of known surface antigens. With the exception of a single protein (NspA) (Martin *et al.*, 1997), *N. meningitidis* surface antigens studied in the last three decades have been found to be highly variable. This problem is illustrated by the results of clinical trials with serogroup B vaccines made from purified outer membrane vesicles (OMV). Two such vaccines have been tested, one in Cuba and one in Norway. While each vaccine has been shown to induce good protection against the homologous strain used in the preparation of the OMV, neither demonstrated effective protection against heterologous strains (reviewed in Rosenstein *et al.*, 2001). The major protective antigen in both these vaccines is an abundant outer membrane protein (PorA), which is, in fact, known to be highly variable across clinical isolates of serogroup B, *N. meningitidis*.

◆◆◆◆◆◆ THE COMPLETE GENOME SEQUENCE OF *N. MENINGIDITIS* SEROGROUP B

To overcome the problems described above, and to identify potential new antigens for use in a meningitis B vaccine, the complete genome sequence of the virulent strain, MC58, of serogroup B, *N. meningitidis* was obtained using the random shotgun strategy (Genbank accession number AE002098) (Tettelin *et al.*, 2000). Strain MC58 has 2 272 351 basepairs with an average content of $G + C$ of 53%. Using open reading frame (ORF) prediction algorithms and whole-genome homology searches, 2158 putative ORFs were identified. Biological roles have been assigned to 1158 ORFs (53.7%) with similarity to proteins of known function according to the classification of Riley (1993). A total of 345 (16%) predicted coding sequences matched gene products of unknown function from other species and 532 (24.7%) had no database match (www.tigr.org/tdb/mdb/mdb.html). The challenge was to design a strategy which would allow the rapid identification of potential candidates among these more than 2000 predicted proteins and to test them for their potential to induce protection against disease.

◆◆◆◆◆◆ CANDIDATE ANTIGEN PREDICTION

Probably the only bias in candidate antigen selection in this approach is based on the belief that protective antigens are more likely to be found among surface exposed or secreted proteins. This belief is based on the fact that, as *N. meningitidis* is essentially an extracellular pathogen, the major protective response relies on circulating antibody. In fact, complement-mediated bactericidal activity is the accepted correlate with *in vivo* protection and, as such, is the surrogate end-point in clinical trials of potential vaccines. Hence the initial selection of candidate analysis is based on computer predictions of secretion or surface location. In this preliminary selection, all *N. meningiditis* predicted ORFs were searched using computer programs, such as PSORT and SignalP, which predict signal peptide sequences. In addition, other indications of surface exposure were identified. For example, proteins containing predicted membrane spanning regions (using TMPRED) predicted lipoproteins, and proteins homologous to surface exposed proteins in other microorganisms were selected. Finally, proteins with homology to known virulence factors or protective antigens from other pathogens were added to the list. The selection procedure is shown schematically in Figure 14.1. Of the 2158 predicted ORFs in the *N. meningitidis* genome, 575 were selected by these criteria. The proportion of each of the predicted protein types is shown in Figure 14.2.

◆◆◆◆◆◆ CANDIDATE SCREENING

The next step in the process was to design a method to screen each of these candidates in a rapid simple assay of their potential to confer protective immunity against *N. meningitidis* infection. As mentioned above, bactericidal activity of antibodies against *N. meningitidis* is known to correlate well with protective immunity. The assay of bactericidal activity is relatively straightforward and can be performed on a large number of samples. Therefore, each candidate antigen was cloned in expression vectors in *E. coli* and the purified recombinant protein was used to immunize mice. Sera were then tested in the bactericidal assay.

Expression in *E. coli*

In order to screen this large number of candidate antigens, it was necessary to use a simple procedure for cloning and antigen purification. To this end, each ORF was amplified by PCR using primers, which eliminated the

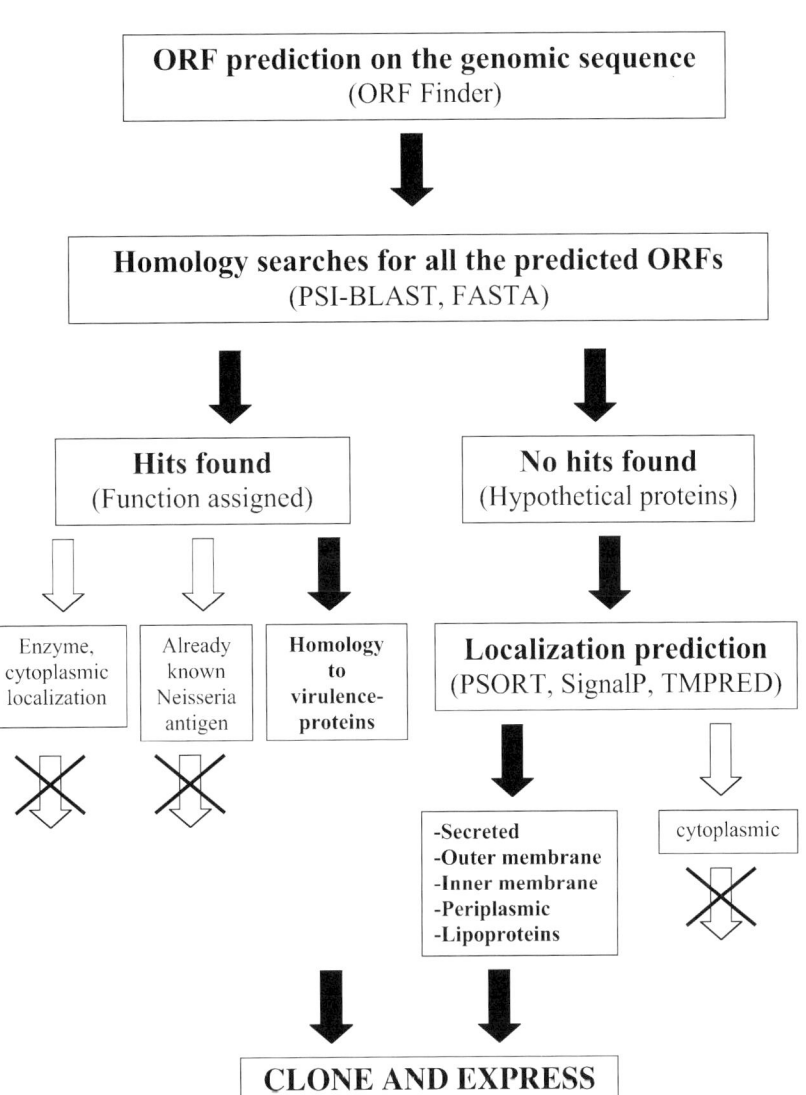

Figure 14.1. Schematic representation of the *in silico* selection of candidate antigens from the complete genome sequence.

signal peptide coding regions and included appropriate restriction sites. These products were then cloned in parallel into vectors containing the necessary promoters and other regulatory sequences and sequences coding for either an amino-terminal six histidine tag or a carboxy-terminal GST fusion protein. These constructs permitted the rapid purification of soluble recombinant protein by simple column chromatography. The parallel cloning in the two different vectors was used to maximize the possibility of obtaining the recombinant protein in a soluble form. Of the 575 selected candidates, 344 were successfully cloned in *E. coli* and soluble recombinant proteins were purified in sufficient quantity for the immunization of mice

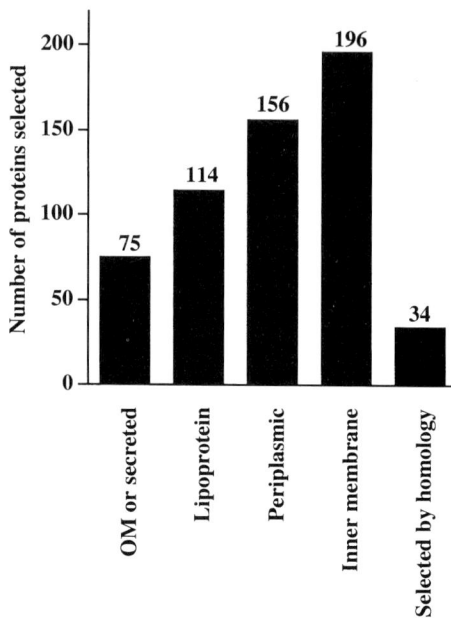

Figure 14.2. Categories and numbers of proteins selected for cloning and expression from the complete genome sequence of a serogroup B strain of *N. meningitidis*.

(see Figure 14.3). Most of the failures, both in cloning and in expression of soluble protein were with proteins predicted to have more than one transmembrane spanning region. This is likely to be due to toxicity to *E. coli* of these proteins or to their inherent insolubility.

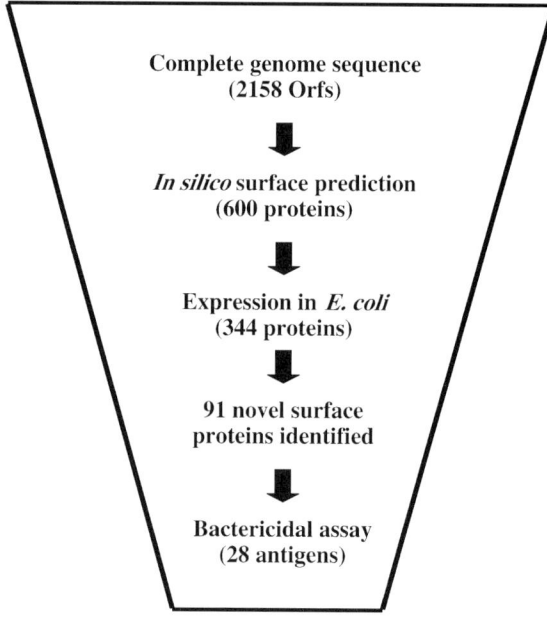

Figure 14.3. Reverse vaccinology. Schematic representation of the process of identifying vaccine candidates, starting with the complete genome sequence.

Immunization and screening of sera

Each purified recombinant protein was used to immunize groups of four CD1 mice. CD1 mice were used as they are outbred and would thus avoid missing potential antigens due to MHC restriction. Sera from each group of mice were pooled and used in a series of assays to characterize the proteins. First, immunoblot analysis was performed in order to demonstrate that the protein is actually expressed in the bacteria. Immunoblots were performed on both total cell extracts and on purified outer membrane vesicles. This latter gives an indication of the subcellular localization of the protein. In addition, surface localization was assessed by flow cytometry of fixed bacteria treated with the mouse antiserum and a fluorescent antimouse immunoglobulin (Ig) secondary antibody. Using this approach, 91 novel surface exposed proteins have been identified in *N. meningitidis* (Pizza et al., 2000).

Assays of the complement-mediated bactericidal activity of the sera identified 28 recombinant proteins capable of inducing a protective immune response. This is a very significant result given that the last four decades of research have identified less than a dozen such proteins. Furthermore, whereas previous research revealed only highly expressed proteins that were highly variable between different isolates, many of the proteins identified by this approach were poorly expressed but, nevertheless, induced high titer bactericidal activity.

◆◆◆◆◆◆ ANTIGENIC VARIATION IN THE PROTECTIVE ANTIGENS

Of the 28 primary candidates identified, seven were chosen for further characterization, based on the bactericidal titers obtained and the expression of the appropriate protein product in *N. meningitidis* (Table 14.1). A

Table 14.1. Characteristics of protective antigens

Antigen (number of amino acids)	Characteristics	FACS*	ELISA[†]	Serum bactericidal activity[‡]
GNA33 (441)	Lipoprotein	+++++	13 000	1/16 000
GNA992 (591)	Outer membrane protein	+++	2 750	1/256
GNA1162	Lipoprotein	++	1 270	1/4
GNA1220 (315)	Membrane protein	+++	1 000	1/256
GNA1946 (287)	Lipoprotein	+++	13 100	1/32
GNA2001 (488)	Outer membrane protein	++	500	1/32
GNA2132 (488)	Lipoprotein	++	1 700	1/16 000

* Relative indirect immunofluorescence of whole fixed bacteria treated with specific antiserum by flow cytometry.
[†] ELISA titers of specific antiserum on whole fixed bacteria.
[‡] Dilution of sera that results in 50% complement-mediated bacterial killing.

major question was to what extent the sequence of these antigens varied among clinical isolates of *N. meningitidis*. To approach this question, a panel of 22 strains of serogroup B of *N. meningitidis* were selected, which represent the major disease-causing lineages as identified by multilocus enzyme electrophoresis and multilocus sequence typing (Maiden et al., 1998). In addition, in the analysis we included: three strains of *N. meningitidis* serogroup A; two strains of serogroup C; one strain each of serogroups Y, X, Z and W135; three strains of *N. gonorrhoeae*; and one strain each of the commensals *N. cinerea* and *N. lactamica*. Each of these strains was analyzed by PCR and dot-blot hybridization for the presence of the genes coding for each of the seven antigens. Remarkably, all seven genes were found to be present in all strains of serogroup B *N. meningitidis* and several were present in the other serogroups and related bacteria (see Table 14.2).

To further characterize the variation in the candidate antigens across *N. meningitidis* isolates, PCR products from the amplifications of each antigen from each of the 31 strains of *N. meningitidis* were sequenced and compared. Among the antigens, GNA992 and GNA2132 had many hypervariable regions, located mostly in the amino-terminal half of the molecules, which suggests that these proteins, like PorA, might induce strain-specific immunity. In marked contrast, GNA33, GNA1162, GNA1220, GNA1946 and GNA2001 were found to be very highly conserved. In fact, the average amino-acid identity with the MC58 sequence was about 99% for each of these antigens (Pizza et al., 2000).

Table 14.2. Conservation of genes coding for protective antigens in isolates of *Neisseria*

Gene	Serogroup B (22 isolates)	ACYXZW (9 isolates)	*N. gonorrhoeae* (3 isolates)	*N. lactamica* (1 isolate)	*N. cinerea* (1 isolate)
gna33	+	+	+	+	+
gna992	+	+	−	+/−	+/−
gna1162	+	+	+	+	+
gna1220	+	+	+	+/−	+/−
gna1946	+	+	+	+	+/−
gna2001	+	+	+	+	+/−
gna2132	+	+	+	+	−

◆◆◆◆◆◆ ARE THE CONSERVED ANTIGENS CROSS-PROTECTIVE?

The presence of the genes for these antigens and their conservation among *N. meningitidis* isolates does not, however, guarantee that they will induce cross-protective immunity. Levels of expression may vary across strains and the availability of the antigens within the capsular polysaccharide capsule may differ. To address these points, the accessibility of the antigens to antibodies was assessed by immunoblot of OMV extracts, enzyme-linked immunosorbent assay (ELISA) of capsulated bacteria and flow cytometry. These experiments provided evidence that each of the seven antigens was present and available in each of the 31 strains tested. Furthermore, bactericidal activity was demonstrated for a subset of the strains for which suitable human complement was available.

The finding of several cross-protective antigens in *N. meningitidis* was remarkable as, with one exception mentioned above, all *N. meningitidis* surface proteins described to date have been found to be highly variable across different strains. It is unclear why conserved surface antigens have not been previously found. It has generally been felt that surface exposed antigens would be under pressure from the immune system to undergo variation. One difference in these antigens may be that they are expressed at relatively low levels. Traditional approaches to antigen identification bias towards relatively abundant proteins. One of the strengths of reverse vaccinology is that the functionality of the antigens is tested directly with the minimum of experimental or ideological bias.

◆◆◆◆◆◆ OTHER PROTECTIVE ANTIGENS

In addition to the antigens described above, the genomic screen identified several other protective antigens with less complete cross-protection, some of which were either unexpected or interesting for other reasons. Two of these deserve mention here.

GNA33: a novel mimetic antigen

GNA33 is a lipoprotein with amino-acid similarity to *E. coli* murein transglycosylase. Recombinant GNA33 induced high-titer bactericidal antibodies against some strains of *N. meningitidis*. Bactericidal titers were as high as those of sera from mice immunized with OMV, and GNA33 was thus considered an attractive antigen. Curiously, an analysis of the surface binding of anti-GNA33 antibodies revealed that surface staining correlated perfectly with the expression of a particular form of the PorA antigen, the P1.2 serosubgroup (Granoff *et al.*, 2001). Immunoblot of various *N. meningitidis* strains, including strains that had been engineered to delete the gna33 gene, revealed that the antibodies in addition to the GNA33 protein, cross-reacted with the serosubgroup P1.2 PorA itself. It was subsequently determined that it was this cross-reaction, which accounted for the bactericidal antibodies induced by GNA33, and that recognition of GNA33 itself was not bactericidal. Epitope mapping of a bactericidal anti-GNA33 monoclonal antibody, identified a short motif (PTF) present in GNA33, which is essential for recognition. This same motif was also found in loop 4 of PorA and was also found to be essential but not sufficient for the binding of the Mab to PorA.

This case of antigen mimicry is not only curious but may have functional value. Although anti-GNA33 antibodies only recognize P1.2 PorA, this serosubgroup accounts for 8% of disease, causing strains of *N. meningitidis* (Tondella et al., 2000) and GNA33 could hence be considered in a combination vaccine. The enormous advantage would be that, unlike PorA, GNA33 could be produced and purified readily in a soluble form from *E. coli*.

A novel *N. meningitidis* adhesin, NadA

During the genomic screen for vaccine candidates, the NMB194 antigen was found to induce bactericidal antibody titers as high as those obtained with OMV preparations. Comparison of the sequence of NMB194 with gene databases revealed sequence motifs highly similar to YadA, an adhesin believed to be involved in virulence of *Yersina*, and UspA, an ubiquitous surface protein of *Moraxella catarrhalis*. These proteins characteristically form oligomeric structures that are resistant to boiling in sodium dodecyl sulfate (SDS) and reducing conditions. NMB194 was also found to form oligomers and to bind to epithelial cells *in vitro*, indicating that it also functions as an adhesin for *N. meningiditis* (Comanducci et al., 2002). For this reason, it has been named NadA (*Neisseria* adhesin A).

NadA was found in only about 50% of *N. meningitidis* isolates, indicating that it may have limited value as a vaccine candidate. Interestingly, however, the *nadA* gene was found in 51 of 53 isolates of the hypervirulent lineages ET35, ET37 and cluster 4, but was not found in any lineage III strains. Hence, this antigen may be very useful for a vaccine for use in outbreaks of these hypervirulent strains.

◆◆◆◆◆◆ OUTLOOK FOR DEVELOPMENT OF A VACCINE

Decades of research have failed to produce an effective vaccine against serogroup B *N. meningitidis*. Attempts to find good protein antigens have identified highly variable abundant antigens capable only of inducing protection against the homologous strain. The reverse vaccinology approach of expressing and testing all potentially surface exposed antigens in the complete genome has identified, in the space of about 2½ years, several conserved proteins capable of inducing high-titer bactericidal antibodies against most clinical isolates of *N. meningitidis*. It is highly likely that a combination of a small number of these antigens will provide an efficacious novel vaccine capable of protecting the population from this devastating disease.

◆◆◆◆◆◆ THE FUTURE OF REVERSE VACCINOLOGY

Progress in DNA sequence technology means that it is now possible to obtain the complete genome sequence of any bacterial pathogen in a very short space of time and at relatively low cost. The success of reverse vaccinology in the development of a novel vaccine against *N. meningitidis* indicates that this approach is likely also to be successful for the development of vaccines against other bacterial diseases. No longer is it necessary to spend years identifying and testing protein antigens one by one only to find that they are highly variable or do not induce the appropriate immune response to give protection against the disease. The single requirement for the success of this approach is the availability of an *in vitro* or *in vivo* model to test the capacity of the antigens to induce protection. Hence, we may expect to see the development of several new vaccines against such bacterial targets in the next few years. Furthermore, there is no reason why the approach should be limited to bacterial pathogens.

Reverse vaccinology of viral pathogens

Many viral pathogens have quite small genomes whose sequence has been available for several years. Conventional approaches to vaccines against disease caused by these viruses have, nevertheless, been based on the preconceived idea that the best antigens will be found among the structural proteins that are present on the surface of the viruses. Frequently, however, these proteins are also highly variable owing to immune system pressure and have limited suitability for a vaccine.

Human immunodeficiency virus (HIV) vaccine research has suffered from this intellectual bias since the beginning, even though the complete sequence of the virus has been known for a long time. Most attempts at designing a vaccine against HIV have been aimed at the envelope glycoprotein in all its forms (gp120, gp140 and gp160) or the GAG antigen (gp55). However, given that the cellular arm of the immune system, in particular cytotoxic T-cells, is probably very important in the protective immune response against these intracellular pathogens, non-structural proteins, which are generally better conserved, may be useful antigens. Recent promising data from experiments with Tat, Nef, Rev and Pol strongly support this hypothesis (Cafaro *et al.*, 1999; Osterhaus *et al.*, 1999; Pauza *et al.*, 2000). Perhaps if a reverse vaccinology approach had been taken to research into an HIV vaccine, progress may have been more rapid.

Eukaryotic pathogens

Although eukaryotic genomes tend to be larger and more complex than bacterial genes, in principle, there is no reason why reverse vaccinology can not be applied to eukaryotic pathogens. Malaria is a good example. The genome of *Plasmodium falciparum* is estimated to contain about 6000 genes. This is only three times more than an average bacterial pathogen. Hence there is no reason why most of these genes cannot be expressed as recombinant proteins and tested. The use of genome-based DNA array technology to identify antigens expressed at different stages of the parasites, life cycle could reduce the number of antigens to be tested and may help to select the best combination of antigens such that an appropriate immune response could be mounted against different stages of the disease. A reverse vaccinology approach to malaria would depend, of course, on the development of a sufficiently straightforward *in vitro* or *in vivo* assay suitable for the testing of a large number of antigens.

◆◆◆◆◆◆ CONCLUSIONS

The availability of complete genome sequence information on many pathogens has led to a new paradigm in vaccine development. If a suitable assay is available, every protein synthesized by the pathogen can be tested as a vaccine candidate without any prior selection based on incomplete knowledge of the pathogenicity and immunogenicity of the organism. The addition of other genome-based technologies, such as DNA microarrays and proteomics, will add power to the approach and permit the refinement of strategies to identify novel antigens (see Grandi, 2001). DNA microarrays containing every gene in the genome of a pathogen can be hybridized with RNA extracted from the pathogen at different stages of infection, permitting the identification of genes important for infection and disease. Similarly, identification of proteins by mass spectrometry based on genomic information can identify antigens expressed at critical stages of infection. Improvements in heterologous expression systems and development of novel models to test antigens will extend the approach to even more human pathogens. Last but not least, defining protective antigens by unbiased selection based on function will add enormously to our knowledge of the protective immune response both for specific pathogens and generally. At least for vaccines, the genome era has come of age.

References

Cafaro, A., Caputo, A., Fracasso, C., Maggiorella, M. T., Goletti, D., Baroncelli, S., Pace, M., Sernicola, L., Koanga-Mogtomo, M. L., Betti, M., Borsetti, A., Belli, R., Akerblom, L., Corrias, F., Butto, S., Heeney, J., Verani, P., Titti, F. and Ensoli, B. (1999). Control of SHIV-89.6P-infection of cynomolgus monkeys by HIV-1 Tat protein vaccine. *Nature Med.* **5**, 643–650.

Comanducci, M., Bambini, S., Brunelli, B., Adu-Bobie, J., Arico, B., Capacci, B., Giuliani, M. M., Masignani, V., Santini, L., Savino, S., Granoff, D. M., Caugunt, D. A., Pizza, M., Rappuoli, R. and Mora, M. (2002). NuclA, a novel vaccine candidate of *Neisseria meningitidis*. *J. Exp. Med.* **195**, 1445–1454.

Grandi, G. (2001). Antibacterial vaccine design using genomics and proteomics. *Trends Biotechnol.* **19**, 181–188.

Granoff, D. M., Moe, G. R., Giuliani, M. M., Adu-Bobie, J., Santini, L., Brunelli, B., Piccinetti, F., Zuno-Mitchell, P., Lee, S. S., Neri, P., Bracci, L., Lozzi, L. and Rappuoli, R. (2001). A novel mimetic antigen eliciting protective antibody to *Neisseria meningitidis*. *J. Immunol.* **167**, 6487–6496.

Hayrinen, J., Jennings, H., Raff, H. V., Rougon, G., Hanai, N., Gerardy-Schahn, R. and Finne, J. (1995). Antibodies to polysialic acid and its N-propyl derivative: binding properties and interaction with human embryonal brain glycopeptides. *J. Infect. Dis.* **171**, 1481–1490.

Maiden, M. C., Bygraves, J. A., Feil, E., Morelli, G., Russell, J. E., Urwin, R., Zhang, Q., Zhou, J., Zurth, K., Caugant, D. A., Feavers, I. M., Achtman, M. and Spratt, B. G. (1998). Multilocus sequence typing: a portable approach to the identification of clones within populations of pathogenic microorganisms. *Proc. Natl Acad. Sci. USA* **95**, 3140–3145.

Martin, D., Cadieux, N., Hamel, J. and Brodeur, B. R. (1997). Highly conserved *Neisseria meningitidis* surface protein confers protection against experimental infection. *J. Exp. Med.* **185**, 1173–1183.

Osterhaus, A. D., van Baalen, C. A., Gruters, R. A., Schutten, M., Siebelink, C. H., Hulskotte, E. G., Tijhaar, E. J., Randall, R. E., van Amerongen, G., Fleuchaus, A., Erfle, V. and Sutter, G. (1999). Vaccination with Rev and Tat against AIDS. *Vaccine* **17**, 2713–2714.

Pauza, C. D., Trivedi, P., Wallace, M., Ruckwardt, T. J., Le Buanec, H., Lu, W., Bizzini, B., Burny, A., Zagury, D. and Gallo, R. C. (2000). Vaccination with tat toxoid attenuates disease in simian/HIV-challenged macaques. *Proc. Natl Acad. Sci. USA* **97**, 3515–3519.

Pizza, M., Covacci, A., Bartoloni, A., Perugini, M., Nencioni, L., De Magistris, M. T., Villa, L., Nucci, D., Manetti, R., Bugnoli, M. *et al.* (1989). Mutants of pertussis toxin suitable for vaccine development. *Science* **246**, 497–500.

Pizza, M., Scarlato, V., Masignani, V., Giuliani, M. M., Arico, B., Comanducci, M., Jennings, G. T., Baldi, L., Bartolini, E., Capecchi, B., Galeotti, C. L., Luzzi, E., Manetti, R., Marchetti, E., Mora, M., Nuti, S., Ratti, G., Santini, L., Savino, S., Scarselli, M., Storni, E., Zuo, P., Broeker, M., Hundt, E., Knapp, B., Blair, E., Mason, T., Tettelin, H., Hood, D. W., Jeffries, A. C., Saunders, N. J., Granoff, D. M., Venter, J. C., Moxon, E. R., Grandi, G. and Rappuoli, R. (2000). Identification of vaccine candidates against serogroup B meningococcus by whole-genome sequencing. *Science* **287**, 1816–1820.

Rappuoli, R. (2001). Conjugates and reverse vaccinology to eliminate bacterial meningitis. *Vaccine* **19**, 2319–2322.

Rosenstein, N. E., Fischer, M. and Tappero, J. W. (2001). Meningococcal vaccines. *Infect. Dis. Clin. North. Am.* **15**, 155–169.

Riley, M. (1993). Functions of the gene products of *Escherichia coli*. *Microbiol. Rev.* **57**, 862–952.

Scholten, R. J., Bijlmer, H. A., Poolman, J. T., Kuipers, B., Caugant, D. A., Van Alphen, L., Dankert, J. and Valkenburg, H. A. (1993). Meningococcal disease in The Netherlands, 1958–1990: a steady increase in the incidence since 1982 partially caused by new serotypes and subtypes of *Neisseria meningitidis*. *Clin. Infect. Dis.* **16**, 237–246.

Summary of notifiable diseases, United States, 1997. (1998). *MMWR Morb. Mortal. Wkly. Rep.* **46**(54), 3–87.

Tettelin, H., Saunders, N. J., Heidelberg, J., Jeffries, A. C., Nelson, K. E., Eisen. J. A., Ketchum, K. A., Hood, D. W., Peden, J. F., Dodson, R. J., Nelson, W. C., Gwinn, M. L., DeBoy, R., Peterson, J. D., Hickey, E. K., Haft, D. H., Salzberg, S. L., White, O., Fleischmann, R. D., Dougherty, B. A., Mason, T., Ciecko, A., Parksey, D. S., Blair, E., Cittone, H., Clark, E. B., Cotton, M. D., Utterback, T. R., Khouri, H., Qin, H., Vamathevan, J., Gill, J., Scarlato, V., Masignani, V., Pizza, M., Grandi, G., Sun, L., Smith, H. O., Fraser, C. M., Moxon, E. R., Rappuoli, R. and Venter, J. C. (2000). Complete genome sequence of *Neisseria meningitidis* serogroup B strain MC58. *Science* **287**, 1809–1815.

Tondella, M. L., Popovic, T., Rosenstein, N. E., Lake, D. B., Carlone, G. M., Mayer, L. W. and Perkins, B. A. (2000). Distribution of *Neisseria meningitidis* serogroup B serosubtypes and serotypes circulating in the United States. The Active Bacterial Core Surveillance Team. *J. Clin. Microbiol.* **38**, 3323–3328.

van Deuren, M., Brandtzaeg, P. and van der Meer, J. W. (2000). Update on meningococcal disease with emphasis on pathogenesis and clinical management. *Clin. Microbiol. Rev.* **13**, 144–166.

Zollinger, W. D. (1997). New and improved vaccines against meningococcal disease. In *New Generation Vaccines* (Levine, M. M., Woodrow, G. C., Kaper, J. B. and Cobon G. S., eds), pp. 469–488. New York: Dekker.

15 Microbial Genomics for Antibiotic Target Discovery

Frank Fan and Damien McDevitt
Microbial, Musculoskeletal and Proliferative Diseases CEDD, GlaxoSmithKline, 1250 South Collegeville Road, Collegeville, PA 19426-0989, USA

◆◆◆

CONTENTS

Introduction
Identification of antibiotic targets from bacterial genomes
Selection of antibacterial targets
Concluding remarks

◆◆◆◆◆◆ INTRODUCTION

Antibiotics have proved to be highly effective in treating and controlling bacterial infection. Countless lives have been saved and morbidity due to these infections has been substantially reduced. Antibiotic screening efforts in the 1940s to the late 1960s yielded many structurally diverse and clinically useful agents. Second- and third-generation antibiotics were later developed that had improved the spectrum of activity, pharmacokinetic characteristics and efficacy in patients (Scholar and Pratt, 2000). The success was such that many pharmaceutical companies closed or severely curtailed their antibiotic research and development programs (Shales *et al.*, 1991). However, antibiotic resistance has steadily increased in the last 25 years (Levy, 1998) and it is now well accepted that the current arsenal of antibiotics is not going to be enough in the fight against pathogenic bacteria. Methicillin-resistant *Staphylococcus aureus* (MRSA), penicillin-resistant pneumococcus and vancomycin-resistant *Enterococci* (VRE) are problem pathogens (Pfaller *et al.*, 1998; Jones *et al.*, 1999) and the emergence of multidrug resistance is a real threat to effective therapy (Swartz, 1994; Baquero, 1997). The response has been a renewed commitment by the pharmaceutical industry to discover and develop new classes of antibiotics to treat these resistant organisms. In 1997, $2.6 billion was spent on anti-infectives research in the pharmaceutical industry in the US (PhRMA, 1997).

These efforts are being propelled by the genomics revolution. The first genome sequence published was of a bacterium *Haemophilus influenzae* in 1995 (Fleischmann *et al.*, 1995). Since then, many bacterial genome sequences have been determined by both public and private groups. In some cases, multiple strains of a given bacterial species such as *S. aureus*, *Streptococcus pneumoniae*, *Helicobacter pylori*, *Neisseria meningitidis* and *H. influenzae* have been sequenced. It is anticipated that there will be over 100 completed microbial genome sequences available within the next year or two as sequencing continues at an exponential rate (for more information, see the following web sites: http://www.tigr.org/tdb/mdb/mdb.html, http://www.sc.doe.gov/production/ober/microbial.html, http://www.hgsc.bcm.tmc.edu/microbial, http://www.sanger.ac.uk/Projects/Microbes).

The challenge is how to exploit all this sequence information to help understand microbial physiology and pathogenesis, and how to apply this information to antibiotic drug discovery, vaccine development and microbial diagnostics. Many pharmaceutical companies have revitalized their antibiotic discovery programs by establishing a genomics-based antibacterial drug discovery process. The enzymes and other macromolecules to be inhibited by new antibacterials are referred to as 'targets' for those antibacterials. First, novel targets are selected and validated using genomics-based technologies. These targets are then studied biochemically and configured into appropriate high-throughput screens of specific biochemical reactions, which are used to run against diverse sets of chemical libraries to identify potent inhibitors. These initial 'hits' are analysed based on their potency, physiochemical properties and antibacterial activities to identify suitable leads for further chemical development. These leads are optimized by chemical structure–activity relationship programs to achieve the desired potency against the target, spectrum of antibacterial activity and selectivity. In addition, the pharmacokinetic properties of the compounds and their antibacterial activities in animal models of infection are studied. Finally, the optimized leads with antibacterial activities are tested in animals for safety and efficacy before being tested in clinical trials in man.

The selection and validation of appropriate targets is key to the success of this antibiotic discovery strategy. In recent years, several exciting new genomics-based technologies have been developed to accelerate and enhance this process. This chapter reviews several of these microbial genomics methodologies used in the pharmaceutical industry to identify, validate and exploit novel antibacterial targets.

◆◆◆◆◆◆ IDENTIFICATION OF ANTIBIOTIC TARGETS FROM BACTERIAL GENOMES

Fully sequenced bacterial genomes provide a platform for identifying all of the potential antibacterial targets. Since the majority of currently used antibiotics only target a limited number of macromolecules involved in critical cellular functions, such as DNA, RNA, protein and cell-wall

biosynthesis, there is now an exciting opportunity to discover additional novel antibacterial targets.

Targets essential for growth versus targets required for virulence

Before beginning the target-hunting process, one needs to decide on what type of target one is seeking to identify and develop antagonists against, namely, the gene products that are essential for bacterial viability *in vitro* or gene products that are required for virulence (the ability to cause infection) but are not essential for bacterial growth *in vitro* (e.g. virulence factors). The molecular targets of most antibiotics are gene products essential for bacterial viability. Therefore, the traditional target selected by the pharmaceutical industry for antibiotic drug discovery is essential for bacterial viability *in vitro* and during infection. Inhibition of the function of these essential gene products usually results in cell growth inhibition and eradication of the bacterial infection. However, the use of antibiotics with such bacterial-killing ability is a key factor in the emergence of widespread antibiotic resistance today. New strategies for the development of antibacterial agents targeting genes that are non-essential for *in vitro* growth, but that are required for virulence are gaining increased attention.

Methods for the random identification of genes relevant to the infection process include signature-tagged mutagenesis (STM) (see Chapters 9 and 10), size-marker identification technology (SMIT), *in vivo* expression technology (IVET) and differential fluorescent induction (DFI). STM and SMIT are technologies that identify genes required for the infection process (Holden and Hensel, 1998; Buysse *et al.*, 1999). IVET and DFI are methods that rely on the identification of random DNA cloned fragments harboring promoters whose transcription is induced during *in vivo* infection (Mahan *et al.*, 1993; Valdivia and Falkow, 1997). Several classes of virulence factors are being considered as putative targets for the intervention of the bacterial infection: quorum-sensing systems (Bassler, 1999), two-component signal transduction systems (Barrett and Hoch, 1998; Throup *et al.*, 2000), type III secretion (Cheng and Schneewind, 2000) and sortase (Mazmania *et al.*, 1999, 2000). A comprehensive review of bacterial virulence as a target for antibacterial chemotherapy can be found in a reference by Alksne and Projan (2000). Although theoretically this approach is very attractive, in practice, many obstacles need to be overcome. In particular, the inability to perform minimum inhibitory concentration (MIC) assays in the laboratory on inhibitors of these targets (as they do not kill bacteria) is a major issue affecting the discovery, development, approval process and ultimate commercialization of potential agents that modulate the infection process. Thus, in this review, we will focus on the methods for identifying and validating traditional targets that are essential for bacterial viability.

The availability of complete genomic sequence data provides the possibility of systematically evaluating the essentiality of each and every gene in the genomes. By performing genomewide 'gene by gene' mutagenesis to isolate knockout or conditional lethal mutants, one can examine the growth behavior of these mutants in a variety of conditions. The resulting information helps to classify and profile each gene, and potentially aid the prediction of the function of unknown genes. In addition, the reproducible inability to isolate knockout mutants, or a resulting conditional lethal phenotype for some genes, would suggest that these genes might be essential for *in vitro* growth. In fact, such projects are already in progress for well-studied model organisms such as *Escherichia coli* and *Bacillus subtilus* (http://www.genome.wisc.edu; http://locus.jouy.inra.fr/cgi-bin/genmic/madbase/progs/madbase.operl).

The 'gene by gene' knockout approach is labor intensive, costly and not easily applied to many clinically important pathogens. Therefore, a number of technologies have been developed to aid this assessment. These methods can be crudely divided into a 'dry' approach of using comparative genomics to perform *in silico* analysis on bacterial genomes and a 'wet' approach using genomewide genetic screening technologies.

Comparative genomics

Aided by accumulating knowledge in bacteriology and advanced bioinformatics, many of the open reading frames (ORFs) in a given bacterium are identified and annotated. Comparative genomic analysis of an appropriate collection of bacterial genomes reveals specific sets of genes responsible for basic housekeeping functions, as well as factors required for survival under extreme conditions and virulence factors.

Bacterial genome sizes vary from 0.6 to 6 million basepairs. By comparing the small genome of the Gram-negative *H. influenzae* with that of the Gram-positive *Mycoplasma genitalium*, Mushegian and Koonin (1996) identified a minimal gene set of 256 genes necessary for cellular life. These genes, not surprisingly, encode proteins and RNAs responsible for transcription, translation, replication, secretion and macromolecule metabolism. In theory, every gene in this minimal set could potentially encode an antibacterial target and could be subjected to the target validation process as discussed below.

When novelty of the targets was most sought after, a similar comparative bioinformatic study between *E. coli* and *M. genitalium* revealed 26 conserved ORFs with unknown function (Arigoni *et al.*, 1998), six of which were demonstrated experimentally to be essential for *E. coli* growth.

Comparative genomic approaches were also successfully used to identify a set of genes only present in the gastric pathogen *Helicobacter pylori*. In this case, as *H. pylori* produces an organ-specific infection and is not associated with coinfection with other pathogens, a narrow-spectrum anti-*H. pylori* agent would be ideal for the treatment of *H. pylori* infection without affecting the normal endogenous flora. A set of these unique genes was identified in two *H. pylori* strains 26695 and J99 for which genome sequences were available. These were subsequently prioritized and tested for essentiality on a 'gene by gene' knockout basis (Chalker *et al.*, 2001).

Genetic screening

Technologies for random genetic screening for essential genes have been developed in many bacteria. The genome sequence information allows quick identification of the essential genes found in these screens. Conceptually, there are two ways to identify essential genes in a bacterial genome. A 'negative' approach refers to the methods that identify all the non-essential genes and presumes that everything else is essential. In contrast, a 'positive' approach refers to the methods, which identify genes that are essential by generating a conditional lethal phenotype.

The negative screening methods include global transposon-insertion mutagenesis or global plasmid-insertion mutagenesis. Saturation transposon mutagenesis, followed by sequencing the insertions and comparison to the genome sequences, enabled the deduction of an essential gene set in *M. genitalium* (Hutchison *et al.*, 1999). This approach is resource intensive, especially in the case of a large-sized genome (e.g. *Pseudomonas aeruginosa*) for which saturation mutagenesis is difficult to achieve. An example of global plasmid insertion mutagenesis was presented by Altieri *et al.* (1999). A library of short DNA fragments (e.g. 500 bp) from *S. pneumoniae* was constructed in an *E. coli* plasmid vector. Individual plasmids with inserts were used to transform *S. pneumoniae*. Sequence analysis of DNA inserts cloned in plasmids unable to transform identified putative essential genes.

In contrast to the negative screens in which no mutants are recovered if the gene is essential, the goal of positive screens is to generate conditional lethal phenotypes by generating temperature-sensitive (TS) mutants or using controlled gene expression. This approach requires less resource yet provides more information on target essentiality. TS mutants are generated by chemical mutagenesis followed by screen of mutants that can grow at permissive temperatures but not at restrictive temperatures. The location of the mutation is mapped by complementation to identify the 'essential' gene (Kaback *et al.*, 1984). However, this approach is not always successful as it can prove impossible to isolate TS mutants in some essential genes and mapping strategies may be difficult or even impossible in many bacteria.

A more efficient and robust way of creating a conditional lethal phenotype is to use tightly regulated gene expression systems (see below). In combination with other technologies, it can be applied on a genomic scale. For example, transposons can be genetically engineered to incorporate an outward-facing inducible promoter at one end and then used for global transposon mutagenesis (Judson and Mekalanos, 2000). Mutants which show inducer-dependent conditional lethal phenotypes usually have transposons inserted upstream of an essential gene or operon. Combining regulated expression systems with antisense technology provides another elegant positive screen to identify essential genes (Ji *et al.*, 2001). In this case, random fragments of the *S. aureus* genome (e.g. 500 bp) were cloned into a plasmid vector containing a regulated expression system. The library was transformed into *S. aureus* and growth of the colonies was compared in the absence or presence of inducer. Clones expressing antisense RNA to essential genes resulted in lethal or growth-defective colonies following antisense induction. This strategy was validated by the identification of many known essential genes but importantly also resulted in the random identification of many novel essential genes of unknown function.

◆◆◆◆◆◆ SELECTION OF ANTIBACTERIAL TARGETS

The several hundred essential genes identified by comparative genomics and the various genetic screening methods outlined above will constitute too large a panel to advance through costly drug discovery efforts and thus they need to be prioritized further. There are several key criteria for consideration: (1) the target should be present in the required *spectrum* of organisms that you plan to make your antibiotic against; (2) the target should be absent in humans or, if present, significantly different to give confidence that *selective* inhibitors can be identified; (3) the target should be *essential* for bacterial growth; (4) the target should be *expressed in vivo* and relevant to the infection process; and (5) the target should be *amenable to assay development* for high-throughput screening to identify inhibitors. The strategies and technologies for determining target spectrum, selectivity, essentiality, expression during infection and assay developability are outlined below.

Spectrum and selectivity

The availability of extensive microbial and eukaryotic genome sequence information enables a rapid *in silico* assessment of a target's potential spectrum. Bioinformatic methods such as basic local alignment search tool (BLAST) allow a rapid identification of all homologous sequences in various organisms (see Chapters 1 and 2). In some special cases, more sophisticated homology searching methods, such as position-specific iterated (PSI)-BLAST can be used (Altschul *et al.*, 1997; Koretke *et al.*, 2000). A phylogenetic analysis of the target provides a more detailed evaluation of the evolutionary relationship among bacterial species and addresses such questions as: 'How related is the target in bacteria?' and 'Is the target only present in Gram-positive or Gram-negative bacteria, and if so, how similar is the target in these organisms?'. This *in silico* analysis is critical at the early stage of target selection for drug discovery because it predicts the widest potential spectrum for a proposed antibacterial agent against the target under study. For the development of broad-spectrum antibiotics, the target should be present in all key pathogens of interest such as *S. aureus, S. pneumoniae, H. influenzae*, etc. In contrast, unique targets with narrow distributions can be exploited for the development of drugs against a specific bacterium such as *Mycobacterium tuberculosis, H. pylori* or *Chlamydia*.

An example that may demonstrate the power of microbial genomics and bioinformatic analysis in defining target spectrum can be found in the case of enoyl-acyl carrier protein (ACP) reductase or FabI. Bacterial fatty-acid biosynthesis is essential for bacterial viability and is significantly different from human fatty-acid biosynthesis. Therefore, many enzymes in this pathway have been proposed as attractive antibacterial drug targets (Payne *et al.*, 2001). Homology searches identified a FabI ortholog in a variety of clinically important bacteria including *S. aureus*, *H. influenzae*, *E. faecalis* and *E. coli*, but it was absent in *S. pneumoniae*. It was subsequently determined that *S. pneumoniae* has an alternative enoyl-ACP reductase called FabK, which does not have significant homology to FabI (Heath and Rock, 2000). Furthermore, FabK is present in several other organisms, and some species such as *E. faecalis* and *P. aeruginosa* have both FabI and FabK. Interestingly, *B. subtilis* has a FabI and a second enoyl-reductase called FabL. This *in silico* analysis is consistent with the fact that FabI inhibitors, such as triclosan, are potent antibacterials against FabI-containing species, such as *S. aureus* and *H. influenzae*, but less effective against FabK-containing bacteria, such as *S. pneumoniae* and *E. faecalis*. It also highlights the difficulty in developing broad-spectrum antibiotics against FabI/FabK as they are so dissimilar.

An *in silico* bioinformatic analysis can also help to assess target selectivity by searching for homologs in the recently available human genome sequence databases (International Human Genome Sequencing Consortium, 2001; Venter *et al.*, 2001). An ideal antibacterial target such as one involved in bacterial protein secretion or cell-wall biosynthesis, will not have a human homolog. However, since all living organisms share some common mechanisms for essential cellular functions such as DNA, RNA and protein synthesis, it is not surprising to find that many bacterial proteins have homologs in humans. This does not necessarily rule out discovery efforts on these targets, since many of the known antibiotics inhibit targets that have human homologs. For these targets, it is important to develop a selectivity screen using the human or other mammalian homologs to direct chemical structure–activity relationship programs to identify bacterial-selective inhibitors.

Gene essentiality

Gene essentiality testing is a critical step in the target validation process because the results predict whether complete inhibition of the target would result in cell death. Whenever possible, putative essential targets identified from comparative bioinformatics or by genetic screening should be tested and confirmed for essentiality in all key pathogens of interest. This is important since there are many examples where target genes are essential in some bacteria but not in others (Throup *et al.*, 2000). Early identification of issues, such as essentiality restricted to some organisms, is critical and helps to avoid costly discovery research efforts.

Many technologies have been developed to test gene essentiality. Most of them were initially developed in well-studied bacteria, such as *E. coli* and *B. subtilis*, and have been subsequently adapted for use in pathogens, such as *S. aureus* and *S. pneumoniae* directly. Again, there are two main approaches. For the negative approaches, such as plasmid-insertion mutagenesis, allelic-replacement mutagenesis and transposon-insertion mutagenesis, no mutants can be isolated in the case of essential genes. In contrast, the isolation of conditional lethal mutants using a regulated gene expression system is an effective positive approach.

Plasmid-insertion mutagenesis

This is the simplest approach. The objective is to disrupt the target gene by inserting a plasmid in the middle of the gene via homologous gene recombination (Figure 15.1). Non-replicating ('suicide') plasmid vectors

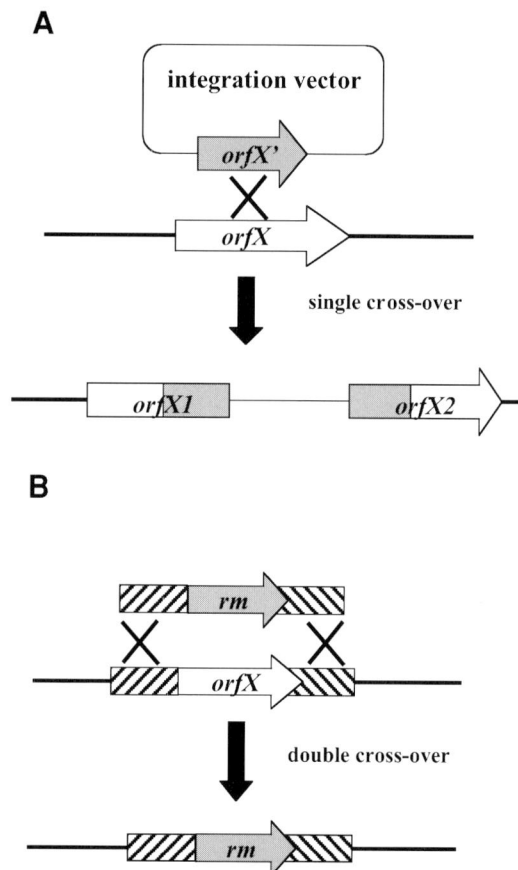

Figure 15.1. Gene essentiality testing using plasmid-insertion mutagenesis and allelic-replacement mutagenesis. (A) Plasmid insertion mutagenesis. An internal fragment (*orfX'*) of the gene of interest (*orfX*) is amplified by PCR and cloned into an integration vector. The resulting plasmid is transformed into the bacterium of interest. *orfX* is disrupted after a single recombination event. (B) Allelic-replacement mutagenesis. Flanking sequences (▨, ▧) of the gene of interest (*orfX*), and an antibiotic resistance marker (*rm*) are PCR amplified separately. These three PCR products are then combined in a second PCR reaction. The resulting mutagenic cassette is transformed into the bacterium of interest to replace *orfX* in a double recombination event.

or vectors with *ts* replicons are most often used. These vectors will contain a DNA internal fragment (*ca* 500 bp) of the target gene and a selective marker for the target bacterium. The resulting plasmid is introduced into the target bacterium. When using a non-replicating plasmid, integration into the host genome occurs and the integrants are selected. When using a vector with a *ts* replicon, it requires growing the transformed strain at a restrictive temperature. If a mutant with an insertion in the gene of interest is isolated, then this gene is non-essential. Alternatively, the inability to isolate mutants for a given gene suggests that it is essential.

Although this approach is simple and rapid, there are several limitations and caveats: (1) for some bacteria (e.g. *S. aureus*), where the efficiency of recombination is low, it would be difficult to apply this method to genes small in size (*ca* 500 bp); (2) the truncated version of the target gene (Figure 15.1A) may still retain partial activities sufficient to sustain cell growth, thus yielding false-negative results on essentiality; (3) the plasmid insertion might disrupt an operon structure and have polarity effects on genes downstream. Therefore, the inability to isolate insertion mutants could be due to the regulation of a distal gene being perturbed, rather than the disruption of the target gene (i.e. false positive); (4) the insertion mutant may not be stable in the absence of selective pressure, so it is inappropriate for studies of these mutants in animal models.

Allelic-replacement mutagenesis

In this case, upstream and downstream sequences for the gene of interest are amplified and combined with a gene that encodes antibiotic resistance designed not to produce polarity effects (Figure 15.1B). Such DNA cassettes can be readily generated using a PCR-based method and introduced into naturally competent bacteria such as *H. pylori*, *H. influenzae* and *S. pneumoniae* (Chalker *et al.*, 2001). A double recombination event between this cassette and the chromosomal DNA results in the replacement of the gene of interest with the selective marker. For some organisms such as *S. aureus*, the DNA cassettes must be cloned into appropriate vectors for mutagenesis (Xia *et al.*, 1999). This strategy has eliminated most of the caveats described in the plasmid inactivation method. Furthermore, allelic-replacement mutants with mutations in *in vitro* non-essential genes, can be used in animal models of infection for virulence testing.

Transposon-insertion mutagenesis

Two methods have been reported using transposon-insertion mutagenesis to glean information on gene essentiality: GAMBIT (genomic analysis and mapping by *in vitro* transposition) and genetic footprinting. For GAMBIT, saturation transposon mutagenesis is performed on PCR-amplified target gene fragments using transposons with small target sequence (e.g. *mariner*). The mutagenized pool of PCR fragments is transformed back into the host, and transposon insertions are mapped. If the gene is essential, no transposon insertions will be recovered in the gene.

This method can only be applied to naturally competent species such as *H. influenzae* and *S. pneumoniae* (Akerley *et al.*, 1998; Reich *et al.*, 1999).

Genetic footprinting was first developed in yeast and later applied to *E. coli* (Smith *et al.*, 1996; Hare *et al.*, 2001). It involves transposon mutagenesis, outgrowth of the mutagenized cell population and analysis of the fate of cells carrying mutations in specific genes. First, random transposon insertional mutagenesis is performed on the whole genome. Then transposition is halted and the mutagenized population is grown under specific conditions. Samples at different time points (e.g. 0, 30 min and 45 min) are taken and the gene of interest is PCR amplified using a set of gene-specific primers. If the gene is essential for growth, the PCR products corresponding to insertions within that gene would be lost with time.

Regulated gene expression

A more convincing way to demonstrate *in vitro* gene essentiality is to isolate mutants with conditional lethal phenotypes using regulated gene expression. Valuable information is gained by modulating gene expression levels and monitoring the effect as the product titrates up or down. Compared with traditional gene inactivation/replacement strategies, which typically result in the total loss of gene function, regulated gene expression can provide quantitative information about the functional importance of a gene product. Such information is valuable for studying gene function and evaluating molecular targets for antibiotic discovery.

Suitable regulated gene expression systems were developed initially in well-studied bacteria, such as *E. coli* and *B. subtilis*, but more recently in pathogens such as *S. pneumoniae* and *S. aureus* (Stieger *et al.*, 1999; Zhang *et al.*, 2000). Because of the authors' expertise on essentiality testing using regulated gene expression systems in *S. aureus*, these methods are described in more detail below.

Two related methods have been developed to generate regulatable target strains in *S. aureus*. In the homologous recombination method (Jana *et al.*, 2000; Zhang *et al.*, 2000), the native promoter for the gene of interest is replaced with a regulated promoter by homologous recombination either via a single cross-over integration of a plasmid construct (Figure 15.2A) or via a double cross-over allelic-replacement process. A fragment (≥ 500 bp) including the 5'-end of the *orfX* together with its own ribosome binding site is amplified by PCR from *S. aureus* strain RN4220. An integration vector incorporating this DNA fragment downstream of a regulated promoter Pr (e.g. Pspac promoter) is constructed. The resulting plasmid, which is obtained in *E. coli* and unable to replicate in *S. aureus*, is used to transform RN4220 and putative mutants containing a regulatable *orfX* are identified by screening erythromycin-resistant colonies using diagnostic PCR analysis, and subsequently confirmed by Southern hybridization analysis.

Although relatively simple, there are two significant limitations to this strategy. First, in cases where the gene of interest is present in an operon, the regulated promoter could potentially co-regulate downstream genes

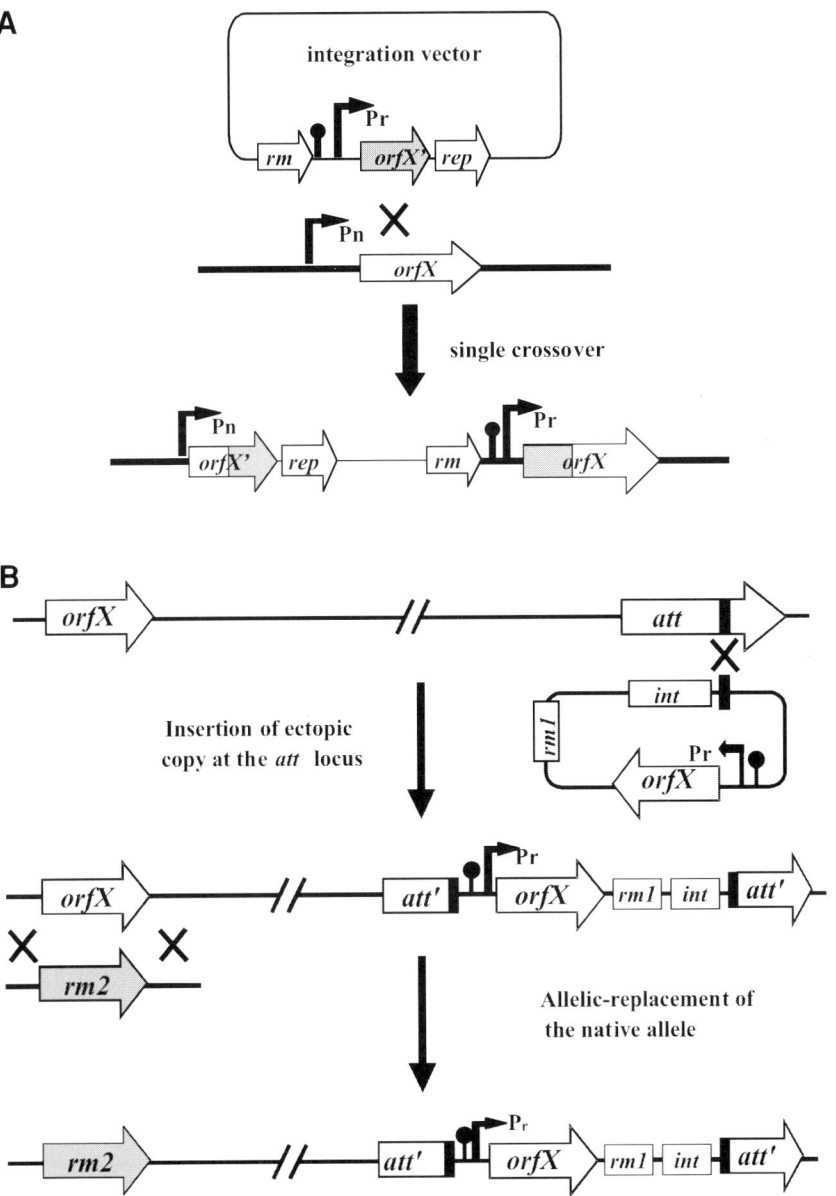

Figure 15.2. Gene essentiality testing using regulated gene expression. (A) Single integration method. A fragment (*orfX'*) containing 5'-end of the gene of interest *orfX* together with a ribosome binding site is cloned downstream of a regulated promoter (Pr) in an integration plasmid. The resulting plasmid is transformed into the bacterium of interest. The native promoter (Pn) for *orfX* is replaced with Pr in a single recombination event. (B) Ectopic expression method. *orfX* is cloned downstream of a regulated promoter (Pr) in an integration vector. Upon transformation into the bacterium of interest, the resulting plasmid specifically inserts into a nonessential locus (*att*) on the chromosome. The strain is then subjected to an allelic-replacement mutagenesis protocol to delete the native allele. Other features shown in the figure are as follows: *rm*, antibiotic resistance marker for the bacterium of interest; *rep*, repressor gene for the Pr; *int*, integrase gene; *att'*, partial *att* gene; ↑, transcriptional terminator.

and complicate the interpretation of essentiality. This is of particular concern for complex operons with multiple essential genes and secondary transcription control mechanisms. Second, this method is difficult to apply to small genes owing to the lack of sufficient homology to drive efficient recombination of targeting constructs at the locus. Fortunately, a second method (ectopic expression method) has been developed in which a gene of interest is placed under the control of a regulated promoter (Pr) and integrated into a *att* locus ectopically, followed by the deletion of the native allele using allelic-replacement mutagenesis (Fan et al., 2001). This method is illustrated in Figure 15.2B. In *S. aureus*, a gene of interest (*orfX*) is fused to a regulated promoter (Pspac or Pxyl/tet) and placed within the integration vector. This vector specifically inserts into the lipase structure gene (*geh*) locus via the *attB*/*attP* φL54a integrase system to produce an *orfX* merodiploid. The strain is then subjected to an allelic-replacement mutagenesis protocol specific for the native allele. Replacement of the native *orfX* requires a two-step procedure that for simplicity is shown here as a double cross-over insertion. The first step involves the generation of a co-integrant between the native locus and the targeting plasmid that contains the *tetA(M)* gene flanked by two blocks (500 bp) of homology to *orfX*. This step is carried out in a wild-type (non-ectopic) strain. Integration could occur at either of the homologous blocks to yield a co-integrant. Phage transduction is used to transfer the co-integrant structure to the *orfX* merodiploid strain in order to increase the frequency of intramolecular recombination and thus resolution to an allelic exchange mutant devoid of targeting plasmid. The final part of Figure 15.2B represents the genomic structure of the strain with the native allele replaced with *tetA(M)* and the ectopic allele at the *geh* locus under control of the regulated promoter. This strategy circumvents the limitations of the homologous recombination method and should be applicable to any *S. aureus* gene regardless of size or genomic structure.

For both methods, the initial mutant strain with a regulatable target gene contains a single copy of the repressor gene (e.g. *lacI* for Pspac promoter) on the chromosome. In order to provide sufficient amounts of LacI for repression, the mutant strain is also transformed with a multicopy LacI-expression plasmid pFF40 (Zhang et al., 2000). Cell growth is examined in liquid or on solid media containing varying amounts of inducer (isopropyl-β-D-thiogalactopyranoside; IPTG).

Three types of growth profile have been observed (Figure 15.3). For type 1, exemplified by *mrs*, which encodes methionyl tRNA synthetase, cell growth showed dependence on inducer even in the absence of additional repressors provided from pFF40. For type 2, exemplified by *pdf*, which encodes polypeptide deformylase, cell growth showed dependence on inducer only in the presence of pFF40. For type 3, exemplified by *polA*, which encodes DNA polymerase, no cell growth dependence on inducer was observed. The growth dependence profile for both types 1 and 2 suggests that the genes of interest are essential. Furthermore, genes of type 1 appear to be more sensitive towards down-regulation and, therefore, potentially more attractive antibacterial drug targets.

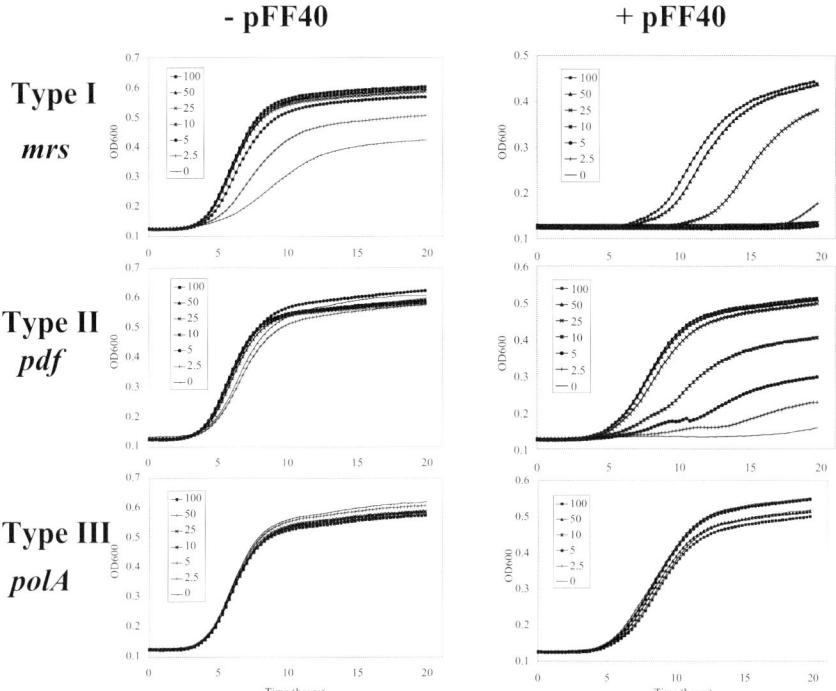

Figure 15.3. Growth dependence on isopropyl-β-D-thiogalactopyranoside (IPTG) for *S. aureus* Pspac-regulated *mrs*, *pdf* and *polA* strains with and without plasmid pFF40. *S. aureus* strains were grown at 37°C overnight in tryptic soy broth with the appropriate antibiotics and inducer IPTG. The cultures were washed and diluted to ~5 × 10^4 cells/ml as starting cultures. A total of 198 µl of the starting culture was placed in a well of a flat-bottom microtiter plate and mixed with 2 µl of IPTG stock solution to make a final IPTG concentration of 0, 2.5, 5, 10, 25, 50 and 100 µM (as shown in the insert of each graph). The cell growth at 37°C was monitored by measuring OD$_{600}$ every 15 min with 1 min mixing before each reading, using a microtiter plate reader SpectraMax250 from Molecular Devices (Sunnyvale, CA, USA).

We have applied these methods to over 40 genes and some results are summarized in Table 15.1. These data correlated very well with the results from independent essentiality testing done by allelic-replacement mutagenesis wherever available (data not shown). As expected, many genes involved in processes, such as DNA replication, translation, cell-wall biosynthesis, fatty acid biosynthesis, two-component signal transduction and secretion, were found to be essential. Virulence factors such as α-toxin and sortase were non-essential. More importantly, the essentiality of some genes encoding proteins with unknown function (e.g. *yneS*, *yeaZ*) was also confirmed. They could, therefore, represent novel targets for antibiotic discovery.

Expression during infection

As discussed earlier, various technologies such as STM and IVET have been developed to identify genes that are expressed and relevant for

Table 15.1. *S. aureus* gene essentiality testing using regulated gene expression

Type 1, essential, sensitive
accD	Acetyl CoA carboxylase subunit D
dnaB	Helicase
dnaE	DNA polymerase
fabD	Malonyl CoA–ACP transacylase
fabF	β-Ketoacyl-acyl carrier protein synthase
fabG	3-Oxoacyl-ACP reductase
ffh	Signal recognition particle
frs	Phenylalanyl tRNA synthetase
gyrA	Gyrase A subunit
pcrA	Helicase
pyrH	UMP kinase
mrs	Methionyl tRNA synthetase
murC	UDP-N-acetylmuranmate:L-alanine ligase
yidC	Membrane insertase
yneS	Unknown
yrs	Tyrosyl tRNA synthetase
yycF	TCSTS* response regulator

Type 2, essential, less sensitive
fabH	β-Ketoacyl-ACP synthetase
pdf	Polypeptide deformylase
rat	tRNA-dependent aminotransferase
topoI	Topoisomerase I
yeaZ	Unknown
yycG	TCSTS histidine kinase

Type 3, non-essential
accCl	Acetyl CoA carboxylase subunit Cl
c509	TCSTS
geh	Lipase
hla	α-Toxin
murA1	UDP-N-acetylglucosamine l-carboxyvinyl-transferase
polA	DNA polymerase
srtA	Sortase
srtA2	Sortase homolog
yeiH	Unknown

* TCSTS, two-component signal transduction system.

bacterial virulence. However, when assessing the expression of given target genes during infection, methods such as transcript analysis are more appropriate. After establishing an infection in a suitable animal model, bacteria are recovered and total RNA is isolated. The RNA samples are then reverse transcribed to make cDNA. The transcription of the target genes can be detected by PCR using gene-specific primer pairs, or more quantitatively by TaqMan analysis (Heid *et al.*, 1996). Coupled with genome sequence information, it is now possible to conduct genomewide global analysis of transcripts during infection. This method can also highlight genes that are selectively induced during infection by comparing transcription at different stages of infection (*in vivo*) versus the transcription occurring during growth in laboratory medium (*in vitro*).

Genes that are essential for bacterial viability in rich laboratory medium are most likely to be also expressed during the infection process. However, there could be exceptions. Even in these cases, it is still possible that these genes are transcribed but their levels are too low for detection. Therefore, the targets should not be considered invalid simply based on the negative results from the *in vivo* expression analysis. Rather, this analysis serves as a prioritization tool to highlight the target genes that are highly expressed during infection.

Amenability to HTS assay development

Once the targets are selected and validated, assays will need to be developed for high throughput screening (HTS) to identify inhibitors of the target. It is prudent to consider the possibility of HTS assay development even at the stage of target selection. Higher priority should be given to those targets with assays readily available. The success of HTS, to a large extent, depends on the quality of the assay. Tremendous advances have been made in the development of HTS assay technologies, especially towards using the fluorescence-based or scintillation proximity assay (SPA)-based formats. The HTS assays can be designed to detect inhibition of enzymatic reactions that the target catalyzes or interactions between the target and other macromolecules. For those targets with unknown function, although they have the merit of being completely novel, it is usually quite difficult to determine their function and develop appropriate assays. This needs to be taken into consideration when prioritizing these targets.

◆◆◆◆◆◆ CONCLUDING REMARKS

The availability of extensive microbial genome sequence information coupled with the development of novel genomics-based technologies has enabled the pharmaceutical industry to revitalize their programs on antibiotic drug discovery. This is very timely given the continued increase in antibiotic resistance worldwide. One of the main strategies that the industry is taking is exploiting entirely novel antibacterial targets. This is being driven by the development of new technologies to rapidly identify and validate gene products as potential targets. Many new validated targets have now been identified and several new classes of antibiotic leads have recently been identified (McDevitt and Rosenberg, 2001). The current challenge is to optimize these and develop them to become part of the next generation of antibiotics. Those companies that succeed will derive significant competitive advantage in creating product portfolios that will address the unmet clinical need.

Acknowledgements

We thank Jing Fan for her contribution in generating the target-regulatable strains in *S. aureus*. We also thank Drs Jianzhong Huang, Paul O'Toole and Jacqueline Cooney for critical reading of this manuscript.

References

Akerley, B. J., Rubin, R. J., Camilli, A., Lampe, D. J., Robertson, H. M. and Mekalanos, J. J. (1998). Systematic identification of essential genes by *in vitro* mariner mutagenesis. *Proc. Natl Acad. Sci. USA* **95**, 8927–8932.

Alksne, L. and Projan, S. J. (2000). Bacterial virulence as a target for antimicrobial chemotherapy. *Curr. Opin. Biotechnol.* **11**, 625–636.

Altieri, M., Camera, M. G., Frandsen, N. M., Falciani, F., Rimini, R., Faggioni, F., Domenici, E. and Mottl, H. (1999). Identification of essential genes in *Streptococcus pneumoniae*. In *99th General Meeting of the American Society for Microbiology*, Abstract H-10.

Altschul, S. F., Madden, T. L., Schaffer, A. A., Zhang, J., Zhang, Z., Miller, W. and Lipman, D. J. (1997). Gapped BLAST and PSI-BLAST: a new generation of protein database search programs. *Nucleic Acids Res.* **25**, 3389–3402.

Arigoni, F., Talabot, F., Peitsch, M., Edgerton, M. D., Meldrum, E., Allet, E., Fish, R., Jamotte, T., Curchod, M. L. and Loferer, H. (1998). A genome-based approach for the identification of essential bacterial genes. *Nature Biotechnol.* **16**, 851–856.

Baquero, F. (1997). Gram positive resistance: challenge for the development of new antibiotics. *J. Antimicrob. Chemother.* **39**, 1–6.

Barret, J. F. and Hoch, J. A. (1998). Two-component signal transduction as a target for microbial anti-infective therapy. *Antimicrob. Agents Chemother.* **42**, 1529–1536.

Bassler, B. L. (1999). How bacteria talk to each other: regulation of gene expression by quorum sensing. *Curr. Opin. Microbiol.* **2**, 582–587.

Buysse, J., Benton, B., Zhang, J. P., Bond, S., Winterberg, K., Burke, C. and Christian, T. (1999). Size-marker identification technology (SMIT) defines *Staphylococcus aureus in vivo* survival genes. In *99th General Meeting of the American Society for Microbiology*, Abstract D/B-214.

Chalker, A. F., Heather, M. W., Hughes, N. J., Koretke, K. K., Lonetto, M. A., Brinkman, K. K., Warren, P. V., Lupas, A., Stanhope, M. J., Brown, J. R. and Hoffman, P. S. (2001). Systematic identification of selective essential genes in *Helicobacter pylori* by genomic prioritization and allelic-replacement mutagenesis. *J. Bacteriol.* **183**, 1259–1268.

Cheng, L. W. and Schneewind, O. (2000). Type III machines of Gram-negative bacteria: delivering the good. *Trends. Microbiol.* **8**, 214–220.

Fan, F., Lunsford, R. D., Sylvester, D., Fan, J., Celesnik, H., Iordanescu, S., Rosenberg, M. and McDevitt, D. (2001). Regulated ectopic expression and allelic-replacement mutagenesis as a method for gene essentiality testing in *Staphylococcus aureus*. *Plasmid* **46**, 71–75.

Fleischmann, R. D., Adams, M. D., White, O., Clayton, R. A., Kirkness, E. F., Kerlavage, A. R., Bult, C. J., Tomb, J. F., Dougherty, B. A., Merrick, J. M. *et al.* (1995). Whole genome random sequencing and assembly of *Haemophilus influenzae* Rd. *Science* **269**, 496–512.

Hare, R. S., Walker, S. S., Dorman, T. E., Greene, J. R., Guzman, L. M., Kenney, T. J., Sulavik, M. C., Baradaran, K., Houseweart, C., Yu, H., Foldes, Z., Motzer, A.,

Walbridge, M., Shimer, G. H., Jr. and Shaw, K. J. (2001). Genetic footprinting in bacteria. *J. Bacteriol.* **183**, 1694–1706.

Heath, R. J. and Rock, C. O. (2000). A triclosan-resistant bacterial enzyme. *Nature* **406**, 145–146.

Heid, C. A., Stevens, J., Livak, K. J. and Williams, P. M. (1996). Real time quantitative PCR. *Genome Res.* **6**, 986–994.

Holden, D. W. and Hensel, M. (1998). Signature tagged mutagenesis. *Methods Microbiol.* **27**, 359–370.

Hutchison, C. A., Peterson, S. N., Gill, S.R., Cline, R. T., White, O., Fraser, C. M., Smith, H. O. and Venter, J. C. (1999). Global transposon mutagenesis and a minimal mycoplasma genome. *Science* **286**, 2165–2169.

International Human Genome Sequencing Consortium (2000). Initial sequencing and analysis of the human genome. *Nature* **409**, 860–921.

Jana, M., Luong, T., Komatsuzawa, H., Shigeta, M. and Lee, C.Y. (2000). A method for demonstrating gene essentiality in *Staphylococcus aureus*. *Plasmid* **44**, 100–104.

Ji, Y., Zhang, B., Van Horn, S. F., Warren, P., Woodnutt, G., Burnham, M. K. and Rosenberg, M. (2001). Identification of critical Staphylococcal genes using conditional phenotypes generated by antisense RNA. *Science* **293**, 2266–2269.

Jones, R. N., Low, D. E. and Pfaller, M. A. (1999). Epidemiological trends in nosocomial and community-acquired infections due to antibiotic resistant Gram-positive bacteria. *Diagn. Mirobiol. Infect. Dis.* **33**, 101–112.

Judson, N. and Mekalanos, J. J. (2000). TnAraOut, a transposon-based approach to identify and characterize essential bacterial genes. *Nature Biotechnol.* **18**, 740–745.

Kaback, D. B., Oeller, P. W., Yde Steensma, H., Hirschman, J., Ruezinsky, D., Coleman, K. G. and Pringle, J. R. (1984). Temperature-sensitive lethal mutations on yeast chromosome I appear to define only a small number of genes. *Genetics* **108**, 67–90.

Koretke, K. K., Lupas, A. N., Warren, P. V., Rosenberg, M. and Brown, J. R. (2000). Evolution of two-component signal transduction. *Molec. Biol. Evol.* **17**, 1956–1970.

Levy, S. (1998). The challenge of antimicrobial resistance. *Scientific America* **XX**, 46–53.

Mahan, M. J., Slauch, J. M. and Mekalanos, J. J. (1993). Selection of bacterial virulence genes that are specifically induced in host tissues. *Science* **259**, 686–688.

Mazmania, S. K., Liu, G., Ton-That, H. and Schneewind, O. (1999). *Staphylococcus aureus* sortase, an enzyme that anchors surface proteins to the cell wall. *Science* **285**, 760–763.

Mazmania, S. K., Liu, G., Jensen, E. R., Lenoy, E. and Schneewind, O. (2000). *Staphylococcus aureus* mutants defective in the display of surface proteins and in the pathogenesis of animal infections. *Proc. Natl Acad. Sci. USA* **97**, 5510–5515.

McDevitt, D. and Rosenberg, M. (2001). Exploiting genomics to discover new antibiotics. *Trends Microbiol.* **9**, 611–617.

Mushegian, A. R. and Koonin, E. V. (1996). A minimal gene set for cellular life derived by comparison of complete bacterial genomes. *Proc. Natl Acad. Sci. USA* **93**, 10268–10273.

Payne, D. J., Warren, P. V, Holmes, D. J., Ji, Y. and Lonsdale, J. T. (2001). Bacterial fatty acid biosynthesis: a genomics driven target for antibacterial drug discovery. *Drug Discovery Today* **6**, 537–544.

Pfaller, M. A. Jones, R. N., Doern, G. V. and Kugler, K. (1998). Bacterial pathogens isolated from patients with blood stream infection: frequency of occurrence and antimicrobial susceptibility patterns from the SENTRY antimicrobial surveillance program (United States and Canada, 1997). *Antimicrob. Agents Chemother.* **42**, 1762–1770.

PhARMA (1997). Pharmaceutical industry spending on anti-infectives. *FDC reports – The ink sheet* **59**, 13–14.

Reich, K. A., Chovan, L. and Hessler, P. (1999). Genome scanning in *Haemophilus influenzae* for identification of essential genes. *J. Bacteriol.* **181**, 4961–4968.

Scholar, E. M. and Pratt W. B. (2000). *The Antimicrobial Drugs*, 2nd edn. New York: Oxford University Press.

Shales, D., Levy, S. and Archer, G. (1991). Antimicrobial resistance – new directions. *ASM News* **57**, 455–458.

Smith, V., Chou, K. N., Lashkari, D., Botstein, D. and Brown, P. O. (1996). Functional analysis of the genes of yeast chromosome V by genetic footprinting. *Science* **274**, 2069–2074.

Stieger, M., Wohlgensinger, B., Kamber, M., Lutz, R. and Keck, W. (1999). Integrational plasmids for the tetracycline-regulated expression of genes in *Streptococcus pneumoniae*. *Gene* **226**, 243–251.

Swartz, M. N. (1994). Hospital-acquired infections: diseases with increasingly limited therapies. *Proc. Natl Acad. Sci. USA* **91**, 2420–2427.

Throup, J. P., Koretke, K. K., Bryant, A. P., Ingraham, K. A., Chalker, A. F., Ge, Y., Marra, A., Wallis, N. G., Brown, J. R., Holmes, D. J., Rosenberg, M. and Burnham, M. K. (2000). A genomic analysis of two-component signal transduction in *Streptococcus pneumoniae*. *Mol. Microbiol.* **35**, 566–576.

Valdivia, R. H. and Falkow, S. (1997). Fluorescence-based isolation of bacterial genes expressed within host cells. *Science* **277**, 2007–2011.

Venter, J. C., Adams, M. D., Myers, E. W., Li, P. W., Mural, R. J., Sutton, G. G. *et al.* (2001). The sequence of the human genome. *Science* **291**, 1304–1351.

Xia, M., Lunsford, R. D., McDevitt, D. and Iordanescu, S. (1999). Rapid method for the identification of essential genes in *Staphylococcus aureus*. *Plasmid* **42**, 144–149.

Zhang, L., Fan, F., Palmer, L. M., Lonetto, M. A., Petit, C., Voelker, L. L., St John, A., Bankosky, B., Rosenberg, M. and McDevitt, D. (2000). Regulated gene expression in *Staphylococcus aureus* for identifying conditional lethal phenotypes and antibiotic mode of action. *Gene* **255**, 297–305.

Part VI
Case Studies – Bacteria

16 *Helicobacter pylori* Functional Genomics

Karen J Guillemin[1] and Nina R Salama[2]

[1]*Institute of Molecular Biology, University of Oregon, Eugene, OR 97403, USA*
[2]*Human Biology Division, Fred Hutchinson Cancer Research Center, 1100 Fairview Avenue North, PO Box 19024, Seattle, WA 98109, USA*

◆◆◆

CONTENTS

Introduction
The genome
The transcriptome
The proteome
Genetic screens
Conclusions

◆◆◆◆◆◆ INTRODUCTION

Helicobacter pylori is a recently described human pathogen and resident of the human stomach, a niche previously believed to be too inhospitable to sustain microbial life (Montecucco and Rappuoli, 2001). *H. pylori* is specialized for life in the gastric mucosa and is fastidious when grown in culture. Strains are highly variable, with no two independently isolated stains exhibiting the same DNA footprints, making cross-strain phenotypic comparisons challenging. The bacteria cause a wide spectrum of diseases. The majority of infected individuals develop asymptomatic gastritis, while approximately 10% will present with more severe diseases including gastric ulcer, duodenal ulcer, gastric adenocarcinoma or mucosa-associated lymphoma (MALT) (Covacci *et al.*, 1999). Unfortunately, none of the animal models recapitulate the complexity of the disease outcome in humans. For these reasons it would seem that this bacterium would be intractable to biomedical research, and yet the references to *H. pylori* in Medline have increased exponentially since its discovery in 1982 (Marshall and Warren, 1984). A notable component of this reference list was the determination of the genome sequence in 1997 (Tomb *et al.*, 1997), at the beginning of the exponential growth phase of fully determined bacterial genome sequences. *H. pylori* was the first bacterium for which a second strain's genome sequence was fully determined, allowing for the first genomic cross-strain comparison (Alm *et al.*, 1999). These genome sequences shed light on the metabolic repertoire of the organism and opened the door for an onslaught of genomic investigations. In many ways *H. pylori* exemplifies the power of the methodologies available to researchers in the postgenomic era. Thanks to the combined approaches outlined in this chapter, a small spiral bacterium which just 15 years ago was greeted with skepticism by the medical community has been transformed into one of the better-studied organisms and a model for systems biology.

The term 'functional genomics' refers to the combined approaches, both experimental and computational, that can be brought to bear on the study of an organism, making use of its genomic sequence. To evaluate the success of a functional genomics approach, one must first define the questions one is trying to answer with the investigation. In the case of bacterial pathogens the most pressing questions initially seem to be of a very applied nature: how can we better prevent, treat or diagnose *H. pylori* infection? Yet to answer these questions we first need to understand how *H. pylori* causes disease. To address this question we need an understanding of the basic bacteriology of this organism: how does it grow and how does it interact with its environment? We need to define its metabolic capacities, its physiology and its ecology. While genomics approaches can be used to answer very specific questions of an applied nature, such as rational vaccine or antibiotic design, their power lies in the potential for generating non-hypothesis-driven exhaustive bodies of information that describe an organism's biology. In many cases these approaches are broadly applicable and do not rely on a specific knowledge of an organism or ability to manipulate it genetically, allowing for the study of non-model systems. The body of information generated by functional genomic approaches can be made publicly available and act as a repository of information to allow individual researchers to focus on more applied questions with a much better chance of successfully manipulating the organism and its environment in a therapeutically beneficial manner.

◆◆◆◆◆◆ THE GENOME

H. pylori genomics resources

A number of resources are available to researchers conducting functional genomic experiments with *H. pylori* (Bjorkholm *et al.*, 2001b). These are summarized in Table 16.1, and their applications are discussed in the following sections. Well-annotated genome sequences and analysis tools are available at the websites of TIGR and AstraZeneca. Further gene function annotation can be found at the PyloriGene website maintained at the Institut Pasteur and useful metabolic pathway annotation can be found at the Metacyc web site. A number of academic groups have constructed whole-genome DNA microarrays from one or both of the genome

Table 16.1. *H. pylori* genomic resources

Resource	Reference	Description	Contact information
Genome sequence information and annotation	Tomb et al. (1997)	26695 genome sequence	http://www.tigr.org/tigr-scripts/CMR2/GenomePage3.spl?database=ghp
	Alm et al. (1999)	J99 genome sequence	http://scriabin.astrazeneca-boston.com/hpylori/
	Institut Pasteur	26695 and J99 genome sequence annotation	http://genolist.pasteur.fr/PyloriGene
	Metacyc	Microbial metabolic encyclopedia	http://ecocyc.org/ecocyc/metacyc.html
DNA microarrays	Salama et al. (2000)	26695 and J99 PCR products spotted on glass slides	Primers available from Sigma Genosys (The Woodlands, TX, USA) http://www.genosys.com/expression/frameset.html
	Ang et al. (2001)	26695 PCR products spotted on nylon membranes	PCR products provided by Invitrogen (San Diego, CA, USA)
	Jenks et al. (2001)	26695 PCR products spotted on nylon membranes	Microarrays prepared by Eurogentec (Seraing, Belgium)
	Allan et al. (2001)	NCTC 11637 genomic clones spotted on nylon membranes	Genomic clones described in Allan et al. (2001)
	MWG-biotech	26695 and J99 50mer oligonucleotides spotted on glass slides	http://ecom.mwgbiotech.com/html/discovery/dna_arrays/heli_pylori.shtml
Protein expression	Jungblut et al. (2000)	Interactive 2D-GEL data for 26695	http://www.mpiib-berlin.mpg.de/2D-PAGE/
Protein interaction maps	Rain et al (2001)	Interactive protein interaction map for 26695	http://pim.hybrigenics.com

sequences, using polymerase chain reaction (PCR) products or genomic DNA clones arrayed on either glass slides or nylon membranes. The primer pairs for one of these microarrays (Salama et al., 2000), as well as fabricated 50mer microarrays, are commercially available. The proteomics community has developed searchable web sites of protein expression and protein–protein interaction data.

Genome sequences

The complete genome sequence of an organism forms the basis of genomic analyses. The small (1.7 Mb) genome of *H. pylori* was one of the earliest for which the sequence was completely determined (Tomb et al., 1997). It yielded some surprises, including the extent of horizontally acquired DNA, the paucity of regulatory factors, the large family of outer membrane proteins (OMPs), and the absence of certain functions such as mismatch repair. *H. pylori* was the first bacterium for which a second strain's genome was determined (Alm et al., 1999), and this shed light on the extent of sequence divergence between strains (see Strain comparison section below) (Alm and Trust, 1999). This strain comparison allowed a more careful analysis of the metabolic repertoire of the organism (Doig et al., 1999; Marais et al., 1999) and it began to define which genes were conserved among strains and possibly essential to the organism.

Another useful metric of comparison besides that of intraspecies genomes is between genomes of closely related organisms. Such analysis was facilitated by the determination of the genome sequence of *Campylobacter jejuni* (Parkhill et al., 2000), which like that of *Helicobacter* is a fastidious, spiral, microaerophilic resident of the human gastrointestinal tract, although it colonizes the intestine and also thrives as an intestinal commensal of many avian species and survives in aquatic environments. While many of the housekeeping genes are highly related between the two organisms, only 55.4% of *C. jejuni* genes have *H. pylori* orthologs, suggesting that these morphologically similar bacteria have very different genetic repertoires when it comes to survival in the different niches they inhabit. In general, *H. pylori* lacks the metabolic and regulatory versatility of *C. jejuni*, consistent with the more diverse environments in which the latter can grow (Kelly, 2001). The genome sequence of an even closer relative of *H. pylori*, *H. hepaticus*, is in the final stages of completion, and here again the theme of environmental adaptation emerges for this bacterium, which like *C. jejuni*, inhabits the small intestine. In certain cases, *H. hepaticus* genes most closely resemble those of *H. pylori*, but other metabolic genes are more closely related to *C. jejuni* (Ge et al., 2001). One sequence that seems to be unique to *H. pylori* among the Helicobacters, and is absent from the *H. hepaticus* genome (Suerbaum et al., 2001), is a horizontally acquired DNA cassette called the *cag* pathogenicity island, which encodes a type IV secretion system that plays an important role in the organism's ability to interact with gastric epithelial cells and cause disease (Censini et al., 1996; Akopyants et al., 1998a; Montecucco and Rappuoli, 2001). These genes, and others unique to *H. pylori*, might be useful targets for *H. pylori*-specific antibiotics that would not harm the host or its resident microflora.

Computational analysis: mining the genome

The genome sequence provides the raw material to ask a myriad of questions about the metabolic capacity of the organism, its regulatory networks and its history. Many of the complexities of the metabolic capacities encoded in the genome can only be revealed by in-depth homology analyses (Doig *et al.*, 1999; Marais *et al.*, 1999) and comparative approaches (Kelly, 2001). The ability to conceptualize complex anabolic and catabolic pathways in a cell is aided by analysis tools such as those provided by the metabolic encyclopedia (Metacyc) at http://ecocyc.org/ecocyc/metacyc.html (Karp *et al.*, 2000). Hypotheses developed by bioinformatics analysis can then be tested experimentally.

Sequence analysis studies have focused on two protein families that are highlights of the *H. pylori* genome: the large family of OMPs (Alm *et al.*, 2000) and the extensive collection of restriction modification (RM) systems (Nobusato *et al.*, 2000). Both of these gene families are highly variable between strains and point to interesting mechanisms of rapid strain evolution by gene duplication, recombination, acquisition and loss. The RM sequence analysis has been complemented by biochemical characterization of these systems, further highlighting the diversity among strains and the extent of non-functional RM systems (Kong *et al.*, 2000; Lin *et al.*, 2001). In both instances of OMPs and RM, the genomic sequence also revealed evidence of an interesting regulator mechanism of protein expression, namely slipped strand repair-mediated phase variation. Genes with tracts of simple nucleotide repeats will incur frame-shift mutations during replication, placing the coding sequence in or out of frame. In the *H. pylori* genome, half of the genes with mononucleotide or dinucleotide repeats encode genes involved in the composition of the cell envelope, suggesting a role for slipped strand repair in antigenic variation. The mechanism has been demonstrated to regulate expression of fucosyltransferases that modulate the expression of Lewis antigens on *H. pylori* lipopolysaccharide (LPS) (Appelmelk *et al.*, 1999; Wang *et al.*, 1999).

Other analyses of the genome sequence have focused on the bacterium's capacity to regulate expression at the transcriptional level. *H. pylori* has a remarkable paucity of transcriptional regulators, with only three complete two-component regulatory systems (as compared with the 118 systems in *Pseudomonas aeruginosa*) and just three sigma factors (Scarlato *et al.*, 2001). Several independent computational approaches were used to define a consensus binding site for this major sigma factor sigma 80 (Vanet *et al.*, 2000), providing the first template for determining transcriptional regulatory networks of *H. pylori*, which can be validated by experimental approaches.

Many of the *H. pylori* genes lack homology to any other genes in the sequence database, but their sequences can still be informative in the context of other microbial genome sequences. Non-homology-based approaches to assigning gene function employ genomic information across multiple genomes, looking for informative instances of co-inheritance of genes across evolution (Pellegrini *et al.*, 1999) and events of gene fusions that can implicate genes encoding closely interacting proteins (Marcotte *et al.*, 1999a). In combinations with other genomewide information, such as protein interaction data and transcript co-regulation, these computational approaches can be used to assign function and build functional connections within genomes (Marcotte *et al.*, 1999b; Marcotte, 2000). With the emerging databases of protein interactions, protein and gene expression studies, and the genome sequences of closely related organisms, *H. pylori* is an excellent subject for this combined computational approach.

Strain comparison

H. pylori strains isolated from different individuals show considerable genetic variation as measured by direct sequencing of specific genes or fingerprinting methods, such as restriction fragment length polymorphisms, random primed PCR, and pulse-field gel electrophoresis (Jiang *et al.*, 1996; Suerbaum *et al.*, 1998; Achtman *et al.*, 1999; Han *et al.*, 2000). This high degree of variation can also be observed by protein fingerprinting methods (Nilsson, 1999). The functional role of this variation, however, is not clear. For example, the high degree of sequence variability of the VacA cytotoxin has been used to establish phylogenetic lineages of strains from different parts of the world (Atherton *et al.*, 1995, 1999). Of particular interest were alleles of the m region, because m2 seemed to be inactive for vacuolating activity (the ability to cause human epithelial cells to fill up with large vacuoles and inhibit growth). Subsequent studies, however, showed that this inability to vacuolate cells was restricted to the HeLa tissue culture cell line used in the assay and that primary gastric epithelial cells show strong vacuolation in response to VacA with the m2 allele (Pagliaccia *et al.*, 1998). Indeed the determination of the second genome sequence of *H. pylori* revealed fewer differences than anticipated at the protein level owing to the fact that most sequence polymorphisms mapped to third basepairs or non-coding regions (Alm *et al.*, 1999). This said, the second sequence revealed more than a hundred genes that were present in only one of the two strains, suggesting that strain comparisons at the level of gene content rather than single basepair polymorphisms could be informative.

To assay for the presence of genes, probes made from genomic DNA of the strain of interest were hybridized to a microarray containing PCR products that cover on average 80% of the coding sequence of each open reading frame present in the two sequenced strains (Salama *et al.*, 2000). Lack of hybridization either meant the gene was not present or differed from the sequence on the microarray by more than 5–10%. This survey of 15 strains, including the two sequenced strains, identified 1281 genes common to all strains while 362 genes were either absent or highly divergent in at least one strain (named strain-specific genes). The strain-specific genes included genes known to vary in their presence among strains, such as the *cag* pathogenicity island genes (Censini *et al.*, 1996; Akopyants *et al.*, 1998a). The largest class of strain-specific genes with homology resembled restriction modification genes, but most of the strain-specific genes had no ascribable function. The invariant genes, which were proposed to represent core *H. pylori* genes, were enriched in homologs of proteins involved in biosynthesis and intermediary metabolism. These invariant genes would be expected to be enriched for essential genes, a hypothesis that could be tested by genetic screens described below.

This work has been extended to correlate changes in genotype measured by microarray analysis with phenotypic differences among strains. In two cases, the major difference between two strains was a deletion of the *cag* pathogenicity island. In one case, this led to a decrease in inflammation caused by the bacteria as measured in tissue culture assays and using a gerbil model of infection (Israel *et al.*, 2001a). In the second case, the strains showed differences in their ability to infect mice carrying a transgene for the expression of the *H. pylori* receptor, Le^b-glycoconjugates, on the pit cell lineage in the stomach (Bjorkholm *et al.*, 2001a). The strain containing the island could infect both germ-free and conventionally raised mice. The strain that had lost the island, but was otherwise highly related, could only infect germ-free mice. Interestingly, in the germ-free animals, there was no observable difference in the degree of inflammation between the two strains, implying multiple functions for the pathogenicity island during infection.

Most recently, this method has been applied to look at subtle changes that occur in the bacterial population within a single infected individual over time (Israel *et al.*, 2001b). Thirteen single-colony isolates were obtained from the same patient from whom the sequenced strain, J99, was isolated 6 years subsequent to the first isolation. Hybridization to the microarray revealed both deletion and acquisition of genes relative to the archival strain. The degree of diversity was tenfold less than seen by comparing 15 unrelated strains (2% *vs.* 22%). Still, the findings highlight that even at the whole gene level, *H. pylori* strains have a remarkable ability to generate genetic diversity during their long sojourn in the human host. Of the newly acquired genes in the recent J99 isolates, some had sequence homology to the other sequenced strain (26695), but others were novel. Clearly there remain unknown *H. pylori* genes beyond those found in the two sequenced genomes. Akopyants and colleagues have described a technique for amplifying unique genes from an unsequenced strain compared to strain 26695 based on subtractive hybridization (Akopyants *et al.*, 1998b). They described 16 unique genes from strain J166, which has been used in monkey infection studies. Seven of these genes had homology to restriction modification genes, two resembled genes involved in metabolism, and four had homology to sequenced *H. pylori* genes of unknown function. Southern blotting revealed these genes were variably present in other strains and thus represent additional strain-specific genes.

Of particular interest to clinical microbiologists is the question of whether genes in addition to the pathogenicity island influence the course of infection and the severity of disease. This will require examination of a large number of strains with known clinical outcome from case controlled studies. Indeed, our ability to distinguish subtle genetic differences among the members of the bacterial populations in a single infection highlights the complexity of the bacterial genetic contributions to disease outcome. Since *H. pylori* can diversify as a population in the stomach, the resulting disease phenotype of the host cannot be ascribed to the genetic make up of any single bacterial isolate. To fully understand *H. pylori* infections, we will need to develop techniques to genotype the entire *H. pylori* community within an individual infection.

◆◆◆◆◆◆ THE TRANSCRIPTOME

H. pylori gene expression

The small number of transcriptional regulators found in the *H. pylori* genome, the lack of operon organization of its genome and the restricted environmental niche it is believed to occupy, led to the notion that *H. pylori* does not exhibit extensive and dynamic regulation of gene expression. Genomic approaches have provided the means to test this idea. DNA microarrays offer the ability to monitor global transcript levels over a large number of growth conditions to determine the extent of transcriptional changes observed. The annotated genome sequence has allowed investigators to delete transcriptional regulators systematically and examine the resulting phenotypes. A combination of these two approaches will allow an assessment of the extent of transcriptional regulation attributable to any given regulator, and will ultimately provide a complete description of the transcriptional networks of *H. pylori*.

Two groups independently undertook the individual deletion of all five of the two-component response regulators in *H. pylori*. These studies revealed several systems to be essential for viability in cultures, and others to have specific roles in motility (Beier and Frank, 2000; McDaniel *et al.*, 2001). Further analysis of the transcriptional profiles of the non-essential regulators should provide candidate target genes that they regulate. Since the environmental stimuli to which these regulators respond is not known, transcriptional profiling of wild-type bacteria under multiple conditions will help determine when the regulators are expressed, suggesting growth conditions under which to scrutinize the gene expression of bacteria containing and lacking a particular regulator. Such profiling experiments have started to reveal examples of extensive co-ordinate regulation of gene expression by *H. pylori*. Microarray analysis of transcript levels have been performed on bacteria at different stages of growth in broth and on cultured epithelial cells, revealing both growth phase and environmental regulation of gene expression (Thompson *et al.*, 2001). Bioinformatics approaches can now be employed to identify potential regulatory sequences in coexpressed genes.

One environmental condition examined in detail, which is of interest given the ability of *H. pylori* to survive in the stomach, is growth in acid. Two groups using either PCR amplified open reading frames (Ang *et al.*,

2001) or genomic DNA fragments (Allan *et al.*, 2001) spotted on nylon have compared the transcript levels of *H. pylori* grown in neutral versus acidic conditions, using slightly different growth conditions (either pH 5.5 *vs.* 7.2 or pH 4.0 *vs.* 7.0, respectively). No overlap exists between these two datasets of acid-regulated genes, nor with the set of protein found to be regulated by two-dimensional gel electrophoresis (see below), highlighting the complexity of acid response studies. Since *H. pylori* survives to varying extents in acid depending on its growth history, and it neutralizes acidic media when grown in the presence of urea, pH titrations, time-course studies of expression and careful monitoring of bacterial viability and media pH should further help refine our picture of *H. pylori* gene expression in acid.

Of great interest to researchers of bacterial pathogenesis is the gene expression of the bacteria in the environment of the host. The measurement of *H. pylori in vivo* transcription has been technically challenging because only small amounts of bacterial mRNA can be isolated from a single infection and this is contaminated with large quantities of host message. One promising approach is the development of reliable quantitative reverse transcription-polymerase chain reaction (RT-PCR) methods that have been used to examine a handful of *H. pylori* genes from both mouse and human stomach biopsies (Rokbi *et al.*, 2001). Advances in RNA amplification and detection methods, along with laser dissection approaches (Mills *et al.*, 2001a) may allow microarray technology to be applied to *in vivo* gene expression studies.

Another approach to studying bacterial gene expression inside the host is to develop *in vivo* reporters of gene expression (Merrell and Camilli, 2000). While the lack of genetic tools currently available for *H. pylori* has made global approaches difficult, several *in vivo* reporters have been used to monitor expression of individual loci including *Pseudomonas putida* catechol 2,3-dioxygenase (encoded by *xylE*) (Karita *et al.*, 1996), green fluorescent protein (Josenhans *et al.*, 1998) and *H. pylori* urease (Joyce *et al.*, 2001). As systematic directed mutagenesis approaches are developed, creation of genomewide libraries of *in vivo* reporters should become feasible. Such libraries would be useful to probe gene expression *in situ* and examine the heterogeneity of microenvironments and metabolic states experienced by the bacteria within the stomach.

Host response

An important aspect of *H. pylori* biology, and of very direct clinical relevance, is the bacterium's effect on the host. This can be studied in clinical samples, modeled in animals or investigated with cultured cells *in vitro*. The genome sequencing efforts for *H. pylori*'s natural host, the human, and one of the best-studied animal models of *H. pylori* infection, the mouse, have paved the way for investigations of host response at a genomic level. Such studies provide a more quantitative analysis of the host response than histopathology, a much higher level of detail than the analysis of a handful of molecular markers, and the discriminatory power to classify strains and study the mechanism of action of individual bacterial genes.

Cultured human gastric epithelial cells derived from gastric adenocarcinomas, including AGS and Kato 3, have been used extensively to study *H. pylori* interactions with host cells. Such studies have shown that infection with bacteria containing the *cag* pathogenicity island (PAI) induces expression of certain cytokines, including interleukin-8 (IL-8), to a much greater extent than strains lacking a functional PAI. Two groups have published studies of the transcriptional response of infected gastric cells using human cDNA microarrays. Maeda and colleagues compared the gene expression of uninfected AGS (American Type Culture Collection number CRL-1739) cells to AGS cells infected for 3 h with a PAI-positive strain. In addition they compared uninfected cells to cells infected with a PAI-defective isogenic mutant bearing an insertional mutation in the *cagE* gene, which encodes a structural component of the secretion apparatus (Maeda *et al.*, 2001). They queried a collection of 2304 named human cDNAs spotted on glass slides, and using the stringent threshold of greater than twofold difference, they found eight genes (*IL-8, IκBα, A20, Kerratin K7, ERF-1, Glutathione peroxidase, MAPKKK12, Stromelysin-3*), which were up-regulated in the PAI+-infected versus uninfected cells, none of which were up-regulated in the PAI– infection. A second study by Cox and colleagues (2001) examined the response of Kato 3 cells to a PAI+ and an unrelated PAI– clinical isolate. They looked at 0.75, 3 and 24 h of infection and measured transcript levels of uninfected and infected cells on a variety of human cDNA collections spotted on nylon membranes, encompassing approximately 57 800 cDNAs. Using a primary cutoff of 1.1-fold change, followed by a rescreening with a 44% false-positive rate, they report identifying 624 cDNAs (encoding 256 unique genes) that were differentially expressed between uninfected and infected cells. In contrast to the previous study, they found more genes up-regulated by the PAI– strain than by the PAI+ strain. The differences in host cell gene expression between the PAI+ and PAI– strains was quite extensive but could not be attributed solely to the presence of the PAI since the two strains were non-isogenic. Indeed expression profiling of AGS cells at 1, 3, 6, 12 and 24 h of infection with wild-type *H. pylori* and isogenic mutants lacking the entire pathogenicity island or individual genes therein has revealed very similar patterns of gene expression. Certain aspects of the gene profile induced by wild-type infection require the presence of the entire pathogenicity island, while others are dependent on individual factors such as CagA (Guillemin, Salama, Tompkins and Falkow, manuscript in preparation).

In contrast to tissue culture models, studies of host-cell responses in animals are complicated by the mixture of cell types present and by the fact that cell census can alter during infection (such as the loss of parietal cells seen in *H. pylori*-induced atrophic gastritis). To address these challenges, Mills and colleagues used lectin panning to isolate pure populations of parietal cells from uninfected and *H. pylori*-infected mouse stomachs, and examined their expression profiles in comparison to the parietal cell depleted gastric cell population (Mills *et al.*, 2001b). The expression pattern of the parietal cells was remarkably constant between the treatments. In contrast, the other cells of the stomach up-regulated a number of genes in response to *H. pylori* infection, including genes involved in interferon signaling, lipopolysaccharide-regulated responses, the cytoskeleton, the extracellular matrix, protein turnover and signal transduction. This study represents the first stage of a mouse Gastric Genome Anatomy Project to characterize the normal gene expression of all the principal gastric cell lineages, as a baseline for studying pathological expression. Indeed, expression profile studies of the development of *H. pylori*-induced MALT lymphoma in a mouse model reveals both changes in cell census, with an increase in lymphoid cell gene expression, as well as the onset of malignancy with the up-regulation of a number of oncogenes (Mueller *et al.*, 2002).

The complexity of host response is even more challenging when studied in human subjects, since genetic diversity makes it impossible to define a clear baseline of expression, such as the one being developed for inbred laboratory mice. In order to develop effective vaccines, however, it is crucial to characterize human immunological responses to *H. pylori*. A study of gene expression from amplified mRNA of gastric biopsies from serially sampled patients before and after *H. pylori* eradication should help to define the intrinsic response to infection (Yang, Tsai, Guillemin and Parsonnet, unpublished results). Another approach has been to screen human sera against *H. pylori* proteins fractionated on two-dimensional gels (McAtee *et al.*, 1998b) (see below). With the increasing annotation of *H. pylori* protein mobilities, such approaches have promising future applications.

◆◆◆◆◆◆ THE PROTEOME

We now know that bacteria accomplish all the complex processes required for life with a repertoire of several hundred (483 in the case of *Mycoplasma genitalium*) to several thousand (6414 for *Pseudomonas aeruginosa*) genes. The challenge to biologists is to understand when the proteins encoded by these genes are expressed and how they interact to direct these processes. Two major approaches have been used to study this in *H. pylori*: two-dimensional gel electrophoresis (2D-GE) and two-hybrid analysis to form protein–protein interaction maps. 2D-GE expands the number of protein species that can be detected from approximately 100 in one-dimensional electrophoresis to over 1000. One can obtain quantitative information for each species and, in contrast to RNA expression analysis, one can distinguish post-translationally modified forms of individual proteins. The number of proteins expressed at any given time is some fraction of the total. *H. pylori*, which contains 1500–1600 open reading frames, likely expresses no more than a thousand genes at any one time. 2D-GE, where the first dimension separates proteins on the basis of charge (isoelectric focusing) and the second dimension separates proteins based on size (polyacrylamide gel electrophoresis; PAGE), can resolve more than 1000 protein species in a complex sample, making it possible to look at global protein expression. This exquisite resolution allows individual spots to be excised free of contamination from other protein species and further analysis. N-terminal sequencing can be performed directly on the whole protein, or the protein can be digested with proteases to allow further sequencing from the N-terminus of these new peptides or analyzed with matrix-assisted laser desorption time-of-flight mass spectrometry (MALDI-TOF MS), which determines the precise mass of the resulting peptides. With the availability of the complete genome sequence, even the mass of a collection of peptides from a protein is sufficient to determine the gene encoding the protein with a very high probability.

Purifying virulence determinants

Even before the genome sequence allowed easy identification of individual spots, 2D-GE was used to characterize a number of virulence determinants and to compare bacterial strains. Dunn and colleagues (1989) used 2D-GE to determine the relationship of *H. pylori* to other *Campylobacter* species, since at the time they were in the same genus. Protein profiles of both whole cell extracts and iodinated surface proteins from a number of *H. pylori* strains were compared. These analyses revealed both variability in protein expression among strains and a set of major outer membrane-associated proteins common to all *H. pylori* strains, but distinct from *C. jejuni* and *C. fetus* outer membrane proteins. Excision of individual spots allowed these researchers to make species-specific antibodies to the major proteins in partially purified urease and flagellar preparations, facilitating *H. pylori* diagnosis and further characterization of these proteins.

Since many bacterial porins show altered mobility on SDS-PAGE depending on whether they are heated, a variation of two-dimensional electrophoresis, where the first dimension of SDS-PAGE was run without heating of the sample and second dimension with heating was used to identify a small family of heat-sensitive proteins in the total membrane fraction (Exner *et al.*, 1995). After further purification based on this assay, N-terminal sequencing revealed these proteins (HopA–D) to be related to each other, but distinct from porins in other bacteria. Antibodies raised against these proteins demonstrated they were specific to *H. pylori* and conversely, antibodies raised against *C. jejuni* porins cross-reacted with an *H. pylori* protein distinct from the Hops. Thus in the pregenomic era, 2D-GE approaches already contributed to the characterization of important *H. pylori* proteins and facilitated strain comparison.

Postgenomic proteome analysis

After the genome sequence of *H. pylori* became available, a number of groups used 2D-GE followed by N-terminal sequencing or MALDI-TOF MS to identify a large number of proteins from whole cell or membrane fractions (Jungblut *et al.*, 2000; Nilsson *et al.*, 2000; Lock *et al.*, 2001) creating a catalog of the most highly expressed *H. pylori* proteins (http://www.mpiib-berlin.mpg.de/2D-PAGE). In order to identify potential vaccine candidates, some groups screened their 2D gels for spots that bound immune sera from multiple (McAtee *et al.*, 1998a, b) or individual (Jungblut *et al.*, 2000) human patients or mouse monoclonal antibodies (Chakravarti *et al.*, 2000). The difficulty in designing good antigen mixes for vaccines is illustrated by the diversity of antigens recognized by the different sera used in these studies (Table 16.2). Thirty-seven proteins (out of 52) were only identified in one of the four studies as being immunogenic. Interestingly, two of these studies used the same pool of sera from 14 patients to probe 2D-GE blots from two different strains of *H. pylori*. Even in this case, only 11/29 protein spots were common to both strains, although many of the proteins identified uniquely in one strain would be expected to be present in both strains, such as flagellin, ATP synthase and a ribosomal protein. Comparison of the sequence differences between the two sequenced strains does not predict a high degree of variation in these proteins to explain the different reactivity of the proteins isolated from different strains. More likely, there were differences in the growth conditions and thus protein expression of the cells, or in some technical aspect of the procedure that masked some antigens in each strain. Of the common antigens (15/52) identified in these studies, many are highly abundant cytosolic proteins, such as the urease enzyme subunits and heat-shock proteins. These antigens have begun to be tested in vaccine studies and in the case of urease have yielded mixed results (Ermak *et al.*, 1998; Lee *et al.*, 2001). Interestingly, many of the proteins identified with human patient sera were also identified by the creation of mouse monoclonal antibodies (Chakravarti *et al.*, 2000), supporting the utility of mouse models for human vaccine development.

Perhaps the most exciting application of 2D-GE technologies is the analysis of protein expression of a single strain subjected to different environmental conditions. Two groups have analyzed changes in protein profiles upon exposure of bacteria to acidic pH using slightly different protocols. When liquid broth grown cultures at pH 5.7 and pH 7.5 were compared, surprisingly few differences were observed (Slonczewski *et al.*, 2000). Only two spots were increased twofold or more at pH 5.7: the UreB subunit of urease and a second spot that corresponds to bovine serum apolipoprotein A-I. *H. pylori* presumably bind this component of the media during growth. Three proteins showed increased expression at pH 7.5: the heat-shock proteins GroES and GroEL, and TsaA, a member of the AhpC antioxidant family. The observation that stress proteins were induced at neutral pH relative to acidic seems to contradict studies in other bacteria and a mRNA expression profiling study (Ang *et al.*, 2001) that found the opposite result. This illustrates a major difficulty in performing such comparative experiments: although both cultures were described as being harvested in mid-log, the growth of *H. pylori* at pH 5.7 is slower than at pH 7.5, raising questions as to the exact growth state of the bacteria in the two cultures. The second study compared plate grown bacteria after 2–5 days of growth at pH 5, 6, 7, and 8 (Jungblut *et al.*, 2000). Two proteins were found consistently to decrease in expression with decreasing pH and had almost undetectable expression at pH 5: HtrA, a

Table 16.2. Antigens identified by two-dimensional gel electrophoresis

ORF	Name	Gene description	McAtee et al. (1998a)	McAtee et al. (1998b)	Jungblut et al. (2000)	Chakravarti et al. (2000)
HP0010	GroEL	Chaperone and heat-shock protein	x	x	x	x
HP0072	UreB	Urease β-subunit	x	x	x	x
HP0875	KatA	Catalase	x	x	x	x
HP0073	UreA	Urease, α-subunit	x	x	x	
HP0912	HopC/omp20/AlpA	Porin, outer membrane protein	x	x		x
HP0027	Icd	Isocitrate dehydrogenase	x	x		
HP0175	–	Cell-binding factor 2		x	x	
HP0243	NapA	Neutrophil-activating protein (bacterioferritin)	x			x
HP0601	FlaA	Flagellin A	x	x		
HP0794	ClpP	ATP-dependent clp protease proteolytic component	x	x		
HP1108	–	Pyruvate ferredoxin oxidoreductase, γ-subunit		x	x	
HP1110	–	Pyruvate ferredoxin oxidoreductase, α-subunit	x	x		
HP1199	Rpl7/l12	Ribosomal protein L7/L12	x		x	
HP1205	TufB	Translation elongation factor EF-Tu	x	x		
HP1350	Prc	Protease	x	x		
HP0011	GroES	Co-chaperone			x	
HP0026	GltA	Citrate synthase			x	
HP0027	Icd	Isocitrate dehydrogenase			x	
HP0109	DnaK	Chaperone and heat-shock protein 70	x			
HP0115	FlaB	Flagellin B		x		
HP0154	Eno	Enolase	x			
HP0192	FrdA	Fumarate reductase, flavoprotein subunit			x	
HP0227	Omp5	Outer membrane protein	x			
HP0254	Omp8	Outer membrane protein	x			
HP0317	Omp9	Outer membrane protein	x			
HP0390	TagD	Adhesin–thiol peroxidase			x	
HP0400	LytB	Penicillin tolerance protein			x	

HP#	Gene	Description		
HP0480	YihK	GTP-binding protein, fusA-homolog	x	x
HP0512	GlnA	Glutamine synthetase	x	x
HP0537	Cag16	*cag* pathogenicity island protein	x	
HP0547	Cag26	*cag* pathogenicity island protein	x	x
HP0561	FabG	3-Ketoacyl-acyl carrier protein reductase	x	x
HP0589	—	Ferredoxin oxidoreductase, α-subunit	x	
HP0653	Pfr	Non-heme iron-containing ferritin		x
HP0657	YmxG	Processing protease	x	x
HP0697	—	Hypothetical protein	x	
HP0706	HopE/omp15	Porin, outer membrane protein		x
HP0779	AcnB	Aconitase B	x	x
HP0825	TrxB	Thioredoxin reductase	x	
HP0896	Omp19	Outer membrane protein	x	
HP0900	HypB	Hydrogenase expression/formation protein	x	x
HP0913	HopB/omp32/AlpB	Porin, outer membrane protein	x	
HP0954	—	Oxygen-insensitive NAD(P)H nitroreductase	x	
HP1018	—	Hypothetical protein	x	x
HP1111	—	Pyruvate ferredoxin oxidoreductase, β-subunit	x	
HP1132	AtpD	ATP synthase F1, β-subunit	x	x
HP1152	Ffh	Signal recognition particle protein	x	
HP1161	FldA	Flavodoxin	x	
HP1177	Omp27	Outer membrane protein	x	
HP1243	Omp28/BabA	Blood group antigen binding protein/outer membrane protein	x	x
HP1562	CeuE	Iron(III) ABC transporter, periplasmic iron-binding protein	x	
HP1563	TsaA	Alkyl hydroperoxide reductase		x

serine protease, and VacA, the major cytotoxin. In these two studies, very few differences in the protein profile were observed at differing pH and there was no overlap in the observed differences.

There are far fewer differences observed in the proteomic experiments than in similar experiments analyzing changes in mRNA expression levels (see earlier). This could mean that expression experiments overestimate the changes in gene expression of *H. pylori* or more likely highlight a lack of sensitivity of the current protocols for observing differences in protein expression using 2D-GE. As methods improve, both for increasing detection of weakly expressed proteins and more uniform methods for culturing the bacteria, 2D-GE will increasingly complement expression array studies in understanding gene expression of *H. pylori*. Comparisons of the transcriptome and proteome from the same cultures will provide insights into the extent to which *H. pylori* uses non-transcriptional mechanisms, such as slipped strand repair or post-translational modifications, to regulate its gene expression.

Protein–protein interaction maps

The yeast two-hybrid screening method (see Chapter 12), originally developed by Fields and Song (1989), has now been used to explore protein–protein interactions on a genomewide scale in *H. pylori* (Rain *et al.*, 2001). This technique relies on the observation that DNA binding and transcriptional activation domains of the *Saccharomyces cerevisiae* Gal4 transcription factor act independently, but must be brought in close proximity to each other in order to drive transcription of a reporter gene. Thus, if the two domains are expressed on separate molecules, no transcription will occur, but if the DNA-binding portion is fused to a protein X and the activation domain is fused to a separate protein Y that can bind to X, transcription of the reporter will occur. To begin to probe global protein–protein interactions in *H. pylori*, Rain and colleagues made 285 bait molecules representing 261 ORFs. Fifty of these ORFs were chosen as proteins known to interact with other proteins in other species or known virulence factors, and the rest were chosen randomly, apparently with some bias towards genes of unknown function. Yeast containing these bait molecules was mated to a library of yeast containing 2 000 000 independent prey molecules. Of 13 962 positive clones, 13 296 were sequenced to determine the interacting domain. These protein interaction map (PIM) data were compiled into a database that can be accessed at http://pim.hybergenics.com.

The researchers computed two very useful statistics to help make sense of the data. The bait molecules on average selected 916 prey molecules. Many of these prey molecules had overlapping sequences. For each interaction they define a selected interaction domain (SID) that is the common sequence in all overlapping prey molecules. In addition, they assign each protein–protein interaction a PIM biological score (PBS) that gives a measure of confidence in the interaction. This score is computed by assessing the number of overlapping prey hits for each individual screen and comparing this to a theoretical distribution of randomly picking fragments from the prey library. This score was then modified by excluding those prey that were actually in non-coding genes and by looking at the whole screen and filtering out prey molecules that selected a very large number of baits, and thus could represent 'sticky' two-hybrid artefacts. The PBS is represented as an E-value between zero and one where values approaching zero represent the highest confidence in the interaction. A review of the data set found that a number of either homodimer or heterodimer interactions reported for *E. coli* proteins were detected for the corresponding *H. pylori* homologs (22/39). The utility of this study in defining new interactions was shown by looking at the *H. pylori* HolB homolog. In *E. coli*, this protein interacts with HolA, which was not thought to be present in *H. pylori*. The HolB bait pulled out a number of prey molecules corresponding to HP1247. Alignment of HP1247 directly with *E. coli* HolA revealed structure similarities that had not been identified by previous bioinformatics. Another exciting example involves the FliA protein, an alternative sigma factor ($\sigma 28$) that regulates flagella biosynthesis. In other bacteria this protein is regulated by an antisigma factor, FlaM. FliA pulled out a bait molecule HP1122 that when overexpressed or underexpressed behaves as an antisigma factor, although it has no homology to FlaM (Colland *et al.*, 2001). Additionally, HP1122 has homologs in *C. jejuni*, *P. aeruginosa* and *Thermotoga maritima*, and thus may define a new class of antisigma factors.

Both 2D-GE and PIM analysis of *H. pylori* are in their infancy. As these studies are expanded to include a greater percentage of the genome, they will give increasing insight into the basic biology of *H. pylori* and have great potential to elucidate molecules and pathways that can be targeted for antimicrobial therapy or vaccine development.

◆◆◆◆◆◆ GENETIC SCREENS

The study of many bacterial pathogens has been facilitated by large genetic screens for genes defective in virulence-associated phenotypes, such as the ability to adhere to or invade host cells or persist in animal models of infection (Finlay and Falkow, 1997). Unfortunately, *H. pylori* do not support the replication of common plasmid or transposon vectors, making such studies difficult. Only in 1998 was there a description of a pair of shuttle plasmids that replicate both in *E. coli* and *H. pylori*. The pHel plasmids contain the *repA* gene from a small cryptic plasmid present in some *H. pylori* strains, a *colE1* origin for replication in *E. coli*, *oriT* to allow mobilization of this plasmid from *E. coli* by conjugation, and either a *cat* gene cassette conferring chloramphenicol resistance (pHel2) or the *aphA3* gene conferring kanamycin resistance (pHel3) in both *E. coli* and *H. pylori* (Heuermann and Haas, 1998). These plasmids can be introduced into *H. pylori* either by natural transformation or conjugation. However, because many strains cannot be transformed with this plasmid owing to variation in restriction modification (Ando *et al.*, 2000) and the instability of some *H. pylori* clones in *E. coli* (Fischer *et al.*, 1999), these plasmids have been used primarily for complementation of gene disruption mutants in known genes, and not for undefined mutants or in promoter trap experiments. Thus far, genetic screens have been limited to screening mutant libraries of insertional mutants made either by transposition of *H. pylori* clones in *E. coli* (shuttle mutagenesis) or random integration of plasmids containing small pieces of *H. pylori* DNA. Both these techniques take advantage of the fact that *H. pylori* will readily integrate homologous DNA by recombination, generally with double cross-overs. Recently, brute force approaches have been used to generate mutants in specific genes that have been identified as interesting by bioinformatics approaches. So far, genetic screens have focused on *in vitro* phenotypes thought to be important for infection of the human stomach, such as motility, acid survival, urease activity, adherence to gastric epithelial cells and the ability to take up exogenous DNA (which may be important for generating genetic diversity). No saturating mutagenesis screens for any phenotype have been preformed to date in *H. pylori*. The combined data of all the screens described below are summarized in Table 16.3.

Searching for virulence genes

The first global mutagenesis study focused on secreted or cell surface genes (Odenbreit *et al.*, 1996). This was accomplished by mutagenesis of *E. coli* clones containing random fragments of *H. pylori* DNA in a plasmid vector with a minitransposon that contained a 5'-truncated BlaM gene. If the transposon landed in a *H. pylori* gene that, when fused to the BlaM gene, would direct the protein to the cell surface, the resulting *E. coli* clones would become resistant to ampicillin. The transposon also encodes a gene conferring chloramphenicol resistance, allowing the reisolated plasmids to be used to introduce an insertional mutant in this gene into *H. pylori*, provided there is sufficient flanking DNA on either side of the transposon (*H. pylori* seem to require on the order of 100–300 bp of flanking homologous DNA to direct homologous recombination into the chromosome). A total of 2400 clones containing 3–6 kb of *H. pylori* DNA behind the *gonococcal* IgA protease promoter Piga, a constitutive, weak promoter in *E. coli* (Kahrs *et al.*, 1995), were isolated in an *E. coli* strain containing a TnMax9 mini-blaM transposon and a source of IPTG-regulated transposase on a separate plasmid vector. This library represents roughly six-fold coverage of the 1.7 Mb genome. Transposition was induced in 198 pools of 11 or 24 clones. Of these 198 pools, 158 yielded ampicillin-resistant colonies. Two colonies from each pool were picked, analyzed by restriction mapping to ensure uniqueness and transformed into *H. pylori*, yielding a mutant library of 135 clones.

These 135 mutants were screened for several phenotypes including motility, production of vacuolating cytotoxin, competence and adherence to the KatoIII gastric epithelial cell line (Table 16.3). Fifteen mutants (13.3%) had reduced or no motility in liquid culture, and all those characterized further had insertions in genes of unknown function. Two mutants that failed to express VacA had independent insertions within

Table 16.3. Summary of genetic screens

Phenotype screened	ORF	Gene name	Gene description	Reference
Acid survival	HP0207	mpr	ATP-binding protein	Bijlsma et al. (2000)
Acid survival	HP0355	lepA	GTP-binding membrane protein	Bijlsma et al. (2000)
Acid survival	HP0642		NAD(P)H-flavin oxidoreductase, putative Oxidoreductase	Bijlsma et al. (2000)
Acid survival	HP0669		Oxidoreductase	Bijlsma et al. (2000)
Acid survival	HP0705	uvrA	Excinuclease ABC subunit A	Bijlsma et al. (2000)
Acid survival	HP0771			Bijlsma et al. (2000)
Acid survival	HP0969	czcA, czcA_1	Cation efflux system protein	Bijlsma et al. (2000)
Acid survival	HP1137	atpF', atpX	ATP synthase F0, subunit b', ATP synthase B'	Bijlsma et al. (2000)
Acid survival	HP1193		Aldo-keto reductase, putative	Bijlsma et al. (2000)
Adherence	HP0912	omp20, hopC	Outer membrane protein/porin	Odenbreit et al. (1999)
Adherence	HP0913	omp21, hopB	Outer membrane protein/porin	Odenbreit et al. (1999)
Competence	HP0038	comB1	DNA transformation competency	Hofreuter et al. (1998)
Competence	HP0039	comB2	DNA transformation competency	Hofreuter et al. (1998)
Competence	HP0040	comB2	DNA transformation competency	Hofreuter et al. (1998)
Competence	HP0042	comB3	trbI protein, DNA transformation competency	Hofreuter et al. (1998)
Competence	HP1527	comH		Smeets et al. (2000)
Motility	HP0080		Conserved hypothetical protein	Bijlsma et al. (1999)
Motility	HP0233		Iron-regulated outer membrane protein	Odenbreit et al. (1996)
Motility	HP0876	frpB, frpB_1		Bijlsma et al. (1999)
Motility	HP0904	pta	Phosphotransacetylase	Bijlsma et al. (1999)
Motility	HP1192		Secreted protein involved in flagellar motility, putative motility protein	Odenbreit et al. (1996)
Motility	HP1274	pflA	Paralyzed flagella protein	Bijlsma et al. (1999)
Motility	HP1462		Secreted protein involved in flagellar motility	Odenbreit et al. (1996)
Motility	HP1465	HI1087	ABC transporter, ATP-binding protein, putative amino acid abc transporter	Bijlsma et al. (1999)
Urease	HP0071	ureI	Urease accessory protein, urea transporter	Bijlsma et al. (1999)
Urease	HP0247	deaD	ATP-dependent RNA helicase, DEAD-box family	Bijlsma et al. (1999)

the *vacA* coding sequence. Two mutants were attenuated in their ability to take up DNA and subsequent characterization identified three consecutive genes, ComB1-3 (HP0038, HP0039-HP0040, HP0042), with sequence homology to conjugation and type IV secretion system genes (Hofreuter *et al.*, 1998). Although polar effects cannot be excluded for ComB1 and ComB2, ComB3 is required for competence. Two mutants that failed to bind KatoIII cells in a fluorescence-activated cell sorting (FACS)-based assay were independent insertions in the same locus encoding two adjacent porin-like outer membrane proteins (Odenbreit *et al.*, 1999), which had been characterized previously as HopB and HopC (Exner *et al.*, 1995). Mutagenesis and antibody inhibition studies indicated that both AlpA (HopB, HP0912) and AlpB (HopC, HP0913) are required for maximal binding to human gastric epithelium (Odenbreit *et al.*, 1999). Thus, this pool of mutants, although biased towards outer membrane proteins and limited in complexity, possibly due to the pooling of clones for transposition, yielded a large number of interesting genes that play a variety of roles in *H. pylori* biology and pathogenesis.

A second random insertional mutagenesis technique again took advantage of the efficient homologous recombination system of *H. pylori* (Bijlsma *et al.*, 1999). This method utilizes a library of small (average size 500 bp) fragments of *H. pylori* genomic DNA in a plasmid vector containing only a *colE1* origin, which allows replication in *E. coli*, but not *H. pylori*, and an *aphA-3* gene conferring kanamycin resistance in *H. pylori*. When transformed into *H. pylori*, a Campbell-like recombination event occurs where a single cross-over directs the integration of the entire plasmid (selected for by kanamycin) resulting in both an N-terminal and C-terminal truncated copy of the gene, neither of which should be functional. Although the starting complexity of the library was not indicated, several rounds of transformation resulted in a library of 1251 *H. pylori* mutant clones.

The entire library, or subsets, were screened for defects in urease activity, motility, competence and growth at low pH (Table 16.3). Insertions in the urease subunit *ureB* (HP0072), the urease transporter *ureI* (HP0071) and a putative RNA helicase *deaD* (HP0247) were found to lack urease activity (although in the case of *deaD*, two genes of unknown function may be inactivated due to polar effects, giving rise to the observed phenotype). The fact that many other genes that affect urease activity, including the *ureA* subunit, were not identified probably reflects the fact that this library would not be expected to cover the genome even once. A total of 180 clones from the library were screened for motility and five (3%) non-motile mutants identified, only one of which had an insert in a gene with putative flagellar function, a homolog of the paralyzed flagellar protein from *C. jejuni* (Yao *et al.*, 1994). All of the mutants had flagella but one mutant, with an insertion in a phosphotransacetylase, had extra flagella at the opposite pole instead of the usual unipolar flagella. Screening for reduced competence yielded three clones attenuated for DNA uptake with identical insertions in HP1527 (*comH*) (Smeets *et al.*, 2000). Complementation of this mutant by integration of the *comH* genes at the *rdxA* locus (which could be selected for by metronidazole resistance conferred by disruption of the *rdxA* gene) confirmed that the phenotype of this mutant was due to disruption of the *comH* gene and not polar effects on upstream or downstream genes. Transformation by electroporation was normal in this mutant, indicating that *comH* is required specifically for DNA uptake during natural transformation. *comH* has a signal sequence and a putative transmembrane domain, but no obvious homologies to genes from other organisms, including DNA-binding domains. An additional screen was performed with this library for mutants with a selective growth defect on solid media at pH 4.8 (Bijlsma et al., 2000). Sixteen mutants were identified of which ten were unique, including two genes, *uvrA* (HP0705) and *atpF* (HP1137) with homology to genes known to be involved in acid tolerance in other organisms. In these screens clones with identical insertion sites were isolated, indicating that the library was of low complexity and thus covered only a fraction of the genome. Still, these screens yielded new genes, highlighting the need for libraries with greater representation to explore fully the genes required for motility, competence and acid survival, as well as the need for new phenotypic screens.

Defining essential genes

Since the types of screens that can be conducted with *H. pylori* genetics are quite limited, two groups have taken another tack and begun systematically to define the set of essential (and non-essential) genes of *H. pylori* using mutagenesis approaches. Chalker and colleagues (2001) used a bioinformatic approach to identify candidate essential genes. To be included in this set, an ORF had to be highly conserved between the two sequenced *H. pylori* strains 26695 and J99, but not conserved among other bacteria defined by having a BLASTP score less than 1.0e-15 in only one or two additional eubacterial genomes excluding the closely related *C. jejuni* genome. A total of 73 genes met these criteria and 29 were randomly chosen for further testing. An additional 13 genes tested were ORFs predicted to encode key enzymes in central intermediary metabolic pathways, four of which had already been described as essential in *H. pylori*. In addition to these 42 genes, three predicted regulatory proteins previously reported to be essential (HP1021, response regulator) or non-essential (HP1027, *fur*, HP1364, a histidine kinase) were included. Non-polar disruption mutants were constructed using a clever vectorless system. PCR primers were designed to amplify separately approximately 500 basepairs of N-terminal and C-terminal sequence. The 3'-primer for the N-terminal portion and the 5'-primer C-terminal portion contained overhangs complementary to a separately amplified chloramphenicol resistance cassette that lacked a transcription terminator such that it would be in the same orientation, thus minimizing polar effects on transcription. A second PCR using the 5'-primer for the N-terminal portion and the 3'-primer of the C-terminal portion with 1 µg each of the three purified products generated the mutagenic cassette used for transformation of *H. pylori*. Genes were considered essential if three consecutive transformations yielded no colonies with the expected insertion as confirmed by Southern blotting and PCR. Insertion-deletion mutants were recovered in 12/45 genes, including the two regulatory genes expected to be non-essential as well as HP1021. This confirms the results of McDaniel and colleagues (2001) who reported HP1021 was not essential and contradicts a previous report (Beier and Frank, 2000). Thus 33/42 or 78% of the genes predicted to be essential did not yield transformants. Contrary to expectation, seven genes with predicted function, but highly divergent from other bacterial homologs were found to be dispensable. This could indicate erroneous annotation of these genes, a requirement for these genes in other growth conditions, or differences in redundancy of function between *H. pylori* and other eubacteria. One major caveat of this study is that no attempt was made rigorously to prove that these genes are in fact essential. This could be accomplished by integrating a second copy of the gene into the genome and showing that now one copy, but not both could be mutated (McDaniel *et al.*, 2001). Definitively identifying those genes that are essential genes but diverge greatly from homologs in other bacteria, as the authors indicate, would highlight potential targets for *H. pylori* antimicrobial therapy.

A second approach to defining essential *H. pylori* genes is to identify all those non-essential genes that can tolerate insertional mutations. One strategy has been described, which uses shuttle mutagenesis combined with hybridization to a DNA array in a technique called RMLA (random mutagenesis and loop amplification) (Jenks *et al.*, 2001). These scientists used a kanamycin resistance marker that preferentially hops into plasmid DNA to mutagenize a library of 1187 individual *E. coli* clones arrayed in 96-well plates, each containing a vector with a random 1.5–2.0 kb *H. pylori* genomic DNA insert. Mutagenized clones from each 96-well plate were pooled and plasmid DNA prepared for transformation into *H. pylori*. Each of the 15 pools yielded 200–1200 kanamycin-resistant colonies. Representation of this library was checked by screening for urease and catalase-deficient mutants. Three independent urease mutants were identified that all mapped in the *ureE* gene (HP0070) and three independent catalase mutants mapped within or upstream of the *katA* gene (HP0875). Although the number of clones screened is not indicated, the fact that multiple insertions were obtained in the *ureE* gene, while none of the other genes involved in urease synthesis and expression were identified, indicated that the Tn*3* insertion was biased or that the original library of inserts was not representative.

To identify non-essential genes, three pools representing a total of 207 original inserts were transformed into *H. pylori* and genomic DNA prepared from each of the transformation pools. This genomic DNA, as well as that of the pooled plasmid DNA used for transformation, was digested with *Hin*dIII, an enzyme that cuts once in the miniTn*3* transposon. The resulting DNA was diluted and ligated to form circles containing the transposition junction. Inverse PCR was performed using primers that anneal within the transposon reading out to amplify the flanking sequence. Radiolabeled probes were prepared from this DNA and hybridized to nylon filters containing each of the open reading frames (ORFs) of the sequenced strain *H. pylori* 26695. Hybridization with a probe prepared from the pooled plasmid used for transformation identified 204 ORFs. Probes prepared from *H. pylori* after transformation hybridized to 72 ORFs. Of the 72 ORFs labeled, many lay adjacent to each other. PCR mapping and sequencing of the insertions defined 43 ORFs with insertions, thus demonstrating them to be non-essential genes. As with the shuttle mutagenesis scheme described above, there seems to be a loss of complexity at the step of transformation into *H. pylori*. This could mean that most genes in *H. pylori* are essential (79%) or that integration of these insertional mutant constructs is inefficient. This could reflect competition of clones for DNA uptake or for the recombination machinery. Alternatively, the transposition events may be biased such that there is insufficient homologous DNA on both sides of the insertion to direct efficient integration of most clones. Beyond the technical limitations of these screening approaches, the ability to experimentally define essential or non-essential genes is restricted to the growth conditions assayed and the strain background used for the insertional mutagenesis. The fact that neither of these variables has been standardized in the *H. pylori* research community represents another challenge for *H. pylori* geneticists.

The future of *H. pylori* genetics

A number of mutagenesis schemes have been devised to construct libraries of *H. pylori* mutants, all of which are rather laborious and none of which have yielded a comprehensive library that saturates the genome. Still screens of these libraries for phenotypes associated with virulence such as motility, adherence to gastric tissue, competence for natural transformation with exogenous DNA or growth under acidic conditions have yielded novel genes whose functions can now be tested further. With *H. pylori*'s small genome size of somewhere between 1500 and 1600 genes, it may be that brute force approaches in which each genes is individually mutated using PCR-based strategies, such as those pioneered in the search for unique *H. pylori* essential genes (Chalker *et al.*, 2001), will precede the discovery of an efficient insertional mutagenesis protocol. Systematically engineered mutants could be screened across panels of growth conditions and the growth rates of different mutants compared using algorithms similar to those employed for transcript profile analysis in order to cluster functionally related genes. To date there have been no direct screens for virulence even though several animal models of infection have been reported. Such screens require methods for identifying from a pool of individual clones that cannot survive in the animal, such as signature-tagged mutagenesis (Shea *et al.*, 1996). The RMLA method described to search for non-essential genes represents an exciting first step in developing such methods. Indeed, a similar strategy has been used to track the presence of pooled transposon insertion clones using glass cDNA microarrays (Salama and Falkow, in preparation). Saturation mutagenesis screens will allow the integration of phenotypic analysis with mRNA and protein expression data by correlating when genes are expressed and when they are required.

◆◆◆◆◆◆ CONCLUSIONS

As we have described in this chapter, the full repertoire of current genomics approaches has been brought to bear on the study of *H. pylori* biology. The real power of functional genomics lies in the combinatorial application of these approaches and this is where the future lies for *H. pylori* research. Comparisons of transcript and protein profiles under various conditions will be used to reveal novel modes of gene regulation. Mutational analysis will be combined with bacterial expression profiling to uncover regulatory circuits. Bacterial gene expression, bacterial genome composition, and genomic analysis of host responses will be used to provide detailed and quantitative phenotypic portraits of new strains. Bioinformatics will be employed for systematic mutagenesis to define gene function. So far research has only just begun in each of these areas with little overlap from the resulting findings but, as more complete data sets are gathered and integrated approaches are applied, we can anticipate complementary and synergistic discoveries. Already with the rapid influx of data from genomic experiments, *H. pylori* is looking less like an intractable but medically important microbe, and more like a model organism for studying a wide range of fascinating biological questions from persistent animal infection to *in vivo* evolution.

Acknowledgements

We would like to thank Stanley Falkow for intellectual contributions and support, and Lucy Thompson, Anne Müller, Jani O'Rourke, Michael Dixon, Hazel Mitchell, Adrian Lee, Shufang Yang, Chiao-Jung Tsia and Julie Parsonnet for sharing unpublished data. This work was supported by a Burroughs Wellcome Fund Career Award in the Biomedical Sciences to KJG.

References

Achtman, M., Azuma, T., Berg, D. E., Ito, Y., Morelli, G., Pan, Z. J., Suerbaum, S., Thompson, S. A., van der Ende, A. and van Doorn, L. J. (1999). Recombination and clonal groupings within *Helicobacter pylori* from different geographical regions. *Mol. Microbiol.* **32**, 459–470.

Akopyants, N. S., Clifton, S. W., Kersulyte, D., Crabtree, J. E., Youree, B. E., Reece, C. A., Bukanov, N. O., Drazek, E. S., Roe, B. A. and Berg, D. E. (1998a). Analyses of the *cag* pathogenicity island of *Helicobacter pylori*. *Mol. Microbiol.* **28**, 37–53.

Akopyants, N. S., Fradkov, A., Diatchenko, L., Hill, J. E., Siebert, P. D., Lukyanov, S. A., Sverdlov, E. D. and Berg, D. E. (1998b). PCR-based subtractive hybridization and differences in gene content among strains of *Helicobacter pylori*. *Proc. Natl Acad. Sci. USA* **95**, 13108–13113.

Allan, E., Clayton, C. L., McLaren, A., Wallace, D. M. and Wren, B. W. (2001). Characterization of the low-pH responses of *Helicobacter pylori* using genomic DNA arrays. *Microbiology* **147**, 2285–2292.

Alm, R. A. and Trust, T. J. (1999). Analysis of the genetic diversity of *Helicobacter pylori*: the tale of two genomes. *J. Mol. Med.* **77**, 834–846.

Alm, R. A., Ling, L. S., Moir, D. T., King, B. L., Brown, E. D., Doig, P. C., Smith, D. R., Noonan, B., Guild, B. C., deJonge, B. L., Carmel, G., Tummino, P. J., Caruso, A., Uria-Nickelsen, M., Mills, D. M., Ives, C., Gibson, R., Merberg, D., Mills, S. D., Jiang, Q., Taylor, D. E., Vovis, G. F. and Trust, T. J. (1999). Genomic-sequence comparison of two unrelated isolates of the human gastric pathogen *Helicobacter pylori*. *Nature* **397**, 176–180.

Alm, R. A., Bina, J., Andrews, B. M., Doig, P., Hancock, R. E. and Trust, T. J. (2000). Comparative genomics of *Helicobacter pylori*: analysis of the outer membrane protein families. *Infect. Immun.* **68**, 4155–4168.

Ando, T., Xu, Q., Torres, M., Kusugami, K., Israel, D. A. and Blaser, M. J. (2000). Restriction-modification system differences in *Helicobacter pylori* are a barrier to interstrain plasmid transfer. *Mol. Microbiol.* **37**, 1052–1065.

Ang, S., Lee, C. Z., Peck, K., Sindici, M., Matrubutham, U., Gleeson, M. A. and Wang, J. T. (2001). Acid-induced gene expression in *Helicobacter pylori*: study in genomic scale by microarray. *Infect. Immun.* **69**, 1679–1686.

Appelmelk, B. J., Martin, S. L., Monteiro, M. A., Clayton, C. A., McColm, A. A., Zheng, P., Verboom, T., Maaskant, J. J., van den Eijnden, D. H., Hokke, C. H., Perry, M. B., Vandenbroucke-Grauls, C. M. and Kusters, J. G. (1999). Phase variation in *Helicobacter pylori* lipopolysaccharide due to changes in the lengths of poly(C) tracts in alpha3-fucosyltransferase genes. *Infect. Immun.* **67**, 5361–5366.

Atherton, J. C., Cao, P., Peek, R. M., Jr, Tummuru, M. K., Blaser, M. J. and Cover, T. L. (1995). Mosaicism in vacuolating cytotoxin alleles of *Helicobacter pylori*. Association of specific *vacA* types with cytotoxin production and peptic ulceration. *J. Biol. Chem.* **270**, 17771–17777.

Atherton, J. C., Sharp, P. M., Cover, T. L., Gonzalez-Valencia, G., Peek, R. M., Jr, Thompson, S. A., Hawkey, C. J. and Blaser, M. J. (1999). Vacuolating cytotoxin (*vacA*) alleles of *Helicobacter pylori* comprise two geographically widespread types, m1 and m2, and have evolved through limited recombination. *Curr. Microbiol.* **39**, 211–218.

Beier, D. and Frank, R. (2000). Molecular characterization of two-component systems of *Helicobacter pylori*. *J. Bacteriol.* **182**, 2068–2076.

Bijlsma, J. J., Vandenbroucke-Grauls, C. M., Phadnis, S. H. and Kusters, J. G. (1999). Identification of virulence genes of *Helicobacter pylori* by random insertion mutagenesis. *Infect. Immun.* **67**, 2433–2440.

Bijlsma, J. J., Lie, A. L. M., Nootenboom, I. C., Vandenbroucke-Grauls, C. M. and Kusters, J. G. (2000). Identification of loci essential for the growth of *Helicobacter pylori* under acidic conditions. *J. Infect. Dis.* **182**, 1566–1569.

Bjorkholm, B., Lundin, A., Sillen, A., Guillemin, K., Salama, N., Rubio, C., Gordon, J. I., Falk, P. and Engstrand, L. (2001a). Comparison of genetic divergence and fitness between two subclones of *Helicobacter pylori*. *Infect. Immun.* **69**, 7832–7838.

Bjorkholm, B. M., Oh, J. D., Falk, P. G., Engstrand, L. G. and Gordon, J. I. (2001b). Genomics and proteomics converge on *Helicobacter pylori*. *Curr. Opin. Microbiol.* **4**, 237–245.

Censini, S., Lange, C., Xiang, Z., Crabtree, J. E., Ghiara, P., Borodovsky, M., Rappuoli, R. and Covacci, A. (1996). cag, a pathogenicity island of *Helicobacter pylori*, encodes type I-specific and disease-associated virulence factors. *Proc. Natl Acad. Sci. USA* **93**, 14648–14653.

Chakravarti, D. N., Fiske, M. J., Fletcher, L. D. and Zagursky, R. J. (2000). Application of genomics and proteomics for identification of bacterial gene products as potential vaccine candidates. *Vaccine* **19**, 601–612.

Chalker, A. F., Minehart, H. W., Hughes, N. J., Koretke, K. K., Lonetto, M. A., Brinkman, K. K., Warren, P. V., Lupas, A., Stanhope, M. J., Brown, J. R. and Hoffman, P. S. (2001). Systematic identification of selective essential genes in

Helicobacter pylori by genome prioritization and allelic replacement mutagenesis. *J. Bacteriol.* **183**, 1259–1268.

Colland, F., Rain, J. C., Gounon, P., Labigne, A., Legrain, P. and De Reuse, H. (2001). Identification of the *Helicobacter pylori* anti-sigma28 factor. *Mol. Microbiol.* **41**, 477–487.

Covacci, A., Telford, J. L., Del Giudice, G., Parsonnet, J. and Rappuoli, R. (1999). *Helicobacter pylori* virulence and genetic geography. *Science* **284**, 1328–1333.

Cox, J. M., Clayton, C. L., Tomita, T., Wallace, D. M., Robinson, P. A. and Crabtree, J. E. (2001). cDNA array analysis of *cag* pathogenicity island-associated *Helicobacter pylori* epithelial cell response genes. *Infect. Immun.* **69**, 6970–6980.

Doig, P., de Jonge, B. L., Alm, R. A., Brown, E. D., Uria-Nickelsen, M., Noonan, B., Mills, S. D., Tummino, P., Carmel, G., Guild, B. C., Moir, D. T., Vovis, G. F. and Trust, T. J. (1999). *Helicobacter pylori* physiology predicted from genomic comparison of two strains. *Microbiol Mol. Biol. Rev.* **63**, 675–707.

Dunn, B. E., Perez-Perez, G. I. and Blaser, M. J. (1989). Two-dimensional gel electrophoresis and immunoblotting of *Campylobacter pylori* proteins. *Infect. Immun.* **57**, 1825–1833.

Ermak, T. H., Giannasca, P. J., Nichols, R., Myers, G. A., Nedrud, J., Weltzin, R., Lee, C. K., Kleanthous, H. and Monath, T. P. (1998). Immunization of mice with urease vaccine affords protection against *Helicobacter pylori* infection in the absence of antibodies and is mediated by MHC class II-restricted responses. *J. Exp. Med.* **188**, 2277–2288.

Exner, M. M., Doig, P., Trust, T. J. and Hancock, R. E. (1995). Isolation and characterization of a family of porin proteins from *Helicobacter pylori. Infect. Immun.* **63**, 1567–1572.

Fields, S. and Song, O. (1989). A novel genetic system to detect protein–protein interactions. *Nature* **340**, 245–246.

Finlay, B. B. and Falkow, S. (1997). Common themes in microbial pathogenicity revisited. *Microbiol. Mol. Biol. Rev.* **61**, 136–169.

Fischer, W., Schwan, D., Gerland, E., Erlenfeld, G. E., Odenbreit, S. and Haas, R. (1999). A plasmid-based vector system for the cloning and expression of *Helicobacter pylori* genes encoding outer membrane proteins. *Mol. Gen. Genet.* **262**, 501–507.

Ge, Z., Feng, Y., White, D. A., Schauer, D. B. and Fox, J. G. (2001). Genomic characterization of *Helicobacter hepaticus*: ordered cosmid library and comparative sequence analysis. *FEMS Microbiol. Lett.* **204**, 147–153.

Han, S. R., Zschausch, H. C., Meyer, H. G., Schneider, T., Loos, M., Bhakdi, S. and Maeurer, M. J. (2000). *Helicobacter pylori*: clonal population structure and restricted transmission within families revealed by molecular typing. *J. Clin. Microbiol.* **38**, 3646–3651.

Heuermann, D. and Haas, R. (1998). A stable shuttle vector system for efficient genetic complementation of *Helicobacter pylori* strains by transformation and conjugation. *Mol. Gen. Genet.* **257**, 519–528.

Hofreuter, D., Odenbreit, S., Henke, G. and Haas, R. (1998). Natural competence for DNA transformation in *Helicobacter pylori*: identification and genetic characterization of the comB locus. *Mol. Microbiol.* **28**, 1027–1038.

Israel, D. A., Salama, N., Arnold, C. N., Moss, S. F., Ando, T., Wirth, H. P., Tham, K. T., Camorlinga, M., Blaser, M. J., Falkow, S. and Peek, R. M., Jr. (2001a). *Helicobacter pylori* strain-specific differences in genetic content, identified by microarray, influence host inflammatory responses. *J. Clin. Invest.* **107**, 611–620.

Israel, D. A., Salama, N., Krishna, U., Rieger, U. M., Atherton, J. C., Falkow, S. and Peek, R. M., Jr. (2001b). *Helicobacter pylori* genetic diversity within the gastric niche of a single human host. *Proc. Natl Acad. Sci. USA* **98**, 14625–14630.

Jenks, P. J., Chevalier, C., Ecobichon, C. and Labigne, A. (2001). Identification of nonessential *Helicobacter pylori* genes using random mutagenesis and loop amplifications. *Res. Microbiol.* **152**, 725–734.

Jiang, Q., Hiratsuka, K. and Taylor, D. E. (1996). Variability of gene order in different *Helicobacter pylori* strains contributes to genome diversity. *Mol. Microbiol.* **20**, 833–842.

Josenhans, C., Friedrich, S. and Suerbaum, S. (1998). Green fluorescent protein as a novel marker and reporter system in *Helicobacter* sp. *FEMS Microbiol. Lett.* **161**, 263–273.

Joyce, E. A., Gilbert, J. V., Eaton, K. A., Plaut, A. and Wright, A. (2001). Differential gene expression from two transcriptional units in the cag pathogenicity island of *Helicobacter pylori*. *Infect. Immun.* **69**, 4202–4209.

Jungblut, P. R., Bumann, D., Haas, G., Zimny-Arndt, U., Holland, P., Lamer, S., Siejak, F., Aebischer, A. and Meyer, T. F. (2000). Comparative proteome analysis of *Helicobacter pylori*. *Mol. Microbiol.* **36**, 710–725.

Kahrs, A. F., Odenbreit, S., Schmitt, W., Heuermann, D., Meyer, T. F. and Haas, R. (1995). An improved TnMax mini-transposon system suitable for sequencing, shuttle mutagenesis and gene fusions. *Gene* **167**, 53–57.

Karita, M., Tummuru, M. K., Wirth, H. P. and Blaser, M. J. (1996). Effect of growth phase and acid shock on *Helicobacter pylori cagA* expression. *Infect. Immun.* **64**, 4501–4507.

Karp, P. D., Riley, M., Saier, M., Paulsen, I. T., Paley, S. M. and Pellegrini-Toole, A. (2000). The EcoCyc and MetaCyc databases. *Nucleic Acids Res.* **28**, 56–59.

Kelly, D. J. (2001). The physiology and metabolism of *Campylobacter jejuni* and *Helicobacter pylori*. *J. Appl. Microbiol.* **90**, 16S–24S.

Kong, H., Lin, L. F., Porter, N., Stickel, S., Byrd, D., Posfai, J. and Roberts, R. J. (2000). Functional analysis of putative restriction-modification system genes in the *Helicobacter pylori* J99 genome. *Nucleic Acids Res.* **28**, 3216–3223.

Lee, M. H., Roussel, Y., Wilks, M. and Tabaqchali, S. (2001). Expression of *Helicobacter pylori* urease subunit B gene in *Lactococcus lactis* MG1363 and its use as a vaccine delivery system against *H. pylori* infection in mice. *Vaccine* **19**, 3927–3935.

Lin, L. F., Posfai, J., Roberts, R. J. and Kong, H. (2001). Comparative genomics of the restriction–modification systems in *Helicobacter pylori*. *Proc. Natl Acad. Sci. USA* **98**, 2740–2745.

Lock, R. A., Cordwell, S. J., Coombs, G. W., Walsh, B. J. and Forbes, G. M. (2001). Proteome analysis of *Helicobacter pylori*: major proteins of type strain NCTC 11637. *Pathology* **33**, 365–374.

Maeda, S., Otsuka, M., Hirata, Y., Mitsuno, Y., Yoshida, H., Shiratori, Y., Masuho, Y., Muramatsu, M., Seki, N. and Omata, M. (2001). cDNA microarray analysis of *Helicobacter pylori*-mediated alteration of gene expression in gastric cancer cells. *Biochem. Biophys. Res. Commun.* **284**, 443–449.

Marais, A., Mendz, G. L., Hazell, S. L. and Megraud, F. (1999). Metabolism and genetics of *Helicobacter pylori*: the genome era. *Microbiol. Mol. Biol. Rev.* **63**, 642–674.

Marcotte, E. M. (2000). Computational genetics: finding protein function by non-homology methods. *Curr. Opin. Struct. Biol.* **10**, 359–365.

Marcotte, E. M., Pellegrini, M., Ng, H. L., Rice, D. W., Yeates, T. O. and Eisenberg, D. (1999a). Detecting protein function and protein–protein interactions from genome sequences. *Science* **285**, 751–753.

Marcotte, E. M., Pellegrini, M., Thompson, M. J., Yeates, T. O. and Eisenberg, D. (1999b). A combined algorithm for genome-wide prediction of protein function. *Nature* **402**, 83–86.

Marshall, B. J. and Warren, J. R. (1984). Unidentified curved bacilli in the stomach of patients with gastritis and peptic ulceration. *Lancet* **1**, 1311–1315.

McAtee, C. P., Fry, K. E. and Berg, D. E. (1998a). Identification of potential diagnostic and vaccine candidates of *Helicobacter pylori* by 'proteome' technologies. *Helicobacter* **3**, 163–169.

McAtee, C. P., Lim, M. Y., Fung, K., Velligan, M., Fry, K., Chow, T. and Berg, D. E. (1998b). Identification of potential diagnostic and vaccine candidates of *Helicobacter pylori* by two-dimensional gel electrophoresis, sequence analysis, and serum profiling. *Clin. Diagn. Lab. Immunol.* **5**, 537–542.

McDaniel, T. K., Dewalt, K. C., Salama, N. R. and Falkow, S. (2001). New approaches for validation of lethal phenotypes and genetic reversion in *Helicobacter pylori*. *Helicobacter* **6**, 15–23.

Merrell, D. S. and Camilli, A. (2000). Detection and analysis of gene expression during infection by *in vivo* expression technology. *Phil. Trans. R. Soc. Lond. B Biol. Sci.* **355**, 587–599.

Mills, J. C., Roth, K. A., Cagan, R. L. and Gordon, J. I. (2001a). DNA microarrays and beyond: completing the journey from tissue to cell. *Nature Cell Biol.* **3**, E175–178.

Mills, J. C., Syder, A. J., Hong, C. V., Guruge, J. L., Raaii, F. and Gordon, J. I. (2001b). A molecular profile of the mouse gastric parietal cell with and without exposure to *Helicobacter pylori*. *Proc. Natl Acad. Sci. USA* **98**, 13687–13692.

Montecucco, C. and Rappuoli, R. (2001). Living dangerously: how *Helicobacter pylori* survives in the human stomach. *Nature Rev. Mol. Cell Biol.* **2**, 457–466.

Mueller, A., O'Rourke, J. L., Guillemin, K., Dixon, M. F., Mitchell, H., Lee, A. and Falkow, S. (2002). Genetic and histopathological changes in *Helicobacter* induced MALT lymphomas are organism specific and time dependent. In *4th Western Pacific Helicobacter Congress ('Two decades of* H. pylori*')*, Perth, Australia.

Nilsson, C. L. (1999). Fingerprinting of *Helicobacter pylori* strains by matrix-assisted laser desorption/ionization mass spectrometric analysis. *Rapid Commun. Mass Spectrom.* **13**, 1067–1071.

Nilsson, C. L., Larsson, T., Gustafsson, E., Karlsson, K. A. and Davidsson, P. (2000). Identification of protein vaccine candidates from *Helicobacter pylori* using a preparative two-dimensional electrophoretic procedure and mass spectrometry. *Analyt. Chem.* **72**, 2148–2153.

Nobusato, A., Uchiyama, I. and Kobayashi, I. (2000). Diversity of restriction-modification gene homologues in *Helicobacter pylori*. *Gene* **259**, 89–98.

Odenbreit, S., Till, M. and Haas, R. (1996). Optimized BlaM-transposon shuttle mutagenesis of *Helicobacter pylori* allows the identification of novel genetic loci involved in bacterial virulence. *Mol. Microbiol.* **20**, 361–373.

Odenbreit, S., Till, M., Hofreuter, D., Faller, G. and Haas, R. (1999). Genetic and functional characterization of the alpAB gene locus essential for the adhesion of *Helicobacter pylori* to human gastric tissue. *Mol. Microbiol.* **31**, 1537–1548.

Pagliaccia, C., de Bernard, M., Lupetti, P., Ji, X., Burroni, D., Cover, T. L., Papini, E., Rappuoli, R., Telford, J. L. and Reyrat, J. M. (1998). The m2 form of the *Helicobacter pylori* cytotoxin has cell type-specific vacuolating activity. *Proc. Natl Acad. Sci. USA* **95**, 10212–10217.

Parkhill, J., Wren, B. W., Mungall, K., Ketley, J. M., Churcher, C., Basham, D., Chillingworth, T., Davies, R. M., Feltwell, T., Holroyd, S., Jagels, K., Karlyshev, A. V., Moule, S., Pallen, M. J., Penn, C. W., Quail, M. A., Rajandream, M. A., Rutherford, K. M., van Vliet, A. H., Whitehead, S. and Barrell, B. G. (2000). The genome sequence of the food-borne pathogen *Campylobacter jejuni* reveals hypervariable sequences. *Nature* **403**, 665–668.

Pellegrini, M., Marcotte, E. M., Thompson, M. J., Eisenberg, D. and Yeates, T. O. (1999). Assigning protein functions by comparative genome analysis: protein phylogenetic profiles. *Proc. Natl Acad. Sci. USA* **96**, 4285–4288.

Rain, J. C., Selig, L., De Reuse, H., Battaglia, V., Reverdy, C., Simon, S., Lenzen, G., Petel, F., Wojcik, J., Schachter, V., Chemama, Y., Labigne, A. and Legrain, P. (2001). The protein–protein interaction map of *Helicobacter pylori*. *Nature* **409**, 211–215.

Rokbi, B., Seguin, D., Guy, B., Mazarin, V., Vidor, E., Mion, F., Cadoz, M. and Quentin-Millet, M. J. (2001). Assessment of *Helicobacter pylori* gene expression within mouse and human gastric mucosae by real-time reverse transcriptase PCR. *Infect. Immun.* **69**, 4759–4766.

Salama, N., Guillemin, K., McDaniel, T. K., Sherlock, G., Tompkins, L. and Falkow, S. (2000). A whole-genome microarray reveals genetic diversity among *Helicobacter pylori* strains. *Proc. Natl Acad. Sci. USA* **97**, 14668–14673.

Scarlato, V., Delany, I., Spohn, G. and Beier, D. (2001). Regulation of transcription in *Helicobacter pylori*: simple systems or complex circuits. *Int. J. Med. Microbiol.* **291**, 107–117.

Shea, J. E., Hensel, M., Gleeson, C. and Holden, D. W. (1996). Identification of a virulence locus encoding a second type III secretion system in *Salmonella typhimurium*. *Proc. Natl Acad. Sci. USA* **93**, 2593–2597.

Slonczewski, J. L., McGee, D. J., Phillips, J., Kirkpatrick, C. and Mobley, H. L. (2000). pH-dependent protein profiles of *Helicobacter pylori* analyzed by two-dimensional gels. *Helicobacter* **5**, 240–247.

Smeets, L. C., Bijlsma, J. J., Boomkens, S. Y., Vandenbroucke-Grauls, C. M. and Kusters, J. G. (2000). comH, a novel gene essential for natural transformation of *Helicobacter pylori*. *J. Bacteriol.* **182**, 3948–3954.

Suerbaum, S., Smith, J. M., Bapumia, K., Morelli, G., Smith, N. H., Kunstmann, E., Dyrek, I. and Achtman, M. (1998). Free recombination within *Helicobacter pylori*. *Proc. Natl Acad. Sci. USA* **95**, 12619–12624.

Suerbaum, S., Josenhans, C., Frosch, M., Bell, M., Braig, K., Brandt, P., Chevreux, B., Dietrich, G., Drescher, B., Droege, M., Fartmann, B., Fischer, H., Ge, Z., Holland, R., Mann, W., Nyakatura, G., Pfannes, O., Schauer, D., Weber, J. and Fox, J. (2001). Determining the whole genome sequence of *Helicobacter hepaticus* ATCC 51449. In *11th International Workshop on Campylobacter, Helicobacter, and Related Organisms*, Vol. 291 (Suppl. 31), abstract M17. Freiburg: Urban & Fischer.

Thompson, L. J., Guillemin, K., Falkow, S. and Lee, A. (2001). Global gene expression of *Helicobacter pylori* during the growth cycle detected on a DNA microarray. In *11th International Workshop on Campylobacter, Helicobacter, and Related Organisms*, Vol. 291 (Suppl. 31), abstract M24. Freiburg: Urban & Fischer.

Tomb, J. F., White, O., Kerlavage, A. R., Clayton, R. A., Sutton, G. G., Fleischmann, R. D., Ketchum, K. A., Klenk, H. P., Gill, S., Dougherty, B. A., Nelson, K., Quackenbush, J., Zhou, L., Kirkness, E. F., Peterson, S., Loftus, B., Richardson, D., Dodson, R., Khalak, H. G., Glodek, A., McKenney, K., Fitzegerald, L. M., Lee, N., Adams, M. D., Venter, J. C. *et al.* (1997). The complete genome sequence of the gastric pathogen *Helicobacter pylori*. *Nature* **388**, 539–547.

Vanet, A., Marsan, L., Labigne, A. and Sagot, M. F. (2000). Inferring regulatory elements from a whole genome. An analysis of *Helicobacter pylori* sigma(80) family of promoter signals. *J. Mol. Biol.* **297**, 335–353.

Wang, G., Rasko, D. A., Sherburne, R. and Taylor, D. E. (1999). Molecular genetic basis for the variable expression of Lewis Y antigen in *Helicobacter pylori*: analysis of the alpha (1,2) fucosyltransferase gene. *Mol. Microbiol.* **31**, 1265–1274.

Yao, R., Burr, D. H., Doig, P., Trust, T. J., Niu, H. and Guerry, P. (1994). Isolation of motile and non-motile insertional mutants of *Campylobacter jejuni*: the role of motility in adherence and invasion of eukaryotic cells. *Mol. Microbiol.* **14**, 883–893.

17 *Streptomyces coelicolor* A3(2): from Genome Sequence to Function

Keith F Chater,[1] Giselda Bucca,[3] Paul Dyson,[2] Kay Fowler,[1] Bertolt Gust,[1] Paul Herron,[2] Andy Hesketh,[1] Graham Hotchkiss,[3] Tobias Kieser,[1] Vassilios Mersinias[3] and Colin P Smith[3]

[1] *John Innes Centre, Norwich Research Park, Colney, Norwich NR4 7UH, UK*
[2] *School of Biological Sciences, University of Wales Swansea, Singleton Park, Swansea, SA2 8PP, UK*
[3] *Department of Biomolecular Sciences, UMIST, PO Box 88, Manchester, M60 1QD, UK*

CONTENTS

Introduction
Microarrays and the transcriptome
The proteome
Back to genetics
Genome-based mutagenesis in practical situations
Summary

◆◆◆◆◆◆ INTRODUCTION

In the mid-1950s a graduate student in Cambridge University's Department of Botany decided to test how far the new genetics of *Escherichia coli* K-12 could be applied to other bacteria. What better choice to make than *Streptomyces*: bacteria so complex that they were still sometimes thought of as fungi, and that were proving astonishingly valuable as a major natural source of antibiotics. The young David Hopwood fairly quickly chose *S. coelicolor* A3(2) as a well-behaved organism that had the added advantage of producing a diffusible blue pigment that might be a valuable genetic marker (see Hopwood, 1999, for a historical account). As we write this chapter, we have just learnt that, under Sir David Hopwood's supervision, and based on his career's classical and molecular analysis of *S. coelicolor* genetics, scientists at the Sanger Centre near Cambridge have completed the sequencing of the A3(2) genome (Bentley et al., 2002; www.sanger.ac.uk/Projects/S_coelicolor). This project, jointly funded by the UK's Biotechnology and Biological Sciences Research Council (BBSRC) and the Wellcome Foundation, took about 3 years to complete. During that time, every new contig of sequence greater than 1 kb was immediately released for use by the general scientific community, and it soon became clear that its scientific interest was even greater than had been anticipated. Accordingly, BBSRC funding was allocated to a substantial effort in *Streptomyces* functional genomics in the UK. In this chapter, we review current progress in this project and, where possible, on related projects in other laboratories. As will become apparent, excellent progress has been made. However, at the time of writing, none of the work has been published in the primary literature, and it is not possible to provide methodological details or protocols.

Streptomyces coelicolor A3(2) is genetically the most studied member of a large genus of saprophytic bacteria that are highly abundant in nearly all soils. Streptomycetes are mycelial organisms (Chater and Losick, 1997): they grow as branched filaments, called hyphae, and they typically obtain nutrients by secreting a battery of extracellular hydrolytic enzymes that can solubilize such abundant but difficult materials as lignocellulose and chitin. Mycelial organisms face a dispersal problem, which is usually solved by converting their biomass into spores. Streptomycetes achieve this by sending up reproductive aerial hyphal branches, which appear to cannibalize the substrate mycelium from which they emerge. Eventually, the aerial hyphae stop growing and metamorphose into chains of uninucleate spores. In the laboratory, each spore can give rise to a colony covered with many millions of spores in 4–5 days. As the aerial branches grow, the substrate mycelium typically produces various antibiotics, one function of which may well be to prevent motile single-celled bacteria in the soil environment from competing with the aerial hyphae for the nutrient resources of the substrate mycelium.

What are the implications of such a lifestyle for the genome? One might anticipate that large numbers of genes will be devoted to extracellular biology – secreted enzymes, transport processes, sensory systems and the like. Indeed, in their analysis of the completed genome sequence, Bentley *et al.* (2002) have predicted more than 800 secreted proteins, and 59 sensor kinase–response regulator gene pairs of the bacterial two-component system type (there are also significant numbers of unpaired members of these systems). Moreover, a recently discovered class of sigma factors, the so-called ECF family (for extracytoplasmic function) has about 50 representatives in *S. coelicolor* (compare this with one in *E. coli* and ten in *Bacillus subtilis*!). Likewise, the complex growth mode might lead one to expect comparatively large numbers of genes to be involved in cell growth and division, as well as in the regulation of morphological differentiation. However, in this case, it seems that most of the 'universal' genes are present in just one version (the exceptions noted being *minD* and a multiplicity of genes for cell wall biosynthesis). In addition the sequence offered the possibility of revealing the extent of the genetic commitment to secondary metabolism – an expectation that has not been disappointed by the revelation of some 400 kb of likely secondary metabolic genes.

Taking all these factors into account, it is hardly surprising that the A3(2) genome is large – at 8.67 Mb about twice the size of the *E. coli* and *B. subtilis* genomes. The 7825 genes present a formidable challenge and a fascinating opportunity for investigation.

◆◆◆◆◆◆ MICROARRAYS AND THE TRANSCRIPTOME

A key objective of the UK *Streptomyces* functional genomics consortium was to produce DNA microarrays [and corresponding primers and PCR products] and provide these resources to the research community. The high gene content of *S. coelicolor* presents a significant technical challenge for producing such microarrays. The difficulties are compounded by the high (72%) G + C composition of the genome and by the presence of large numbers of gene families that could potentially create serious problems with cross-hybridization between paralogous sequences on an array.

To deal with all these requirements and problems, a relatively expensive but comparatively versatile two-stage PCR-based strategy was adopted at UMIST to generate the 'CoelicolorArray' (Figure 17.1; see also http://www.arrays.umist.ac.uk). The probe size designed for each open reading frame (ORF) was typically 150–500 nt. This relatively small size was selected to maximize the chances of generating a unique product that would not cross-hybridize to other sequences from the genome, and to facilitate the PCR amplification of the G + C-rich DNA. Long, *ca* 40 nt primers were used for the 'first-round' PCR cycle. The 3' halves of each primer pair are specific for the target gene and the 5'-halves represent

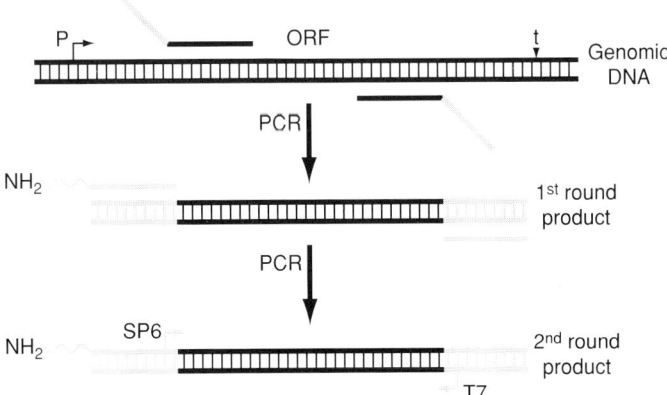

Figure 17.1. Microarray probe production using PCR amplification from genomic DNA. A unique portion of each open reading frame (ORF) was amplified using gene-specific primers tailed with universal sequences. A second round of PCR using the universal primer pair enables the incorporation of 5'-end modifications, such as the amine group indicated that allows covalent attachment to specific (e.g. aldehyde-coated) surfaces. The universal tag sequences also contain bacteriophage promoters for *in vitro* transcription to generate either sense or antisense RNA.

'universal tag' sequences. A very small quantity of each first-round PCR product was then amplified using a single primer pair corresponding to the respective universal tag sequences. This approach has several important advantages over the more typical single round amplification cycle:

1. The two-round amplification minimizes the amount of genomic template DNA carried across on to the array.
2. The entire set of probes can be reamplified using a single primer pair. Thus, a small aliquot of the diluted first-round PCR product and a sample of the universal primers can be provided to other laboratories that have a high demand for microarrays.
3. One of the universal primers can be derivatized (e.g. 5'-amidated) to allow covalent attachment of the probe set to specific slide coatings.
4. Useful features can be built into the universal tag sequences. For this project we have incorporated bacteriophage T7 and SP6 polymerase-recognized promoters in the respective tags. This allows the generation of labelled or unlabelled RNA of either strand, and provides routes to normalizing and validating array data.
5. The presence of the extraneous universal 'tails' on the PCR products allows them to be used directly in nuclease-protection assays for validating signal ratio results from specific probes on the array (the tails allow the distinction to be made between probe–probe reannealing and protection by RNA, when the non-homologous tail is removed).

S. coelicolor microarrays (5000–7001 ORFs) have been independently produced at Stanford University (Huang *et al.*, 2001). In this case, the probes were generally much larger and were generated from a single PCR amplification using 20mer gene-specific primers. About 10% of probes in the Stanford arrays were duplicates or overlapping sequences. At UMIST the gene-specific sequences of the primers were designed using automated scripts which made full use of the emerging *S. coelicolor* genome sequence (Figure 17.2; G. Hotchkiss and C. P. Smith, unpublished data). In each case an optimal primer pair was selected by the primer3 programme (http://www-genome.wi.mit.edu/cgi-bin/primer/primer3_www.cgi), and the deduced DNA sequence of each PCR product was compared against a non-redundant contig of the genome sequence using the Blast program. Where necessary, this automated process was conducted iteratively until a product was identified that received a Blast score below a certain threshold (and hence would not cross-hybridize with other sequences). In parallel with this process the selected primer sequences were also checked against the genome sequence using the FastA program to identify, and reject, any primer pairs that could mis-prime elsewhere.

To date, arrays have been produced at UMIST that represent *ca* 75% (*ca* 6000 ORFs) of the predicted *S. coelicolor* ORFs (Figure 17.3; G. Hotchkiss, V. Mersinias, G. Bucca and C. P. Smith, unpublished data). It is envisaged that near-whole genome arrays will be printed by the time this manuscript goes to print. Despite the advantages of the PCR-based approach outlined above, it is possible that these arrays will eventually be superseded by 'long oligonucleotide' arrays as the oligo-chip technology

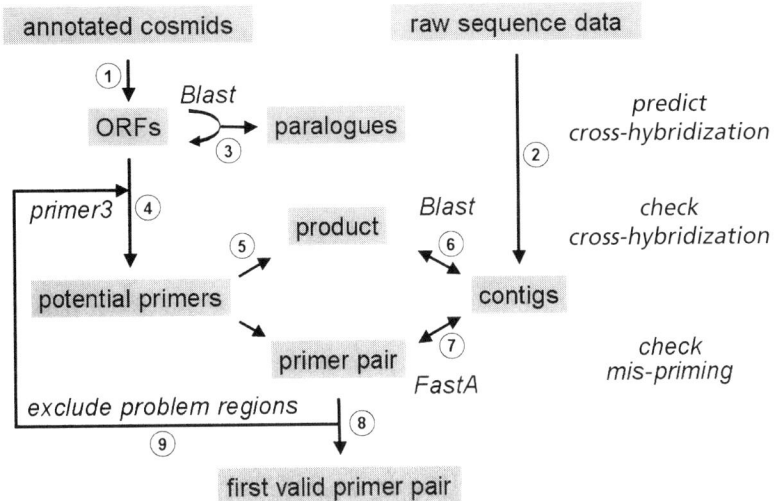

Figure 17.2. Primer design schema. Predicted open reading frames (1) and contigs (2) were generated from the emerging *S. coelicolor* sequence data at the Sanger Centre. An automated primer design programme predicted genes expected to cross-hybridize with any probe from a particular ORF using Blast similarity searches against all ORFs (3). The primer3 primer design software then selected five potential primer pairs (4) and the virtual products were generated (5). A Blast search of the product against the contig library revealed if it would cross-hybridize to another portion of the genome (6). A FastA search of the primer pair against the contigs was used to identify potential annealing sites that might give rise to additional PCR products (7). The first product that only picked up itself, and not any predicted paralogues, during the Blast search and whose primers were not expected to give extraneous bands in the FastA search was selected as the microarray probe for that ORF (8). If all five potential primer pairs were rejected, the programme repeated the procedure from step 4, excluding regions that would cross-hybridize to the predicted product and preventing potentially promiscuous primers from being selected (9).

matures and costs fall (see Chapter 7). The ease of production of the probes, ability to generate complete coverage, and certainty of the purity and identity of the DNA being spotted are some of the major advantages of this new technology.

The initial experiments conducted with the '75% array' assessed its specificity by hybridizing sets of labelled cosmids from the 'Redenbach' collection (Redenbach *et al.*, 1996); only probes corresponding to ORFs on the cosmids produced signals on the array. The heat-shock response was also used to evaluate the reliability of the array. The HspR protein is known negatively to regulate the *dnaK* operon of *S. coelicolor* (comprising *dnaK-grpE-dnaJ-hspR*) (Bucca *et al.*, 1997). Microarray analysis of wild-type and *hspR* knockout mutants demonstrated that *dnaK*, *grpE* and *dnaJ* were significantly up-regulated in the *hspR* mutant as expected, as were at least 20 other genes. Subsequent S1 nuclease protection assays on some of these newly identified genes have confirmed the array results (G. Bucca, V. Mersinias, G. Hotchkiss and C. P. Smith, unpublished data).

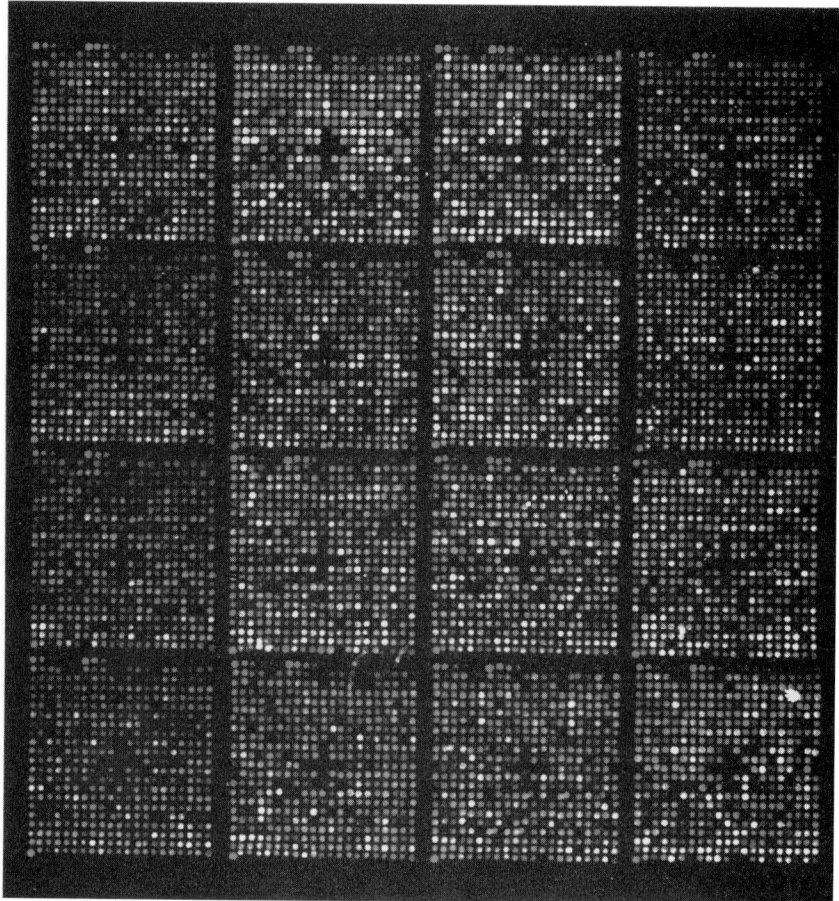

Figure 17.3. Version 1 of the 'CoelicolorArray' representing *ca* 6000 ORFs. Cy5-labelled cDNA (red) generated from RNA isolated from the prototrophic *S. coelicolor* MT1110 grown on cellophane-coated Oxoid Nutrient Agar for 25 h was co-hybridized with Cy3-labelled *S. coelicolor* M145 genomic DNA (green). In each subarray the first row contains a rRNA dilution series and negative control (*Pseudomonas* ORF) spots; a number of carry-over negative controls are also included and empty spots form a cross in the middle of each subarray. Reproduced in colour between pages 178 and 179.

A key application of the microarrays, also addressed by Huang *et al.* (2001), will be to analyse dynamic changes in transcript profiles associated with developmental (metabolic and morphological) transitions during the life cycle of *S. coelicolor*. This will reveal relevant genetic/metabolic networks and dependencies. This is particularly critical in an organism like *S. coelicolor*, which possesses a significant level of genetic redundancy. Clearly, mutational analyses will also play an important role (see later) but interpretation of the resulting phenotypes will probably require substantial information from transcriptome and proteome studies, particularly when gene knockouts result in a 'silent' phenotype. Even when gene knockouts cause clearly visible pleiotropic effects on phenotype – perturbing both sporulation and antibiotic production, for example – the genes targeted for mutagenesis will often have no direct role in these processes. Instead, the complex phenotype may result from the disruption of an interconnected web of genetic/metabolic networks. Thus, mutagenesis will need to be integrated with the '-omic' technologies and data modelling, if we are to unravel the biological roles of a significant proportion of the genes.

The current status of the array programme, and information on access to the resources can be found at www.arrays.umist.ac.uk.

◆◆◆◆◆◆ THE PROTEOME

Although the term proteome came into common use when the electrophoretically separated proteins of diverse organisms could be related to their genome sequences, the principle of observing changes in the total profile of cellular proteins accompanying physiological and/or genetic changes was established much earlier. In work on *S. coelicolor*, the laboratory of C. J. Thompson began to carry out such two-dimensional (2D) gel electrophoretic analysis about 10 years ago, helped by numerical analytical methods developed by J. Vohradsky (e.g. Puglia *et al.*, 1995; Vohradsky *et al.*, 1997). In a recent large series of experiments, Vohradsky *et al.* (2000) showed that, both in liquid and during surface growth on glutamate-maltose medium, four defined stages in the growth cycle (including a brief hiatus between two stages of biomass accumulation before entry into stationary phase) are associated with dramatic changes in the radio-labelled protein profile obtained by 2D gel analysis after pulse-labelling with ^{35}S-cysteine and ^{35}S-methionine. Twenty-two sets of proteins showing characteristic profiles were recognized. Some of these could be equated with sets of proteins whose rate of synthesis was changed by such stresses as heat, cold, changed osmotic pressure and changed pH. The two most remarkable conclusions were that several stress regulons are expressed in the stage between the two growth phases, even at a fixed temperature, and that the proteins induced by different stresses show virtually no overlap with each other, in sharp contrast to observations in *B. subtilis* and *E. coli*.

Now, with the availability of the *S. coelicolor* genome sequence, a substantial number of these spots have been identified by matrix-assisted laser desorption ionization time-of-flight mass spectroscopy (MALDI-TOF MS) on tryptic digests (see Chapter 11). The accumulating data can be accessed at http://www.proteon.biomed.cas.cz. More extensive spot identification by the MALDI-TOF MS procedure has been carried out as part of the BBSRC-funded functional genomics project (A. Hesketh, unpublished data) and is documented at http://www.qbab.aber.ac.uk/S_coeli. These data have been obtained using a different culture medium from that of the Basel group, as well as a different strain (the M145 prototroph, instead of J1501, a double auxotroph). The SMM 'enriched' minimal medium used at Norwich is that used in extensive physiological studies of antibiotic production by the group of M. J. Bibb (see Kieser *et al.*, 2000), and does not reveal a lag between two growth phases, such as is seen on glutamate–maltose medium. It is, therefore, not straightforward to superimpose the 2D gel patterns obtained in Norwich and Basel, as displayed on the two web sites.

In the Norwich study, initial studies of proteins of pI 4–7 were given higher resolution by the use of 'zoom gels', with a narrower pI range (e.g. 4.5–5.5, 5.0–6.0, 5.5–6.7). Zoom gels permit higher loading as well as better separation, so far more spots can be detected. Colloidal Coomassie Blue was found to be about five times more sensitive than standard Coomassie Blue staining and is fully compatible with downstream mass spectrometric analysis. Silver staining is several times more sensitive still but interferes with further analysis. Virtually any spot visible with colloidal Coomassie Blue staining could be unambiguously identified by MALDI-TOF MS. About 770 spots have been identified so far, including some regulatory proteins and some enzymes of antibiotic biosynthesis. The spots correspond to 651 genes. The finding of, on average, 1.2 spots per gene shows the value of proteomics in revealing aspects of regulation invisible to transcriptome analysis, whether involving covalent modifications, such as phosphorylation and nucleotidylation, or specific proteolytic cleavage. Specific examples of post-translational changes in proteins revealed by proteomics are now being analysed in more detail, in part by the additional application of quadrupole time-of-flight mass spectrometry (Q-TOF MS MS) (A. Hesketh, unpublished data). It is clear that protein activity does not necessarily follow either quantitatively or qualitatively from the transcription of a gene.

Another important way in which proteome analysis has special value is in the study of subproteomes, in which cellular subfractions are analysed. Some obvious examples include proteins that fractionate with membranes, extracellular proteins, large protein assemblies (e.g. those involved in macromolecular syntheses, or in 'degradosomes'); and immuno-coprecipitates.

◆◆◆◆◆◆ BACK TO GENETICS

Annotation of the genome sequence has given us 7825 genes to study. Transcriptome and proteome analyses are beginning to reveal expression patterns for these genes in diverse physiological states and transitions. The analysis and modelling of these vast amounts of data can be expected to provide many new and important insights. However, the allocation of causality or biological function will also require genetic intervention. What happens if a gene is inactivated or overexpressed? Are there obvious phenotypic changes that indicate that the gene is important for some particular area of physiology (e.g. sporulation, antibiotic production, stress responses, primary metabolism, etc.)? And what are the consequences of genetic changes for global gene expression as revealed by transcriptome and proteome analysis?

Although classical approaches to the isolation and study of mutants will no doubt continue to complement the '-omic' technologies, the availability of genome sequences has revolutionized mutant isolation by permitting the rapid precise mapping of transposon insertions and the highly efficient construction of precisely defined mutations by oligonucleotide-targeted mutagenesis.

Transposon mutagenesis

There have been several moderately successful attempts over a period of about 15 years to use transposons in *Streptomyces* but their impact has been less than in most other organisms. This can, at least in part, be attributed to: the difficulty of finding good transposon delivery systems; the use of protoplast transformation to deliver transposons (after protoplasting and regeneration, unwanted 'spontaneous' mutations often occur at much increased frequency); and the absence of an efficient way to test whether a transposon insertion is linked to a particular phenotype – the latter problem is potentially ameliorated by the recent description of a generalized transduction system for *S. coelicolor* and other species (Burke *et al.*, 2001), although there is some way to go before this procedure can be considered efficient enough for high throughput.

Recently, several laboratories have made important progress in applying transposon mutagenesis to *S. coelicolor*. Gehring *et al.* (2000, 2001) used *in vitro* transposition with derivatives of both Tn5 and *mariner*, in each case carrying an apramycin resistance gene. The target for the transposition reactions was a library of *S. coelicolor* DNA shotgun cloned into an *E. coli* plasmid. After recovery of mutated plasmids in *E. coli*, the mutations were transferred into *S. coelicolor* by either protoplast transformation or conjugation, for the latter relying on an origin of transfer in the cloning vector, and then selecting for marker replacement. Linkage of the transposon marker to interesting phenotypes was checked by transformation with denatured linear DNA (Oh and Chater, 1997; this procedure does not lend itself to high throughput). As a result of this pioneering work, several new genes involved in morphological development were identified. More recently, a different Tn5 derivative (Tn5062, containing an *egfp* reporter gene; Figure 17.4) has been used by P. Herron and P. Dyson (unpublished data; http://www.swan.ac.uk/biosci/strepTn/; Figure

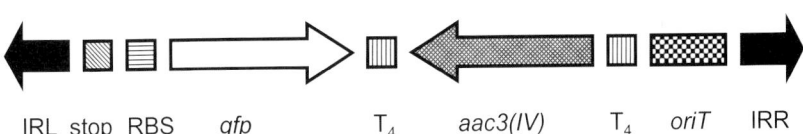

IRL stop RBS   gfp   T$_4$ aac3(IV)   T$_4$ oriT   IRR

Figure 17.4. Organization of Tn5062. The transposon consists of two inverted repeats (IRL and IRR) flanking an internal region containing the following elements: (1) a sequence consisting of translational stop codons in all three reading frames (stop); (2) a consensus streptomycete ribosome site (RBS); (3) a promoterless copy of a fluorescence-enhanced and codon-optimized version of the green fluorescent protein gene (*gfp*); (4) an apramycin-resistance gene [*aac3(IV)*] flanked itself by two T4 transcriptional terminators (T$_4$); and (5) the RP4 origin of transfer (*oriT*).

17.5) for *in vitro* mutagenesis of cosmids taken from the ordered *S. coelicolor* cosmid library of Redenbach *et al.* (1996), which was used for the *S. coelicolor* genome project (the complete, end-to-end sequences of these are available at the Sanger web site, ftp.sanger.ac.uk/pub/S_coelicolor/cosmid_inserts). Because a wide host range origin of transfer has been inserted into Tn5062, mutated cosmids can be transferred into *Streptomyces* by conjugation from an *E. coli* host carrying a suitable *tra* operon. The *in vitro* mutated cosmids are, therefore, introduced into a suitable mobilizing host, using electroporation. Mating of such transformed strains with *Streptomyces* spp. by growth together on a Petri dish results in large numbers of cosmid transfers. The cosmids are based on Supercos 1, which has no replication origin capable of functioning in *Streptomyces*, so rescue of the transposon marker in exconjugants depends on integration into the genome. Because of the long flanking regions in the

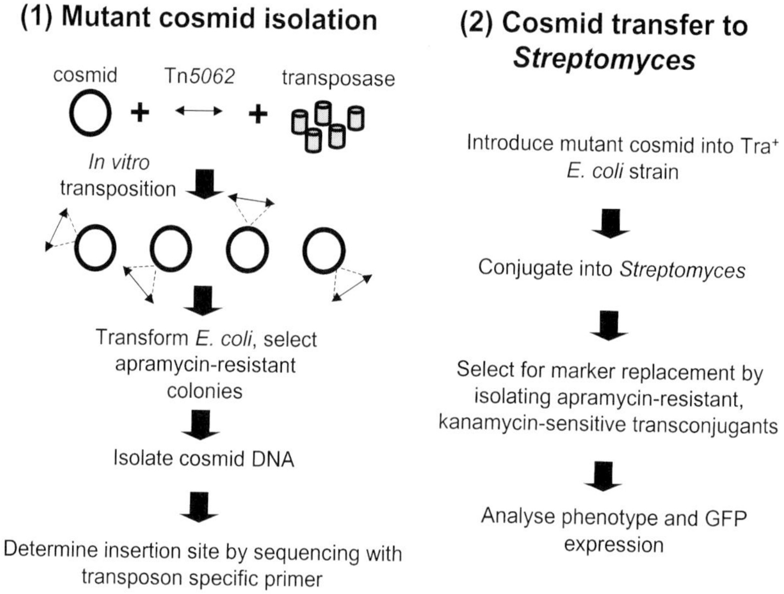

(1) Mutant cosmid isolation

cosmid + Tn5062 ↔ + transposase

In vitro transposition ↓

Transform *E. coli*, select apramycin-resistant colonies
↓
Isolate cosmid DNA
↓
Determine insertion site by sequencing with transposon specific primer

(2) Cosmid transfer to *Streptomyces*

Introduce mutant cosmid into Tra⁺ *E. coli* strain
↓
Conjugate into *Streptomyces*
↓
Select for marker replacement by isolating apramycin-resistant, kanamycin-sensitive transconjugants
↓
Analyse phenotype and GFP expression

Figure 17.5. *In vitro* transposon mutagenesis combined with marker replacement in *Streptomyces*. (1) An efficient *in vitro* reaction generates a population of mutated cosmids, each containing Tn5062 insertions in different locations. The mutated cosmids are isolated after electroporation of *E. coli*, selecting for the apramycin-resistance gene carried by the transposon. The location and orientation of an insertion can then be identified by limited DNA sequencing using an 'outward-facing' Tn5062 primer. (2) To transfer a specific mutation into *Streptomyces*, a mutated cosmid is mobilized from a Tra⁺ *E. coli* strain that is also deficient in DNA modification. Transconjugant apramycin-resistant *Streptomyces* colonies are screened for the absence of the kanamycin-resistance gene carried by the cosmid, indicative of homologous recombination-dependent marker replacement and loss of the non-replicating cosmid molecule. A clear indication of gene function is provided if several independently isolated transconjugants exhibit the same mutant phenotype. Depending on the orientation of Tn5062, fluorescence microscopy and/or immunological detection can subsequently be used to analyse the temporal and spatial regulation of the mutated gene and the results correlated with data from transcriptome and/or proteome analysis.

cosmid inserts, integration is readily obtained, and double cross-overs allowing the transposon-induced mutation to replace the wild-type allele occur in more than 10% of colonies. An attractive feature of this approach is the ability to mobilize mutations into different genetic backgrounds and even into the genomes of closely related species. Similarly, as mutations are introduced into the cosmids *in vitro*, an inability to recover viable exconjugants carrying a mutation in a given gene is indicative of the essential nature of that gene.

Following electroporation, each *in vitro* transposition reaction typically generates several thousand independent *E. coli* clones. To date, a randomly selected sample of only 96 clones from each reaction has been sequenced to determine individual insertion sites. The vast majority of clones contain a single insertion and data from the first 20 mutagenized cosmids has indicated a frequency of one insertion every 612 bp, with 71% of predicted genes containing one or more insertions, representing 1300 insertions in 462 genes. Mutagenesis of small genes (< 600 bp) that are often 'missed' by this limited sampling can be pursued either by sequencing more clones from each transposition reaction, or by employing PCR-based screening of the total mutant library for each cosmid, or by an alternative strategy of targeted mutagenesis of cosmids (see later).

The transposon Tn*5062* developed for this strategy contains a promoterless copy of the *egfp* reporter gene that is preceded by translational stops in all three reading frames. One consequence of insertion is that the temporal and spatial regulation of expression of a gene can be investigated. For example, a Tn*5062* insertion has revealed up-regulation of expression of a cell division gene in aerial hyphae that undergo multiple septation to form spores. Another outcome, a consequence of translation termination, is the formation of truncated proteins. This will be particularly important in analysing the functions and interactions of multidomain polypeptides. In addition, by selecting an insertion in a cosmid in a region adjacent to a gene of interest, cloning of this gene for further studies, for example, for complementation analysis, is greatly eased by use of the antibiotic resistance marker carried on the transposon. These features, together with the high throughput of the *in vitro* system, will enable coverage of the entire genome in a relatively short period and will provide an invaluable resource for investigating streptomycete gene function.

In an independent approach, the first large-scale high-quality global library of *in vivo* transposon mutants to be described for *S. coelicolor* was recently made by K. Fowler and T. Kieser (unpublished data). They used Tn*4560*, originally derived from the phenotypically cryptic *Streptomyces fradiae* transposon Tn*4556* (Simieniak *et al.*, 1990). Over a number of years, this transposon has been valuable on a small scale, but its full value was realized only when it was delivered by conjugation from *E. coli* on a highly effective suicide vector. The result was a 'megalibrary' of more than 10^5 independent highly random simple insertions, preserved in the form of various levels of clonally separate pools to permit the use of sib selection techniques. Individual mutants (e.g. chosen on phenotype criteria) are readily characterized at the sequence level by ligation-mediated PCR (Figure 17.6; Mueller and Wold, 1989). In addition, gene-specific PCR

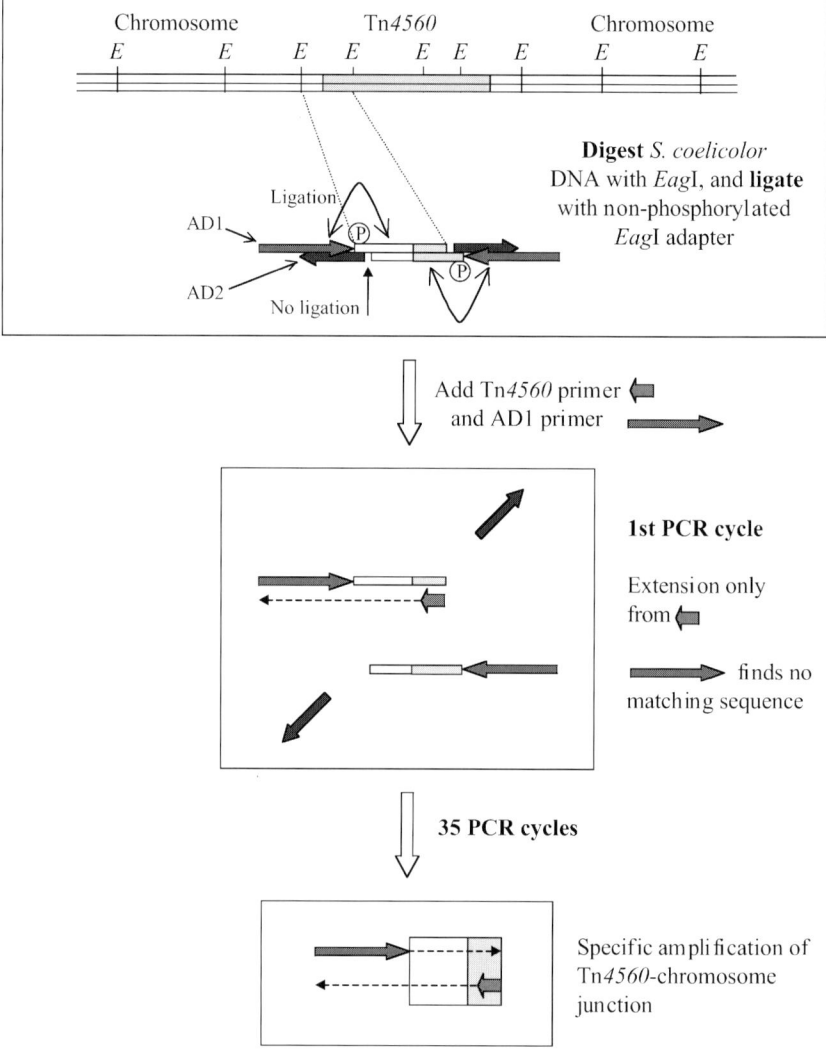

Figure 17.6. Use of ligation-mediated PCR to detect a Tn*4560* insertion in a specific gene (applicable to 10^5-complexity libraries). Reproduced in colour between pages 178 and 179.

has been successfully used even on the 'megalibrary' to detect the presence of insertions in particular genes, indicating that the relevant gene function was non-essential under the conditions in which the library was produced. Mutants for specific genes can be isolated from the megalibrary by successive evaluation of pools of 10^4 and 10^3 complexity, followed by colony hybridization (Figure 17.7).

Targeted mutagenesis

In the last few years, it has proved possible to obtain homologous recombination into microbial genomes via very short stretches of sequence

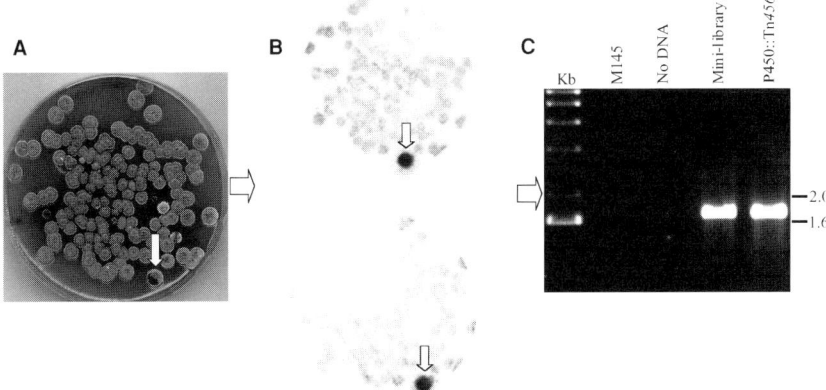

Figure 17.7. Isolation of a specific Tn*4560* insertion mutant in the putative cytochrome P450 gene, 7E4.20, by colony hybridization. (A) After PCR had shown that minilibrary 3 (10^3-complexity) contained an insert in 7E4.20, mini-library 3 (100 μl inoculum of 5×10^5-fold dilution) was spread on medium selecting for Tn*4560* and counterselecting any residual *E. coli* colonies and incubated for 7 days. (B) Autoradiographs of the duplicate colony lift filters, after hybridization at 55°C with a ^{32}P-labelled 'P450-TN4560' oligonucleotide probe, and final washes with 2 × SSC, 0.1% SDS at 55°C. Arrows indicate the strongly hybridizing spot on each autoradiograph and the corresponding discrete, isolated colony. (C) The identity of the positive colony was confirmed by PCR (35 cycles), using liquid-grown mycelium as template, and Tn*4560* and gene-specific primers. Kb, 1 Kb DNA ladder.

identity. The first example of this was in budding yeast, in which a selectable resistance gene could be targeted to replace a chromosomal segment by flanking sequences as short as 35 bp (Baudin *et al.*, 1993). This opened the way for generalized 'PCR targeting', in which the 5'-ends of long oligonucleotide primers supply the flanking targeting sequences, while the 3'-ends prime on to a selectable resistance cassette, permitting its amplification by PCR.

In wild-type bacteria, unlike yeast, homologous recombination at experimentally useful frequencies usually requires much more than a few tens of bases of sequence identity, and linear DNA is often degraded rapidly. However, several authors have exploited basic knowledge of *E. coli* and phage λ recombination to overcome DNA degradation and to enhance the basal recombination frequency greatly up to a level at which PCR targeting is possible. In a well-developed example of this (Datsenko and Wanner, 2000), recombination potential is elevated by placing the λ *red* genes under the control of the P_{araBAD} promoter, and exposing cells to arabinose, before introducing a suitable PCR product. With *ca* 40 bp of flanking targeting sequence to guide recombination, Datsenko and Wanner successfully isolated a number of *E. coli* chromosomal mutants by this technique. Recently, B. Gust (unpublished data) has PCR-amplified resistance genes selectable in both *E. coli* and *Streptomyces*, using primers whose 39-base 5'-ends are homologous to *Streptomyces* DNA. The amplicons are then used to transform λ Red-expressing *E. coli* cells containing the targeted *Streptomyces* gene as part of a *ca* 40 kb insert in a Supercos 1 cosmid (using the precisely defined cosmids used to sequence the genome: Redenbach *et al.*, 1996; Sanger Centre Home Page, ftp.sanger.ac.uk/pub/S_coelicolor/cosmid_inserts). Typically, many tens of resistant transformants are obtained, all with the expected gene replacement. The mutations are easily reintroduced into *Streptomyces* by conjugation, because the resistance gene templates also carry an *oriT* region. The rest of the procedure, leading to replacement of the equivalent region in *S. coelicolor*, is essentially as described above for Tn*5062*.

In a trial experiment, replacement of 50 distinct chromosomal genes by resistance cassettes was successfully carried out in cosmids in two weeks and 30 of the disruptions were introduced into the *S. coelicolor* chromosome in a further 6 weeks (B. Gust, unpublished data). As a result, functional attribution was possible to a number of genes, including some for secondary metabolism and morphological differentiation. Previous methods were so labour intensive that only a few disruptions could be made in parallel, and the time-scale for both the *E. coli* and the *S. coelicolor* manipulations was typically several times longer. The new procedure makes it feasible to construct many mutations in parallel in a short time and without conventional cloning steps. Indeed, one might now contemplate targeted mutagenesis of thousands of genes.

◆◆◆◆◆◆ GENOME-BASED MUTAGENESIS IN PRACTICAL SITUATIONS

The combination of efficient transposon-mediated local and global mutagenesis with analysis based on information from the complete genome raises the possibility of randomly defining the role and/or importance of every gene in the genome. PCR targeting permits the opposite – defining the role and/or importance of preselected genes. Examples of the latter include genes whose annotation indicates to the skilled interpreter that they are involved in the biosynthesis of secondary metabolites. Thus, G. L. Challis (personal communication) predicted that either or both of two gene sets containing possible determinants of sesquiterpene cyclases might be involved in making geosmin, the volatile compound famously responsible for the earthy odour of *Streptomyces*. A mutation in one of these cyclases was discovered by chance among the first 100 Tn*4560* insertions sequenced by K. Fowler and T. Kieser (unpublished data), and a mutation in the other was constructed by PCR targeting (B. Gust, unpublished data). Mutants carrying either or both mutations were screened by GCMS (gas chromatography mass spectrometry), and the double mutant and one of the single mutants were, indeed, found to lack geosmin production.

The PCR targeting technique allows multiple mutants to be constructed sequentially, with considerable ease. One aid to this is the availability of several different resistance genes in a form suitable for use as templates for PCR to generate disruption cassettes. A second approach is the elimination of the disrupting resistance cassette to leave a phenotypically silent 81 bp scar. This is achieved by the inclusion of target sites (FRT) for the yeast FLP recombinase on either side of the resistance gene-*oriT* templates. Expression of FLP from the pCP20 plasmid (Datsenko and Wanner, 2000) gives efficient and reliable excision in *E. coli* of sequences flanked by direct repeats of FRT, to leave an 81-bp scar with no stop codons. The 'silent' construct can then be used in the rather easy sequential build-up of multiple mutants.

◆◆◆◆◆◆ SUMMARY

The huge genome (8.7 Mb) and developmental complexity of *Streptomyces coelicolor* A3(2) present significant challenges for functional genomics. Nevertheless, effective microarrays of PCR-generated gene fragments have been generated in two laboratories, and significant information is accruing from proteome analysis. Genomewide transposon mutagenesis and highly efficient PCR-targeted gene disruption are making genetic intervention straightforward.

Acknowledgements

We thank Jason Hinds, Joe Mangan and Philip Butcher for their helpful advice on microarray technology, Mike Naldrett and Mervyn Bibb for crucial help in establishing proteomics, and Jason Hinds for advice on related software and for spotting the array products. Work on functional genomics in our laboratories was funded by the BBSRC's Technologies for Functional Genomics and Investigating Gene Function initiatives, and by a BBSRC Special Studentship.

References

Baudin, A., Ozier-Kalegeropoulos, O., Denouel, A., Lacroute, F. and Cullin, C. (1993). A simple and efficient method for directed gene deletion in *Saccharomyces cerevisiae*. *Nucleic Acids Res.* **21**, 3329–3330.

Bentley, S. D., Chater, K. F., Cerdeno-Tarraga, A. M., Challis, G. L., Thomson, N. R., James, K. D., Harris, D. E., Quail, M. A., Kieser, H., Harper, D., Bateman, A., Brown, S., Chandra, G., Chen, C. W., Collins, M., Cronin, A., Fraser, A., Goble, A., Hidalgo, J., Hornsby, T., Howarth, S., Huang, C. H., Kieser, T., Larke, L., Murphy, L., Oliver, K., O'Neil, S., Rabbinowitsch, E., Rajandream, M. A., Rutherford, K., Rutter, S., Seeger, K., Saunders, D., Sharp, S., Squares, R., Squares, S., Taylor, K., Warren, T., Wietzorrek, A., Woodward, J., Barrell, B. G., Parkhill, J. and Hopwood, D. A. (2002). Complete genome sequence of the model actinomycete *Streptomyces coelicolor* A3(2). *Nature* **417**, 141–147.

Bucca, G., Hindle, Z. and Smith, C. P. (1997). Regulation of the *dnaK* operon of *Streptomyces coelicolor* A3(2) is governed by HspR, an autoregulatory repressor protein. *J. Bacteriol.* **179**, 5999–6004.

Burke, J., Schneider, D. and Westpheling, J. (2001). Generalised transduction in *Streptomyces coelicolor*. *Proc. Natl Acad. Sci. USA* **98**, 6289–6294.

Chater, K. F. and Losick, R. (1997). Mycelial lifestyle of *Streptomyces coelicolor* A3(2) and its relatives. In *Bacteria as Multicellular Organisms* (Shapiro, J. A. and Dworkin, M., eds), pp. 149–182. New York: Oxford University Press.

Datsenko, K. A. and Wanner, B. L. (2000). One-step inactivation of chromosomal genes in *Escherichia coli* K-12 using PCR products. *Proc. Natl Acad. Sci. USA* **97**, 6640–6645.

Gehring, A. M., Nodwell, J. R., Beverley, S. M. and Losick, R. (2000). Genomewide insertional mutagenesis in *Streptomyces coelicolor* reveals additional genes involved in morphological differentiation. *Proc. Natl Acad. Sci. USA* **97**, 9642–9647.

Gehring, A. M., Yoo, N. J. and Losick, R. (2001). RNA polymerase sigma factor that blocks morphological differentiation by *Streptomyces coelicolor*. *J. Bacteriol.* **183**, 5991–5996.

Hopwood, D. A. (1999). 40 years of genetics with *Streptomyces*: from *in vivo* through *in vitro* to *in silico*. *Microbiology* **145**, 2183–2202.

Huang, J., Lih, C-J., Pan, K-H. and Cohen, S. N. (2001). Global analysis of growth phase responsive gene expression and regulation of antibiotic biosynthetic pathways in *Streptomyces coelicolor* using DNA microarrays. *Genes Development* **15**, 3183–3192.

Kieser, T., Bibb, M. J., Buttner, M. J., Chater, K. F. and Hopwood, D. A. (2000). *Practical Streptomyces Genetics*. Norwich: John Innes Foundation.

Mueller, P. R. and Wold, B. (1989). In vivo footprinting of a muscle specific enhancer by ligation mediated PCR. *Science* **246**, 780–786 (Erratum in *Science* **248**, 802).

Oh, S-H. and Chater, K. F. (1997). Denaturation of circular or linear DNA facilitates targeted integrative transformation of *Streptomyces coelicolor* A3(2): possible relevance to other organisms. *J. Bacteriol.* **179**, 122–127.

Puglia, A. M., Vohradsky, J. and Thompson, C. J. (1995). Developmental control of the heat-shock stress regulon in *Streptomyces coelicolor*. *Mol. Microbiol.* **17**, 737–746.

Redenbach, M., Kieser, H. M., Denapaite, D., Eichner, A., Cullum, J., Kinashi, H. and Hopwood, D. A. (1996). A set of ordered cosmids and a detailed physical and genetic map for the 8 Mb *Streptomyces coelicolor* A3(2) genome. *Mol. Microbiol.* **21**, 77–96.

Siemieniak, D. R., Slightom, J. L. and Chung, S. T. (1990). Nucleotide sequence of *Streptomyces fradiae* transposon Tn4556: a class II transposon related to Tn3. *Gene* **86**, 1–9.

Vohradsky, J., Li, X. M. and Thompson, C. J. (1997). Identification of prokaryotic developmental stages by statistical analysis of two-dimensional gel patterns. *Electrophoresis* **18**, 1418–1428.

Vohradsky, J., Li, X. M., Dale, G., Folcher, M., Nguyen, L., Viollier, P. H. and Thompson, C. J. (2000). Developmental control of stress stimulons in *Streptomyces coelicolor* revealed by statistical analyses of global gene expression patterns. *J. Bacteriol.* **182**, 4979–4986.

18 Functional Analysis of the *Bacillus subtilis* genome

Colin R Harwood[1], Anil Wipat[2] and Zoltán Prágai[1]

[1]*School of Cell and Molecular Biosciences* and [2]*School of Computing Science, Newcastle University, Newcastle upon Tyne NE2 4HH, UK*

CONTENTS

Introduction
The *Bacillus subtilis* 168 genome
Genome management
Analysis of genes of unknown function
Analysis of the transcriptome and proteome

◆◆◆◆◆◆ INTRODUCTION

Bacillus subtilis and close relatives

Members of the genus *Bacillus* are generally aerobic endospore-forming Gram-positive bacteria that are widely distributed in nature. With the principal exception of members of the *B. cereus* group (*B. anthracis*, *B. cereus* and *B. thuringiensis*), the genus comprises non-pathogenic species that are of proven value for both traditional (e.g. fermented soybeans) and newer biotechnologies (e.g. antibiotics, industrial enzymes, insecticides and vitamins). The genus has undergone extensive revision in recent years, with many species being diverted to newly established genera (e.g. *Alicyclobacillus, Brevibacillus, Paenibacillus*). *Bacillus subtilis* is the type species for the genus and, following the discovery that strain 168 could be transformed with naked DNA (Spizizen, 1958), this bacterium has been developed as a highly tractable model for Gram-positive bacteria and for the study of cellular differentiation processes. The accumulation, over 40 years, of knowledge of the biochemistry, genetics and physiology of *B. subtilis* has been enhanced during the last 10 years by systematic analyses of its genome, transcriptome and proteome. This is reflected by the publication of books on the methodology (Harwood and Cutting, 1990; Schumann *et al.*, 2001) and biology (Sonenshein *et al.*, 1993, 2002) of *B. subtilis* and its relatives.

Most species of *Bacillus* will grow at mesophilic temperatures on commercially prepared nutrient media, although for some species it is necessary to modify the pH or salt concentration. Obligate thermophilic species, such as *B. stearothermophilus*, are usually grown at 60°C while moderately thermophilic species, such as *B. coagulans*, are grown between 45°C and 50°C. *B. subtilis* and many other species are able to grow in simple salts media containing glucose or other simple sugars as sources of carbon, and ammonium salts or certain amino acids as sources of nitrogen (Harwood and Archibald, 1990). However, since *B. subtilis* encodes substrate-induced catabolic pathways for a number of amino acids (e.g. arginine, glutamine, glutamate, asparagine and aspartate), care needs to be taken to ensure that the growth of strains that are auxotrophic for these amino acids do not become limited by their availability. It is also worth noting that *B. subtilis* strain 168, on which most studies are performed, is a tryptophan auxotroph and requires tryptophan for growth even in media with acid-hydrolysed casein as the main source of nitrogen. A widely used minimal medium for *B. subtilis* is the chemically defined Spizizen's minimal medium (Spizizen, 1958). Some species of *Bacillus* have specific growth requirements, such as *B. stearothermophilus*, which requires additional calcium and iron, and *B. pasteuri*, which requires urea. The more fastidious insect pathogens, *B. larvae* and *B. popilliae*, require thiamine for growth and are usually grown at or below 30°C.

Although *B. subtilis* forms endospores in response to nutrient deprivation (e.g. carbon, nitrogen or phosphate), special media and growth protocols are required for efficient sporulation. Schaeffer's sporulation medium is widely used for efficient spore production in the laboratory (Schaeffer *et al.*, 1965). The production of a pink pigment during sporulation provides a good visual indication of the formation of spores both in liquid and on solid media.

Bacillus subtilis survives well on agar plates, either at room temperature or at 4°C, although it is recommended to subculture on a weekly basis. Viable cells may even be recovered from severely dehydrated agar plates that encourage sporulation. For long-term preservation, *B. subtilis* is stored as glycerol or lyophilized cultures. Strains that sporulate well (i.e. not sporulation mutants) may also be preserved as spore suspensions in sterile water at 4°C. Provided they have been extensively washed to remove nutrients that may cause them to germinate, spores are generally stable for many years. *Bacillus subtilis* can be transported on the surface of freshly inoculated nutrient agar slopes or as spore suspensions spotted on sterilized filter paper discs and encased in aluminium foil.

Cultures of *B. subtilis* and close relatives, such as *B. licheniformis* and *B. amyloliquefaciens*, are available from a variety of international culture collections and a comprehensive list has been published previously (Claus and Fritze, 1989). In addition, the *Bacillus* Genetic Stock Center (http://bacillus.biosci.ohio-state.edu) has an extensive collection of mutant *B. subtilis* strains, bacteriophages and plasmids. The collection also includes strains of *B. cereus*, *B. licheniformis*, *B. megaterium*, *B. pumilus*, *B. stearothermophilus* and *B. thuringiensis*. BGSC produces a catalogue that can be requested by e-mail at 'zeigler@osu.edu'.

♦♦♦♦♦♦ THE *BACILLUS SUBTILIS* 168 GENOME

Genome sequencing

The genome of *Bacillus subtilis* strain 168 was sequenced by a joint European/Japanese consortium. The resulting sequence, completed in June 1997, was the first for both a Gram-positive bacterium and a differentiating bacterium (Kunst *et al.*, 1997). Extensive pre-existing genetical (Anagnostopoulos *et al.*, 1993) and physical maps (Itaya and Tanaka, 1991) meant that specific regions of the chromosome could be allocated to individual research groups who employed a variety of sequencing strategies. Library-based approaches used phage λ-derived vectors (Wipat *et al.*, 1996) or an ordered collection of yeast artificial chromosomes (YAC) (Azevedo *et al.*, 1993; Sorokin *et al.*, 1996), in combination with high copy-number pUC- or M13-based sequencing vectors. In both cases, particular fragments of *B. subtilis* DNA exhibited instability, probably because of the tendency of *E. coli* to overexpress AT-rich DNA. The problem was usually overcome by use of low copy number vectors or an *E. coli* host that maintains ColE1-based vectors at a low copy number.

An alternative vector-based chromosome walking technique exploited the ease with which integration vectors can be directed to specific locations in the *B. subtilis* chromosome via a single cross-over recombination event (see later). The vector, containing a 200–400 bp fragment of sequenced DNA (the target site) adjacent to the region to be sequenced, is transformed into the *B. subtilis* host. After selecting for transformants in which the vector has integrated at the target site (i.e. integrants), chromosomal DNA is extracted and digested with restriction enzymes that cut neither the vector nor target site, but which do cut adjacent chromosomal DNA. After dilution and ligation, recombinants containing DNA adjacent to the target site are recovered in *E. coli* and sequenced (Glaser *et al.*, 1993).

Inverse and long-range PCR techniques, developed during the latter stages of the sequencing programme, were ultimately responsible for increasing the rate of sequencing. The amplified DNA was either sequenced directly (Rose and Entian, 1996) or used to generate shotgun libraries of fragments ranging from 1 to 1.5 kb (Wipat *et al.*, 1996).

In silico analysis of the *B. subtilis* genome

The genome of *B. subtilis* is currently estimated to be 4 214 630 in size, comprising 4106 protein encoding genes, 86 tRNA genes, 30 rRNA genes and three stable RNA genes. These are organized into approximately 1500 operons that are controlled by some 200 regulatory proteins. The annotation of the genome, carried out primarily at the Institut Pasteur (Moszer, 2002), was facilitated by the relative ease with which ribosome binding sites (RBS) are identified: the absence of a homologue of the *E. coli* S1 ribosomal protein appears to be compensated by a strong interaction between the RBS and the cognate sequence close to the 3′-end of the 16S rRNA. Importantly, the genome continues to be checked for errors and the SubtiList database continues to be updated, both with respect to detected errors and information on gene names and function (Moszer et al., 1995).

Slightly less than 90% of the genome encodes proteins or stable RNA. Most of the remaining DNA appears to be structural or to be involved in gene regulation. However, a few grey holes exist, two of which (0.7 kb and 2.2 kb) are associated with genes involved in teichoic acid synthesis (Lazarevic et al., 2002). At least 5% of the chromosome appears to have been acquired recently by horizontal gene transfer (Rocha et al., 1999). Although B. subtilis encodes ten transposon/insertion sequence-like proteins, there is an apparent absence of transposable genetic elements. There are, however, ten copies of a 190 bp repeated element that appears to code for an RNA molecule of unknown function (Kunst et al., 1997). An analysis of long repeats (>24 bp) found 170 repeats of strict homology in B. subtilis, the lowest level for any microbial genome analysed to date. When repeats in prophages and rRNA were removed (about 40%), the remaining repeats were clustered on the genome with an average separation distance of about 10 kb. This pattern of repeats has been tentatively suggested to be related to the acquisition of foreign DNA, since they are associated with DNA with characteristics that are typical of horizontally transferred genes (Rocha et al., 1999).

Just over half of the genome is required for cell processes, intermediary metabolism and for macromolecular synthesis, while a significant proportion of the remaining genome is required for growth and survival in the environment (Sonenshein et al., 2002). The ability to transport substrates and to remove toxic substances is clearly important to survival, as illustrated by the presence of about 400 transporters, including approximately 50 ABC transporters. Finally, B. subtilis senses the environment via 34 signal transducers, one of which (YycF/G) is essential for growth.

The B. subtilis is a low %G + C bacterium (43.5%). Nevertheless, variation in G + C content and codon usage are found throughout the chromosome. One group of genes, believed to have been acquired by horizontal gene transfer, is particularly A+T-rich. They include prophage SPβ and defective prophages PBSX and the *skin* element, which were identified prior to the sequence determination, and seven additional prophage-like elements, which together account for about 10% of the chromosome.

Under conditions of nutrient deprivation or stress, B. subtilis initiates a series of responses designed to maintain or restore growth. These include the induction of macromolecular hydrolases (e.g. proteases and carbohydrases), chemotaxis and motility, transport systems and competence (i.e. the ability to take up DNA from the environment). For the most part these responses are transitional; if they fail to re-establish growth, sporulation is induced. This ability of *Bacillus* species to form resistant endospores imparts an enormous competitive advantage in environments such as soil, where long periods of drought and nutrient deprivation are common. About 5% of the B. subtilis genome is dedicated to the processes of sporulation and germination.

In addition to specific stress responses, the B. subtilis genome encodes more than 200 general stress proteins, members of the sigma-B (σ^B) general stress regulon, that provide a non-specific resistance to stress by protecting DNA, membranes and proteins from the damage. These proteins include ATPase-dependent chaperones and associated proteases, enzymes involved in protection from oxidative stress, and proteins involved in drug efflux and the uptake of osmoprotectants.

Sigma-B is one of 17 sigma factors involved in the transcription cycle of B. subtilis (Helmann and Moran, 2002). The principal sigma factor, Sigma-A (σ^A), exhibits a similar consensus sequence to that of E. coli σ^{70}. Of the remaining 15 sigma factors, four are required for sporulation specific (σ^E, σ^F, σ^G, σ^K), one for transition-phase activities, including sporulation initiation and competence (σ^H), one for levanase and amino acid catabolism (σ^L), one for chemotaxis and motility (σ^D), one for salt resistance (σ^M), while the remaining seven sigma factors are members of the extracellular function (ECF) group.

B. subtilis can utilize a wide range of substrates and analysis of the genome reveals the presence of a large number of transporter proteins for the uptake or efflux of particular substrates (Saier et al., 2002). These include six voltage-gated ion channel proteins, 185 secondary active transporters, and more than 70 ABC transporters (or components thereof).

As with E. coli, the analysis of the B. subtilis chromosome has revealed high-level organization and relationships. Highly expressed genes are typically found close to the origin of replication, presumably because of the resulting gene dosage effect, while prophages and genes that have been acquired via horizontal transfer are typically found near the replication termination region. Highly expressed genes and genes expressed during growth tend to be oriented in the direction of the replication fork, presumably to reduce conflicts between replication and transcription (Kunst et al., 1997). Another high-level relationship results from the different modes of replication of the leading and lagging strands and consequential differences in mutational biases. This leads to differences in the distribution of C and G nucleotides on the two strands that invert at the replication terminus. In B. subtilis this leads to an over-representation of valine, glutamic acid, arginine and aspartic acid on the leading strand and serine, phenylalanine, leucine, isoleucine and histidine on the lagging strand (Rocha et al., 2001).

◆◆◆◆◆◆ GENOME MANAGEMENT

Competence and recombination

The ability to introduce isolated DNA into B. subtilis and, where necessary, to have it recombine with sequences in the chromosome are at the heart of its extreme genetic amenability. The most widely used method for introducing DNA into B. subtilis is the transformation of naturally competent cells, although protoplasts of *B. subtilis* and several other *Bacillus* species can be efficiently transformed by naked DNA and their cell walls regenerated (Chang and Cohen, 1979). Electrotransformation (electroporation) is possible, but usually results in low efficiencies (Bron, 1990). DNA can also be introduced in *B. subtilis* by transducing phages, although the use of transduction is nowadays limited to very specific purposes.

The transformation of *B. subtilis* was first described in 1958 by Spizizen (Spizizen, 1958; Anagnostopoulos and Spizizen, 1961). The development of competence to take up DNA is one of several post-exponential phase phenomena that are a characteristic of this bacterium's life cycle, and which includes the production of peptide antibiotics, the secretion of macromolecular hydrolases and sporulation. Competence develops naturally shortly after the transition from exponential to stationary phase and is, in part, promoted by a quorum-sensing mechanism involving secreted oligopeptides (Dubnau and Lovett, 2002). Maximally, only about 10% of the cells in a population become competent. Competence is best documented for *B. subtilis* strain 168 and has been confirmed for only a limited number of other *Bacillus* species. The size of DNA fragments that can be taken up is about 20–30 kb (Dubnau, 1993). With saturating amounts of DNA (> 1 μg/ml of competent cells), transformation frequencies of up to 5% of the cells in a population can be achieved with homologous chromosomal DNA. Under these conditions, the co-transfer of unlinked genetic markers is possible, a phenomenon referred to as congression. Congression can be used for the introduction of non-selectable genes into the chromosome since approximately 1% of recombinants will also contain the required non-selected gene.

Transformation of plasmid DNA into *B. subtilis* by natural transformation is possible, although the frequency at which the plasmid becomes established is usually relatively low; between 0.001% and 0.01% for intact plasmids, and one or two orders of magnitude lower for ligation mixtures. This is because transforming DNA is converted into a single-stranded form and randomly fragmented during entry in the competent cell. Consequently, only multimeric plasmid DNA (present in most *Bacillus* plasmid preparations) or monomers containing the internal repeats required for recircularization, are effective in plasmid-mediated transformation.

Plasmid-based host–vector systems

Most plasmids from Gram-positive bacteria use the rolling-circle mode of replication, which is characterized by the uncoupled synthesis of leading and lagging strands. A characteristic for rolling circle replication is the formation of single-stranded DNA intermediates that are subsequently converted to double-stranded DNA. This requires a secondary replication origin known as the single-strand origin (SSO). SSOs are often active in a limited number of strains and, although dispensable, affect the efficiency of replication and plasmid stability. The copy number of rolling circle plasmids in *B. subtilis* can vary from about 5 to 200 per chromosome.

A number of theta replicating plasmids have been developed as vectors for *B. subtilis*, including the enterococcal plasmid pAM81 (Bruand *et al.*, 1991; Janniere *et al.*, 1993) and the endogenous *B. subtilis* plasmid pLS20 (Meijer *et al.*, 1995b). Despite the fact that these plasmids tend to be more stable than rolling circle plasmids, they are still not widely used as cloning vectors.

The absence of native antibiotic resistance plasmids meant that vectors for *B. subtilis* were initially developed using plasmids from other Gram-positive bacteria, such as *Staphylococcus aureus* and *Lactococcus lactis* (Bron, 1990; Janniere *et al.*, 1993). The replication functions and/or antibiotic resistance genes from several of these plasmids are still in common use (e.g. pUB110, pC194, pE194 and pWVO1). More recently, vectors based on endogenous *Bacillus* plasmids have been developed.

S. aureus-derived plasmids (e.g. pUB110, pC194, and pE194), although still widely used, are relatively unstable in *B. subtilis*, particularly when carrying large inserts (Bron, 1990). pWVO1, from *L. lactis*, is a small (2178 bp) cryptic rolling circle plasmid (Leenhouts and Venema, 1993) that replicates stably in *B. subtilis*, a variety of other Gram-positive bacteria and even in the Gram-negative *E. coli*. pWVO1 (which has a copy number of about five in *B. subtilis* but 50–100 in *E. coli*) has been used as the basis for a variety of special-purpose vectors. The plasmid-encoded RepA function of this plasmid can be provided in *trans*, which means that, with the provision of a suitable *E. coli* helper strain, this plasmid can even be used as the basis of a *B. subtilis* integration system (see Figure 18.4) (Leenhouts, 1995; Leenhouts *et al.*, 1996).

More recently, a number of cryptic *Bacillus* rolling circle plasmids have been developed into vectors (Meijer *et al.*, 1998). pTA1060, a rolling circle plasmid from *B. subtilis*, is 8.6 kb in size and has a copy-number of five per chromosome. Its SSO is very efficient in *B. subtilis* (Meijer *et al.*, 1995a) and derivatives of pTA1060 can stably carry inserts up to at least 30 kb.

Derivatives of the enterococcal plasmid pAMB1, a theta replicating plasmid, have been used as the basis of a series of vectors for *B. subtilis* (Janniere *et al.*, 1990). pHV1431 is a *B. subtilis*/*E. coli* shuttle plasmid, carrying the pBR322 replication functions for maintenance in *E. coli* and the pAMB1 replication functions for *B. subtilis*. This plasmid has a copy number of 200 in *B. subtilis* and long inserts are generally maintained stably.

For many purposes, stable cloning into *B. subtilis* is best achieved using an integrative vector. The methods, relative advantages and applications of integrative plasmids are discussed later in relation to the *B. subtilis* functional analysis programme.

Promoters for controlled gene expression

Relatively few systems have been developed for controlled high-level expression of genes in *B. subtilis*. Industrial applications tend to use native promoters (e.g. promoters from α-amylase genes) capable of directing the sustained synthesis of extracellular proteins during stationary phase to concentrations of about 20 g/l. The following controllable promoters are widely used for research purposes.

- P_{spac} promoter. A widely used promoter formed by fusing the 5'-sequences of a promoter from the B. subtilis phage SPO1 and the 3'-sequences of the E. coli lac promoter, including the operator region (Yansura and Henner, 1984). The controllability of P_{spac} is dependent on the repressor encoded by the E. coli lacI gene, modified to facilitate expression in B. subtilis. Although not a particularly strong promoter, P_{spac} is inducible 50-fold with 1–10 mM isopropyl-β-D-thiogalactopyranoside (IPTG).
- XylR-controlled promoters. The xylose-inducible promoter/operator elements of B. subtilis have been used without modification to control gene expression (Gartner et al., 1992). This system is regulated by the XylR repressor that is inactivated in the presence of xylose. If the system is used on a high-copy-number expression vector, a copy of the xylR gene needs to be included to maintain a balance between the number of repressor molecules and operator sites. XylR-controlled promoters direct moderately high levels of expression.
- sacB promoter. The inducible expression of sacB, encoding levansucrase, is controlled positively by sucrose, the SacY antiterminator and the products of degQ and sacU, and negatively by SacX (Crutz et al., 1990). This promoter can be induced during exponential growth and is not subject to catabolite repression. The level of induction can be modulated by using levels of sucrose from 1 to 30 mM.

Reporter genes

Gene fusion has proved to be an effective means of studying gene expression and protein location in B. subtilis and several reporter genes have been used in this bacterium.

- The lacZ from E. coli is a widely used reporter gene for monitoring gene expression in Bacillus, including the ability to detect gene expression at the single cell level using combined cytochemical and video microscopy techniques (Lewis et al., 1994). lacZ expression can be detected on solid media using the chromogenic substrate X-gal or the fluorogenic substrate MUG (4-methylumbelliferyl-β-D-galactopyranoside) (Youngman, 1990). The lacZ reporter is not suitable for studying gene expression during heat shock, since E. coli β-galactosidase is degraded rapidly under these conditions. In its place, the β-galactosidase encoded by the B. stearothermophilus bgaB gene has been adapted for this purpose (Yuan and Wong, 1995).
- The xylE gene from Pseudomonas putida, specifying catechol-2, 3-dioxygenase, is a useful reporter gene in Bacillus sp. for analysing expression from strong promoters (Zukowski et al., 1983). Catechol-2, 3-dioxygenase converts colourless catechol to the yellow compound 2-hydroxymuconic-semialdehyde, which can be detected in colonies or spectrophotometrically.
- The luxAB genes from Vibrio harvei encode a luciferase which emits light when exposed to a suitable substrate (e.g. Decanal). A chromosomally located

luxAB gene fusion has been used as a reporter for tracking *B. subtilis* in soil (Cook *et al.*, 1993).
- The *gfp* gene from the jellyfish *Aequorea victoria* encodes the green fluorescent protein (GFP), one of the most versatile reporters (Prasher *et al.*, 1992). This small protein (27 kDa) fluoresces as a result of the autocatalytic cyclization of amino acids Ser-65 and Tyr-67, and their subsequent oxidation (Heim *et al.*, 1994). Wild-type GFP is excited at 395 nm and emits green light at 590 nm. However, the isolation of blue- and red-shifted fluorescent derivatives has facilitated the use of GFP for dual-labelling experiments. GFP has been used extensively in *B. subtilis* for the localization of protein involved in sporulation (Webb *et al.*, 1995) and cell morphogenesis (Jones *et al.*, 2001).

◆◆◆◆◆◆ ANALYSIS OF GENES OF UNKNOWN FUNCTION

The 4.2 Mbp genome encodes 4106 proteins, more than a third of which are currently of unknown function (Table 18.1). Although the targeted sequencing approach was superseded by TIGR's shotgun cloning and sequence assembly strategy (Fleischmann *et al.*, 1995), it had the advantage of producing a cohesive research community that subsequently initiated a programme for the systematic analysis of genes of unknown function. Indeed, the absence of extensive functional analysis programmes for many sequenced organisms is a major bottleneck in the exploitation of sequenced bacterial genomes. The systematic approach adopted for the genome sequencing has been extended to include: (1) a joint European/Japanese functional analysis programme for the construction and phenotypic characterization of a set of isogenic mutants for virtually every gene of unknown function; (2) a programme aimed at defining the *B. subtilis* secretome (Tjalsma *et al.*, 2000) and adapting it for high-level production of heterologous proteins (http://www.ncl.ac.uk/ebsg/); (3) a genome minimizing programme designed to maximize the efficiency of *B. subtilis* as a host for the synthesis of proteins and fine biochemicals; and (4) a programme designed to model regulatory networks in *B. subtilis* through the analysis of global gene expression in response to environmental stress (http://www.ncl.ac.uk/bacellnet/).

Data from the *B. subtilis* genomic and postgenomic consortia have been compiled into databases that are available over the Internet. These include:

- Subtilist (http://genolist.pasteur.fr/SubtiList/) (Moszer *et al.*, 1995), a dedicated DNA sequence database;
- Micado (http://locus.jouy.inra.fr/cgi-bin/genmic/madbase/progs/madbase.operl/) (Biaudet *et al.*, 1997; Samson *et al.*, 2001) and JAFAN (http://bacillus.genome.ad.jp/), which include data on the characterization of mutants generated during the functional analysis programme;
- Sub2D (http://microbio2.biologie.uni-greifswald.de:8880/) (Bernhardt *et al.*, 2001), a dedicated proteome database;

Table 18.1. Categorization of the ORFs of the genome of B. subtilis strain 168; data extracted from Appendix 2 of Sonenshein et al. (2002)

Category	Area of metabolism	Number of genes (%)
1	*Cell envelope and cellular processes*	928 (22.6%)
1.1	Cell wall (89)	
1.2	Transport and lipoproteins (400)	
1.3	Sensor proteins (39)	
1.4	Membrane bioenergetics (82)	
1.5	Mobility and chemotaxis (55)	
1.6	Protein secretion (26)	
1.7	Cell division (22)	
1.8	Sporulation (164)	
1.9	Germination (26)	
1.10	Competence and transformation (25)	
2	*Intermediary metabolism*	774 (18.8%)
2.1	Carbohydrates (271)	
2.2	Amino acids (201)	
2.3	Nucleotides and nucleic acid (92)	
2.4	Lipids (89)	
2.5	Co-enzymes and prosthetic groups (103)	
2.6	Phosphate (10)	
2.7	Sulphur (8)	
3	*Information pathways*	528 (12.9%)
3.1	DNA replication (26)	
3.2	DNA restriction, modification and repair (42)	
3.3	DNA recombination (19)	
3.4	DNA packaging and segregation (11)	
3.5	RNA synthesis (254)	
3.6	RNA modification (28)	
3.7	Protein synthesis (101)	
3.8	Protein modification (35)	
3.9	Protein folding (12)	
4	*Other functions*	334 (8.1%)
4.1	Adaptation to stress (81)	
4.2	Detoxification (89)	
4.3	Antibiotic production (35)	
4.4	Phage-related functions (87)	
4.5	Transposon and IS-related (10)	
4.6	Miscellaneous (30)	
5	*Similar to unknown proteins*	921 (22.4%)
6	*No similarity to other proteins*	623 (15.2%)

- Subscript, a database currently under development for the storage and analysis of transcriptomic data (Harwood and Moszer, 2002);
- DBTBS (http://elmo.ims.u-tokyo.ac.jp/dbtbs/), a database dedicated to promoters and transcription factors;

- SPID (http://www-mig.versailles.inra.fr/bdsi/SPiD), a database dedicated to protein–protein interactions (Hoebeke et al., 2001).

Databases that are not specific for B. subtilis but which contain relevant information are:

- KEGG (http://www.genome.ad.jp) contains current information on the metabolic pathways of a range of organisms, including B. subtilis;
- TIGR microbial database (http://www.tigr.org/tdb/mdb/mdb.html), which has genomic information on species related to B. subtilis (e.g. B. anthracis).

Bacillus subtilis functional analysis programme

Despite extensive knowledge of its biochemistry and physiology, a little under a third of B. subtilis genes have functions that have been confirmed experimentally, while the function of a similar proportion can be identified on the basis of their close similarity to genes of known function in other organisms. The remaining genes (approximately 1550) either show homology to genes of unknown function in other organisms or are (currently) unique to B. subtilis (Table 18.1). The genetic amenability of B. subtilis has resulted in this bacterium being used as the basis for a functional analysis programme in which ORFs of unknown function have been investigated systematically. The B. subtilis Function Analysis (BFA) project, in which more than 30 European and Japanese groups jointly participated, was initiated in 1996. The project, coordinated by S. D. Ehrlich (Jouy en Josas, France) and N. Ogasawara (Nara, Japan), was divided into two components, a *resource consortium* and a function consortium. The *resource consortium* was responsible for: (1) the construction and initial characterization of a set of isogenic mutants in target genes of unknown function; (2) the construction of transcription maps; (3) the analyses of cellular proteins and cell composition; and (4) the development of databases. The *function consortium* was responsible for allocating genes to specific areas of metabolism or cellular processes: (1) the metabolism of small molecules and inorganic molecules (e.g. carbon, nitrogen and sulphate); (2), macromolecular metabolism (DNA, RNA and proteins); (3), cell structures and mobility (e.g. cell envelope, motility); (4), stress and stationary phase; and (5), cell processes (e.g. cell cycle, competence, sporulation and germination). Mutants were characterized at up to three levels of detail. At the primary level all mutants were screened using relatively simple high-throughput tests. Mutants showing relevant characteristics were subjected to secondary and sometimes tertiary level tests of increasing complexity by groups with specific expertise. The data are available to the scientific community through a public domain on the Micado database (see earlier).

Gene regulation can also assist in the identification of gene function, although not necessarily to a level that will allow the precise activity of the encoded protein to be described. Genes in the same operon, and genes and operons in the same regulon, usually contribute to the same function or metabolic pathway. Thus, if the function of a gene in an operon or regulon is known, it provides clues as to the function of the other genes with which it is co-ordinately regulated.

Generation of chromosome mutations

The identification of genes currently of unknown function is a major challenge of postgenomics, and gene disruption is an important approach to studying the activity of such genes. The BFA mutant collection was constructed using a common technological platform that exploited the ease with which *B. subtilis* takes up and recombines homologous DNA. The BFA collection of isogenic mutants is based around the pMUTIN series of integration/reporter gene vectors (Vagner et al., 1998). Chromosome integration systems provide powerful tools for gene technology because *B. subtilis* is:

- readily transformed with isolated plasmid or chromosomal DNA;
- not able to sustain the replication of *E. coli*-based plasmids with ColE1 or pBR322 origins of replication;
- extremely recombinagenic, allowing for cross-over events between small regions of homology (ca 200 bp) at a relatively high frequency.

Transformation of competent cells with homologous DNA fragments cloned on to plasmids that are not able to replicate in *B. subtilis* is the preferred method for generating targeted mutations. However, plasmids with temperature-sensitive Gram-positive origins of replication have also been used and may be necessary for *Bacillus* species that are less recombinagenic. Integration can occur by either single or double cross-over recombination events.

Single cross-over recombination

In single cross-over recombination, also referred to as 'Campbell-type' integration, the vectors are usually based on *E. coli* plasmids such as the pUC series. In cases where the *Bacillus* DNA sequences are toxic when overexpressed, low-copy-number plasmids (e.g. based on pSC101) can be used. The integration vector must encode an antibiotic-resistance gene that can be selected in *B. subtilis* in the single-copy state, and genes encoding chloramphenicol, kanamycin/neomycin, erythromycin and tetracycline resistance are available for this purpose. A fragment of homologous *B. subtilis* DNA, no smaller than about 150 bp but preferably a little larger to increase the efficiency of integration, is cloned in the vector. After propagation in *E. coli*, competent *B. subtilis* cells are transformed with the construct and integrants in which the entire plasmid has integrated into the chromosome are selected using the cognate antibiotic resistance marker. The chromosome of the integrant contains a copy of the vector and a duplication of the cloned fragment (Figure 18.1). Integrants are relatively stable, with excision occurring at a frequency of about 10^{-5} per cell generation. The outcome of a single cross-over event with respect to the functionality of target genes is dependent on the structure of the cloned fragment. If the fragment carries an intact gene, two functional copies will be present on the chromosome (Figure 18.1A). If the fragment contains one or other of the ends of the target gene, then one functional and one deleted copy will be present on the chromosome (Figure 18.1B). However, if the fragment contains sequences that are internal to the target gene, no functional copies will be present on the chromosome after integration (Figure 18.1C).

A specialized use of single cross-over recombination is the generation of 'clean' deletions (Figure 18.2), in which target sequences are removed from the chromosome without their replacement with a marker gene (Leenhouts et al., 1996). One such system, which can also be used for unlabelled gene replacements, is based on a derivative of the lactococcal plasmid pWV01 that lacks the *repA* gene encoding the replication initiation protein. The pORI series of vectors can, therefore, only replicate in Gram-positive or Gram-negative helper strains that provide RepA in *trans*. One such integration vector, pORI240 (Figure 18.2) encodes a tetracycline resistance gene, *lacZ* reporter and multiple cloning site (MCS). Sequences flanking the target gene are cloned together into the MCS on the integration vector and amplified in a host able to synthesize RepA. After transformation into *B. subtilis*, integrants resulting from a single cross-over recombination are selected on tetracycline plates and checked for their production of blue colonies in the presence of the chromogenic β-galactosidase substrate X-gal (5-bromo-4-chloro-3-indoylgalactoside). Although Figure 18.2 shows a cross-over between the 5'-flanking ends, it can occur with equal frequency between the 3'-flanking ends. The final step is to screen for an excision event between the flanking ends not involved in the original cross-over recombination, in the example shown in Figure 18.2, between the 3'-flanking ends. Since this occurs at a low frequency, and currently no mechanism exists to select for these directly in *B. subtilis*, the excision event is screened on plates containing X-gal, looking for white rather than blue colonies. Strains resulting from an excision event will also be sensitive to tetracycline. Since the excision event could equally involve a simple reversal of the original integration event, which would leave the target gene in place on the chromosome, the strains need to be tested for the loss of the target gene function and/or by a diagnostic polymerase chain reaction (PCR) across the flanking sequences.

Single cross-over recombination has been used in the genome minimalization project to generate clean deletions of specific genes and even larger regions of the chromosome, including prophages and the genes encoding polyketide antibiotics.

Double cross-over recombination

In contrast to single cross-over recombination, double cross-over recombination (also known as 'replacement recombination') results in only one copy of the target DNA fragment. Typically, a region of chromosomal DNA is replaced with either foreign DNA or mutationally altered homologous DNA. Sequences either side of the chromosome target are incorporated

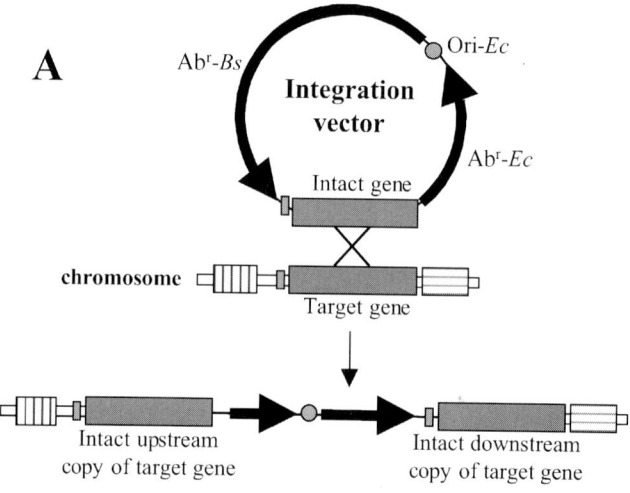
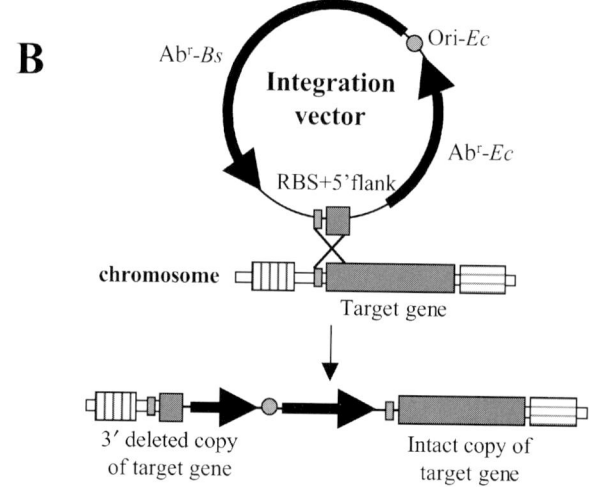
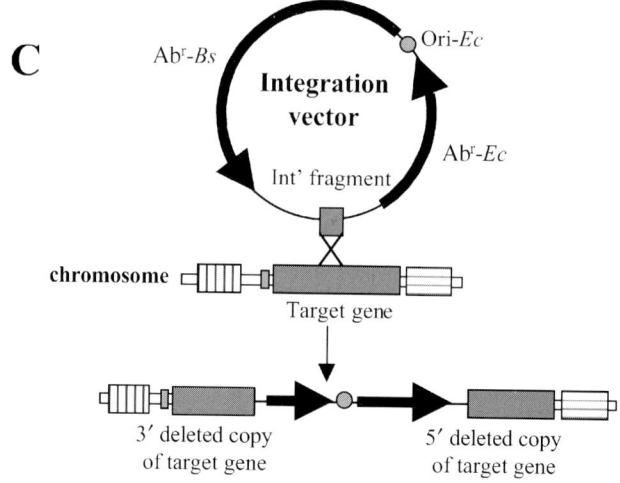

into the integration vector and all or part of the target gene sequence is replaced with a selectable marker gene and, if required, additional genes. A double cross-over recombination between the vector and target sequence on the chromosome leads to the integration only of the sequences between the flanking regions (Figure 18.3). Linearizing the vector before transformation forces the selection of integrants that have resulted from a double rather than single cross-over recombination, since the latter would be lethal to the host.

A special application of double cross-over recombination is shown in Figure 18.3, where the homologous fragments of the vector encode fragments of the front (5′) and back (3′) ends of the *B. subtilis amyE* gene specifying α-amylase. Integration places the cloned fragment, which may be native or foreign DNA, within the *amyE* gene. The inactivation of the latter provides a useful phenotype that can be screened on starch plates stained with iodine. The method is used for the stable introduction of a single copy of native and foreign DNA, for instance, for gene expression or complementation studies.

Construction of BFA mutants

During the BFA programme target genes were mainly inactivated via single cross-over recombination (Figure 18.1) using the pMUTIN series of integration vectors (Figure 18.4). pMUTIN has the following properties:

- a ColEI replication origin that is functional in *E. coli* but not *B. subtilis*;
- two antibiotic resistance genes, one (Ap^R) functional in *E. coli* and the other (Em^R) in *B. subtilis*;
- a *lacZ* reporter gene preceded by a *B. subtilis spoVG* ribosome binding site (Perkins and Youngman, 1986) and stop codons in all three reading frames to avoid translational fusions with upstream genes;
- an IPTG-inducible P_{spac} promoter (Yansura and Henner, 1984) with an RNA polymerase recognition sequence from phage SPO1 and an *E. coli* OI *lac* operator;
- a *lacI* gene under the control of a *B. licheniformis* promoter – the LacI protein interacts with the *lac* operator of P_{spac} unless inactivated with IPTG.

DNA fragments containing either an internal region of the target gene (knockout mutant) or the ribosome binding site and the 5′-end (RBS-fusion mutant) were amplified by PCR, cloned into pMUTIN and propagated in *E. coli* (Figure 18.5A). The integrity of the resulting plasmids was confirmed by diagnostic PCR (Figure 18.5A, C) prior to transforming into *B. subtilis*. Integrants were selected for their ability to grow in the presence of erythromycin, the antibiotic resistance that is elaborated in *B. subtilis*.

Figure 18.1. The outcome of single cross-over recombination using an integration vector and various gene fragments. (A) Whole gene fragment, including ribosome binding site. (B) 5′-Flanking region and associated ribosome binding site. (C) Internal fragment. Ab^r-Bs, antibiotic resistance gene functional in *B. subtilis*; Ab^r-Ec, antibiotic resistance gene functional in *E. coli*; Ori-Ec, origin of replication functional in *E. coli*; RBS, ribosome binding site.

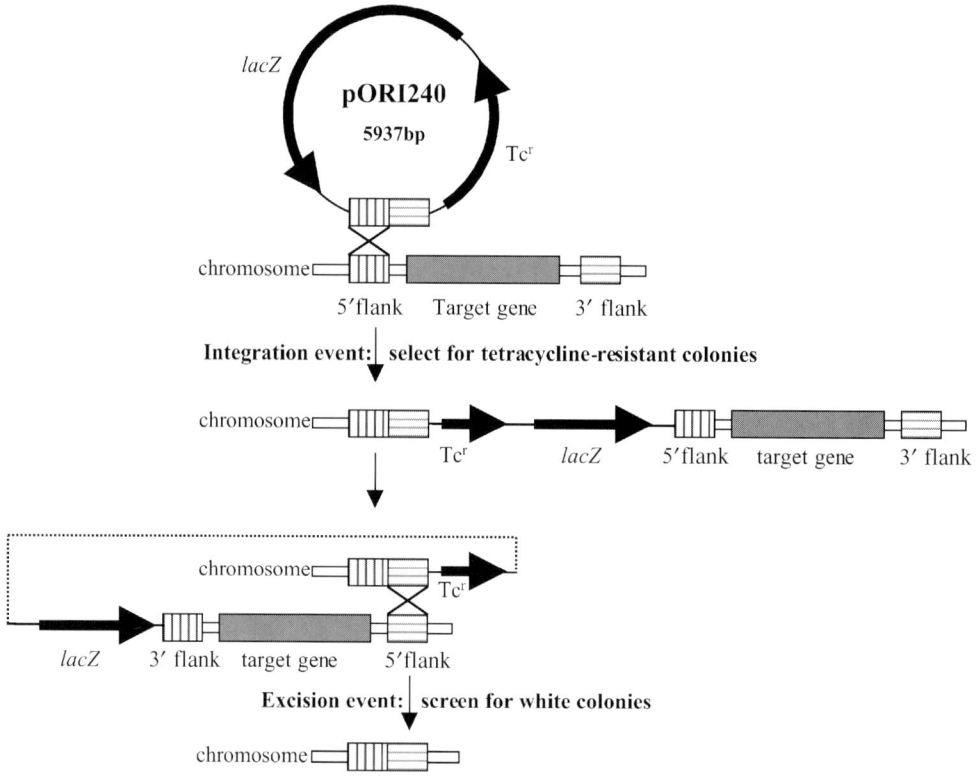

Figure 18.2. The generation of 'clean' chromosome deletions using the pVWO1-based vector pORI240. *lacZ*, encoding β-galactosidase from *E. coli*; Tc^r, tetracycline resistance gene.

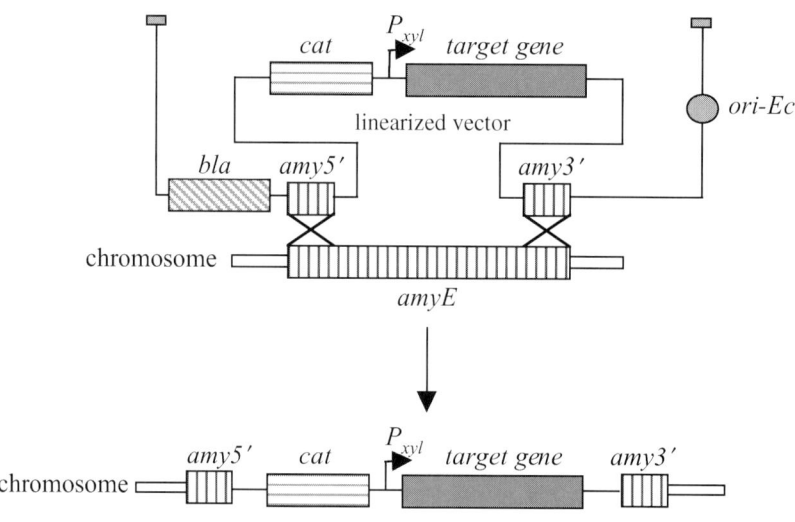

Figure 18.3. The use of the *B. subtilis amyE* gene and double cross-over recombination for replacement recombination. *amyE*, chromosomal gene encoding α-amylase; *cat*, chloramphenicol resistance gene; Ori-*Ec*, origin of replication functional in *E. coli*; P$_{xyl}$, xylose-inducible promoter.

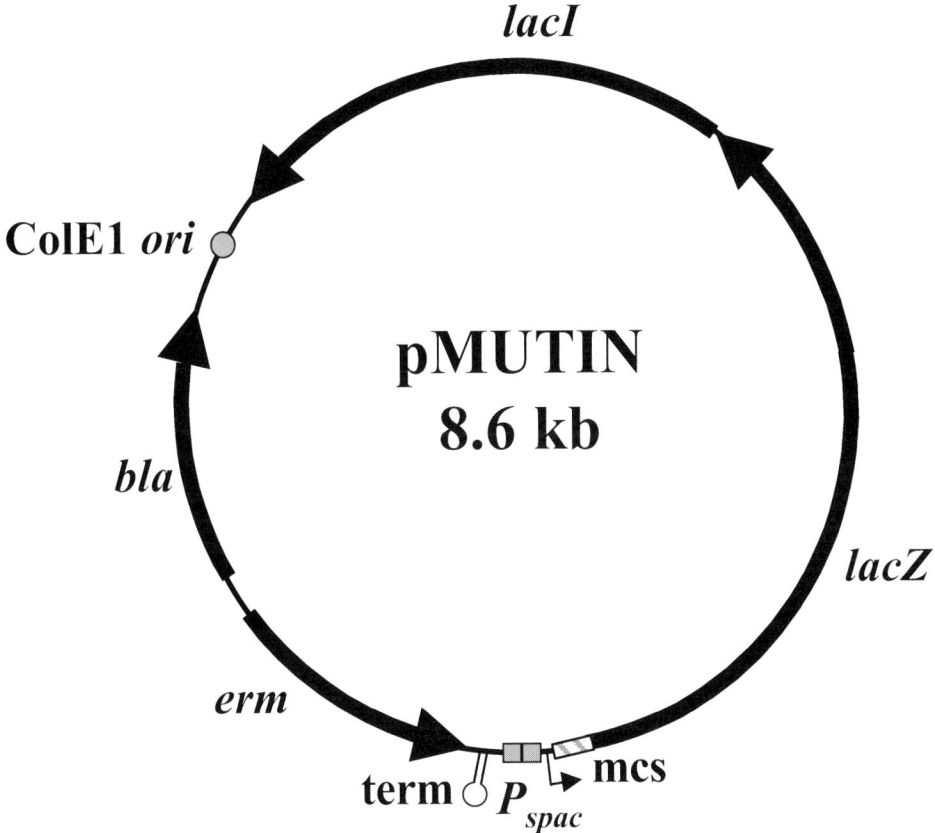

Figure 18.4. General structure of the pMUTIN series of integration/reporter vectors (adapted from Vagner et al., 1998). ColE1 ori, origin of replication from the ColE1 plasmid; lacI, encoding the Lac repressor; lazZ, encoding β-galactosidase; m cs, multiple cloning site; P_{spac}, IPTG-controllable promoter; erm, lincosamide/erythromycin resistance gene; bla, penicillin resistance gene encoding β-lactamase.

The integrants were confirmed by diagnostic PCR (Figure 18.5B, C). The integration places the lacZ gene within the transcriptional unit of the target gene, enabling it to be used to monitor the activity of the target gene's promoter via the synthesis of β-galactosidase. The IPTG-controllable P_{spac} promoter is located so that, in the case of polygenic operons, any genes downstream of the target gene can be induced by the addition of IPTG to the growth medium. This latter feature reduces the potential polar effects of the insertion in the upstream target gene, and allows pMUTIN to be used even when downstream genes are essential for growth and viability. If the target gene itself is essential, as determined by an inability to isolate integrants, its expression can be made conditional by generating an RBS-fusion rather than a knockout mutant. In which case the presence of IPTG is used to control the expression of the downstream (and intact) copy of the target gene.

In a limited number of instances, the target 'y' genes were too small to use a single cross-over recombination to generate a knockout mutant. In

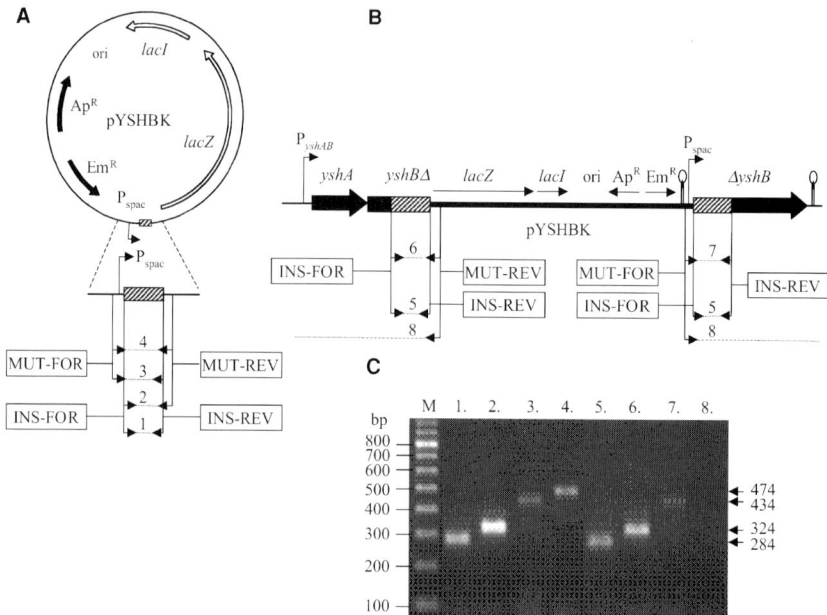

Figure 18.5. Construction of a knockout *B. subtilis* mutant of *yshB*. Schematic representations of (A) pYSHBK, a pMUTIN4-based integration plasmid, and (B) the *yshB* region of BFA2406 after integration of pYSHBK. Filled thick and open arrows indicate structural genes and putative ρ-independent terminators are shown with stem-loop structures. A promoter upstream of *yshA* (P_{yshAB}) is marked with a fine arrow. Striped boxes show the tandem duplication of the internal part of *yshB*. *yshB*Δ is the 5′-end of *yshB* and Δ*yshB* is the 3′-end of *yshB*. Plasmid pMUTIN4 is shown as a thick line. The *lacZ* reporter gene, *lacI*, ampicillin resistance (Ap^R) and erythromycin resistance (Em^R) genes are marked with fine arrows, and promoter P_{spac} is marked with a fine arrow above that of pYSHBK. The region in pYSHBK used for replication in *E. coli* is labelled 'ori', and three terminators ($t_1 t_2 t_0$) upstream of P_{spac} are indicated with stem-loop structures. The arrows below the genes indicate the location and orientation of the primers, while the dashed lines indicate their expected PCR products. Positions of the primers specific for pMUTIN4 were as follows: MUT-FOR, 147–165, and MUT-REV, 361–379 (GenBank accession number AF072806; Vagner *et al.*, 1998). The numbers above the dashed lines correspond to the lanes of the agarose electrophoresis gel shown in panel C. (C) Diagnostic PCR confirming the correct integration of pYSHBK into *yshB* in mutant BFA2406. Lane M, 100 base-pair ladder. The PCR reaction was performed with pYSHBK plasmid DNA (lanes 1–4) or BFA2406 chromosomal DNA (lanes 5–8). The primers (and expected product lengths) were as follows: lanes 1 and 5, INS-FOR and INS-REV (284 bp); lanes 2 and 6, INS-FOR and MUT-REV (324 bp); lanes 3 and 7, MUT-FOR and INS-REV (434 bp); lane 4, MUT-FOR and MUT-REV (474 bp); and lane 8, MUT-FOR and MUT-REV (no PCR product expected since these primers are orientated away from each other).

these cases a double cross-over recombination was used. Sequences either side of the target gene were incorporated into integration vector, with part or all of the target gene sequence replaced with a selectable antibiotic resistance marker. Careful design of the vector allowed a *lacZ* reporter gene to be transcriptionally linked to the target gene promoter, as with the pMUTIN vectors.

Verified BFA mutants were routinely cultured in specific nutrient and minimal media (with or without IPTG), and their growth and production of β-galactosidase monitored in parallel. The mutants were then classified according to the level and the timing of target gene expression. In a few cases, the presence of IPTG revealed operons that were autoregulated.

The construction of the BFA mutant collection is virtually complete. Not only have virtually all genes of unknown function ('*y* genes') been knocked out, but so too (thanks to the efforts of the Japanese consortium, JAFAN) have the majority of genes of known function. BFA mutants were not generated in 120 genes encoding components of the translation apparatus (e.g. rRNA or ribosomal protein genes) or 304 genes encoded by the major prophage genomes (e.g. *skin* element, SPβ or PBSX). The BFA mutant collection, comprising some 2880 strains, currently represents an almost unique microbiological resource for postgenomic studies. It was not possible to construct mutants for 7 '*y*' genes and 267 genes have been designated as essential on the basis of their IPTG dependency or reports from the literature, including 82 genes currently classified as '*y*' genes.

On the basis of the phenotypic tests, many '*y*' genes have been ascribed preliminary phenotypes (Biaudet-Brunaud *et al.*, 2001b; Sonenshein *et al.*, 2002). In a number of cases, their precise role has been determined experimentally. However, the existence of paralogues and the limited range of conditions that can be applied in high-throughput analyses has limited the extent to which phenotypes and/or functions can be ascribed to particular genes and alternative approaches (both bioinformatical and experimental) have subsequently been applied.

The data from the European and Japanese BFA programmes are publicly available in dedicated databases. The *Micado* (MICrobial Advanced Database Organization) database (http://locus.jouy.inra.fr/cgi-bin/genmic/madbase/progs/madbase.operl/) is dedicated to the European BFA programme (Samson *et al.*, 2001) and includes information on:

- contigs and gene names (hyperlinked to Swiss-Prot for access to DNA or amino acid sequences);
- the co-ordinates of the fragment used for insertional mutagenesis;
- the verification protocol for individual mutants;
- growth, expression and phenotypic data;
- links to the SubtiList and Sub2D proteomic database.

The corresponding public database of the Japanese consortium is the BSORF-DB database (http://bacillus.genome.ad.jp/). The general structure is similar to Micado, but it contains some additional features, such as gene category classification.

Complementation analyses

Complementation studies, even across the species divide, can be carried out using a combination of a pMUTIN and double cross-over integration vectors. For example, we have used this system to study the activities of three *B. anthracis* homologues of a gene that is essential in *B. subtilis*.

Placing the B. subtilis gene under the control of the pMUTIN P_{spac} promoter (Figure 18.5B) produces a strain that is IPTG dependent. Integrating the homologous B. anthracis genes, under the control of a xylose-dependent promoter, in turn at the amyE locus (Figure 18.3) allows complementation studies to be carried out. Complementing genes allow these strains to grow in the absence of IPTG in a xylose-dependent manner.

Analysis of essential genes

Genes were identified as essential, if they were required for the growth and viability of B. subtilis on a standard rich medium used throughout the functional analysis project (Dervyn and Ehrlich, 2001). Essential genes were identified on the basis that: (1) it was not possible to isolate gene insertions with pMUTIN (seven genes); or (2) that pMUTIN insertions could only be constructed in the presence of IPTG and when an intact copy of the gene was present (see Figure 18.1). However, we have found caveats to the above. Even when interrupted with pMUTIN4, the version of pMUTIN with the tightest control of the P_{spac} promoter, certain genes that are known to be essential (e.g. secA) do not exhibit a lethal phenotype unless additional copies of the lacI gene are supplied on a multicopy plasmid such as pMAP65 (Petit et al., 1998). Secondly, care needs to be taken in the selection and handling of BFA mutants to avoid the accumulation of suppressor mutations. We have observed the accumulation of suppressor mutants in liquid cultures at low (µM) IPTG concentrations (Figure 18.6). In the case of the gene encoding the essential GTPase, ysxC, the majority of the suppressors mapped to a single mutation (Figure 18.7)

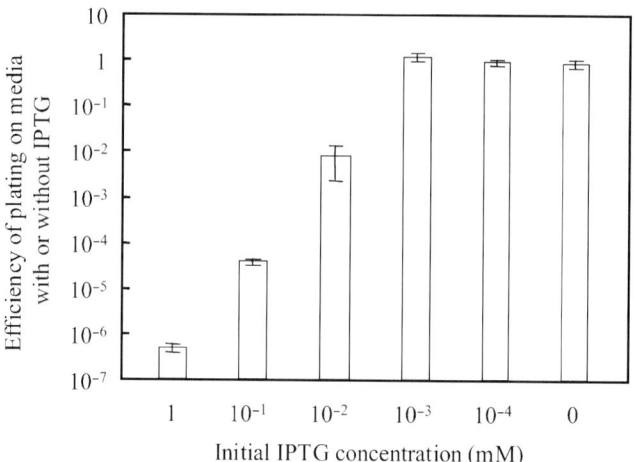

Figure 18.6. The ratio of IPTG-independent (suppressor mutants) to total colony-forming units (cfu) from stationary phase cultures grown in Luria Bertani medium containing various initial concentrations of IPTG. The data are the means of triplicate samples ± standard deviation.

	-35	-10	+1	'Oid' *lac* operator C→T	*Hind*III

```
BFA2414     TTGACTTTATCTACAAGGTGTGGCATAATGTGTGGAATTGTGAGCGCTCACAATTAAGCTT
BFA2414SUP  TTGACTTTATCTACAAGGTGTGGCATAATGTGTGGAATTGTGAGTGCTCACAATTAAGCTT
```

Figure 18.7. Sequence of the P$_{spac}$ promoter region in BFA2414 and BFA2414SUP. The –35 and –10 regions of P$_{spac}$ promoter are indicated. +1 is both the transcriptional initiation point and the first nucleotide of the 'oid' *lac* operator (bold). In BFA2414SUP, the spontaneous C→T transition is indicated at nucleotide 10 of the *lac* operator.

in the *lac* operator sequence of pMUTIN4, reducing its ability to bind LacI (Pragai and Harwood, 2000).

Attempts were made to categorize essential genes to three broad areas of metabolism, namely DNA, RNA or protein synthesis, on the basis of depletion studies. Conditional BFA mutants, grown in the presence of IPTG, were washed free of this gratuitous inducer and then grown in medium containing RNA-([^{14}C]-uridine), DNA-([^{3}H]-thymidine) and protein-([^{35}S]-methionine)-specific radiolabels. Although macromolecular synthesis is interdependent, it was hypothesized that the first precursor to be affected by the depletion of the target gene product was likely to correspond to a defect in that class of macromolecule. This was confirmed experimentally for a number of well-characterized genes (Dervyn and Ehrlich, 2001). Additional useful information can be obtained by monitoring changes in the cell cycle, shape and the partitioning of nucleoids using various microscopy techniques, e.g. immunofluorescence, GFP translational fusions, fluorescent *in situ* hybridization (FISH) (Seror and Errington, 2001).

◆◆◆◆◆◆ ANALYSIS OF THE TRANSCRIPTOME AND PROTEOME

Whilst the completion of bacterial genomes represents a major achievement in our understanding of the molecular biology of bacteria, revealing the extent of the transcriptome and proteome represents a much larger challenge. This is because, unlike the genome, the transcriptome and proteome are almost infinitely variable, and data representing these '-omes' are likely to be orders of magnitude larger than that of the genome.

Transcriptome

Transcriptomics attempts to catalogue the RNA content of a cell using tools such as reporter gene technology, Northern blotting, S1 or primer extension mapping, reverse transcriptase (RT)-PCR and DNA arrays. Transcriptome analysis is complex since cellular RNA content (messenger RNA, transfer RNA, ribosomal RNA and various other species) is dependent on the cell's physical and chemical environment, its stage in growth, cell or differentiation cycles, and the patterns of transcription and RNA processing. Analysis of the transcriptome can be carried out for a variety of purposes:

- to convert ORFs that are defined by bioinformatical techniques into genes, by confirming they are capable of being transcribed and identifying transcription start and end-points;
- to determine transcription and RNA processing patterns that can be overlaid on the genome to identify transcriptional units or operons;
- to identify whole-cell mRNA responses to environmental changes and to identify regulatory proteins and their cognate target genes;
- to identify high-level regulatory networks.

Reporter gene technology

The *lacZ* gene, transcriptionally fused to the target gene of each of the BFA mutants, has been used to determine the level and timing of target expression in minimal and nutrient media. The result are published on the Micado and BSORF databases (see earlier). An analysis of the reporter gene transcription patterns of between 400 and 700 of the BFA 'y' gene mutants is shown in Table 18.2. The data show that surprisingly few of the 'y' genes are not expressed and that a higher proportion are expressed in nutrient medium. Interestingly, a higher proportion of 'y' genes is expressed in minimal medium during the exponential phase, and in nutrient medium during the stationary phase. By comparing the reporter gene transcription patterns of genes adjacent to each other on the chromosome, putative assignments can be made to operons (Biaudet-Brunaud *et al.*, 2001a; Prágai *et al.*, 2001).

Table 18.2. Expression profiles of a set of BFA mutants; data extracted from Biaudet-Brunaud *et al.* (2001a)

	Culture medium	
	Minimal (%)	Nutrient (%)
Expression		
None	16	5
Constant	23	20
Regulated	60	74
Expression level		
Low	18	15
Medium	53	64
High	28	21
Growth phase		
Exponential	69	37
Stationary	31	63

BFA reporter genes have also been used in screening programmes to identify new members of stimulons and regulons, confirming the assignment with the use of null-mutations in the cognate regulatory proteins (Prágai and Harwood, 2002).

Promoters

Bacillus subtilis encodes 17 sigma factors (Table 18.3) that direct the synthesis of sets of genes in response to specific physical and chemical environmental conditions. With the exception of σ^L, all are members of the *E. coli* σ^{70} family with consensus sequences centred around −35 and −10. In contrast, σ^L is a member of the *E. coli* $\sigma^{N/54}$ family with recognition sequences centred around −24 and −12. The principal sigma factor, σ^A, has a similar consensus sequence to that of *E. coli* σ^{70}. The secondary sigma factors control or contribute to the expression of more than 500 genes or at least 12% of the genome. A number of genes, including *sigA*, are controlled by several sigma factors. Additionally, at least two sigma factors, σ^E and σ^K involved in directing compartment-specific transcription in spore mother cells, are synthesized as inactive pro-sigma factors that are activated by proteolytic cleavage in response to signals from the forespore.

Although the consensus sequences have been determined for 12 sigma factors, the use of bioinformatical techniques to search the sequence database for putative promoter sequences usually generates many false positives.

Table 18.3. Sigma factors of *Bacillus subtilis* and their consensus sequences; data extracted from Helmann and Moran (2002)

Sigma factor	Gene	Function	'−35'	Spacer	'−10'
σ^A	*sigA*	Housekeeping	TTGaca	14	tgnTAtaat
σ^B	*sigB*	General stress response	rGGwTTrA	12–15	GGgtAt
σ^D	*sigD*	Chemotaxis/autolysin/flagella	TAAA	14–16	gCCGATAT
σ^E	*sigE*	Sporulation, early mother cell	ATa	16–18	cATAcanT
σ^F	*sigF*	Sporulation, early forespore	GywTA	15	GgnrAnAnTw
σ^G	*sigG*	Sporulation, late forespore	gnATr	15	cAtnnTA
σ^H	*sigH*	Competence/early sporulation	RnAGGAwWW	11–12	RnnGAAT
σ^K	*sigK*	Sporulation, late mother cell	AC	16–18	CATannnT
σ^L	*sigL*	Degradative enzymes	TGGcA	5	CTTGCAT
σ^M	*sigM*	Salt resistance	TGCAAC	16–17	CGTGta
σ^V	*sigV*	Unknown	Unknown		Unknown
σ^W	*sigW*	Antimicrobial resistance	TGAAAC	16–17	CGTa
σ^X	*sigX*	Cell surface properties	tGtAAC	16–17	CGwC
σ^Y	*sigY*	Unknown	Unknown		Unknown
σ^Z	*sigZ*	Unknown	Unknown		Unknown
σ^{ykoZ}	*ykoZ*	Unknown	Unknown		Unknown
σ^{ylaC}	*ylaC*	Unknown	Unknown		Unknown

Transcription mapping

Extensive attempts have been made to map transcription patterns onto the B. subtilis genome map (Yoshida et al., 2000). Northern blotting, which detects full-length transcripts, allows the transcription pattern of newly identified genes to be determined and whether they are transcribed singly or as part of a polygenic operon. The isolation of mRNA from developing endospores is difficult at the later stages of sporulation and additional treatments are required. RT-PCR allows for the measurement of weakly expressed genes, but is not quantitative and does not provide full-length transcripts (Mellado, 2001).

Regulatory proteins

A wide variety of transcription regulatory proteins has been identified in B. subtilis on the basis of multiple protein alignments and hidden Markov modelling (Ishii et al., 2001). Data on B. subtilis promoters and transcription factors are available at the DBTBS website (http://elmo.ims.u-tokyo.ac.jp/dbtbs/) and the classes of regulatory proteins are summarized in Table 18.4.

DNA arrays

The completion of the B. subtilis genome sequence has provided the opportunity to use DNA genome arrays to study whole-cell patterns of transcription in B. subtilis. To date there are published studies using oligonucleotide-based microarrays (Affymetrix GeneChip technology) (Lee et al., 2001), and ORF-based nylon substrate macroarrays (Fawcett et al., 2000; Petersohn et al., 2001) and glass substrate microarrays (Yoshida et al., 2001). These analyses have covered various aspects of central metabolism (riboflavin, biotin and thiamine synthesis), stress responses (sigB-dependent general stress pathway), catabolite repression and sporulation. The use of null mutants has been particularly useful for identifying members of specific regulons.

Proteome

Proteome analysis facilitates the grouping of genes into regulatory classes, and provides information of post-translational modifications and protein stability. A theoretical two-dimensional protein map has been constructed from the calculated pI and molecular mass values of each of the polypeptides encoded by the genome of B. subtilis strain 168 (Bernhardt et al., 2001). A plot of the pI-frequency distribution shows a bimodal distribution, with 65% of the proteins apparently having an acid pI: YvcF is the most acidic (calculated pI value of 2.79) and RpmH the most basic (pI 13.0). A high proportion of secreted proteins have a pI close to 7.0, which may facilitate their passage through the acidic cell wall (Jensen et al., 2000). The largest protein is PksK, involved in polyketide synthesis, with a molecular mass of 496 047 Da.

Table 18.4. Classes of transcription regulators in *Bacillus subtilis*; data extracted from Ishii et al. (2001)

Classes of transcription factor	Number
Sigma factors	
Sigma-70	9
Sigma-54	1
Sigma-70 ECF	7
Helix-turn-helix	
MarR	23
LacI	11
GntR	21
LysR (HTH_1)	20
ArsR (HTH_5)	7
DeoR	7
AraC (HTH_2)	12
GerE	10
Crp	2
Xre (HTH_3)	17
MerR	10
TetR	19
AsnC	7
LexA	1
HTH_6	2
Other families	
C family regulators	13
Response regulators	35
Fur	3
Sigma-54-related factors	5
Bgl-antiterminators	8
Cold shock-dependent (CSD)	3
IclR	1
GreA/B	1
Fe-dependent repressors	1
HrcA	1
Arg repressor	1
Not assigned	35

Two-dimensional protein index

Two-dimensional polyacrylamide gel electrophoresis (2D-PAGE) has been used to catalogue the polypeptides synthesized by *B. subtilis* under various growth and stress regimes by the group of Hecker at the University of Griefswald in Germany. 2D-PAGE has been of particular value for the identification of components of stimulons (proteins encoded by genes induced by a specific physical or chemical stress) or regulons (proteins encoded by genes induced by a specific regulatory protein)

(Eymann *et al.*, 1996; Bernhardt *et al.*, 1997). In recent years, 2D-PAGE protocols have been modified and adapted to detect secreted proteins (Antelmann *et al.*, 2001) and those with a low pI (Ohlmeier *et al.*, 2000).

Proteins may be radiolabelled (^{35}S-methionine) or stained with Coomassie Brilliant Blue or silver. Images can be false coloured so that direct comparisons can be made between separate gels. In many cases the polypeptides have been excised from the gel and identified, initially by N-terminal sequencing but more recently by mass spectrometry, for example, matrix-assisted laser desorption ionization time-of-flight (MALDI-TOF) or quadrupole/orthogonal-acceleration time-of-flight (Q-TOF), after reference to the *B. subtilis* protein database. Currently more than 500 polypeptide spots have been identified.

The resulting data have been stored as co-ordinates that allow individual polypeptides from independent gels to be compared and expression profiles to be constructed (Antelmann *et al.*, 1997; Bernhardt *et al.*, 1999). The data, together with clickable representative gels, are available at the Sub2D website at: http://microbio2.biologie.uni-greifswald.de:8880/sub2d.htm. Sub2D includes data on house-keeping proteins, secreted proteins, proteins induced following phosphate starvation, heat shock, oxidative shock, ethanol stress as well as the effects of null mutations in genes encoding specific regulatory proteins. Sub2D is cross-linked to other *B. subtilis* databases, including SubtiList, BSORF, Micado, as well as SwissProt and Medline. Calculated molecular mass and p*I* values have allowed predicted and actual migration positions in 2D gels to be compared and aberrant migration, owing to post-transcriptional modification (processing or derivitization), to be investigated. A remaining challenge is to integrate the transcriptomic and proteomic data to allow high-level analyses for mRNA and protein stability, and of post-translational effects.

Protein–protein interactions

Protein–protein interactions are a potential source of information about the biological role of both characterized and uncharacterized proteins. *B. subtilis* protein–protein interactions have been studied using the yeast two-hybrid system. SPiD is a newly developed online database for the graphical representation of two-hybrid protein interactions in *B. subtilis*. The interactions stored in SPiD are either obtained directly from yeast two-hybrid experiments or from the literature (http://www-mig.versailles.inra.fr/bdsi/SPiD) (Hoebeke *et al.*, 2001).

References

Anagnostopoulos, C. and Spizizen, J. (1961). Requirements for transformation in *Bacillus subtilis*. *J. Bacteriol.* **81**, 741–746.

Anagnostopoulos, C., Piggot, P. J. and Hoch, J. A. (1993). The genetic map of *Bacillus subtilis*. In Bacillus subtilis *and Other Gram-positive Bacteria: Biochemistry, Physiology, and Molecular Genetics* (Sonenshein, A. L., Hoch, J. A. and Losick, R., eds), pp. 425–461. Washington DC: American Society for Microbiology.

Antelmann, H., Bernhardt, J., Schmid, R., Mach, H., Völker, U. and Hecker, M. (1997). First steps from two-dimensional protein index towards a response regulation map for *Bacillus subtilis*. *Electrophoresis* **18**, 1451–1463.

Antelmann, H., Tjalsma, H., Voigt, B., Ohlmeier, S., Bron, S., van Dijl, J. M. and Hecker, M. (2001). A proteomic view on the genome-based signal peptide predictions. *Genome Res.* **11**, 1484–1502.

Azevedo, V., Alvarez, E., Zumstein, E., Damiani, G., Sgaramella, V., Ehrlich, S. D. and Serror, P. (1993). An ordered collection of *Bacillus subtilis* DNA segments cloned into yeast artificial chromosomes. *Proc. Natl Acad. Sci. USA* **90**, 6047–6051.

Bernhardt, J., Völker, U., Völker, A., Antelmann, H., Schmid, R. M. H. and Hecker, M. (1997). Specific and general stress proteins in *Bacillus subtilis* – a two dimensional protein electrophoretic study. *Microbiol. UK* **143**, 999–1017.

Bernhardt, J., Buttner, K., Scharf, C. and Hecker, M. (1999). Dual channel imaging of two-dimensional electropherograms in *Bacillus subtilis*. *Electrophoresis* **11**, 2225–2240.

Bernhardt, J., Buttner, K., Coppee, J. Y., Lelong, C., Ogasawara, N., Scharf, C., Vagner, V., Schmid, R., Volker, U. and Hecker, M. (2001). The contribution of the European Community consortium to the two-dimensional protein index of *Bacillus subtilis*. In *Functional Analysis of Bacterial Genes* (Schumann, W., Ehrlich, S. D. and Ogasawara, N., eds), pp. 63–74. Chichester: John Wiley and Sons.

Biaudet, V., Samson, F. and Bessieres, P. (1997). Micado: a network-orientated database for microbial genomes. *Comp. Appl. Biosci.* **13**, 431–438.

Biaudet-Brunaud, V., Samson, F., Gas, S., Dervyn, E., Gallezot, G., Duchet, S., Ehrlich, S. D. and Bessieres, P. (2001a). Phenotype responses and reporter gene activity from the systematic functional analysis of *Bacillus subtilis* unknown genes. In *Functional Analysis of Bacterial Genes* (Schumann, W., Ehrlich, S. D. and Ogasawara, N., eds), pp. 53–61. Chichester: John Wiley and Sons.

Biaudet-Brunaud, V., Samson, F., Gas, S., Dervyn, E., Gallezot, G., Duchet, S., Ehrlich, S. D. and Bessieres, P. (2001b). List of *Bacillus subtilis* genes with a phenotype determined by the functional analysis project. In *Functional Analysis of Bacterial Genes* (Schumann, W., Ehrlich, S. D. and Ogasawara, N., eds), pp. 283–292. Chichester: John Wiley and Sons.

Bron, S. (1990). Plasmids. In *Molecular Biological Methods for* Bacillus (Harwood, C. R. and Cutting, S. M., eds), pp. 75–174. Chichester: John Wiley and Sons.

Bruand, C., Ehrlich, S. D. and Janniere, L. (1991). Unidirectional theta replication of the stable *Enterococcus faecalis* plasmid pAM81. *EMBO J.* **10**, 2171–2177.

Chang, S. and Cohen, S. N. (1979). High frequency transformation of *Bacillus subtilis* protoplasts by plasmid DNA. *Mol. Gen. Genet.* **168**, 111–115.

Claus, D. and Fritze, D. (1989). Taxonomy of *Bacillus*. In *Biotechnology Handbooks 2*: Bacillus (Harwood, C. R., ed.), pp. 5–26. New York: Plenum Publishing Corp.

Cook, N., Silcock, D. J., Waterhouse, R. N., Prosser, J. I., Glover, L. A. and Killham, K. C. (1993). Construction and detection of bioluminescent strains of *Bacillus subtilis*. *J. Appl. Bacteriol.* **75**, 350–359.

Crutz, A.-M., Steinmetz, M., Aymerich, S., Richter, R. and LeCoq, D. (1990). Induction of levansucrase in *Bacillus subtilis*: an antitermination mechanism negatively controlled by the phosphotransferase system. *J. Bacteriol.* **172**, 1043–1050.

Dervyn, E. and Ehrlich, S. D. (2001). Analysis of essential genes. In *Functional Analysis of Bacterial Genes* (Schumann, W., Ehrlich, S. D. and Ogasawara, N., eds), pp. 25–32. Chichester: John Wiley and Sons.

Dubnau, D. (1993). Genetic exchange and homologous recombination. In Bacillus subtilis *and Other Gram-positive Bacteria: Biochemistry, Physiology, and Molecular Genetics* (Sonenshein, A. L., Hoch, J. A. and Losick, R., eds), pp. 555–584. Washington, DC: American Society for Microbiology.

Dubnau, D. and Lovett, C. M. (2002). Transformation and recombination. In Bacillus subtilis *and Its Closest Relatives: from Genes to Cells* (Sonenshein, A. L., Hoch, J. A. and Losick, R., eds), pp. 453–471. Washington, DC: American Society for Microbiology.

Eymann, C., Mach, H., Harwood, C. R. and Hecker, M. (1996). Phosphate starvation-inducible protein in *Bacillus subtilis* a two-dimensional protein electrophoretic study. *Microbiol. UK* **142**, 3163–3170.

Fawcett, P., Eichenberger, P., Losick, R. and Youngman, P. (2000). The transcriptional profile of early to middle sporulation in *Bacillus subtilis*. *Proc. Natl Acad. Sci. USA* **97**, 8063–8068.

Fleischmann, R. D., Adams, M. D., White, O. *et al.* (1995). Whole genome random sequencing and assembly of *Haemophilus influenzae* Rd. *Science* **269**, 496–512.

Gartner, D., Degenkolb, J., Ripperger, J. A. E., Allmansberger, R. and Hillen, W. (1992). Regulation of the *Bacillus subtilis* W23 xylose utilization operon: interaction of the Xyl repressor with the *xyl* operator and the inducer xylose. *Mol. Gen. Genet.* **232**, 415–422.

Glaser, P., Kunst, F., Arnaud, M., Coudart, M-P., Gonzarles, W., Hullo, M-F., Ionescu, M., Lubochinsky, S., Marcelino, L., Mozser, I., Presecan, E., Santana, M., Schneider, E., Schweizer, J., Vertes, A., Rapoport, G. and Danchin, A. (1993). *Bacillus subtilis* genome project: cloning and sequencing of the 97 kb region from 325° to 333°. *Mol. Microbiol.* **10**, 371–384.

Harwood, C. R. and Archibald, A. R. (1990). Growth, maintenance and general techniques. In *Molecular Biological Methods for* Bacillus (Harwood. C. R. and Cutting, S. M., eds), pp. 1–26. Chichester: John Wiley and Sons.

Harwood, C. R. and Cutting, S. M. (eds) (1990). *Molecular Biological Methods for* Bacillus. Chichester: John Wiley and Sons.

Harwood, C. R. and Moszer, I. (2002). From gene regulation to gene function: regulatory networks in *Bacillus subtilis*. *Comp. Functional Genomics* **3**, 37–41.

Heim, R., Prasher, D. C. and Tsien, R. Y. (1994). Wavelength mutations and post-translational autooxidation of green fluorescent protein. *Proc. Natl Acad. Sci. USA* **91**, 12501–12504.

Helmann, J. D. and Moran, C. P. (2002). RNA polymerase and sigma factors. In Bacillus subtilis *and its Closest Relatives: from Genes to Cells* (Sonenshein, A. L., Hoch, J. A. and Losick, R., eds), pp. 289–312. Washington, DC: American Society for Microbiology.

Hoebeke, M., Chiapello, H., Noirot, P. and Bessieres, P. (2001). SPID: a *Bacillus subtilis* protein interaction database. *Bioinformatics* **17**, 1209–1212.

Ishii, T., Yoshida, K., Terai, G., Fujita, Y. and Nakai, K. (2001). DBTBS: A database of *Bacillus subtilis* promoters and transcription factors. *Nucleic Acids Res.* **29**, 278–280.

Itaya, M. and Tanaka, T. (1991). Complete physical map of the *Bacillus subtilis* 168 chromosome constructed by a gene-directed mutagenesis method. *J. Mol. Biol.* **220**, 631–648.

Janniere, L., Bruand, C. and Ehrlich, S. D. (1990). Structurally stable *Bacillus subtilis* cloning vectors. *Gene* **87**, 53–59.

Janniere, L., Gruss, A. and Ehrlich, S. D. (1993). Plasmids. In Bacillus subtilis *and Other Gram-positive Bacteria: Biochemistry, Physiology, and Molecular Genetics* (Sonenshein, A. L., Hoch, J. A. and Losick, R., eds), pp. 625–644. Washington, DC: American Society for Microbiology.

Jensen, C. L., Stephenson, K., Jørgensen, S. T. and Harwood, C. R. (2000). Cell-associated degradation affects yield of secreted engineered and heterologous proteins in the *Bacillus subtilis* expression system. *Microbiol. UK* **146**, 2583–2594.

Jones, L. J., Carballido-Lopez, R. and Errington, J. (2001). Control of cell shape in bacteria: helical, actin-like filaments in *Bacillus subtilis*. *Cell* **104**, 913–922.

Kunst, F., Ogasawara, N., Moszer, I. *et al.* (1997). The complete genome sequence of the Gram-positive bacterium *Bacillus subtilis*. *Nature* **390**, 249–256.

Lazarevic, V., Abellan, F., Möller, S. B., Karamata, D. and Mauël, C. (2002). Comparison of ribitol and glycerol teichoic acid genes in *Bacillus subtilis* W23 and 168: identical function, similar divergent organization, but different regulation. *Microbiol. UK* **148**, 815–824.

Lee, J.-M., Zhang, S., Saha, S., Anna, S. S., Jiang, C. and Perkins, J. B. (2001). RNA expression analysis using an antisense *Bacillus subtilis* genome array. *J. Bacteriol.* **183**, 7371–7380.

Leenhouts, K. (1995). Integration strategies and vectors. In *Genetics of Streptococci, Enterococci and Lactococci* (Ferretti, J. J., Gilmore, M. S., Klaenhammer, T. R. and Brown, F., eds), pp. 523–530. Basel: Karger.

Leenhouts, K. J. and Venema, G. (1993). Lactococcal plasmid vectors. In *Plasmids, A Practical Approach* (Hardy, K. G., ed.), pp. 65–94. New York: Oxford University Press.

Leenhouts, K., Buist, G., Bolhuis, A., ten Berge, A., Kiel, J., Mierau, I., Dabrowska, M., Venema, G. and Kok, J. (1996). A general system for generating unlabelled gene replacements in bacterial chromosomes. *Mol. Gen. Genet.* **253**, 217–224.

Lewis, P. J., Nwoguh, C. E., Barer, M. R., Harwood, C. R. and Errington, J. (1994). Use of digitized video microscopy with a fluorogenic enzyme substrate to demonstrate cell and compartment-specific gene expression in *Salmonella enteritidis* and *Bacillus subtilis*. *Mol. Microbiol.* **13**, 655–662.

Meijer, W. J. J., Venema, G. and Bron, S. (1995a). Characterization of single strand origins of cryptic rolling-circle plasmids from *Bacillus subtilis*. *Nucleic Acids Res.* **23**, 612–619.

Meijer, W. J. J., de Boer, A., van Tongeren, S., Venema, G. and Bron, S. (1995b). Characterization of the replication region of the *Bacillus subtilis* pLS20: a novel type of replicon. *Nucleic Acids Res.* **23**, 3214–3223.

Meijer, W. J. J., Wisman, G. B. A., Terpstra, P., Thorsted, P. B., Thomas, C. M., Holsappel, S., Venema, G. and Bron, S. (1998). Rolling-circle plasmids from *Bacillus subtilis*: complete nucleotide sequences and analyses of genes of pTA1015, pTA1040, pTA1050 and pTA1060, and comparisons with related plasmids from Gram-positive bacteria. *FEMS Microbiol. Rev.* **21**, 337–368.

Mellado, R. P. (2001). Transcription analysis of large regions of the bacterial genome. In *Functional Analysis of Bacterial Genes* (Schumann, W., Ehrlich, S. D. and Ogasawara, N., eds), pp. 33–35. Chichester: John Wiley and Sons.

Moszer, I. (2002). *Bacillus subtilis* genome, genes and function. In *Bacillus subtilis and its Closest Relatives: from Genes to Cells* (Sonenshein, A. L., Hoch, J. A. and Losick, R., eds), pp. 7–11. Washington, DC: American Society for Microbiology.

Moszer, I., Glaser, P. and Danchin, A. (1995). SubtiList: a relational database for the *Bacillus subtilis* genome. *Microbiol. UK* **141**, 261–268.

Ohlmeier, S., Scharf, C. and Hecker, M. (2000). Alkaline proteins of *Bacillus subtilis*: first steps towards a two-dimensional alkaline master gel. *Electrophoresis* **21**, 3701–3709.

Perkins, J. B. and Youngman, P. (1986). Construction and properties of Tn*917-lac*, a transposon derivative that mediates transcriptional gene fusions in *Bacillus subtilis*. *Proc. Natl Acad. Sci. USA* **83**, 140–144.

Petersohn, A., Brigulla, M., Haas, S., Hoheisel, J., Volker, U. and Hecker, M. (2001). Global analysis of the general stress response of *Bacillus subtilis*. *J. Bacteriol.* **183**, 5617–5631.

Petit, M. A., Dervyn, E., Rose, M. *et al.* (1998). PcrA is an essential DNA helicase of *Bacillus subtilis* fulfilling functions both in repair and rolling circle replication. *Mol. Microbiol.* **29**, 261–273.

Prágai, Z. and Harwood, C. R. (2000). YsxC, a putative GTP-binding protein essential for the growth of *Bacillus subtilis* 168. *J. Bacteriol.* **184**, 6819–6823.

Prágai, Z. and Harwood, C. R. (2002). Regulatory interactions between the Pho and σB-dependent general stress regulons of *Bacillus subtilis*. *Microbiol. UK* **148**, 1593–1602.

Prágai, Z., Eschevins, C., Bron, S. and Harwood, C. R. (2001). *Bacillus subtilis* NhaC, a Na+/H+ antiporter, influences the expression of the *phoPR* operon and the production of alkaline phosphatases. *J. Bacteriol.* **184**, 6819–6823.

Prasher, D. C., Eckenrode, V. E., Ward, W. W., Prendergast, P. G. and Cormier, M. J. (1992). Primary structure of the *Aequorea victoria* green-fluorescent protein. *Gene* **111**, 229–233.

Rocha, E. P. C., Danchin, A. and Viari, A. (1999). Analysis of long repeats in bacterial genomes reveals alternative evolutionary mechanisms. *Mol. Biol. Evol.* **16**, 1219–1230.

Rocha, E., Moszer, I., Klaerr-Blanchard, M., Sekowska, A., Medigue, C., Viari, A. and Danchin, A. (2001). *In silico* genome analysis. In *Functional Analysis of Bacterial Genes* (Schumann, W., Ehrlich, S. D. and Ogasawara, N., eds), pp. 6–19. Chichester: John Wiley and Sons.

Rose, M. and Entian, K. D. (1996). New genes in the 170 degrees region of the *Bacillus subtilis* genome encode DNA gyrase subunits, a thioredoxin, a xylanase and an amino acid transporter. *Microbiol. UK* **142**, 3097–3101.

Saier, M. H. J., Goldman, S. R., Maile, R. R., Moreno, M. S., Weyler, W., Yang, N. and Paulsen, I. T. (2002). Overall transport capabilities of *Bacillus subtilis*. In Bacillus subtilis *and its Closest Relatives: from Genes to Cells* (Sonenshein, A. L., Hoch, J. A. and Losick, R., eds), pp. 113–128. Washington DC: American Society for Microbiology.

Samson, F., Biaudet-Brunaud, V., Gas, S., Dervyn, E., Gallezot, G., Duchet, S., Ehrlich, S. D. and Bessieres, P. (2001). Micado, an integrative database dedicated to the functional analysis of *Bacillus subtilis* and microbial genomes. In *Functional Analysis of Bacterial Genes* (Schumann, W., Ehrlich, S. D. and Ogasawara, N., eds), pp. 45–52. Chichester: John Wiley and Sons.

Schaeffer, P., Millet, J. and Aubert, P-J. (1965). Catabolite repression of bacterial sporulation. *Proc. Natl Acad. Sci. USA* **54**, 704–711.

Schumann, W., Ehrlich, S. D. and Ogasawara, N. (2001). *Functional Analysis of Bacterial Genes: A Practical Manual*. Chichester: John Wiley and Sons.

Seror, S. and Errington, J. (2001). Cell cycle gene analysis. In *Functional Analysis of Bacterial Genes* (Schumann, W., Ehrlich, S. D. and Ogasawara, N., eds), pp. 189–196. Chichester: John Wiley and Sons.

Sonenshein, A. L., Hoch, J. A. and Losick, R. (eds) (1993). Bacillus subtilis *and Other Gram-positive Bacteria: Biochemistry, Physiology, and Molecular Genetics*. Washington, DC: American Society for Microbiology.

Sonenshein, A. L., Hoch, J. A. and Losick, R. (eds) (2002). Bacillus subtilis *and its Closest Relatives: from Genes to Cells*. Washington, DC: American Society for Microbiology.

Sorokin, A., Azevedo, V., Zumstein, E., Galleron, N., Ehrlich, S. D. and Serror, P. (1996). Sequence analysis of the *Bacillus subtilis* chromosome region between the *serA* and *kdg* loci cloned in a yeast artificial chromosome. *Microbiol. UK* **142**, 2005–2016.

Spizizen, J. (1958). Transformation of biochemically deficient strains of *Bacillus subtilis* by deoxyribonucleate. *Proc. Natl Acad. Sci. USA* **44**, 1072–1078.

Tjalsma, H., Bolhuis, A., Jongbloed, J. D. H., Bron, S. and van Dijl, J. M. (2000). Signal peptide-dependent protein transport in *Bacillus subtilis*: a genome-based survey of the secretome. *Microbiol. Mol. Biol. Rev.* **64**, 515–547.

Vagner, V., Dervyn, E. and Ehrlich, S. D. (1998). A vector for systematic gene inactivation in *Bacillus subtilis*. *Microbiol. UK* **144**, 3097–3104.

Webb, C. D., Decatur, A., Teleman, A. and Losick, R. (1995). Use of green fluorescent protein for visualization of cell-specific gene expression and subcellular protein localization during sporulation in *Bacillus subtilis*. *J. Bacteriol.* **177**, 5906–5911.

Wipat, A., Carter, N., Brignell, S. B., Guy, B. J., Piper, K., Saunders, J., Emmerson, P. T. and Harwood, C.R. (1996). The region *dnaB–pheA* (256º–240º) of the *Bacillus subtilis* chromosome encoding genes responsible for stress responses, the utilisation of plant cell walls and primary metabolism. *Microbiol. UK* **142**, 3067–3078.

Yansura, D. G. and Henner, D. J. (1984). Use of the *Escherichia coli lac* repressor and operator to control gene expression in *Bacillus subtilis*. *Proc. Natl Acad. Sci. USA* **81**, 439–443.

Yoshida, K., Ishio, I., Nagakawa, E., Yamamoto, Y., Yamamoto, M. and Fujita, Y. (2000). Systematic study of gene expression and transcription organization in the *gntZ–ywaA* region of the *Bacillus subtilis* genome. *Microbiol. UK* **146**, 573–579.

Yoshida, K., Kobayashi, K., Miwa, Y., Kang, C. M., Matsunaga, M., Yamaguchi, H., Tojo, S., Yamamoto, M., Nishi, R., Ogasawara, N., Nakayama, T. and Fujita, Y. (2001). Combined transcriptome and proteome analysis as a powerful approach to study genes under glucose repression in *Bacillus subtilis*. *Nucleic Acids Res.* **29**, 683–692.

Youngman, P. (1990). Use of transposons and integrational vectors for mutagenesis and construction of gene fusions in *Bacillus* species. In *Molecular Biological Methods for* Bacillus (Harwood, C. R. and Cutting, S. M., eds), pp. 221–266. Chichester: John Wiley and Sons.

Yuan, G. and Wong, S-L. (1995). Isolation and characterisation of *Bacillus subtilis* regulatory mutants: evidence for *orf39* in the *dnaK* operon as a repressor gene in regulating the expression of both *groE* and *dnaK*. *J. Bacteriol.* **177**, 6462–6468.

Zukowski, M., Gaffney, M., Speck, D. F., Kauffmann, D., Findeli, M., Wisecup, A. and Lecocq, J. P. (1983). Chromogenic identification of genetic regulatory signals in *Bacillus subtilis* based on expression of a cloned *Pseudomonas* gene. *Proc. Natl Acad. Sci. USA* **80**, 1101–1105.

Part VII
Case Studies – Parasites

19 *Plasmodium falciparum* DNA Microarrays and Interpretation of Data

Adam A Witney,* Robert M Anthony, Trevor R Jones and Daniel J Carucci

Malaria Program, Naval Medical Research Center, Silver Spring, MD 20910, USA

CONTENTS

Introduction
Preparation of chromosome-specific DNA microarrays
Preparation of *P. falciparum* total RNA
Performing the microarray hybridization
Data acquisition and analysis
Examples of applications of DNA microarrays to the study of *P. falciparum* biology
Summary

◆◆◆◆◆◆ INTRODUCTION

Plasmodium falciparum is one of four species of the genus *Plasmodium* that cause malaria in humans. The completion of the *P. falciparum* genome sequence will allow the identification of the entire set of genes for this organism. The next challenge is to make use of this vast amount of information to better understand the parasite's complex biochemistry and metabolism with the ultimate aim of identifying novel drug and vaccine targets for use in the fight against malaria. DNA microarray technology allows the analysis of gene expression at the whole-genome level and is an appealing approach for this purpose. The construction of DNA microarrays for any organism can be achieved by several approaches:

1. the arraying of libraries of DNA, e.g. cDNA and mung bean libraries;
2. the arraying of oligonucleotides synthesized to target specific regions of gene sequences;

* Current address: Department of Medical Microbiology, St George's Hospital Medical School, Cranmer Terrace, London SW17 0RE, UK.

3. the arraying of PCR products targeted to specific gene sequences.

The first approach is readily accessible as there is no prerequisite for genomic sequence; all that is required is that plasmid clones from a DNA library be transferred to multiwell plates and then arrayed on to the array surface of choice. In *Plasmodium*, this approach has been successfully applied using *P. falciparum* mung bean (Hayward *et al.*, 2000) and cDNA libraries (Ben Mamoun *et al.*, 2001). The drawbacks of this approach are that the library used may not be a complete representation of the genome and the analysis process can be more labour intensive as any clones shown to be interesting by the array must be sequenced for identification. The second and third approaches avoid some of the limitations of random library approaches, but require whole-genome sequence information.

In this chapter, we discuss the construction of *P. falciparum* gene-specific polymerase chain reaction (PCR) product DNA microarrays (the third approach listed above), emphasizing particularly the approaches taken to overcome specific problems encountered with this organism and the requisite application of statistical methods for their interpretation.

◆◆◆◆◆◆ PREPARATION OF CHROMOSOME-SPECIFIC DNA MICROARRAYS

Exon selection

The design of a gene-specific PCR product involves the prediction of PCR primers targeting all or some of the selected gene sequence. When selecting a region of a gene to design specific primers to, it is important to remember the complex structure of a eukaryotic gene compared to its processed messenger RNA counterpart. Thus, it is essential to design PCR products targeting only exon sequences of the gene. It is also necessary to bear in mind the manner in which cDNA synthesis will be performed: oligo-dT primed synthesis is often 3′-biased, therefore, it would be beneficial to select exons closer to the 3′-end of the genes.

Primer design

Once candidate exons have been identified, the selection of primer sequences for microarrays follows similar design principles for any standard single PCR reaction. However, we have found that, not only is it beneficial to select 3′ for most exons, but higher signal intensities can be observed if the predicted products are targeted to the 3′-end of the selected exons. We design primers to have melting temperatures between 60°C and 65°C consisting of between 18 and 35 nucleotides that amplify products up to 1500 kb specifically biased to the 3′-end of target genes. In order to achieve this, we use the primer prediction program Primer3 (available from the Whitehead Institute; http://www-genome.wi.mit.edu/genome_software/other/primer3.html) to select the primer sequences. However, we use custom Perl (http://www.perl.com) scripts to control Primer3 in such a way as to bias the products to the 3'-end of the target genes.

PCR amplification

Many PCR amplification kits are available. However, we have found the HotStar™ kit (Qiagen, USA) to be a robust and reproducible method for amplifying large numbers of PCR products simultaneously. Fifty microlitre reactions using the standard manufacturers' conditions with a final primer concentration of 1 μM provide sufficient yield of product for microarray analysis. Also, reducing the extension temperature from the recommended 72°C to 60°C improves the overall success rate for amplification; this most likely accommodates for the highly A-T-rich nature of *P. falciparum* genomic DNA. The PCR conditions are as follows:

95°C for 15 min   { 94°C for 30 s   60°C for 10 min
     x40 cycles { 50°C for 1 min
                 { 60°C for 1 min

PCR reactions can be performed in 96-well plate format in commercial PCR machines, e.g. GeneAmp® 9700 (Applied Biosystems, USA). Following amplification, the products should be precipitated and diluted 1:1 with dimethyl sulphoxide (DMSO) ready for arraying. The procedure for precipitation of amplified PCR products ready for arraying is given below.

1. Add 1/10 vol. of 3 M sodium acetate and an equal volume of isopropanol directly to the amplified PCR reactions in the 96-well plates.
2. Store at −20°C for approximately 1–2 h.
3. Centrifuge at 1500g for 15 min and discard the supernatant.
4. Wash with 70% ethanol, centrifuge at 1500g for 5 min, decant and allow to air dry.
5. Resuspend products in 50 μl ddH$_2$O.
6. Dilute 5 μl of PCR products from the master plates with 5 μl DMSO into working plates ready for arraying.
7. PCR products with and without DMSO can be stored at −20°C.

Arraying

Standard microscope slides must be pretreated in order to provide a positively charged surface to allow the negatively charged DNA to bind; such a charge can be achieved by pretreatment with poly-L-lysine. The method described is the DeRisi method (http://www.microarrays.org/pdfs/PolylysineSlides.pdf) (University of California, San Diego, USA) adapted by Lance Miller (NIH, Bethesda, MD, USA). Glass microscope slides (Corning Costar, USA) are cleaned in 3.5 M sodium hydroxide/70% ethanol for 2 h, rinsed with water and then soaked in 10% poly-L-lysine/phosphate-buffered saline solution for 1 h. After coating, the slides are rinsed again in water, dried and stored for 2 weeks in the dark at room temperature.

Several companies manufacture arraying robots; the microarrays developed by the Naval Medical Research Center were printed by The Institute for Genomic Research using an arraying robot (IAS; Intelligent Automation Systems, Cambridge, MA, USA). The arrays were printed at a constant temperature of 22°C and a relative humidity of approximately 45%. Following printing, the arrays are allowed to dry and then the DNA covalently bound by ultraviolet (UV) cross-linking at 90 mJ using a Stratalinker (Stratagene, USA). The amount of UV irradiation (90 mJ) required was titrated to provide optimal binding and minimal damage to the DNA. Until use, the slides should be protected from light and stored in a desiccator at room temperature.

◆◆◆◆◆◆ PREPARATION OF P. FALCIPARUM TOTAL RNA

Much of the *P. falciparum* life cycle cannot yet be cultured. Blood stage asexual parasites, however, can be generated in relatively large numbers and to a high degree of synchrony using current techniques (Haynes and Moch, 2002). In addition, an asexual stage parasite culture can be induced to produce sexual stage parasites under certain conditions, which can thus be purified (Carter *et al.*, 1993). Good signal intensities following hybridization usually require approximately 10 µg total RNA per labelling reaction. The inaccessibility of the exo-erythrocytic stages of the malaria parasite and the lack of any good culture systems for these stages means that they are not yet suitable for microarray analysis. However, emerging strategies of limited amplification of small amounts of RNA may allow the application of microarrays to further more extensive study of the *P. falciparum* life cycle in the future.

Purifications of total RNA from all blood stage forms of the parasite are similar and are described below.

1. Centrifuge the parasite culture material at 3000*g* for 10 min at room temperature and discard the supernatant.
2. Resuspend the pellet in an equal volume of ice-cold 1% acetic acid and incubate on ice for 5 min.
3. Centrifuge the cells at 3000*g* for 10 min at 4°C and discard the supernatant.
4. Resuspend the pellet in the appropriate volume of Trizol® Reagent (Life Technologies, Gaithersburg, MD, USA) and extract total RNA according to standard manufacturer's protocols.
5. Total RNA can be stored in 70% ethanol, 0.3 M sodium acetate at –70°C. When required, reprecipitate the RNA, wash with 70% ethanol and resuspend in the appropriate amount of deionized water.

◆◆◆◆◆◆ PERFORMING THE MICROARRAY HYBRIDIZATION

Synthesis of fluorescently labelled *P. falciparum* cDNA

Several methods exist for the synthesis of fluorescently labelled cDNA from total RNA. The choice of method centres around three issues: the priming method of cDNA synthesis; the choice of fluor-labelled nucleotide; and the method of incorporation of the fluorescent dye. The choices made are more critical for *P. falciparum* because of its highly A + T-rich genome, with coding sequences composed of up to 70% A + T.

P. falciparum, like most other eukaryotes, processes transcribed RNA by adding a poly-A tail, allowing us to prime cDNA synthesis with an oligo-dT primer, and thus providing much more specific labelling of messenger RNA without a preceding mRNA purification step, which would undoubtedly reduce our yield. Indeed, we have found oligo-dT priming to give much greater fluorescence intensities following hybridization to our 3'-biased DNA microarray when compared to random hexamer priming (unpublished data).

The use of a fluorescently labelled dCTP in labelling *P. falciparum* cDNA produces higher intensities than fluorescently labelled dUTP (unpublished data), an observation that may be a consequence of the highly A + T-rich genome of *P. falciparum*, as the use of dUTP would result in an increase in the labelling density of the resulting molecule. Randolph and Waggoner (1997) described how such an increase in labelling density may indeed lead to a reduction in the sensitivity of a labelled probe. Also, Zhu and Waggoner (1997) have shown that a reduced incorporation of Cy-labelled dUTP occurs in a linear PCR (PCR with a single-stranded template and a single primer), when the sequence demands that multiple dyes be inserted at adjacent sites.

Indirect labelling techniques are often preferred as they obviate the problem of any interference between the fluorophore and the enzyme extending the cDNA molecule. However, we found that the commercial direct labelling kit using dCTP, CyScribe (Amersham Biosciences, Sweden) gave consistently better results in our hands.

An adaptation of the CyScribe protocol used to generate fluorescently labelled cDNA from *P. falciparum* total RNA is described below. (N.B. All reagents are supplied in the CyScribe kit unless otherwise stated.)

1. To 10 μg total RNA, add 2 μl oligo-dT in 11 μl dH$_2$O.
2. Mix thoroughly and spin down for 30 s in a microcentrifuge.
3. Incubate at 70°C for 5 min.
4. Incubate on ice for 10 min, and then spin down for 30 s.
5. Add in order, 4 μl 5× CyScribe buffer, 2 μl 0.1 M dithiothreitol (DTT), 1 μl dCTP dNTP mix, 1 μl Cy3- or Cy5-labelled dCTP and 1 μl CyScript RT enzyme (keep in freezer until required).
6. Mix thoroughly and spin down for 30 s.
7. Incubate at 42°C for 1.5 h.
8. Add 2 μl 500 mM NaOH (Sigma, USA).
9. Mix and incubate at 37°C for 15 min.
10. Neutralize with 2 μl 500 mM HCl (Sigma, USA).
11. Purify with GFX™ PCR DNA and Gel Band Purification columns (Amersham Biosciences, Sweden) according to standard manufacturer's protocols.
12. Resuspend probe in 10 μl diethylpyrocarbonate (DEPC)-treated dH$_2$O.

Prehybridization

Prior to hybridization, we must first block non-specific DNA-binding sites on the glass to reduce background fluorescence in the scanning stage of the experiment. The procedure for prehybridization of printed slides, which is adapted from the method described by Hegde et al. (2000), is given below.

1. Locate the printed region of slide (this can be done by simply breathing gently on the slide) and mark the location with a diamond-tipped pen.
2. Preheat 50 ml of stock prehybridization buffer (5 × SSC, 0.1% SDS) to 42°C before adding 0.5 g bovine serum albumin (1% final concentration). Filter sterilize and remove all foam and bubbles with a 1 ml pipette.
3. Prehybridize the printed slides in a Coplin jar at 42°C for 45 min with the fresh prehybridization buffer.
4. Wash slides by dipping five times in ddH$_2$O at room temperature.
5. Dip slides in isopropanol at room temperature and allow to air dry. (Breathing on the slides can show how clean they are. Repeat washing until the slides are very clean.)

Hybridization

The method describes the use of two differently fluorescently labelled (Cy3 or Cy5) cDNA samples in a competitive hybridization. One should note that we have only discussed the use of fluorescently labelled cDNA samples in this chapter. However, the following hybridization procedure could equally be used to compare labelled cDNA with labelled genomic DNA (used as a common reference), or labelled genomic DNA with labelled genomic DNA (for genomic comparison studies). The procedure for hybridization of printed slides with two fluorescently labelled nucleic acid samples described below is adapted from the method described by Hegde et al. (2000).

1. Combine each purified Cy3- and Cy5-labelled sample, mix well and add 2 µl sonicated salmon sperm DNA to further prevent non-specific hybridization.
2. Incubate at 95°C for 3 min.
3. Spin for 1 min in a microcentrifuge.
4. Mix with an equal volume of 2× hybridization buffer (50% formamide, 10× SSC, 0.2% SDS) preheated to 42°C.
5. Pipette mixture on to the slide and cover with a 22 mm × 60 mm polyethylene hydrophobic cover slip (PGC Scientific). (Spread the labelled sample mix along one edge of the coverslip and slowly lower the coverslip down to ensure an even distribution of the solution).
6. Position slide in a sealed hybridization chamber (Corning Costar, USA) and pipette 10 µl dH$_2$O along each side to humidify the chamber.
7. Incubate overnight in a water bath at 42°C.

8. Carefully remove the slide from the chamber leaving the coverslip in place. The coverslip should be removed by gentle agitation in a low stringency wash buffer (1 × SSC, 0.2% SDS) at 42°C.
9. Wash the slide for a further 4 min in low-stringency wash buffer with gentle shaking.
10. Wash the slide at higher stringency (0.1 × SSC, 0.2% SSC) at room temperature for 4 min.
11. Wash the slide in 0.1 × SSC for 2 min and repeat once with fresh 0.1 × SSC.
12. Air dry slides and store in a dark place.

◆◆◆◆◆◆ DATA ACQUISITION AND ANALYSIS

The competitive nature of the hybridization means that we can deduce the relative expression of a particular gene by comparison of the intensities measured at the wavelengths corresponding to the particular fluorophore used (543 nm – Cy3; 633 nm – Cy5). The fluorescence intensities generated by the hybridization of the two labelled nucleic acid samples are measured using a confocal laser scanner. Several scanners applicable to this task are available. One such scanner, the ScanArray 3000 (GSI Lumonics, now Packard Biosciences, USA) measures both channels (representing Cy3 and Cy5) sequentially, generating a separate image for each channel. These images can be easily combined, assigning red and green to represent each channel individually, thus producing the false colour images that are generally reported in the literature. Once scanned, the intensity data for each spot on the array must then be extracted from each image. Image analysis software is available, e.g. ImaGene™ (Biodiscovery, USA), to facilitate in this process. ImaGene™ can automatically locate the spots within the image but will also allow manual manipulation for fine-tuning of the spot identification. The output of the image analysis software is a text file containing the average (mean, median and mode) and standard deviation of the signal and background intensities for each individual spot for each channel scanned. In addition, spot co-ordinates and measurements of spot quality are included.

The next challenge is to analyse the data to extract information that may be of biological significance. The subject of data analysis of microarrays would constitute a book in itself. However, a good review of current thinking on the subject can be found in Quackenbush (2001). One of the most critical steps in this task is data normalization. This is the process by which we try to remove systematic errors from the data that exist by virtue of the way in which the experiment is performed. The end-product, therefore, should be only the biologically relevant information. Systematic errors can arise in microarray experiments owing to, for example, different starting quantities of RNA or different labelling efficiencies between the two samples being tested, or different excitation properties of the Cy3 and Cy5 fluorophores. One approach to data normalization is to assume that the total amount of mRNA is the same in both of the competing RNA samples.

If we make this assumption, then we can simply scale one channel by the ratio of the means of the two channels. In essence, if the Cy3 channel is twice as bright as the Cy5 channel, then we can simply scale all of the Cy5 measurements by a factor of two. One should note, however, that part of this assumption is that some genes are overexpressed and some underexpressed in one RNA sample relative to the other. Such an assumption may be applicable when we are comparing a mutant with a wild-type population or a population with and without drug treatment, where we would not expect a large amount of overall difference between the gene expression patterns of the two populations. However, if we were comparing two different stages of the parasite life cycle where we may expect a larger diversity in the genes being expressed, then this assumption may not be valid.

In general, two basic types of gene expression microarray experiments are performed: (1) a pairwise comparison of two states, e.g. comparing two life-cycle stages, wild-type versus mutant, or with and without drug treatment; or (2) a time series experiment.

Traditionally, pairwise analysis of microarray experiments have been performed by simply identifying genes, which have signal intensity ratios between the two channels of greater than two in replicate experiments. However, gene expression ratios can be misleading, as they do not account for the absolute intensity of the measurements. One would prefer to be able to assign some level of statistical significance to a gene expression ratio. A level of confidence can be obtained by printing replicate spots on an array. This can be enhanced by analysis in conjunction with replicate arrays, such that a paired t-test can then be used to assign a P-value to a particular expression ratio.

Time-series experiments, in which we are interested in patterns of gene expression over time, are a very powerful technique for identifying related genes, which may be co-regulated or may be members of the same metabolic pathways. Clustering algorithms can be used to pull out such information from this kind of data. In order to perform clustering on a data set, one must first decide on a distance metric to use as a measure of similarity between two gene expression profiles, e.g. Euclidean distance, Pearson correlation coefficient or Spearman rank-order correlation coefficient. One must then select a clustering algorithm, e.g. hierarchical, k-means or self-organizing maps. The choice of metric and algorithm is dependent on the type of question one is trying to ask and will, in general, give different results. For example, using a Euclidean distance metric will tend to cluster genes with similar magnitudes as well as similar profiles. However, a centred Pearson correlation coefficient metric will cluster genes with similar profiles but is not so dependent on the magnitudes. Hierarchical clustering is an agglomerative clustering technique where every gene starts in an independent cluster and, through an iterative process, clusters are built up until a single cluster remains. k-Means clustering, however, is a divisive clustering technique where all genes start in one group and are iteratively assigned to a user-defined number of independent clusters. A detailed description of clustering is beyond the scope of this book, but reviews can be found in Quackenbush (2001) and

Sherlock (2000). Although clustering is a very powerful technique for drawing out gene expression patterns, clustering algorithms will always produce clusters of genes from data, even if these clusters are not biologically relevant. Unfortunately, no clustering techniques are currently available that will assign any level of statistical significance to the joining of a gene to a cluster. Therefore, one must be careful when trying to deduce biological relevant information from clustered data.

One issue that is of paramount importance to the whole microarray process is that concerning the methods used to manipulate, document and archive the whole microarray system. A huge amount of diverse information is derived from a single microarray experiment and it is vital that everything is organized and maintained to allow easy querying and manipulation of the data. The use of relational databases for this purpose facilitates the development of very powerful tools for microarray data storage, analysis and visualization, and several systems are publicly available (Gardiner-Garden and Littlejohn, 2001). An international effort (Microarray and Gene Expression Database group; http://www.mged.org) was started in 1998 to devise and develop a set of standards to describe the kinds of information that are required to fully document a microarray experiment (Brazma et al., 2001). The standards introduced will also allow the development of compatible database systems for the easy exchange and comparison of microarray data generated by different laboratories.

◆◆◆◆◆◆ EXAMPLES OF APPLICATIONS OF DNA MICROARRAYS TO THE STUDY OF *P. FALCIPARUM* BIOLOGY

P. falciparum has a complex multistage life cycle, passing through many different environmental conditions within the mammalian and mosquito hosts. Tight control of the transcriptional machinery must, therefore, be maintained in order to sustain such complexity. Also, the diverse environmental conditions experienced require different gene expression profiles to produce a proteomic repertoire capable of sensing and responding to the changing conditions. Little is known of the specific processes involved in this complex life cycle; however, microarray technology is an ideal technique for identifying global expression profiles throughout the life cycle.

To date, only two studies using microarray technology to analyse *P. falciparum* biology have been published. This probably reflects the difficulty experienced when trying to isolate enough *P. falciparum* material for microarray experiments. Hayward et al. (2000) produced an array containing 3648 PCR amplified inserts from a *P. falciparum* mung bean nuclease library and used it to compare blood stage sexual and asexual stages of the parasite. By simply calculating fluorescence ratios for the two RNA samples for each spot, they generated lists of spots potentially up-regulated and down-regulated in each stage of the life cycle. Following subsequent sequencing of the products corresponding to the spots that showed the greatest difference, they observed a good correlation between the genes that they identified and genes previously known to be stage-specifically expressed. A slight variation on this approach was taken by Ben Mamoun et al. (2001). Instead of using a genomic DNA-derived library for arraying, a stage-specific cDNA library was printed. In this study, the array was used to investigate changing gene-expression profiles throughout asexual stage development in erythrocytes. Gene lists were generated in a similar way to those of Hayward et al. (2000), and clustering methods were also used to try to group genes by their expression profiles.

The simple analysis methods employed in both studies lack statistical rigour, and so the lists of up-regulated and down-regulated genes that were generated may include false positives. However, they have produced many new potential leads for further validation and investigation by independent methods. Indeed, a gene encoding phosphoenolpyruvate carboxylase, whose stage-specific expression was identified by microarray experiments (Hayward et al., 2000), has already been further investigated and characterized, and shown to play an important role in the transition of the parasite from the asexual to the sexual stage (Hayward, 2000). A greater understanding of this transition process may identify molecules that could be potential drug and vaccine targets, and thus used to prevent transmission of the parasite and the disease.

◆◆◆◆◆◆ SUMMARY

Microarray analysis is a very powerful technique that can examine the gene expression of thousands of genes simultaneously. The completion of the *P. falciparum* genome allows a targeted approach to the analysis of global gene expression, by allowing the design of gene-specific PCR product arrays. Such arrays can be used in a multitude of experimental designs, including comparing gene expression between different stages of the parasite life cycle; the examination of the effect of drugs on gene expression as a single-dose event or by following a time course of parasite growth in the presence of drug; and genomic analysis can be performed by comparing the hybridization patterns of fluorescently labelled DNA from different strains of the parasite. Such an approach has been shown to be an effective way to identify gene deletions in *Campylobacter jejuni* (Dorrell et al., 2001) (see Chapter 7).

The explosion of whole-genome sequencing in the last 10 years has played a major role in the popularity of microarray technology in gene expression studies. Currently, the practical technologies for running a microarray are relatively well defined; however, the methods for data analysis are still in their infancy. Therefore, until more robust methods for data analysis are available, microarrays should only be considered as an exploration experiment where useful leads can be identified for more specific examination by other better-defined techniques.

Emerging technologies are further pushing the bounds of microarray technology. The introduction of three-, four- or more colour scanners will allow even more complex analysis of biological samples, but this will pose further questions to the computational biologists and bioinformaticists, who will have to develop techniques for analysing these complex data. In addition, the development of techniques allowing the application of microarrays to much smaller amounts of biological samples will vastly expand their use into the microbiological arena, where the limiting amount of testing material is very often an issue.

Acknowledgements

The authors wish to thank John Quackenbush of The Institute for Genomic Research (TIGR) for much guidance and advice when setting up the malaria microarray project, and also to Priti Hegde, Cheryl Gay and Rong Qi, also from TIGR, for printing the arrays used in this project. We would also like to thank David Haynes and Kathy Moch for supplying the cultured parasite material. The work described was funded by Naval Medical Research Center Work Units 61102AA0101BFX and 611102A0101BCX, and with funding from the Burroughs Wellcome Fund. The opinions and assertions herein are the private ones of the authors and are not to be construed as official or as reflecting the views of the US Navy or the naval service at large.

References

Ben Mamoun, C., Gluzman, I. Y., Hott, C., MacMillan, S. K., Amarakone, A. S., Anderson, D. L., Carlton, J. M., Dame, J. B., Chakrabarti, D., Martin, R. K., Brownstein, B. H. and Goldberg, D. E. (2001). Co-ordinated programme of gene expression during asexual intraerythrocytic development of the human malaria parasite *Plasmodium falciparum* revealed by microarray analysis. *Mol. Microbiol.* **39**, 26–36.

Brazma, A., Hingamp, P., Quackenbush, J., Sherlock, G., Spellman, P., Stoeckert, C., Aach, J., Ansorge, W., Ball, C. A., Causton, H. C., Gaasterland, T., Glenisson, P., Holstege, F. C., Kim, I. F., Markowitz, V., Matese, J. C., Parkinson, H., Robinson, A., Sarkans, U., Schulze-Kremer, S., Stewart, J., Taylor, R., Vilo, J. and Vingron, M. (2001). Minimum information about a microarray experiment (MIAME) – toward standards for microarray data. *Nature Genet.* **29**, 365–371.

Carter, R., Ranford-Cartwright, L. and Alano, P. (1993). The culture and preparation of gametocytes of *Plasmodium falciparum* for immunochemical, molecular, and mosquito infectivity studies. In *Methods in Molecular Biology*, Vol. 21, *Protocols in Molecular Parasitology* (Hyde, J. E., ed.), pp. 67–88. Totowa, NJ: Humana Press.

Dorrell, N., Mangan, J. A., Laing, K. G., Hinds, J., Linton, D., Al-Ghusein, H., Barrell, B. G., Parkhill, J., Stoker, N. G., Karlyshev, A. V., Butcher, P. D. and Wren, B. W. (2001) Whole genome comparison of *Campylobacter jejuni* human isolates using a low-cost microarray reveals extensive genetic diversity. *Genome Res.* **11**, 1706–1715.

Gardiner-Garden, M. and Littlejohn, T. G. (2001). A comparison of microarray databases. *Brief Bioinform.* **2**, 143–158.

Haynes, J. D. and Moch, J. K. (2002). Automated synchronization of *P. falciparum* malaria parasites by culture in a temperature-cycling incubator. In *Malaria Methods and Protocols* (Doolan, D. L., ed.). *Methods in Molecular Medicine*. Totowa, NJ: Humana Press (in press).

Hayward, R. E. (2000). *Plasmodium falciparum* phosphoenolpyruvate carboxykinase is developmentally regulated in gametocytes. *Mol. Biochem. Parasitol.* **107**, 227–240.

Hayward, R. E., Derisi, J. L., Alfadhli, S., Kaslow, D. C., Brown, P. O. and Rathod, P. K. (2000). Shotgun DNA microarrays and stage-specific gene expression in *Plasmodium falciparum* malaria. *Mol. Microbiol.* **35**, 6–14.

Hegde, P., Qi, R., Abernathy, K., Gay, C., Dharap, S., Gaspard, R., Hughes, J. E., Snesrud, E., Lee, N. and Quackenbush, J. (2000). A concise guide to cDNA microarray analysis. *Biotechniques* **29**, 548–562.

Quackenbush, J. (2001). Computational analysis of microarray data. *Nature Rev. Genet.* **2**, 418–427.

Randolph, J. B. and Waggoner, A. S. (1997). Stability, specificity and fluorescence brightness of multiply-labelled fluorescent DNA probes. *Nucleic Acids Res.* **25**, 2923–2929.

Sherlock, G. (2000). Analysis of large-scale gene expression data. *Curr. Opin. Immunol.* **12**, 201–205.

Zhu, Z. and Waggoner, A. S. (1997). Molecular mechanism controlling the incorporation of fluorescent nucleotides into DNA by PCR. *Cytometry* **28**, 206–211.

20 Functional Analysis of the *Plasmodium falciparum* Genome Using Transfection

Alan F Cowman, Manoj Duraisingh, Rebecca A O'Donnell, Tony Triglia and Brendan S Crabb

The Walter and Eliza Hall Institute of Medical Research, Victoria 3050, Australia

◆◆◆

CONTENTS

Introduction
Growing parasites for and transfection of *P. falciparum*
Plasmid vectors for transient and stable transfection
Analysis of transfected parasites
Summary

◆◆◆◆◆◆ INTRODUCTION

Plasmodium falciparum is an important pathogen of humans causing the most lethal form of malaria. This protozoan parasite causes over 300 million infections per year in humans and it has been estimated that over two million of these individuals die as a result of this disease. Antimalarial drugs are used to control and treat this disease; however, the parasite has developed the ability to evade the lethal effect of most of these drugs. Important goals are the development of vaccines and novel antimalarial drugs. The genomic sequence of *P. falciparum* is rapidly being completed, and this is providing large amounts of new information that will assist in identifying potential vaccine and drug targets (Gardner et al., 1998; Bowman et al., 1999).

In order to help define good candidate proteins to serve as vaccines and drug targets, it is important to have at least some understanding of their function and potential role in the survival of the parasite and its pathogenesis in the human host (Cowman, 2001). The ability to construct loss-of-function and gain-of-function mutants using transfection of *P. falciparum* is an approach that will allow, in combination with more biochemical and cell biological strategies, the elucidation of the role of these proteins (Crabb and Cowman, 1996; Wu et al., 1996; Crabb et al., 1997a). Importantly, construction of parasites with gene knockouts will help determine if particular proteins are essential for normal parasite growth, and is an approach to validate them as drug targets or vaccine molecules (Cowman, 2001).

In this chapter, we discuss strategies and methods to transfect *P. falciparum* in order to generate specific loss-of-function and gain-of-function mutants. This has broad applications to analyse many different aspects of the parasites biology such as drug resistance, invasion of erythrocytes by the merozoite form, protein trafficking, cytoadherence of parasite-infected erythrocytes and sexual stage reproduction.

◆◆◆◆◆◆ GROWING PARASITES FOR TRANSFECTION OF *P. FALCIPARUM*

The process that we routinely use for the stable and transient transfection of *P. falciparum* is summarized below.

1. Synchronize *P. falciparum* parasites at approximately 1% ring stages using 5% sorbitol 2 days before transfection. It is important to use fresh human erythrocytes to ensure that they support growth of parasites during the lengthy initial selection process. On the day of transfection, parasites should be approximately 1–3% parasitaemia, although they can be a higher density for transient transfection.
2. Prepare plasmid DNA for transfection by ethanol precipitating at least 50 µg of the vector (usually 50–100 µg). Allow the pellet to dry for 5 min in a laminar flow hood. Resuspend DNA in 15 µl of sterile TE (10 mM Tris–HCl, pH 7.5, 1 mM EDTA). It is essential that the DNA be fully dissolved in the buffer before adding further solutions. Add 385 µl of sterile Cytomix to each plasmid DNA pellet. Cytomix consists of 120 mM KCl, 0.15 mM $CaCl_2$, 2 mM EGTA, 5 mM $MgCl_2$, 10 mM K_2HPO_4/KH_2PO_4, 25 mM Hepes, pH 7.6.
3. Centrifuge 5 ml of culture/transfection at 1500g for 5 min and remove supernatant.
4. Add the Cytomix/plasmid mixture to the parasitized erythrocyte pellet and pipette up and down gently to mix.
5. Transfer the parasitized erythrocyte/DNA mixture to a GenePulser (BioRad) cuvette (0.2 cm gap). Electroporate at 0.310 kV and 950 µF. The resulting time constant should be between 7 and 12 ms.
6. Immediately add the electroporated sample to a labelled 10 ml petri dish containing 3% erythrocytes in complete RPMI/Hepes medium with 5% serum and 5% albumax. Grow parasites at 37°C in a gas mixture of 5% carbon dioxide, 1% oxygen and 94% nitrogen.
7. After 1 day post-transfection, change the medium.
8. At 2 days post-transfection add WR99210 to 2.5–10 nM when using the human *dihydrofolate reductase (dhfr)* gene as the selectable marker or 0.2 µM pyrimethamine when using the *Toxoplasma gondii dhfr* gene as

the selection system (see other positive selectable markers later). Fresh media and the appropriate drug are added to cultures for a further 48 h, then every 2 days until parasite establishment.

9. Parasitized erythrocytes should be detectable in Giemsa-stained smears of erythrocytes after 14–40 days.
10. To select parasites with the plasmid vector integrated by homologous recombination, grow the parasites for 3 weeks without drug selection then reapply drug pressure and continue to culture until parasites appear in blood smears. The parasites should be analysed by pulsed-field gel electrophoresis (PFGE) and Southern blotting to determine if integration into the relevant gene has been obtained (see details later).
11. For selection of transfected parasites using vectors containing the thymidine kinase gene for negative selection, the parasites should be transfected as described earlier and selected with WR99210. Once stable transfected parasites are established on WR99210, add ganciclovir to 4 mM for 9 days. During this period there may be substantial parasite death. On day 9, subculture to ~1% parasitaemia and add WR99210 at 10 nM for 2 days, followed by ganciclovir at 4 nM for 1–2 weeks until parasites reappear. Analyse chromosomes and genomic DNA of the parasites using PFGE and Southern blotting to determine if integration into the appropriate gene has occurred in these parasites.

This is the standard procedure for both transient (Wu *et al.*, 1995; Crabb and Cowman, 1996) and stable (Crabb and Cowman, 1996; Wu *et al.*, 1996; Fidock and Wellems, 1997) transfection, although higher initial parasitaemias are used for transient expression. The method shown below uses electroporation of parasite-infected erythrocytes; however, an alternative method of transfection involving spontaneous uptake of plasmid DNA by *P. falciparum* has also been described (Deitsch *et al.*, 2001). Some plasmid constructs lead to the establishment of transfected parasite populations much faster than others. For example, parasites transfected with plasmids designed to express green fluorescent protein (GFP) grow considerably slower than those transfected with other vectors (Waller *et al.*, 2000; Wickham *et al.*, 2001), especially those containing Rep 20 sequence (see below) (O'Donnell *et al.*, 2002). Plasmids should be transfected as undigested circular DNA and will replicate episomally in parasites following transfection and drug selection. Gene targeting by single cross-over recombination to disrupt genes (Crabb *et al.*, 1997a; Baldi *et al.*, 2000) or for allelic replacement (Triglia *et al.*, 1998; O'Donnell *et al.*, 2000) can be achieved using a drug cycling procedure. Furthermore, gene knockouts using double recombination cross-over can be generated using the pHTK plasmid (Duraisingh *et al.*, 2002) (see later). This transfection plasmid possesses a *thymidine kinase* gene cassette that allows negative selection against the presence of the plasmid backbone and hence selects for parasites that have integrated the positive selectable marker cassette via double cross-over recombination. It also greatly shortens the length of time required to derive parasites that have integrated the appropriate portion of the plasmid vector.

◆◆◆◆◆◆ PLASMID VECTORS FOR TRANSIENT AND STABLE TRANSFECTION

Vector construction

Construction of appropriate vectors for stable or transient transfection using some of the available plasmids can be problematic owing to instability and poor growth in *E. coli* (Coppel and Black, 1998). The main reason for this instability appears to be the high AT composition of the genes and, in particular, the extragenic region that can be >90% AT. Interestingly, once a construct is obtained, it usually remains stable henceforth. Therefore, the problems encountered can usually be overcome by testing a number of *E. coli* strains with different genetic backgrounds to identify one that provides a stable vector. The *E. coli* strains PMC103 and XL10-Gold have proven to be very useful for this problem but can provide poor yields of the plasmid. This can frequently be overcome by transferring the plasmid to another *E. coli* strain with a different genetic background, such as XL1-Blue.

Subcloning of fragments into the various transfection vectors, such as pHH1, pHH2 and pHTK, can be very inefficient and it may be necessary to screen large numbers of *E. coli* colonies to identify those that contain the correct plasmid in an unrearranged state. This can more easily be achieved by using screening by the polymerase chain reaction (PCR), picking a portion of each colony directly into the PCR reaction mixture. This facilitates screening of large numbers of colonies to identify those that have the appropriate structure required.

Common *P. falciparum* transfection plasmids

Stable transfection of *P. falciparum* has used primarily two types of vectors containing either *Toxoplasma gondii dhfr* (Crabb and Cowman, 1996; Wu *et al.*, 1996) or human *dhfr* (Fidock and Wellems, 1997) as the gene for selection of transfected parasites. More recently, other genes, such as blastocidin, neomycin and puromycin resistance genes, have been used successfully for selection of *P. falciparum* transfectants (Ben Mamoun *et al.*, 1999; de Koning-Ward *et al.*, 2001). The structure of some commonly used vectors is shown in Figure 20.1.

The pHH1 vector, its parent vector pHC1 and derivatives have been useful for gene targeting (Triglia *et al.*, 1998; Reed *et al.*, 2000) and for transgene expression (Crabb *et al.*, 1997b; Baldi *et al.*, 2000; Waller *et al.*, 2000; Triglia *et al.*, 2001) to analyse protein trafficking, merozoite invasion and drug resistance. This vector allows integration of the plasmid into the genome of *P. falciparum* by single cross-over recombination (Figure 20.2). Although very useful, this strategy has a major drawback in that it does not allow selection of gene disruptions that are not lethal but are deleterious to parasite growth. This is because of the persistence of episomal plasmid in some parasites despite growth of transfectants in the absence of drug selection. Reapplication of drug pressure selects for

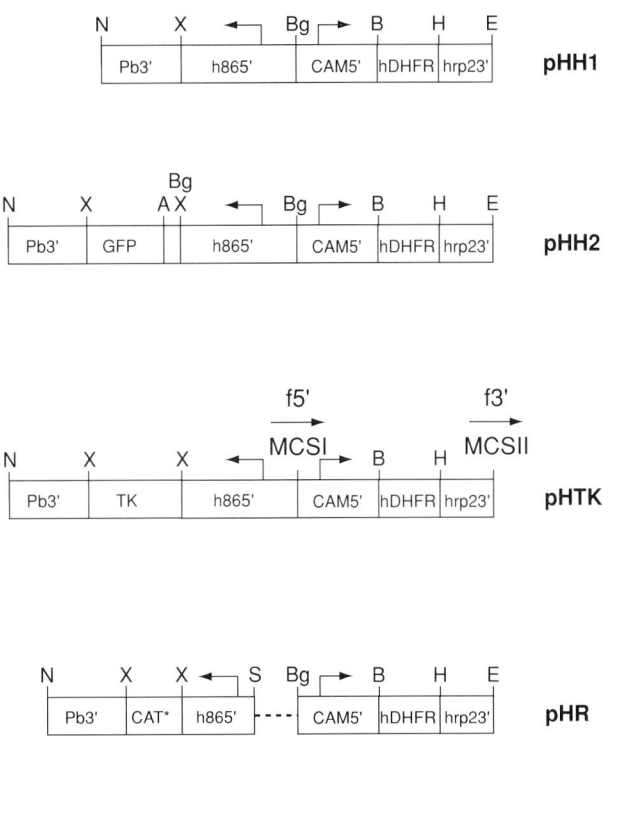

Figure 20.1. Vector maps for cloning. In all plasmids the h*DHFR* cassette is comprised of 1.0 kb of CAM5' UTR, 0.56 kb h*dhfr* gene and 0.6 kb hrp2 3' UTR. pHH1 is a single cross-over vector used for targeted disruption and gene replacement. The target sequence is inserted either as a *Xho*I (X), *Xho*I fragment or a *Bgl*II (Bg), *Xho*I fragment, if the hsp86 5' UTR is not required to drive the expression of the target gene. pHH2 (Waller et al., 2000; Wickham et al., 2001) is similar to pHH1 (Reed et al., 2000), although it contains the gene encoding GFP. The plasmid as shown in this figure contains an acyl carrier protein leader sequence that can be removed and replaced with a target sequence (Waller et al., 2000; Wickham et al., 2001). An ATG needs to be provided within the target sequence and fused in frame to GFP. pHTK is a double cross-over plasmid used for gene disruption (Duraisingh et al., 2002). The 5'- and 3'-target sequences need to be cloned into the multicloning sites (MCSI and MCSII), respectively, in the same direction as expression of the hDHFR cassette, as shown by the arrows. From 5' to 3' the unique restriction sites within MCSI are *Sac*II, *Spe*I, *Bgl*II, *Hinc*II, *Hpa*I, and those within MCSII are: *Eco*RI, *Cla*I, *Nco*I, *Avr*II, *Kas*I and *Sfo*I. Two Rep20-containing plasmids have also been included in this figure, both containing a CAT gene within the *Xho*I site (O'Donnell et al., 2002). pHR has 509 bp of Rep20 sequence (dashed line) inserted into the *Sac*II (S), *Bgl*II (Bg) sites of pHH1, while pHHC*/DR0.28 has a 280 bp Rep20 sequence within the *Not*I site of pHH1. Other restriction sites labelled on plasmids include *Hind*III (H) and *Bam*HI (B).

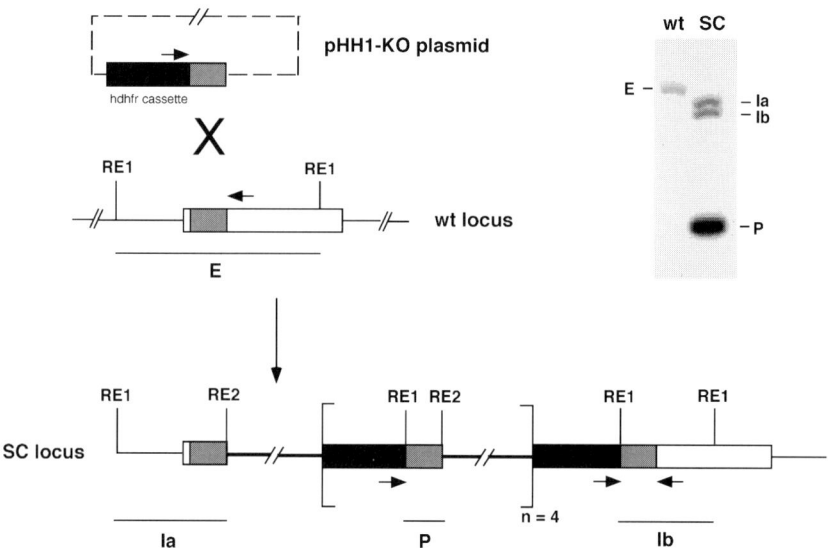

Figure 20.2. A typical single cross-over homologous recombination event in *P. falciparum*. (A) Diagrammatic representation of a single homologous recombination in *P. falciparum*. The pHH1-KO plasmid contains a homologous targeting sequence (grey box) that allows integration into the *wt* locus. The black region corresponds to the rest of the gene. Frequently more than one copy of the plasmid can integrate. (B) A typical Southern blot is shown that has been probed with the homologous targeting sequence (grey box). This detects a single DNA band in *wt* parasite genomic DNA but in the transfected parasites, where the plasmid has integrated, a number of DNA bands are observed. Importantly, in the SC parasite genomic DNA there is no longer a band of equivalent size as that seen in *wt* parasites, proving that homologous integration has occurred. The DNA band labelled P represents the multiple copies of plasmid that have been incorporated during the recombination event as revealed after digestion with the restriction enzymes RE1 and RE2. Other genomic bands produced include 1a and 1b.

parasites that have the plasmid integrated but also for those that contain the plasmid as an episome. If the parasites containing the integrated form of the plasmid grow more slowly, they will be lost in the parasite population and parasites with episomal copies of plasmid will predominate.

To overcome the problem of persisting episomal plasmid in transfected *P. falciparum*, we developed a new vector (pHTK and derivatives) that utilizes the *thymidine kinase* gene to select negatively against its maintenance (Duraisingh et al., 2002). This has been very successful as it allows disruption of genes not previously obtained using pHH1 and also significantly decreases the length of time required to select the *P. falciparum* parasites that have integrated the plasmid. Importantly, this vector allows selection of parasites that have integrated a region of the transfection plasmid by double cross-over recombination (Figure 20.3). This is an important advance for reasons described above, but also allows more defined deletions and mutations in the *P. falciparum* genome. It will also facilitate the production of double mutations and knockouts into the genome.

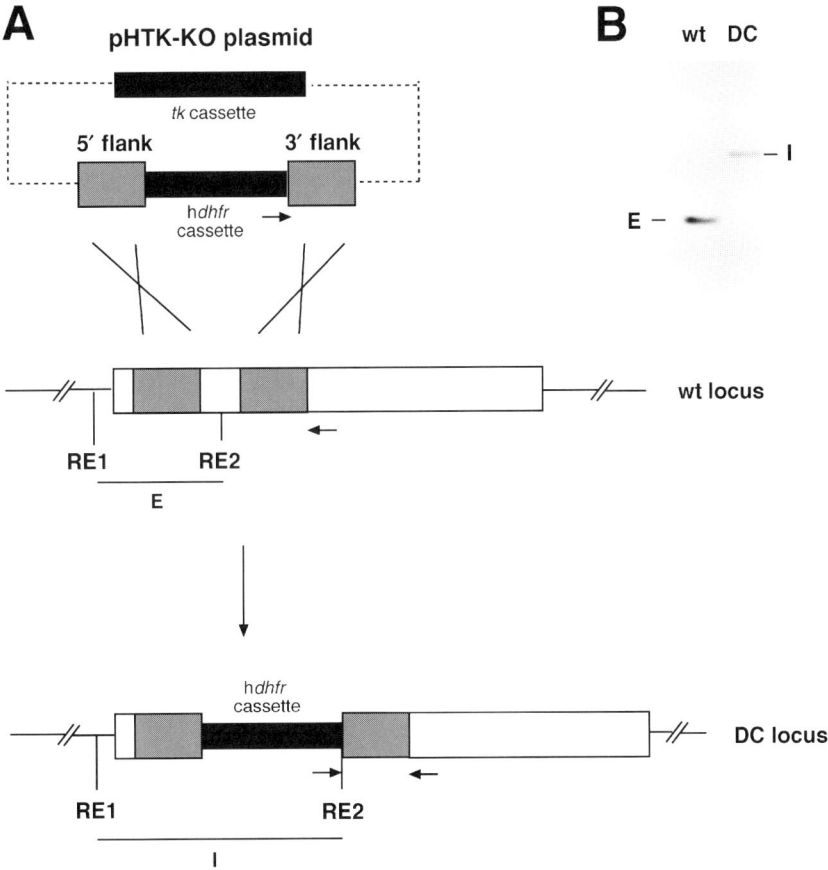

Figure 20.3. A double cross-over homologous recombination event using the pHTK plasmid in *P. falciparum*. (A) Diagrammatic representation of a typical double homologous recombination cross-over event using pHTK vector. The 5'- and 3'-flanks (grey boxes) are two regions from a gene (*wt* locus) that allow a double recombination event as shown to insert the h*dhfr* gene into the endogenous locus. (B) Integration via a double recombination event is confirmed by Southern blot hybridization on genomic DNA from the parental and transfected parasite strains. Hybridization with the 5'-flank sequence allows identification of the *wt* locus (fragment E), whereas integration of the h*dhfr* cassette by double recombination yields an altered hybridizing DNA fragment (fragment I).

The use of transfection vectors for transgene expression is proving to be an important approach to analyse many aspects of *P. falciparum* biology, including merozoite invasion and particularly protein trafficking (Waller et al., 2000; Triglia et al., 2001; Wickham et al., 2001). The use of GFP-tagged proteins has been an important application to follow the trafficking pathway of proteins in live *P. falciparum*-infected erythrocytes (Waller et al., 2000; Wickham et al., 2001). The transfection vector pHH2 allows cloning of sequences into a gene cassette to obtain expression of proteins fused to GFP. This vector uses the promoter from the *hsp86* gene, which allows a broad expression of the GFP in *Plasmodium* blood stages.

Rep20 plasmids

It has recently been shown that the inclusion in transfection plasmids of stretches of the *P. falciparum* subtelomeric repeat sequence Rep20 confers improved plasmid maintenance in transfected parasites (O'Donnell et al., 2002). This occurs because Rep20 sequence allows transfected plasmids to tether to *P. falciparum* chromosomes and, as a result, plasmids are segregated efficiently between daughter merozoites. The primary advantage of this for transfection technology is that drug-resistant parasite populations are established much more rapidly if Rep20 is included in the transfection plasmid: some 1–2 weeks before the appearance of parasites transfected with control plasmids.

A number of different Rep20-containing plasmids have been reported. All possess the human *dhfr* selectable marker and three plasmids (pHHC*/DR1.4, pHHC*/DR0.28 and pHHMC*/3R0.5) possess a transgene expression cassette under the control of the PfHSP86 5' promoter and PbDHFR-TS 3' transcriptional termination sequence (O'Donnell et al., 2002). The *chloramphenicol acetyl transferase* (*CAT*) gene in this expression cassette, which is inactive in these plasmids owing to a mutation in *CAT*, can be excised with *Xho*I to allow insertion of a transgene of interest.

Other positive selectable markers

In addition to the *dhfr* selectable markers described earlier, three other positive selectable markers have been successfully used to derive drug-resistant parasite populations. These markers, blasticidin S deaminase (BSD) (Ben Mamoun et al., 1999), neomycin phosphotransferase II (NEO) (Ben Mamoun et al., 1999) and puromycin-*N*-acetyltransferase (PAC) (de Koning-Ward et al., 2001), confer resistance to blasticidin S, geneticin (G418) and puromycin, respectively. Although all three selectable markers have been used to derive drug-resistant parasite populations harbouring episomally replicating plasmids, to date only PAC has been successfully used for gene targeting.

◆◆◆◆◆◆ ANALYSIS OF TRANSFECTED PARASITES

Characterization of transient transfectants

Because of the low efficiency of *P. falciparum* transfection, highly sensitive reporter systems are required for use in transient transfection. Two such reporter genes, *CAT* and luciferase (*LUC*), have been successfully used for promoter analysis following transient transfection (Wu et al., 1995; Crabb and Cowman, 1996; Horrocks and Kilbey, 1996). Dechering et al. (1999) have also reported characterized two sexual-stage-specific promoters using CAT, LUC and GFP. In our laboratories, the *CAT* reporter is most commonly used. To detect CAT expression following transient transfection, extract parasites at day 2–3 post-transfection and assay for CAT activity as follows.

1. Centrifuge parasites (from 10 ml culture) ~1500g for 5 min.
2. Discard supernatant and resuspend pellet in ~1.5 vols of 0.15% saponin (dissolved in RPMI-Hepes).
3. Place on ice for 10 min.
4. Centrifuge ~2800g for 10 min.
5. Carefully remove and discard supernatant.
6. Wash the pellet by resuspending in 500 µl TEN buffer (40 mM Tris–HCl, pH 7.6, 1 mM EDTA, pH 8.0, and 150 mM NaCl) and centrifuging at 14 000g for 1–2 min. Repeat once.
7. Resuspend pellet in 120 µl 0.25 M Tris–HCl (pH 7.6).
8. Freeze–thaw ×3 in dry ice/ethanol bath.
9. Heat for 10 min at 65°C (to destroy endogenous CAT activity).
10. Centrifuge at 14 000g for 1 min and transfer 114 µl of the supernatant to a fresh tube.

To assay supernatant for CAT activity:

11. Establish a substrate mix of 5 µl acetyl (or buteryl) CoA (5 mg/ml in ddH$_2$O) and 0.5 µl ^{14}C-chloramphenicol per sample.
12. Add 5.5 µl of substrate mix to 114 µl of supernatant.
13. For negative control, add substrate mix to 114 µl 0.25 M Tris–HCl (pH 7.6).
14. Positive control is as for point 13 with the addition of 0.5 µl bacterial CAT enzyme (Promega) prediluted 1/500 in 0.25 M Tris–HCl (pH 7.6).
15. Incubate reactions overnight at 37°C.
16. To extract ^{14}C-chloramphenicol species, add 500 µl ethyl acetate to each tube and vortex for 1 min.
17. Centrifuge at 14 000g for 2 min and collect the top phase.
18. Evaporate using a vacuum centrifuge or by leaving open tubes in a fume hood.
19. Add 10 µl ethyl acetate to each tube and spot on to a TLC plate.
20. Resolve with 97% chloroform/3% methanol in a TLC chamber.
21. Allow to dry and expose to film or a phosphoimaging plate to detect and quantify unacetylated and acetylated ^{14}C-chloramphenicol spots.

Characterization of stable transfectants: genetic analysis

It is particularly important to monitor stable transfectants genetically once a drug-resistant population emerges post-transfection and during the drug cycling process. It is important to do this for three key reasons:

1. To ensure that transfected populations do not represent naturally drug-resistant mutants, such as those with a mutation in the endogenous DHFR-TS gene, but are instead transformed with the desired plasmid.
2. To determine if the transfected populations possess episomally replicating and/or integrated copies of the transfection plasmid.
3. To examine the nature of the integration event (i.e. homologous vs. non-homologous; single vs. double cross-over recombination).

A combination of three approaches is used for this analysis.

Polymerase chain reaction

This approach can be used for all requirements. However, for a number of reasons we believe that its use is limited and that the technique should be used as a guide only. For example, detection of the transfection plasmid by PCR using oligonucleotides specific for a unique sequence (such as a targeting sequence) is confounded by the presence of residual DNA left over from the original transfection. This DNA can be destroyed by pre-digestion of the gDNA with *Dpn*I, a restriction enzyme with a frequently found recognition sequence that cleaves only methylated (such as that replicated in *E. coli*) and not unmethylated (parasite-replicated) DNA, although *Dpn*I digestion is unlikely to be 100% efficient. PCR is particularly useful to detect the presence of homologous integration events using a combination of a plasmid-specific oligonucleotide (not specific to the gene-targeting sequence) and one directed to the genomic sequence located immediately outside of the gene-targeting fragment found in the plasmid. The presence of such a product (which should be sequenced for confirmation) demonstrates that homologous integration has indeed occurred. Using this approach, however, it is not possible to determine the proportion of the parasites that possess integrated forms of the plasmid.

Pulsed-field gel electrophoresis

The separation of *P. falciparum* chromosomes by PFGE followed by Southern blotting is a powerful approach to monitor the genotype of transformants (Crabb *et al.*, 1997a; Baldi *et al.*, 2000). By hybridizing identical blots with a probe to detect the presence of the transfected plasmid (e.g. plasmid backbone or selectable marker gene) and a probe to detect the targeting sequence present in the plasmid, the progress of episomally replicating to integrated plasmid can be followed. While this approach demonstrates plasmid integration and the chromosome into which this has occurred, it does not reveal the specific nature of the integration event. A good example of the use of PFGE to analyse transfectants is shown in a paper by Baldi *et al.* (2000).

Southern blotting of digested gDNA

To determine if the plasmid has integrated into the intended locus by homologous recombination, gDNA (~1 µg) is digested by appropriate restriction endonucleases, separated by agarose gel electrophoresis and Southern blotted to a nitrocellulose membrane. The enzymes to be used should be ones that are intended to reveal a distinct difference in size of the fragments representing a wild-type locus, an integrated locus and an episomal plasmid when the blot is hybridized to a targeting sequence probe. Plasmids that integrate by single cross-over recombination event often insert a number of head to tail plasmid copies into the locus. If this event has occurred, a band corresponding to that expected for the episomal plasmid will be observed. Southern blot analysis of integration by single and double cross-over recombination is represented diagrammatically in Figures 20.2 and 20.3.

Characterization of stable transfectants: protein analysis

A combination of Western blotting, immune-precipitation and immunofluorescence assays (IFA) are commonly used to confirm that the gene-targeting or transgene expression procedure had the expected effect on protein expression and localization. The nature of gene targeting in *P. falciparum* often means that it is possible for truncated products, usually representing the N-terminal portion of the relevant protein, to be synthesized. If possible, antibodies that recognize this region should be utilized to test if this is the case. The use of appropriate control antibodies that ensure both correct protein loading and stage specificity of the parasites are essential, particularly when analysing gene-disruption mutants.

Characterization of stable transfectants: phenotyping

Of course the type of experiments employed to investigate the phenotype of transformed lines depends heavily on the experimental design and hence the expected effect. However, in most circumstances, it is beneficial to perform blood-stage growth rate and/or red blood cell (RBC) invasion assays, especially if a gene encoding a predicted or known blood-stage molecule has been manipulated by gene targeting. It is possible to use a few different approaches here. For these analyses, it is necessary to first culture the transfectants in the absence of the selecting agent for at least 1 week prior to performing the assay (although this is not appropriate for analysing transfectants with episomally replicating plasmids as these will be lost).

Five-day growth rate assay

For the growth rate assay, parental/control and transfected parasite lines (clonal) are synchronized by sorbitol lysis twice, at 4-h intervals, and then plated in duplicate at 0.5% parasitaemia in medium containing 4% haematocrit. Thin blood smears are made every 8 h. Fresh media are added daily and every 48 h cultures are diluted 1 : 5 with fresh medium containing 4% haematocrit. It is important to analyse control and transfected parasite lines in parallel (Baldi *et al.*, 2000; O'Donnell *et al.*, 2000).

Co-cultivation assay

This analysis can be used if the parental/control and transfected parasite lines can be differentiated using distinct antibodies (O'Donnell *et al.*, 2000, 2001). Following double synchronization as described earlier, parent and transfected ring-stage parasites are mixed at four different ratios and maintained in medium containing 4% haematocrit. Parasites are smeared at the trophozoite/schizont stage at day 1, day 3 and day 5. These smears are analysed by IFA using a mixture of the two relevant antibodies, which must have originated from different species if secondary antibodies are to be used for detection. A mixture of the relevant FITC and rhodamine-labelled secondary antibodies (which should be chosen such that they do not recognize each other) are used to detect parasites. Parasites in 10–20 individual fields each containing at least 20 fluorescent parasites should be counted. Results can then be expressed as a ratio of parasites expressing red and green fluorescence.

RBC invasion assay

Ring-stage parasites are synchronized twice as above and then allowed to mature through to trophozoite/schizont stages. Adjust to 2% haematocrit with 0.5–2% infected erythrocytes and add aliquots of 100 µl into the wells of a 96-well tray. Incubate cultures for approximately 26 h to allow for schizont rupture and merozoite invasion of fresh erythrocytes.

There are a number of ways to quantify invasion rate, although microscopic analysis to detect ring-stage parasites should be used (O'Donnell et al., 2001; Triglia et al., 2001). Here, smears made of the duplicate/triplicate wells are stained with Giemsa and the number of ring-stage parasites per 2000 RBC determined for each well. The mean parasitaemia from duplicate/triplicate wells is calculated and expressed as a percentage of the mean parasitaemia observed in parallel cultures of each parasite line in the presence of control non-immune sera.

Another approach to assay for invasion is to measure [^{3}H]-labelled hypoxanthine uptake (Reed et al., 2000; O'Donnell et al., 2001). Although strictly an assay that measures growth, when used within the one growth cycle and in combination with microscopic analysis, this approach does give a meaningful estimate of invasion rate. For this assay, the parasites are washed in hypoxanthine-free media and the assay set up as above in this media. At ~24 h post-cultivation, 90 µl of each culture is transferred to a fresh microtitre tray containing 20 µl of hypoxanthine-free media supplemented with [^{3}H]-labelled hypoxanthine (10 µCi/ml). A further 24 h later, ~100 µl of water is added to each well, which is then frozen and thawed to lyse infected erythrocytes. Samples are transferred to glass-fibre filters via a cell harvester and quantified using a scintillation counter.

◆◆◆◆◆◆ SUMMARY

Since its development in 1996 (Crabb and Cowman, 1996; Wu et al., 1996), stable transfection technology in *P. falciparum* has been successfully employed to address a number of key biological questions in this most important human pathogen. However, it remains a relatively inefficient system that appears to be sensitive to apparently minor variations in methodology. Hence, the technology has not transferred well to many laboratories around the world. We hope that this review contributes to the effective and widespread use of this powerful technology to study *P. falciparum* gene function.

Acknowledgements

The authors acknowledge the contribution of many colleagues in our own laboratories and elsewhere who have contributed to the development of *P. falciparum* transfection technology. We are grateful to the Australian Red Cross Blood Bank for the generous donation of human blood that is central to this technology and to the National Health and Medical Research Council of Australia for financial support. RAO is the recipient of an Australian Postgraduate Research Award, and AFC and BSC are International Research Scholars of the Howard Hughes Medical Institute.

References

Baldi, D. L., Andrews, K. T., Waller, R. S., Roos, D., Crabb, B. S. and Cowman, A. F. (2000). RAP1 controls rhoptry targeting of RAP2 in the malaria parasite *Plasmodium falciparum*. *EMBO J.* **19**, 1–9.

Ben Mamoun, C., Gluzman, I. Y., Goyard, S., Beverley, S. M. and Goldberg, D. E. (1999). A set of independent selectable markers for transfection of the human malaria parasite *Plasmodium falciparum*. *Proc. Natl Acad. Sci. USA* **96**, 8716–8720.

Bowman, S., Lawson, D., Basham, D., Brown, D., Chillingworth, T., Churcher, C. M., Craig, A., Davies, R. M., Devlin, K., Feltwell, T., Gentles, S., Gwilliam, R., Hamlin, N., Harris, D., Holroyd, S., Hornsby, T., Horrocks, P., Jagels, K., Jassal, B., Kyes, S., McLean, J., Moule, S., Mungall, K., Murphy, L., Oliver, K., Quail, M. A., Rajandream, M-A., Rutter, S., Skelton, J., Squares, R., Squares, S., Sulston, J. E., Whitehead, S., Woodward, J. R., Newbold, C. and Barrell, B. G. (1999). The complete nucleotide sequence of chromosome 3 of *Plasmodium falciparum*. *Nature* **400**, 532–538.

Coppel, R. L. and Black, C. G. (1998). Malaria parasite DNA. In *Malaria, Parasite Biology, Pathogenesis, and Protection* (Sherman, I. W., ed.), pp. 185–202. Washington, DC: ASM Press.

Cowman, A. F. (2001). Functional analysis of drug resistance in *Plasmodium falciparum* in the post-genomic era. *Int. J. Parasitol.* **31**, 871–878.

Crabb, B. S. and Cowman, A. F. (1996). Characterization of promoters and stable transfection by homologous and nonhomologous recombination in *Plasmodium falciparum*. *Proc. Natl Acad. Sci. USA* **93**, 7289–7294.

Crabb, B. S., Cooke, B. M., Reeder, J. C., Waller, R. F., Caruana, S. R., Davern, K. M., Wickham, M. E., Brown, G. V., Coppel, R. L. and Cowman, A. F. (1997a). Targeted gene disruption shows that knobs enable malaria-infected red cells to cytoadhere under physiological shear stress. *Cell* **89**, 287–296.

Crabb, B. S., Triglia, T., Waterkeyn, J. G. and Cowman, A. F. (1997b). Stable transgene expression in *Plasmodium falciparum*. *Mol. Biochem. Parasitol.* **90**, 131–144.

Dechering, K. J., Kaan, A. M., Mbacham, W., Wirth, D. F., Eling, W., Konings, R. N. and Stunnenberg, H. G. (1999). Isolation and functional characterization of two distinct sexual-stage-specific promoters of the human malaria parasite *Plasmodium falciparum*. *Mol. Cell Biol.* **19**, 967–978.

Deitsch, K., Driskill, C. and Wellems, T. (2001). Transformation of malaria parasites by the spontaneous uptake and expression of DNA from human erythrocytes. *Nucleic Acids Res.* **29**, 850–853.

de Koning-Ward, T. F., Waters, A. P. and Crabb, B. S. (2001). Puromycin-N-acetyltransferase as a selectable marker for use in *Plasmodium falciparum*. *Mol. Biochem. Parasitol.* **117**, 155–160.

Duraisingh, M. T., Triglia, T. and Cowman, A. F. (2002). Negative selection of *Plasmodium falciparum* reveals targeted gene deletion by double crossover recombination. *Int. J. Parasitol.* **32**, 81–89.

Fidock, D. A. and Wellems, T. E. (1997). Transformation with human dihydrofolate reductase renders malaria parasites insensitive to WR99210 but does not affect the intrinsic activity of proguanil. *Proc. Natl Acad. Sci. USA* **94**, 10931–10936.

Gardner, M. J., Tettelin, H., Carucci, D. J., Cummings, L. M., Aravind, L., Koonin, E. V., Shallom, S., Mason, T., Yu, K., Fujii, C., Pederson, J., Shen, K., Jing, J., Aston, C., Lai, Z., Schwartz, D. C., Pertea, M., Salzberg, S., Zhou, L., Sutton, G. G., Clayton, R., White, O., Smith, H. O., Fraser, C. M., Adams, M. D., Venter, J. C. and Hoffman, S. L. (1998). Chromosome 2 sequence of the human malaria parasite *Plasmodium falciparum*. *Science* **282**, 1126–1132.

Horrocks, P. and Kilbey, B. J. (1996). Physical and functional mapping of the transcriptional start sites of *Plasmodium falciparum* proliferating cell nuclear antigen. *Mol. Biochem. Parasitol.* **82**, 207–215.

O'Donnell, R. A., Saul, A., Cowman, A. F. and Crabb, B. S. (2000). Functional conservation of the malaria vaccine antigen MSP-1_{19} across distantly related *Plasmodium* species. *Nature Med.* **6**, 91–95.

O'Donnell, R. A., de Koning-Ward, T. F., Burt, R. A., Bockarie, M., Reeder, J. C., Cowman, A. F. and Crabb, B. S. (2001). Antibodies against merozoite surface protein (MSP)-1(19) are a major component of the invasion-inhibitory response in individuals immune to malaria. *J. Exp. Med.* **193**, 1403–1412.

O'Donnell, R. A., Freitas-Junior, L. H., Preiser, P. R., Williamson, D. H., Duraisingh, M., McElwain, T. F., Scherf, A., Cowman, A. F. and Crabb, B. S. (2002). A genetic screen for improved plasmid segregation reveals a role for Rep20 in the interaction of *Plasmodium falciparum* chromosomes. *EMBO J.* **21**, 1231–1239.

Reed, M. B., Saliba, K. J., Caruana, S. R., Kirk, K. and Cowman, A. F. (2000). Pgh1 modulates sensitivity and resistance to multiple antimalarials in *Plasmodium falciparum*. *Nature* **403**, 906–909.

Triglia, T., Wang, P., Sims, P. F. G., Hyde, J. E. and Cowman, A. F. (1998). Allelic exchange at the endogenous genomic locus in *Plasmodium falciparum* proves the role of dihydropteroate synthase in sulfadoxine-resistant malaria. *EMBO J.* **17**, 3807–3815.

Triglia, T., Healer, J., Caruana, S. R., Hodder, A. N., Anders, R. F., Crabb, B. S. and Cowman, A. F. (2001). Apical membrane antigen 1 plays a central role in erythrocyte invasion by *Plasmodium* species. *Mol. Microbiol.* **41**, 47–58.

Waller, R. F., Reed, M. B., Cowman, A. F. and McFadden, G. I. (2000). Protein trafficking to the plastid of *Plasmodium falciparum* is via the secretory pathway. *EMBO J.* **19**, 1794–1802.

Wickham, M. E., Rug, M., Ralph, S. A., Klonis, N., McFadden, G. I., Tilley, L. and Cowman, A. F. (2001). Trafficking and assembly of the cytoadherence complex in *Plasmodium falciparum*-infected human erythrocytes. *EMBO J.* **20**, 1–14.

Wu, Y., Sifri, C. D., Lei, H-H., Su, X-S. and Wellems, T. E. (1995). Transfection of *Plasmodium falciparum* within human red blood cells. *Proc. Natl Acad. Sci. USA* **92**, 973–977.

Wu, Y., Kirkman, L. A. and Wellems, T. E. (1996). Transformation of *Plasmodium falciparum* malaria parasites by homologous integration of plasmids that confer resistance to pyrimethamine. *Proc. Natl Acad. Sci. USA* **93**, 1130–1134.

21 Chromosome Fragmentation as an Approach to Whole-genome Analysis in Trypanosomes

John M Kelly* and Samson Obado
Department of Infectious and Tropical Diseases, London School of Hygiene and Tropical Medicine, London WC1E 7HT, UK

CONTENTS

Introduction
Genetic manipulation of trypanosomatids
Techniques required for chromosome fragmentation
Generation of parasites with truncated chromosomes
Applications and limitations of the approach

◆◆◆◆◆◆ INTRODUCTION

Diseases caused by the trypanosomatid parasites *Trypanosoma cruzi* (Chagas' disease or American trypanosomiasis), *Trypanosoma brucei* (African sleeping sickness) and *Leishmania* sp. (visceral and cutaneous leishmaniasis) are major public health problems in many regions of the world. For example, *T. cruzi* infects up to 20 million people in South America, 30% of whom will develop the debilitating chronic form of the disease, often many years after the initial infection. Vaccine development against African sleeping sickness and Chagas' disease has been hampered by a number of factors. The *T. brucei* genome contains a repertoire of up to 1000 variant surface glycoprotein (*vsg*) genes and constant switching of the expressed antigen type within the population prevents elimination of the parasite by the host immune system (Vanhamme et al., 2001). With *T. cruzi*, there is a possible association with autoimmunity (Leon and Engman, 2001). As a result, the major long-term focus of laboratory-based research has been to improve the range and effectiveness of anti-trypanosomatid drugs. At present, most chemotherapeutic regimes are

* Corresponding author.

unsatisfactory because of their toxicity and limited efficacy. One of the goals of the trypanosomatid genome projects (Table 21.1) is to address these problems through the identification of novel chemotherapeutic targets.

Trypanosomatids are early branching eukaryotes and their study has led to the discovery of many novel genetic and biochemical phenomena ranging from RNA editing (Stuart et al., 1997) to the glycosylphosphatidylinositol (GPI) moieties that anchor some surface proteins to the cell membrane (Ferguson, 1999). In addition, genome organization and expression in these organisms display several unusual features. There are no recognizable RNA polymerase II-specific promoters for protein coding genes, transcription is polycistronic, almost all genes lack introns and each mRNA is modified post-transcriptionally by addition of a spliced leader RNA to its 5'-end (Graham, 1995; Horn, 2001). Trypanosome genomes also contain an extensive range of repetitive elements (Melville et al. 1999; Aguero et al., 2000) and many genes are organized in large tandem arrays. In *T. cruzi*, the surface antigens are encoded by several extremely large and complex gene families that are widely dispersed in the genome (Frasch, 2000). Within trypanosomatids, there is significant variation in the size of chromosomes between different parasite strains (Henriksson et al., 1996; Melville et al., 1999) and, although their genomes are generally diploid, the sizes of homologous chromosomes can differ considerably. Sequence elements within chromosomes that have a role in mitotic stability and replication have not been identified. Genome projects for each of the three main parasitic trypanosomatids are now well advanced (Degrave et al., 2001; Table 21.1) and the completion of fully annotated sequences is expected between 2003 and 2005.

Table 21.1. The trypanosomatid genome project strains

Species	Genome size (Mb) (haploid content)	Number of chromosomes	Further information
Trypanosoma cruzi (CL Brener clone)	45	30–35	http://www.tigr.org/tdb/mdb/tcdb/ http://cruzi.genpat.uu.se/
Trypanosoma brucei (TREU927/4)	35	11[a]	http://parsun1.path.cam.ac.uk
Leishmania major (Friedlin strain)	35	36	http://www.ebi.ac.uk/parasites/leish.html http://www.genome.sbri.org/lmjf

[a] In addition, *T. brucei* also contains a variable number of intermediate size (200–900 kb) and minichromosomes of unknown ploidy. These contain *vsg* expression sites and non-transcribed copies of the *vsg* repertoire, respectively.

◆◆◆◆◆◆ GENETIC MANIPULATION OF TRYPANOSOMATIDS

The availability of whole-genome sequences and recent developments in 'postgenome' technologies, including microarray (Lockhart and

Winzeler, 2000) and proteomic procedures (Pandey and Mann, 2000), are revolutionizing the approaches that we use to study trypanosomatid parasites. However, these advances have also highlighted *the need to develop techniques that will facilitate high-throughput analysis of the genome at the functional level.* In trypanosomatids, integration of transfected DNA occurs almost exclusively by homologous recombination and targeted gene deletion has been the main approach used to investigate gene function (for reviews, see Kelly, 1997; Clayton, 1999). Over the past 10 years a number of additional genetic tools have also been produced. These include plasmid and cosmid shuttle vectors (Kelly, 1995), inducible expression systems (Wirtz and Clayton, 1995), negative selectable vectors (Le Bowitz *et al.*, 1995), artificial chromosomes (Lee *et al.*, 1995), transposon-mediated mutagenesis (Gueiros-Filho and Beverley, 1997), and antisense/RNA interference approaches (Ngo *et al.*, 1998; Wang *et al.*, 2000; Zhang and Matlashewski, 2000).

In *T. cruzi*, however, several factors currently restrict our ability to utilize the genome information and to investigate function fully at a 'global' level. The procedure of targeted gene deletion is time consuming and, since the genome is generally diploid, at least two rounds of transfection are required to obtain a 'functional knockout' (each round takes about 3 months). Furthermore, the parasite has a complex chromosome repertoire (karyotype), many genes are present in dispersed multiple copies and the availability of drug-selectable markers is limited. New transfection-based approaches are needed to circumvent these problems. To explore techniques that could be used to exploit the *T. cruzi* genome sequence and to investigate the function of genes, both individually and collectively, we have developed a transfection-mediated chromosome fragmentation system. This technique can be used to create parasites in which defined regions of the genome have been rendered haploid (partial monosomy), so that null mutants can then be produced using a 'single-knockout' approach. In addition, the procedure has the potential to contribute to our understanding of *T. cruzi* chromosome structure by providing a means of delineating the elements necessary for chromosome stability and maintenance. In this chapter we will outline the principle of the approach and discuss possible applications.

◆◆◆◆◆◆ TECHNIQUES REQUIRED FOR CHROMOSOME FRAGMENTATION

We designed the vector pTEX-CF to facilitate transfection-mediated site-directed chromosome fragmentation in *T. cruzi* as outlined in Figure 21.1. The aim was to achieve this through integration of the vector by single cross-over at a site specified by a targeting fragment. A *T. cruzi* ribosomal DNA promoter, which has been shown to mediate high-level expression in transfected cells (Martinez-Calvillo *et al.*, 1997) was included in the vector to ensure expression of the drug-selectable marker irrespective of the direction of pol II transcription at the site of integration. An array of 21

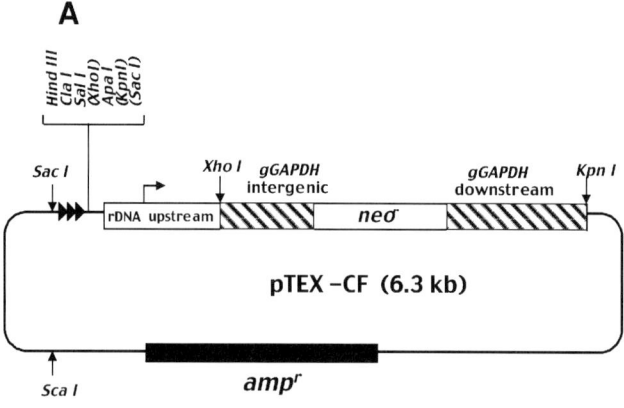

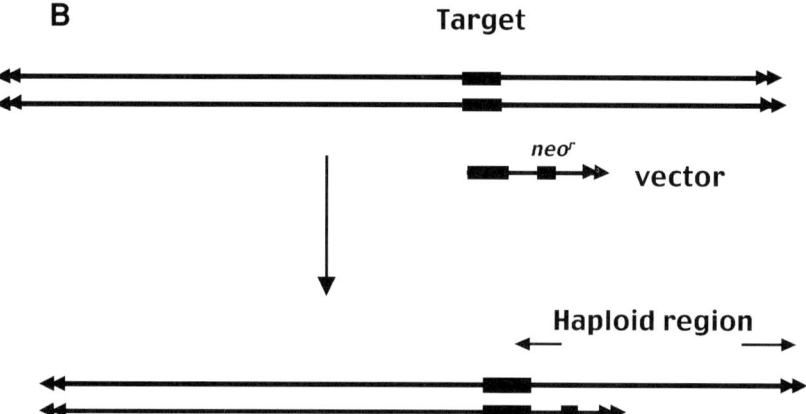

Figure 21.1. How to create *T. cruzi* mutants with restricted haploidy. (A) The chromosome fragmentation vector pTEX-CF. The vector was constructed by modifying the expression construct pRIBOTEX (Martinez-Calvillo et al., 1997). Expression of the drug resistance gene (*neo^r*) is driven by a ribosomal promoter (flagged) and the glycosomal glyceraldehyde-3-phosphate dehydrogenase (*gGAPDH*) flanking sequences (hatched box) ensure correct processing of the transcript (Kendall et al., 1990). Targeting fragments can be cloned in both orientations into the MCS. Bracketed restriction sites in the MCS are not unique. The location of the repeated telomeric hexamers is indicated by the horizontal arrowheads. The plasmid DNA (thin line) is derived from pBluescript and contains an ampicillin resistance gene (*amp^r*). (B) Site-directed chromosome fragmentation. Following insertion of the target sequences into the MCS of pTEX-CF, restriction digestion with the appropriate restriction enzyme produces a linear DNA fragment with the target sequences at one end and telomeric repeats at the other. The vector is designed to integrate by a single cross-over event to produce a truncated chromosome with the new telomere supplied by the vector. Transformants can be selected on the basis of G418 resistance. By cloning the target DNA into pTEX-CF in the opposite orientation, it is feasible to delete the other arm of the chromosome from that shown here. Partially monosomic parasites can be isolated provided that the deletion does not give rise to haploid insufficiency owing to the presence of dose-dependent genes. A 'single knockout' approach can then be used to screen the haploid region for essential genes.

copies of the telomeric repeat hexamer (5'-TTAGGG) was placed in the vector to supply a new telomere for the truncated chromosome (Figure 21.1). We cannot exclude the possibility that integration might also arise from a double cross-over event involving participation of the short telomeric repeat in the vector and the telomere of the targeted chromosome. However, both mechanisms should result in the deletion of all of the genes between the site of insertion and the telomere. The pTEX-CF vector system has now been used to truncate chromosomes in both the CL Brener and Sylvio X10.6 clones of *T. cruzi*, which belong to distinct genetic lineages of this diverse parasite species. We have found no evidence in any of these experiments that the DNA deleted following integration of the vector is maintained in the parasite population. This could be a consequence of the loss of telomeric or centromeric sequences.

To achieve site-specific integration, it is crucial to ensure that the target DNA is single copy in the genome and we recommend that this be confirmed by Southern blotting. We have used fragments ranging from 600 to 2000 bp to target integration to specific chromosomal sites and have observed no significant differences in transformation efficiency. In situations where the whole chromosomal sequence is available, a number of integration sites can be selected to generate a panel of truncated products. In the absence of this information, known loci within defined chromosomes can be targeted in both orientations. The results in these cases should immediately identify the positions of the targeted loci within the chromosome (see Figure 21.2 as example) and reveal the direction of polycistronic transcription at the integration site.

Electroporation is the procedure of choice for transfection experiments involving trypanosomes. We have previously described protocols for electroporators, which operate by the capacitor discharge method (e.g. Bio-Rad Gene Pulser) or by the use of a DC power supply to generate a square wave for the duration of the pulse (e.g. Hoefer Progenitor) (Kelly et al., 1995). Cells electroprated with linearized pTEX-CF constructs (1–2 μg of DNA are sufficient) are selected on the basis of G418 resistance. With the CL Brener clone we use 100 μg/ml, although the optimal level for selection can vary between strains. Drug-resistant cells are usually evident after approximately 14 days, but it often takes an additional 14–21 days to establish a cell line. At this stage the parasites can be cloned by limiting dilution or by growth on agarose plates in the presence of G418 (Mondragon et al., 1999). Stocks of each clone should be cryopreserved and stored long term in liquid nitrogen.

For analysis of transformed clones, we routinely use two approaches: the polymerase chain reaction (PCR), and contour-clamped homogeneous electric field electrophoresis (CHEFE). For PCR analysis we use a primer derived from the rDNA promoter region and from a sequence immediately upstream of the integration site (Figure 21.1). This should yield an amplified fragment of a predicted size and sequence, if integration has occurred at the targeted site. The technique has the advantage that it can be used early during the selection process to confirm the presence of transformants within the population. In addition, many clones can be checked simultaneously. We use CHEFE analysis to confirm

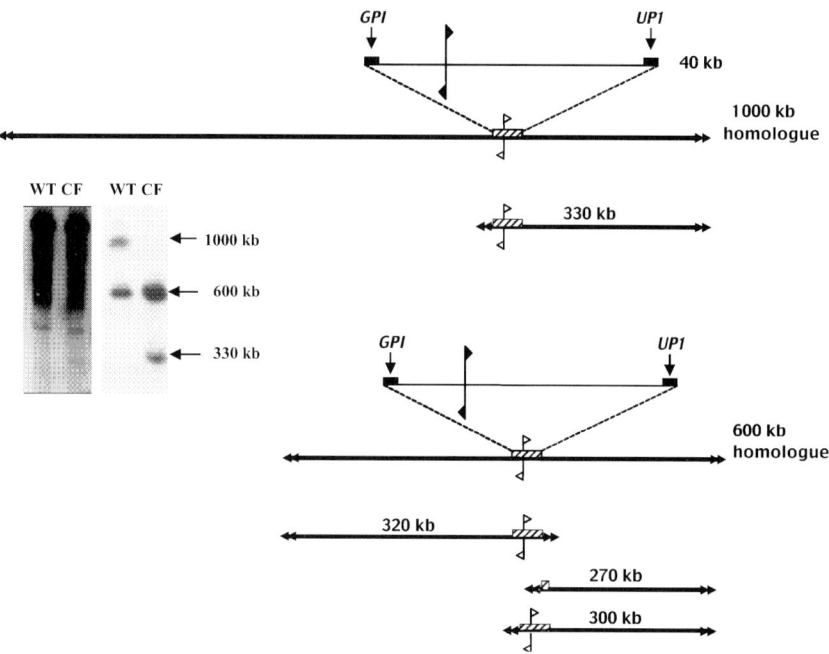

Figure 21.2. Truncated products generated following fragmentation of chromosome III homologues using linearized pTEX-CF constructs targeted to the *GPI* (330-kb, 300-kb products) and *UP1* (320-kb, 270-kb products) loci. The hatched boxes identify a 40-kb sequence that flanks the transcription 'strand switch' region and the flags indicate the direction of polycistronic transcription. The inset shows an ethidium bromide-stained CHEFE gel (left) containing chromosomes from wild-type cells (WT) and cells transfected with a construct targeted to the *GPI* locus (CF). The photograph has been overexposed to allow visualization of the 330-kb truncated product. The inset also shows the autoradiograph (right) corresponding to this gel following hybridization with a *GPI* probe.

the disappearance of the targeted chromosome and to identify the size of the truncated product (Figure 21.2).

❖❖❖❖❖❖ GENERATION OF PARASITES WITH TRUNCATED CHROMOSOMES

For our initial experiments, we targeted *T. cruzi* chromosome III. This chromosome exists as homologues of 600 and 1000 kb in the CL Brener clone and substantial amounts of sequence data are available, including a 93-kb contig containing a transcriptional 'strand-switch' region (Andersson *et al.*, 1998). In chromosome III most of the genes occur in two extremely long 'head-to-head' clusters that are transcribed in opposite directions. Similar types of organization have been observed in chromosomes of *T. brucei* and *Leishmania*, suggesting that the arrangement of genes in large transcription units is a common feature of the trypanosomatids (Myler and Stuart, 2000).

Figure 21.2 serves to illustrate the type of data that can be produced by the chromosome fragmentation approach. In these experiments we used single-copy targeting sequences located proximal to the 'strand-switch' region. By targeting the glucose phosphate isomerase (*GPI*) locus, we were able to delete approximately 700-kb from the left arm of the larger homologue and to generate a 330-kb product as shown in the autoradiograph. The same construct generated a 300-kb product when it integrated into the smaller homologue. It can be concluded from these results that the left arm of the 1000-kb chromosome contains most of the additional 400-kb of the sequence that is responsible for the size heterogeneity between the two homologues. Although the nature of these additional sequences is unknown, their deletion had no phenotypic consequences for the parasite. We were also able to fragment the 600-kb homologue at the *UP1* (unidentified protein 1) locus in both orientations to generate products of 320-kb and 270-kb, thus demonstrating the feasibility of deleting all of the sequences in a chromosome in parallel experiments (Figure 21.2).

The chromosome fragmentation approach also offers a method for mapping regions that are essential for mitotic stability. Parasites containing the truncated chromosomes shown in Figure 21.2 were grown in the absence of G418 (each truncated chromosome contains a *neo*^{r} gene; see Figure 21.1) for approximately 3 months. The 300-kb, 320-kb and 330-kb products were found to be stable for at least 100 generations in the absence of selective pressure. In contrast, the 270-kb product, which lacks the region adjacent to the transcriptional 'strand-switch', was lost from the population within ten generations. This experiment provides preliminary evidence on the location of the sequences necessary for the regulated segregation of chromosome III and further analysis should allow the extent and nature of the centromere to be determined.

◆◆◆◆◆◆ APPLICATIONS AND LIMITATIONS OF THE APPROACH

In the postgenome era, the yeast research community has been an example to other fields. They have taken the long-term view and recognized that new, more expansive approaches are now required to address the study of function. To this end, there has been a co-ordinated attempt to produce deletion mutants covering all 6000 yeast genes and to analyse the resulting phenotypes. Because of technical and logistical constraints (see earlier) this approach has not yet been proposed for *T. cruzi*, or for other parasitic protozoa. The procedures outlined here could circumvent many of these problems and provide the resources to make the study of function at the 'global' level a feasible proposition. The reasons for doing this are clear; in all eukaryotic genomes so far sequenced, approximately 30% of the genes encode unique products of no known biochemical or biological function. In *T. cruzi*, the estimated 3000 parasite-specific *FUN* (Function Unknown) genes may encompass a large number that have important roles in the clinically relevant mammalian forms of the parasite.

With the completion of the genome sequence, the chromosome fragmentation technique will have the potential to create a bank of partially monosomic *T. cruzi* mutants covering a large percentage of the parasite genome. This will facilitate the systematic deletion of genes from the regions of restricted haploidy, either individually or collectively, and provide a rapid method tentatively to discriminate between genes that are essential (knockouts are not obtainable) and those that are non-essential (knockouts obtainable). Since these transfection experiments are carried out on the non-infectious epimastigote (insect) stage of the life cycle (Kelly et al., 1995), it should then be possible to screen the resulting null mutants for those that have lost the ability to differentiate into the infectious metacyclic stage, or that have become deficient in some other aspect of virulence. Proof of function could then be established using an episomal copy of the relevant gene(s) to complement the phenotype.

A possible constraint on this approach to high-throughput functional analysis could be the frequency of haploid insufficiency in *T. cruzi*. We have observed that chromosome truncation in some instances is accompanied by an increase in the copy number of the corresponding chromosome homologue, suggesting the presence of dose-dependent genes in the deleted region. The extent to which this will restrict whole-genome analysis remains to be determined. However, our initial experiments suggest that the chromosome fragmentation is a promising new approach for exploiting the genome sequence and could represent the basis of an integrated framework for addressing the question of function at a more 'global' level.

Acknowledgements

This work was supported by the Wellcome Trust and the UNDP/World Bank/WHO Special Programme for Research and Training in Tropical Diseases (TDR).

References

Aguero, F., Verdun, R. E., Frasch, A. C. and Sanchez, D. O. (2000). A random sequencing approach for the analysis of the *Trypanosoma cruzi* genome: general structure, large gene and repetitive families, and gene discovery. *Genome Res.* **10**, 1996–2005.

Andersson, B., Aslund, L., Tammi, M., Tran, A-N., Hoheisel, J. D. and Pettersson, U. (1998). Complete sequence of a 93.4-kb contig from chromosome 3 of *Trypanosoma cruzi* containing a strand-switch region. *Genome Res.* **8**, 809–816.

Clayton, C. E. (1999). Genetic manipulation of kinetoplastida. *Parasitol. Today* **15**, 372–378.

Degrave, W. M., Melville, S., Ivens, A. and Aslett, M. (2001). Parasite genome initiatives. *Int. J. Parsitol.* **31**, 532–536.

Ferguson, M. A. (1999). The structure, biosynthesis and functions of glycosylphosphatidylinositol anchors, and the contributions to trypanosome research. *J. Cell Sci.* **112**, 2799–2809.

Frasch, A. C. (2000). Functional diversity in the trans-sialidase and mucin families in *Trypanosoma cruzi. Parasitol. Today* **16**, 282–286.

Graham, S. V. (1995). Mechanisms of stage-regulated gene expression in the Kinetoplastida. *Parasitol. Today* **11**, 217–223.

Gueiros-Filho, F. J. and Beverley, S. M. (1997). Trans-kingdom transposition of the *Drosophila* element mariner within the protozoan *Leishmania. Science* **276**, 1716–1719.

Henriksson, J., Aslund, L. and Pettersson, U. (1996). Karyotype variability in *Trypanosoma cruzi. Parasitol. Today* **12**, 108–114.

Horn, D. (2001). Nuclear gene transcription and chromatin in *Trypanosoma brucei. Int. J. Parasitol.* **31**, 1157–1165.

Kelly, J. M. (1995). Trypanosomatid shuttle vectors: new tools for the functional dissection of parasite genomes. *Parasitol. Today* **11**, 447–451.

Kelly, J. M. (1997). Genetic transformation of parasitic protozoa. *Adv. Parasitol.* **39**, 227–270.

Kelly, J. M., Taylor, M. C., Rudenko, G. and Blundell, P. A. (1995). Transfection of the African and American trypanosome. In *Methods in Molecular Biology*, Vol. 47: *Electroporation Protocols in Microorganisms* (Nickoloff, J. A., ed.). Totowa, NJ: Humana Press Inc.

Kendall, G., Wilderspin, A. F., Ashall, F., Miles, M. A. and Kelly, J. M. (1990). *Trypanosoma cruzi* glycosomal glyceraldehyde-3-phosphate dehydrogenase does not conform to the 'hotspot' topogenic signal model. *EMBO J.* **9**, 2751–2758.

Le Bowitz, J. H., Cruz, A. and Beverley, S. M. (1992). Thymidine kinase as a negative selectable marker in *Leishmania major. Mol. Biochem. Parasitol.* **51**, 321–326.

Lee, M. G-S., Yaping, E. and Axelrod, N. (1995). Construction of a trypanosome artificial chromosome. *Nucleic Acids Res.* **23**, 4893–4899.

Leon, J. S. and Engman, D. M. (2001). Autoimmunity in Chagas heart disease. *Int. J. Parasitol.* **31**, 555–561.

Lockhart, D. J. and Winzeler, E. A. (2000). Genomics, gene expression and DNA arrays. *Nature* **405**, 827–836.

Martinez-Calvillo, S., Lopez, I. and Hernandez, R. (1997). pRIBOTEX expression vector: a pTEX derivative for rapid selection of *Trypanosoma cruzi* transfectants. *Gene* **1999**, 71–76.

Melville, S. E., Gerrard, C. S. and Blackwell, J. M. (1999). Multiple causes of size variation in the diploid megabase chromosomes of African trypanosomes. *Chromosome Res.* **7**, 191–203.

Mondragon, A., Wilkinson, S. R., Taylor, M. C. and Kelly, J. M. (1999). Optimization of conditions for growth of wild-type and genetically transformed *Trypanosoma cruzi* on agarose plates. *Parasitology* **118**, 461–467.

Myler, P. J. and Stuart, K. D. (2000). Recent developments from the *Leishmania* genome project. *Curr. Opin. Microbiol.* **3**, 412–416.

Ngo, H., Tschudi, C., Gull, K. and Ullu, E. (1998). Double-stranded RNA induces mRNA degradation in *Trypanosoma brucei. Proc. Natl Acad. Sci. USA* **95**, 14687–14692.

Pandey, A. and Mann, M. (2000). Proteomics to study genes and genomes. *Nature* **405**, 837–846.

Stuart, K., Allen, T. E., Kable, M. L. and Lawson, S. (1997). Kinetoplastid RNA editing: complexes and catalysts. *Curr. Opin. Chem. Biol.* **1**, 340–346.

Vanhamme, L., Lecordier, L. and Pays, E. (2001). Control and function of the bloodstream variant surface glycoprotein expression site in *Trypanosoma brucei. Int. J. Parasitol.* **31**, 523–531.

Wang, Z., Morris, J. C., Drew, M. E. and Englund, P. T. (2000). Inhibition of *Trypanosoma brucei* gene expression by RNA interference using an integratable vector with opposing T7 promoters. *J. Biol. Chem.* **275**, 40174–40179.

Wirtz, E. and Clayton, C. (1995). Inducible gene expression in trypanosomes mediated by a prokaryotic repressor. *Science* **268**, 1179–1183.

Zhang, W. W. and Matlashewski, G. (2000). Analysis of double stranded RNA downregulation of A2 protein expression in *Leishmania donovani*. *Mol. Biochem. Parasitol.* **107**, 315–319.

Index

ACT (Artemis Comparison Tool) 76
β-actin 94
ADH1 (alcohol dehydrogenase 1) promoter 213
adhesin protein 16
Adi proteins 232
Aequorea victoria gfp gene 345
affinity chromatography 200
Affymetrix 123, 428
Affymetrix GeneChip™ 68, 70, 85, 128, 177
agarose gel electrophoresis 89
Alicyclobacillus 337
amidosulphobetaine-14 (ASB-14) 197
amino-acid usage 5
aminoallyl dUTP, coupling of RNA 144–5, 147
3-aminotriazole (3-AT) 216, 220, 222
amplification of differences 116–18
annotation of genomes 3–22
antibiotic target discovery 271–85
    amenability to HTS assay development 285
    antibacterial targets 276–85
    expression during infection 283–5
    gene essentiality 277–83
        allelic-replacement 279
        regulated gene expression 280–3
        plasmid-insertion mutagenesis 278–9
        transposon-insertion mutagenesis 279–80
    comparative genomics 274
    genetic screening 275–6
    identification from bacterial genomes 272–6
    spectrum and selectivity 276–7
    targets essential for growth vs. virulence 273–4
anti-GNA33 265
antisense RNA runoff transcripts (aRNA) 146–7
antisigma factor FLaM 307
array construction 87
Array-Ready Oligo Set™ 123
Artemis 6, 7, 33, 38, 80
Artemis Comparisons Tool (ACT) 21
Aspergillus fumigatus,
    restriction enzyme mediated insertion 156
AT content 56
    percentage, in *Proteobacter* genomes 49, 50, 51
atlas visualization of genomewide information 49–61
ATP-binding proteins 200
auxiliary bait 227

BAC libraries 248–50
bacillary dysentery 154
Bacillus
    plasmid pC194 247
Bacillus amyloliquefaciens 338
Bacillus anthracis 55, 337, 355, 356
Bacillus cereus 55, 337
    BceT diarrhoeal toxin 37
    food poisoning 38
Bacillus coagulans 338
Bacillus larvae 338
Bacillus licheniformis 338
Bacillus megaterium 338
Bacillus pasteuri 338

Bacillus popilliae 338
Bacillus pumilus 338
Bacillus stearothermophilus 338
Bacillus subtilis 41, 198, 274, 277
    categorization of ORFs 346
    plasmid pAM81 343
    plasmid pLS20 343
    see also *Bacillus subtilis strain* 168 genome
Bacillus subtilis Function Analysis (BFA) project 347
Bacillus subtilis strain 168 genome 337–62
    genes of unknown function 345–57
        complementation analyses 355–6
        construction of BFA mutants 351–5
        double cross-over recombination (replacement recombination) 349–51
        essential genes 356–7
        functional analysis programme 347–8
        generation of chromosome mutations 348–51
        single cross-over recombination (Campbell-type) integration 348–9
    genome management 341–5
        competence and recombination 341–2
        plasmid-based host–vector systems 342–3
        promoters for controlled gene expression 343–4
        reporter genes 344–5
    genome sequencing 339
    in silico analysis 339–41
    ORFs 346
    proteome 360–2
        protein–protein interactions 362
        two-dimensional protein index 361–2
    transcriptome 357–60
        DNA arrays 360
        promoters 359
        regulatory proteins 360
        reporter gene technology 358–9
        transcription mapping 360
Bacillus thuringiensis 337, 338
Bacillus thuringiensis subsp. *israelensis*
    GenomeAtlas for 55
    pBtoxis plasmid 54
bacterial artificial chromosomes (BAC) vectors 244–5
bacterial two-hybrid systems 230
basic local alignment search tool (BLAST) 8, 29, 35, 52, 53, 54, 59, 79, 196, 276
    BLASTP 15, 35
    BLASTX comparison 5
    position-specific iterated (PSI)-BLAST 28, 35, 36, 39, 276
bead-beating 190
β-propellers 34
bioinformatics analysis 29
Biomek 224
BioRobot 224
BioRobotics MicroGridII arrayer 93
biotin rNTPs 147
biotinylated antisense oligonucleotides 143–4
BLAST 8, 29, 35, 52, 53, 54, 59, 79, 196, 276
blasticidin S 390
blasticidin S deaminase (BSD) 390

BLASTP 15, 35
BLASTX comparison 5
Bordetella pertussis 257
Brevibacillus 337
Brucella suis 181
BSORF database 355, 358, 362
bZIP (basic domain leucin zipper motif) 211–12

Caenorhabditis elegans 59
Campbell-type integration 348–9
Campylobacter jejuni 68, 83, 87, 120, 133, 181, 200, 294
   1318 family 38–40
   11168 genome project 181
   capsule biosynthesis locus 125–6
   low-cost whole-genome arrays 88
   microarray analysis 125
   NCTC 11168 123, 125, 126
   ORF prediction in 4
   Penner serotyping system 126
   reference 2DGE map of soluble proteins 201
Candida albicans 156
Candida glabrata 156
candidate antigen prediction 260
CAST™ Slides 94
cDNA RDA 106
cDNA representational difference analysis 110–19
   bacterial strains and plasmids 110
   cDNA synthesis 111–12
   cloning of difference products 119
   generation of difference products 116–18
   generation of representations 113–15
   isolation of *S. uberis* 16S and 23S rRNA genes 112–13
   isolation of total RNA 110–11
   preparation of driver and tester components 115–16
   subtractive hybridization and amplification of differences 116–18
cDNA RNA 116
cDNA synthesis 111–12
CDP-Star 162
Chlamydia 121, 276
   cryptic plasmid 125
chromosome fragmentation of trypanosomes 397–404
   applications and limitations 403–4
   generation of parasites with truncated chromosomes 402–3
   techniques required 399–402
c-Jun amino-terminal kinase (JNK) 231
classification of gene products, functional 21–2
cloning of difference products 109
CLUSTALW 36
clusters of orthologous groups (COGs) 15
coding sequences (CDS) 4, 75
codon preference 5, 20
codon usage 18
CoelicolorArray 323, 326
coiled-coil domain 34
COILS 34
ComB1-B3 310
comet tails 95
community genomics 242
comparative genomics 84
complex repeats 34

conserved hypothetic proteins 17
contour-clamped homogenous electric field electrophoresis (CHEFE) 401–2
Coomassie stains 192
culture supernatant (CSN) 198
CURVATURE program 52
cycloheximide 223
Cy-dNTPs, direct incorporation 144
cytokeratin 230

DBTBS 346
3'-degenerate oligonucleotides 147
degradosomes 328
denaturing gradient gel electrophoresis 249
dexamethasone 229
diethyl pyrocarbonate (DEPC)-treated water 111
difference products, cloning of 109
differential display proteomics 201–2
differential fluorescence induction (DFI) 138, 273
differential gene expression 84
dinucleotide frequency 18
dinucleotide signature 20
direct repeat 53
dithiothreitol (DTT) 192
DNA activation domain (AD) 209
DNA binding domain (BD) 209
DNA chips 84, 85
DNA curvature 51, 52
DNA fingerprinting 84
DNA labelling 96
DNA microarrays
   applications 125–32
      genome diversity 125–8
      genome plasticity 128–30
      genomes from related species 130–2
      pathological outcomes of infection 130
   in comparative and evolutionary genomics 121–34
   future perspectives 133–4
   limitations 132–3
   types 122–5
DNS reannealing 104
domain/motif searching 13–14
double cross-over recombination (replacement recombination) 349–51
dried droplet method 194
driver:driver hybrids 108
driver:tester hybrids 108
Drosophila, protein–protein interaction in 225
dual-bait systems 227–8
dual-laser confocal microarray scanner 97

electroporation 219, 401
electrospray-ionization tandem MS (ESI-MS/MS) 194
EMBL files 74, 75
enoyl-acyl carrier protein (ACP) reductase 276
Enterobacter aerogenes insect toxin 17
Entrez protein 29
environmental genomics 242
equilibration 192
errors in annotation 29
Escherichia coli 37, 40, 41, 121, 129, 133, 154, 274, 277
   2DGE/MS approach 190
   AIDA-1 adhesin 10
   comparative genomics 274

expression of candidate genes in 260–2
interaction map 226
libraries 168, 170
neural networks 58, 59
OMPs 197
pO157 plasmid 54
PCR product selection 72
Piga 308
protein-folding chaperone GroEL 17
sequences stain MG 1655 127
S17-1 *gammapir* 158
stable RNA species 6
Top10F' 106, 110
Escherichia–coli-based two-hybrid systems 230
Escherichia coli strains
    0157:H7 128
    536 132
    19851 donor strain 172
    DH10B 245
    F632, WaaZ from 39
    HB101 223
    JW366 245
    K12–131
        RfaZ from 39
    K12 strain MG1655 132
Escherichia faecalis 277
ethidium bromide staining 89
European Molecular Biology Laboratory 28
ExpressionAtlas 57–8
extracellular signal-regulated kinase (ERK) 231

FAST™ slides 94
FASTA 8, 9, 10, 12, 13, 15, 38, 75, 76, 78, 79
fingerprinting technique 194
5' to 3' bias 74
fluorescence-based assay 285
frameshift errors in sequences 33
frameshifts 33
functional analysis consortium 41
functional genomics 154
functional proteomics 189, 203–4

G + C content 5, 18–20
GAL4 213, 215, 217–18
Gal4p transcription factor 211
GAL80 218
GAMBIT 279
GAPSII™ slide 93
gas chromatography mass spectometry (GCMS) 334
GC skew 53, 56
GenBank files 74, 75
gene by gene knockout 273, 274
gene by gene mutagenesis 273, 274
gene clusters 36
gene deletion mapping 84
Gene Ontology 21
gene prediction 4–6, 75
   programs 4, 5
gene regulation 84
GeneChip *E. coli* Genome Array 123
GeneChip Yeast Genome S98 Array 124
gene-for-gene hypothesis 231
gene-specific oligonucleotide primers 88–9
GeneSpring 98
genetic footprinting 279, 280

GeneWiz 51
genome plasticity 84
genome sequence divergence 70
GenomeAtlases 52–3
   base composition 53
   custom-made 55–7
   repeat elements 53
   of pathogenicity plasmids 54–7
   structural parameters 52–3
genome-directed primers (GDPs) 145–6
   with T7 tags 147
genome-independent measures 5
genomewide RNA expression, atlases for 57–8
Genomic-Tip 100/G system 106
glass slide microarrays 83–98
Glimmer 33, 75
global prediction of protein function, atlases for 59–61
glycerol phosphate dehydrogenase (GAPDH) 94
GNA33 265–4
Google 29
Gradiflow 197
Gramicidin S 194
GRAVY 198
guanidine isothiocyanate (GTC) solution 140–2

Haemophilus influenzae 68, 87, 272, 276, 277
   comparative genomics 274
Helicobacter hepaticus genome 294
Helicobacter pylori 272, 274, 276, 294–5
   comparative genomics 274
   composite DNA microarray 126
   computational analysis 295–6
   functional genomics 291–314
   gene expression 298–9
   genetic screens 307–13
      defining essential genes 311–13
      future of *H. pylori* genetics 313
      virulence genes 308–11
   genome sequences 294–5
   genomics resources 292–4
   host response 299–301
   interaction map 225, 226
   proteome 301–7
      postgenomic analysis 302–6
      protein–protein interaction maps 306–7
      purifying virulence determinants 302
   signature-tagged allele replacement 175–81
   strain comparison 296–8
   survival rates of mutants 168
   transcriptome 298–301
   unsuitability for 2DGE/MS approach 189
Helicobacter pylori strains 83, 121, 129, 133
   J99 129
   B128 130
   G1.1 130
   26695 132, 177
hairpin fold-back structures 71
4-HCCA 194
heat shock 95
hepatitis C virus 225
herpes simplex virus (HSV) virion protein 16 (VP16) 213
hidden Markov model (HMM) 14
high-performance liquid chromatography (HPLC)-
   purified adapter oligonucleotides 107

high throughput screening (HTS) 285
HIV vaccine research 267
HMMER package 14, 35
horizontal gene transfer 18–21
host–pathogen interactions,–gene expression during 137–49
HotStar+TM+ kit 373
human immunodeficiency virus (HIV) vaccine research 267
hybrid two hybrid systems 218
hybridization analysis of DNA and RNA 95–8
    hybridization 97
    image acquisition and analysis 97–8
    washing 97
hypothesis formulation 36–7

ImaGene™ 97, 377
immobilized pH gradient (IPG) technology 190
in silico hypothesis generation and testing 27–41, 276
in vivo expression technology (IVET) 138, 273, 283
insertion sequence (IS)
    element mediated inversion 20
    elements 54
    location 84
integration vector pORI240 349
interacting domain profile pairs (IDPP) 226
interaction mating 224
InterPro 14, 29
intimin 230–1
inverse PCR mutagenesis (PICRM) 176, 181
inverted repeat 53
Invitrogen 40
isoelectric focusing 190
isopropyl-β-D-thiogalactopyranoside (IPTG) 106, 110, 282
isotope-coded affinity tags (ICAT) 202–3

JalView 36, 38
J-Bgl adapter set 115, 118

KEGG 347
Kelnow DNA polymerase 96
Klebsiella pneumoniae 342 131
Kyoto Encyclopaedia of Gene and Genomes (KEGG) 22
Kyte–Doolittle value 198

Lactococcus lactis-derived plasmids 343
lateral gene transfer microarray 133–4
Leishmania major GenomeAtlas of chromosome 1 55–7
Leishmania sp. 397
LexA BD protein 211, 213, 215, 216, 217, 218
lipopolysaccharide (LPS) core biosynthesis proteins 39
liquid chromatography (LC)–MS 196
lithium acetate method (LiAc) 219–20
low-complexity sequences 34
Luria Bertani (LB) media 158

macroarrays 85
MAGE-ML 87
major groove compressibility 51
major outer membrane protein (MOMP) 200
major polymorphic tandem repeats (MPTRs) 73

malaria 267
MAPK kinase superfamily (MKKs) 231
mass spectrometry 188
matrix approach 224–6
matrix-assisted laser desorption ionization time-of-flight (MALDI-TOF) 362
matrix-assisted laser desorption ionization time-of-flight mass spectrometry (MALDI-TOF MS) 194, 200, 302, 327, 328
Medline 362
Melanie 3 193
melt depletion 115
membrane arrays 85
meningitis B 258–9
Met25 promoter 227
metabolic reconstruction 21–2
Metacyc website 292
metagenomics 241–51
    applications 243
    host cells 247
    host characterizing metagenomic libraries 248–50
        functional analysis 250
        phylogenetic analysis 248–9
        sequence analysis 249–50
    isolation of environmental DNA 243–4
    libraries 242–3
    storage and lone management 248
    vectors 244–7
        expression vectors 245–6
        large-insert 244–5
        shuttle vectors 246–7
methicillin-resistant *Staphylococcus aureus* (MRSA) 271
Micado (MICrobial Advanced Database Organization) 345, 355, 358, 362
microarray design for bacterial genomes 67–81
    general considerations 68–74
    genome sequence and annotation 69–70
    genomic features 70–1
    multiple strain coverage 75–8
    oligonucleotide primers for PCR 71
    PCR product selection 71–4, 79
    primer design 78–9
    quality control of primer prediction 79–80
    sequence acquisition of genomic features 74–5
microarray scanning equipment 97
microarrays, application in microbiology 84
microelectronic chip arrays 124–5
MIG1 repressor 218
minimal cross-hybridizing array 72, 73
minimum inhibitory concentration (MIC) assays 273
mini-Tn5 156, 162
Minnesota Biocatalysis Biodegradation database 22
modification of the transcription response 228
molecular epidemiology 84
Moraxella catarrhalis, UspA in 266
motif searching 13–14
mRNA
    amplification of 146–7
    rapid stabilization 140–3
    separation from ribosomal RNA 143–4
    stability 138–40
MUG 344
Multicoil 34, 39
multidimensional LC-MS (MuD LC-MS) 202–3

multiple strain comparison 75–8
multiplex PCR 84
Multiscreen96 93
mung bean nuclease 104
Mycobacterium bovis
    BCG 141–2
    BCG RNA preparations 144
    BCG vaccine strain 128
    dnaK mRNA 141
Mycobacterium. tuberculosis 68, 83, 87, 133, 198, 276
    deletion polymorphisms 124
    2DGE/MS approach 190
    ESAT-6 protein 37
    genome 123
    metabolic reconstruction 22
    microarray 90, 92
    PCR product array 72, 73
    PGRS proteins 20
    PPE and PGRS proteins 10, 12
    RNA extraction 138–49
    transposons for 156
Mycobacterium. tuberculosis strain H37Rv 123, 128, 129, 130
Mycoplasma genitalium–275, 301
    comparative genomics 274

Nanogen NanoChip™ system 124–5
National Center for Biotechnology Information 28
N-Bgl adaptors 118
Neisseria cinerea 264
Neisseria gonorrhoeae 264
Neisseria lactamica 264
Neisseria meningitidis 121, 162, 272
    candidate antigen prediction 260
    candidate screening 260–3
    complete genome sequence of serogroup B 259–60
    meningitis B and 258–9
    protective antigens in 263–4
    TN*10* transposition 156
neomycin-*N*-acetyltransferase II (NEO) 390
neural networks 58, 59
non-cross-contamination (NCC) design 89
non-similarity-based methods 17–18
Northern blotting 40, 138, 141
nuclear factor kappa-B (NF-kappaB) signalling pathway 231
nuclear localization signal (NLS) 215
nucleotide plots, anomalies in 19, 21
Nytran SuPerCharge 94

O-antigen biosynthesis 175
oligonucleotide arrays 68, 69
oligonucleotide-based DNA microarrays 122–4
oligonucleotide primers for PCR 71
one and a half hybrid system 229
one-colour technology 85
open reading frames (ORFs) 4, 73, 145–6, 155–6, 187, 224–5, 274
    ORF-specific 3′ primer pool with T7 tags 147
Orcbp protein 229
orphan proteins 59
Orpheus 33
orthology 14–17
outer membrane protein (OMPs) 197–8
outer membrane vesicles (OMV) 259

p38 mitogen-activated protein kinase (MAPK) signalling pathway 231
pACT2 vector 216
Paenibacillus 337
pan-pathogen arrays 84
pan-species arrays 75
pan-strain arrays 75
paralogy 14–17
parsing proteins into domains 30–6
    homology searches 35–6
    quality control 33–4
    structural features 34–5
pathogenicity island (PAI) 157
pBeloBAC11 245
pBridge system 227
PBSX prophage 340
pCR2.1 TOPO 163
PD-Quest 193
Penicillin resistant-pneumococcus 271
peptide mass mapping (PMM) 194, 195
Perl scripts 75, 80, 373
PFAM 14, 35, 38–9
pGAD424 vector 213, 216
pGBT9 vector 213
PGRSs 73
phospholipase A 175
phospholipase PldA 175
phosphoproteomics 199
phosphoribosylamino-imidazole (AIR) 221
phosphotyrosine binding domains 227
phosphotyrosinephosphatase 227
PIM analysis of *H. pylori* 307
Plasmodium falciparum
    chloramphenicol acetyl transferase (CAT) gene 390
    luciferase gene (*LUC*) 390
    thymidine kinase gene–385, 388
Plasmodium falciparum DNA microarrays 371–81
    applications to study of *P. falciparum* biology 379–80
    chromosome-specific 372–4
        arraying 373–4
        exon selection 372
        PCR amplification 373
        primer design 372–3
    data acquisition and analysis 377–9
    microarray hybridization 374–7
        hybridization 376–7
        prehybridization 376
        synthesis of fluorescently labelled cDNA 374–5
    preparation of total RNA 374
Plasmodium falciparum genome, functional analysis 267, 383–95
    analysis of transfected parasitres 390–4
    growing parasites for transfection 384–5
    plasmid vectors for transient and stable transfection 386–90
        common plasmids 386–9
        rep20 plasmids 390
        vector construction 386
    stable transfectants: phenotyping 393–4
        co-cultivation 393–4
        five-day growth rate assay 393
        RBC invasion assay 394
    stable transfectants: protein analysis 393

Plasmodium falciparum genome, functional analysis (cont.)
  stable transfectants: genetic analysis 391–3
    PCR 392
      pulsed-field gel electrophoresis 392
      Southern blotting of digested gDNA 392–3
    transient transfectants 390–1
pMAP65 356
pMUTIN 348, 351, 353, 354, 355–7
polyA tails 74
polyethylene glycol (PEG) 219
poly-L-lysine slides 93
polymerase chain reaction (PCR) 38, 104, 168, 401
  amplification of array elements 89–92
  cycling robotics 89
  labelling of tags 160–1
  product-based arrays 68, 69, 122
  products, preparation for printing 92–3
  targeting 333
polysaccharide biosynthesis 175
position preference 53
position-specific iterative basic local alignment sequence tool (PSI-BLAST) 28, 35, 36, 39, 276
post-print processing 94–5
post-translational modifications (PTMs) 59
  in yeast two-hybrid system 226–7
prediction of bacterial protein sequences, problems in 33
Primer3 71, 372, 373
printing
  arrays 93–4
  preparation of PCR products for 92–3
probe 86–7
ProDom 14
PROSITE 13–14
protein interaction map (PIM) 306
  PIM biological score 306
protein–ligand detection systems 229
protein–nucleic acid interactions, detection 229
protein–protein interaction maps 224–6
protein–protein interaction using two-hybrid systems 209–32
  history and evolution 210–12
  methodology 212–26
    analysis of positive clones 223
    autoactivation 216–17
    genomics-bioinformatics 223–4
    independent confirmation 226
    interaction mating 224
    library plasmid rescue 222–3
    library plasmids 217
    library transformation 219–20
    promoter 213–15
    selecting a yeast strain 217–18
    selection procedures 220–2
    two-hybrid vectors 215–16
  variations on 226–32
    post-translational modifications 226–7
protein secretion pathway 6
proteomics 187
ProtFun 52
pseudogene 33, 75
Pseudomonas aeruginosa 275, 277, 295
  CSN proteins 198
  gentamicin resistance gene 133
  OMPs 198
Pseudomonas putida 299
Pseudomonas syringae pv tomato 231
PSI-BLAST 28, 35, 36, 39, 276
PSORT 34, 260
P_{spac} promoter 344
Pto protein 231–2
Pto-interacting proteins (Pti) 231–2
PubMed 29
PUREGENE DNA Isolation Kit 106
purification of DNA and RNA labelled samples 96–7
puromycin 390
puromycin-*N*-acetyltransferase (PAC) 390
PyloriGene website 292

Q-Bot 224
Qiagen MinElute PCR Purification 96
Qiagen Operon systems 123
qrna 6
quadrupole time-of-flight mass spectrometry (Q-TOF MS) 328
quadrupole/orthogonal-acceleration time-of-flight (Q-TOF) 362
quorem-sensing systems 273

RACK1 protein 232
random hexamers 96
random shotgun strategy 259
Ras recruitment system (RRS) 228
R-Bgl-12/24 adaptors, ligation of 113
reference mapping by 2DGE 200–1
regions of difference (RD) 127
regulon 189
RepA 349
replacement recombination 349–51
representational difference analysis (RDA) 103–20
  bacterial strains and plasmids 106
  cloning of difference products 109
  hybridization/amplification steps in 105
  isolation chromosomal DNA 106
  production of input material 106–7
  subtractive hybridization 108–9
  see also cDNA representational difference analysis
restriction fragment polymorphism analysis 249
reverse transcriptase 96
reverse transcriptase polymerase chain reaction (RT-PCR) 40, 119, 138, 146–7, 299
reverse vaccinology 257–68
  antigenic variation in protective antigens 263–4
  candidate antigen prediction 260
  candidate screening 260–3
    expression in E. coli 260–2
    immunization and screening of sera 263
  cross-protective in conserved antigens 264–5
  future of 266–8
    viral pathogens 267
    eukaryotic pathogens 267–8
  meningitis B and 258–9
  vaccine development 266
ribosome binding sites (RBS) 5, 339–40
RMLA (random mutagenesis and loop amplification) 312, 313
RNA extraction from *in-vitro* cultures of M. tuberculosis 148–9

RNA genes, stable 7–7
RNA labelling 96
RNA:DNA hybrids, antibody detection 145
RNA*later*™ 142
RoboAmp 4200 89
Rotofor 197
rRNA
    intergenic spacer analysis 249
    separation from mRNA 143–4

sacB promoter 344
Saccharomyces cerevisiae 59, 306
    ExpressionAtlas for 57–8, 60
    FunctionAtlas of chromosome VIII 60
    genomewide tagged knockouts 156
    interaction map 226
    mutagenesis project 168
    polymorphisms 124
Salmonella typhi
    G/C bias plot 21
Salmonella typhimurium
    AvrA 231
    RfaZ from 39
    STM use in 154, 155
SAP-1 229
scintillation proximity assay (SPA) 285
SDS-PAGE 192, 196, 199
selected interaction domain (SID) 306
self-annealing 71
sequence homology 35
sequence repeats 34
sequence similarity 35
serum response element (SRE) 212
serum response factor (SRF) 229
Shigella flexneri 154
    5a virulence plasmid pWR501
    application of STM to 156–7
    GenomeAtlas for 50
    mutant library construction 158–9
    strain M90T 158
    virulence plasmid pWR501 54
    yeast two-hybrid system 231
shigellosis, selecting an animal model for 157–8
sigma-A 341
sigma-B 341
signal peptide 34
signalP 34, 260
signature tags 154–5
signature-tagged allele replacement (STAR) in *Helicobacter pylori* 175–81
    growth studies under *in vitro* conditions 178–9
    growth studies under *in vitro* stress conditions 179
    mutagenesis and tagging strategy 176–7
    tag detection 177–8
signature-tagged mutagenesis (STM) 153–63, 273, 283
    application to *Shigella flexneri* 156–7
    bacterial strains, vectors and culture conditions 158
    extraction of genomic DNA 159–60
    identification of attenuated mutants 161–2
        hybridization of dot blots 161–2
        plasmid dot blot preparation 161
    identification of transposon insertion sites 162–3
    inoculation of rabbit ligated loop 159
    limitations 167
    mutant library construction 158–9
    PCR labelling of tags 160–1
    preparation of inoculum 159
    selection of animal model 157–8
similarity searching 8–13
single cross-over recombination (Campbell-type) integration 348–9
single nucleotide polymorphisms (SNPs) 123
single-strand origin (SSOs) 343–4
SipA protein 230
SipC protein 230
Sir4p 211
size-marker identification technology (SMIT) 273
skin element 340
SMART 14, 35, 38
sortase 273
SOS recruitment system (SRS) 228
Southern blot of DNA from *Y. pseudotuberculosis* random mutants 172–3
Southern hybridization 119
SPβ prophage 340
spheroplast method 219
SPID 347, 362
spotted arrays 85–6
spotted PCR product microarrays 68
    construction of 87–95
    sequence-defined clone array 87–8
src family 227
SRF accessory protein-1 (SAP-1) 212
stacking energy 52–3
Staphylococcus aureus 121, 129, 133, 198, 272, 275, 276, 277
    derived plasmids 343
    DNA microarray analysis 127
    regulated gene expression systems in 280
    sortase gene 28
    Tn5 156
    Tn*916* 156
Staphylococcus aureus strains
    ET234 130
    N315 37
    RN4220 280
statistical significance, measures 8
stimulon 189
Streptococcus pneumoniae 68, 87, 272, 275, 276, 277
    insertion–duplication in 156
    R6 76
    regulated gene expression systems in 280
    TIGR4 76
Streptococcus sanguis
    adhesin protein 16
Streptococcus uberis
    application of cDNA RDA to 119
    cDNA representational difference analysis 110
    isolation of 16S and 23S rRNA genes 112–13
Streptomyces coelicolor A3(2)
    genetics 328–34
        targeted mutagenesis 332–4
        transposon mutagenesis 329–32
    genome sequence 321–35
        proteome 327–8
        transcriptome 323
    ORF prediction in 4–5
    practical situations 334–5
    transposon Tn*5062* 331

Streptomyces fradiae 331
Sub2D 345
Subscript 346
Substance P 194
SubtiList 340, 345, 362
subtractive hybridization 108–9, 116–18, 132
superBAC1 245
superCos1 245
swapped systems 228
SwissProt 29, 54, 362
synteny 14–17
Sypro Ruby dye 192

Taq polymerase 90
TaqMan analysis 284
temperature-sensitive (TS) mutants 275
tester:tester hybrids 108
TetR-family transcriptional regulators 8
three-hybrid system 227
TIGRFAM 14
TIGR 292, 347
TMPRED 260
tmRNA 6
toxic shock syndrome 127, 130
T-plastin 230
transcription fusion 40
transcription-induced artefacts 138–40
transcriptome analysis 84, 137
translocated intimin receptor (tir) 230–1
transmembrane (TM) domains 34
tributyl phosphine (TBP) 192
tRNAs 6
tRNA-SE 6
Trypanosoma brucei 397
Trypanosoma cruzi 397
trypanosomes
 applications and limitations 403–4
 chromosome fragmentation 397–404
 generation of parasites with truncated chromosomes 402–3
 genetic manipulation of trypanosomatids 398–9
 techniques required for chromosome fragmentation 399–402
trypticase soy (TCS) broth 158
two-colour technology 86
two-component signal transduction systems 273
two-dimensional gel electrophoresis (2DGE) 40, 187, 189–200, 301, 307
 2DGE/MS approach 200–1
 components of buffers 191
 differential display proteomics 201–2
 identification 193–6
 prefractionation 196–200
 reference mapping 200–1
 sample preparation 189–90
 separation 190–2
 visualization 192–3

two-hybrid systems 209–32
 commonly used 213
 history and evolution 210–12
 vectors 214, 215–16
 yeast selection procedures 220–2
typeIII secretion 273
tyrosine kinase 227

ubiquitin-based split protein sensor (USPS) 228–9
UHP water 111
upstream activating sequences (UAS) 211, 217

vacuolating cytotoxin (VacA) protein 232
validation
 biological, of arrays 95
 of PCR 92
vancomycin-resistant *Enterococci* (VRE) 271
Vibrio cholerae 133
 ampicillin-resistant exconjugants 172
 strain N16861 129
Vibrio harvei, luxAB 344–5
virulence determinants 126
ViruloGenome 28, 33, 35, 39
visualization software 51–2

Western blots 40
Wigglesworthia glossinidia 130–2
WIT 22

Xanthomonas campestris pv. vesicatori AvrRxv 231

X-gal 106, 110, 211, 218, 222, 344, 349
XylR-controlled promoters 344

y genes 355
yeast two-hybrid system 203
 at the bacterial–eukaryote interface 230–2
Yersinia entercolitica 181
Yersinia outer membrane proteins (Yops) 231
Yersinia pestis 68, 87, 174
 G/C bias plot 21
 IS-element mediated inversions in 20
 Muk8 protein 10, 13
Yersinia pseudotuberculosis
 microarray-based STM 168–75
  conjugal DNA transfer method for transposon mutagenesis 172
  double-tagged mini-Tn5 library construction 168–2
  quantitative analysis of tagged mutants 173–4
  sequencing of transposon insertion sites 174–5
  testing of tagged mutant pools in murine infection model 172–3
YukA family of membrane-associated ATPases 37

Z3 193
Zip-Tips™ 194
zoom gels 328